PLEASE STAMP DATE DUE

Nonlinear Physical Oceanography

ATMOSPHERIC AND OCEANOGRAPHIC SCIENCES LIBRARY

VOLUME 28

The titles published in this series are listed at the end of this volume.

Nonlinear Physical Oceanography

A Dynamical Systems Approach to the Large Scale Ocean Circulation and El Niño

2nd Revised and Enlarged Edition

Henk A. Dijkstra

Institute for Marine and Atmospheric Research Utrecht,
Department of Physics and Astronomy,
Utrecht University, Utrecht, The Netherlands
and
Department of Atmospheric Science,
Colorado State University,
Fort Collins, CO, U.S.A.

 Springer

A C.I.P. Catalogue record for this book is available from the Library of Congress.

ISBN-10 1-4020-2262-X (HB) Springer Dordrecht, Berlin, Heidelberg, New York
ISBN-10 1-4020-2263-8 (e-book) Springer Dordrecht, Berlin, Heidelberg, New York
ISBN-13 978 1-4020-2262-3 (HB) Springer Dordrecht, Berlin, Heidelberg, New York
ISBN-13 978 1-4020-2263-0 (e-book) Springer Dordrecht, Berlin, Heidelberg, New York

Published by Springer,
P.O. Box 17, 3300 AA Dordrecht, The Netherlands.

Printed on acid-free paper

Printed in the Netherlands.

To Julia, Joost, Bas
and Anneloes, my
companions over large
distances and long
periods of time

Contents

Preface   xiii

Acknowledgments   xv

1. INTRODUCTION   1

  1.1 Past Climate Variability   2

    1.1.1 The last 2.5 million years   3

    1.1.2 The Younger Dryas   5

    1.1.3 The Little Ice Age   6

    1.1.4 Causes of past climate variability   8

    1.1.4.1 External forcing: orbital and solar variations   9

    1.1.4.2 Internal variability   11

  1.2 The Present Ocean Circulation   12

    1.2.1 Surface forcing   12

    1.2.2 Ocean circulation patterns   14

    1.2.3 Heat and freshwater transport   18

    1.2.4 Ocean circulation and past climate variability   19

  1.3 Present Climate Variability   21

    1.3.1 ENSO   21

    1.3.2 The Atlantic Multidecadal Oscillation   24

  1.4 Physics of Climate Variability   26

    1.4.1 The system view   26

    1.4.2 Central questions   27

    1.4.3 Approach   28

  1.5 Exercises on Chapter 1   30

2. BACKGROUND MATERIAL   37

  2.1 Basic Equations   38

    2.1.1 Coordinate free   38

    2.1.2 Spherical coordinates   40

    2.1.3 Dissipative processes   42

    2.1.4 Boundary conditions   43

    2.1.5 Integral constraints   45

  2.2 Vorticity transport   47

    2.2.1 Vortex stretching and vortex tilting   49

		2.2.2	Baroclinic vorticity production	50
	2.3	Potential Vorticity (PV)		50
		2.3.1	The Ertel theorem	51
		2.3.2	PV conservation	52
	2.4	Stability		53
	2.5	Exercises on Chapter 2		56
3.	A DYNAMICAL SYSTEMS POINT OF VIEW			63
	3.1	An Elementary Problem		64
		3.1.1	The Stommel two-box model	64
		3.1.2	Equilibrium solutions	67
		3.1.3	Stability of steady solutions	69
		3.1.4	The presence of symmetry	71
		3.1.5	Imperfections	74
	3.2	Dynamical Systems: Fixed Points		75
		3.2.1	Elementary concepts	76
		3.2.2	Codimension-1 bifurcations	80
		3.2.2.1 A single zero eigenvalue		80
		3.2.2.2 A single complex conjugate pair of eigenvalues		83
		3.2.3	Imperfection theory	85
		3.2.4	Codimension-2 bifurcations	88
		3.2.4.1 The cusp bifurcation		88
		3.2.4.2 The Bogdanov-Takens bifurcation		89
	3.3	Periodic Solutions and their Stability		90
		3.3.1	Poincaré section and Poincaré map	90
		3.3.2	Floquet theory	93
	3.4	Bifurcations of Periodic Orbits		96
		3.4.1	The cyclic-fold bifurcation	96
		3.4.2	The period-doubling bifurcation	96
		3.4.3	The Naimark-Sacker bifurcation	98
	3.5	Global bifurcations		100
		3.5.1	Homoclinic orbits	100
		3.5.2	The Lorenz (1963) dynamical system	101
	3.6	Resonance phenomena: frequency locking		102
	3.7	Physics of Bifurcation Behavior		106
		3.7.1	Physical constraints	106
		3.7.2	Qualitative versus quantitative sensitivity	107
		3.7.3	Instability mechanisms	109
		3.7.3.1 Saddle-node bifurcation		109
		3.7.3.2 Transcrictal and pitchfork bifurcation		110
		3.7.3.3 Hopf bifurcation		112
	3.8	Exercises on Chapter 3		113
4.	NUMERICAL TECHNIQUES			119
	4.1	An Example Problem		122
		4.1.1	Introduction	122
		4.1.2	Model	124

	4.1.3	Motionless solution	125
	4.1.4	Dimensionless equations	126
4.2		Computation of Steady Solutions	126
	4.2.1	Discretization	127
	4.2.2	Pseudo-arclength continuation	130
	4.2.3	The Euler-Newton method	132
4.3		Detection and Switching	134
	4.3.1	Detection of bifurcations	134
	4.3.2	Branch switching	136
	4.3.3	Finding isolated branches	138
4.4		Linear Stability Problem	139
	4.4.1	The simultaneous iteration method	141
	4.4.2	The Jacobi-Davidson QZ-method	144
4.5		Implicit Time Integration	147
4.6		Linear System Solvers: Direct Methods	148
	4.6.1	Basic principle	148
	4.6.2	Pivoting	150
4.7		Linear System Solvers: Iterative Methods	151
	4.7.1	Relaxation methods	151
	4.7.2	Projection techniques	152
	4.7.2.1	The GMRES technique	154
	4.7.2.2	The BICGSTAB technique	155
4.8		Application to the Example Problem	158
4.9		Exercises on Chapter 4	163
5. THE WIND-DRIVEN CIRCULATION			169
5.1		Phenomena	170
	5.1.1	Gulf Stream	170
	5.1.2	Kuroshio	174
	5.1.3	Central questions	175
5.2		Models of the Midlatitude Ocean Circulation	176
	5.2.1	The homogeneous model	176
	5.2.2	Dominant balances	178
	5.2.3	The multi-layer model	180
5.3		Shallow-water and Quasi-geostrophic Models	181
	5.3.1	The spherical shallow-water model	182
	5.3.2	The β-plane model	183
	5.3.3	Quasi-geostrophic β-plane models	185
	5.3.4	Overview of the SW and the QG models	188
5.4		Classical Results	190
	5.4.1	The Sverdrup-Munk-Stommel theory	190
	5.4.2	Temporal variability	194
	5.4.2.1	Rossby waves	194
	5.4.2.2	Rossby basin modes	196
	5.4.2.3	Basic instability mechanisms	197
5.5		Regimes of Double-Gyre QG Flows	202
	5.5.1	Equivalent barotropic flows	203

5.5.1.1 Basic bifurcation diagrams   203
5.5.1.2 Transient flows   212
5.5.2   Baroclinic flows   216
5.5.2.1 Basic bifurcation diagrams   216
5.5.2.2 Transient flows   219

5.6   Regimes of Double-Gyre SW Flows   225
      5.6.1   The equivalent barotropic case   226
      5.6.2   Connection: SW- and QG-models   229
      5.6.3   The baroclinic case   232

5.7   Continental Geometry   235
      5.7.1   Continents within a β-plane SW model   235
      5.7.2   Continents on the sphere   237
      5.7.2.1 North Atlantic basin   238
      5.7.2.2 North Pacific domain   240
      5.7.3   Summary   242

5.8   High-Resolution Ocean Models   245
      5.8.1   Typical results   246
      5.8.2   Analysis of POCM results   249
      5.8.2.1 Gulf Stream   250
      5.8.2.2 Kuroshio   250

5.9   Synthesis   253
      5.9.1   Summary   253
      5.9.2   Interpretation framework   255
      5.9.2.1 Multiple mean paths?   255
      5.9.2.2 Modes of variability?   256

5.10 Exercises on Chapter 5   259

6. THE THERMOHALINE CIRCULATION   267

6.1   North Atlantic Climate Variability   268
      6.1.1   Observations   268
      6.1.2   Central questions and approach   272

6.2   Potential Mechanisms   273
      6.2.1   What drives the THC?   273
      6.2.2   Advective feedback   274
      6.2.3   Convective feedback   275
      6.2.4   The flip-flop oscillation   278
      6.2.5   The loop oscillation   280
      6.2.6   Models of the THC   281

6.3   Two-dimensional Boussinesq Models   282
      6.3.1   Formulation   282
      6.3.2   Nondimensional equations   284

6.4   Diffusive Thermohaline Flows   285
      6.4.1   Basic bifurcation diagram   285
      6.4.2   Physical mechanisms   288
      6.4.2.1 Symmetry breaking   288
      6.4.2.2 Transition to time-dependence   290
      6.4.3   Model-model comparison   293

6.5	Convective Thermohaline Flows	298
	6.5.1 Basic bifurcation diagrams	300
	6.5.2 Imperfections	302
	6.5.2.1 Coupled model	304
	6.5.2.2 Asymmetric air-sea interaction	305
	6.5.2.3 Regime diagram	309
6.6	Zonally Averaged Models	310
	6.6.1 Scaling of the equations	310
	6.6.2 Zonal averaging	312
	6.6.2.1 Procedure	313
	6.6.2.2 Convective adjustment	316
	6.6.3 Bifurcation diagrams	317
6.7	Three-Dimensional Models	320
	6.7.1 The SH configuration: thermal flows	320
	6.7.1.1 The multidecadal mode	321
	6.7.1.2 Finite-amplitude flows	327
	6.7.2 The SH configuration: thermohaline flows	329
	6.7.2.1 Multiple equilibria and new internal modes	330
	6.7.2.2 Finite-amplitude flows	332
	6.7.3 The DH configuration: thermohaline flows	337
	6.7.3.1 Bifurcation diagrams	337
	6.7.3.2 Finite-amplitude flows	341
	6.7.4 Multi-basin and global models	342
	6.7.4.1 Bifurcation diagrams	342
	6.7.4.2 Finite-amplitude flows	344
6.8	Coupled ocean-atmosphere models	350
6.9	Synthesis	355
	6.9.1 Different mean thermohaline flows?	358
	6.9.2 Temporal variability through internal modes?	359
6.10	Exercises on Chapter 6	363
7.	THE DYNAMICS AND PHYSICS OF ENSO	369
7.1	Basic Phenomena	370
	7.1.1 The annual-mean state	371
	7.1.2 The seasonal cycle	373
	7.1.3 Interannual variability	373
	7.1.4 Low-frequency variability of ENSO	378
	7.1.5 Central questions and Approach	378
7.2	Models of the Equatorial Ocean	379
	7.2.1 Constant density ocean model	379
	7.2.2 The reduced gravity model	380
	7.2.3 Equatorial waves	381
	7.2.4 Forced response in a basin	387
7.3	Physics of Coupling	392
	7.3.1 Atmospheric response to diabatic heating	392
	7.3.2 Adjustment of the ocean	397
	7.3.3 Processes determining the SST	400

7.3.4	Feedbacks	403
7.3.4.1	Thermocline feedback	403
7.3.4.2	Upwelling feedback	404
7.3.4.3	Zonal advection feedback	404
7.3.4.4	Strength of the feedbacks	404
7.4	The Zebiak-Cane Model	405
7.4.1	Formulation	406
7.4.2	Results	408
7.5	Towards the Delayed Oscillator	411
7.5.1	Coupled modes: periodic ocean basin	411
7.5.2	Coupled modes: bounded basin	415
7.5.2.1	The near equatorial behavior	416
7.5.2.2	The fast wave limit	421
7.5.2.3	The weak-coupling limit	424
7.5.3	Modes in the full problem	426
7.5.4	Conceptual models of the ENSO oscillation	430
7.5.4.1	The two-strip model	431
7.5.4.2	The delayed oscillator	434
7.5.4.3	The coupled wave oscillator	436
7.5.4.4	The recharge oscillator	437
7.6	Coupled Processes and the Annual-Mean State	438
7.6.1	Constructed versus coupled mean states	439
7.6.2	Demise of multiple equilibria	440
7.6.3	The position of the cold tongue	443
7.7	Unifying Mean State and Variability	448
7.7.1	The warm pool/cold tongue state	448
7.7.2	The ENSO mode	449
7.7.3	Model reduction	454
7.8	Presence of the Seasonal Cycle	455
7.8.1	Coupled processes and the seasonal cycle	455
7.8.2	Interaction of seasonal cycle and ENSO	457
7.8.3	The irregularity of ENSO	461
7.9	ENSO in General Circulation Models	463
7.10	Synthesis	468
7.10.1	A summary of ICM results	469
7.10.2	Multi-scale physics	471
7.10.3	The future of ENSO	476
7.11	Exercises on Chapter 7	478
Bibliography		483
Copyright Acknowledgements		511
Index		513
Color Plates		517

Preface

In the first edition of this book (published by Kluwer Academic in November 2000) the methodology of dynamical systems theory was introduced and applications of this theory to the large-scale ocean circulation and El Niño were provided. Surprised by the favorable reactions, I decided to make a second edition of the book which could be more easily used as a textbook for a graduate (700-level) course. The first edition has undergone a substantial rewrite on three aspects:

(i) the text has been adapted at many locations to improve clarity and readability,

(ii) many recent results on the wind-driven ocean circulation, the thermohaline circulation and El Niño have been included, and

(iii) a number of exercises have been added at the end of each chapter.

In chapter 1, the description of what is known from observations on the global ocean circulation has been improved by including, for example, recent estimates of transport quantities. Both the chapters 2 and 3 have only slightly changed; in chapter 3, the text on homoclinic orbits has been extended as these type of phenomena have now clearly been found in the wind-driven double-gyre ocean circulation (as presented in chapter 5). In chapter 4, I have added a paragraph on the computation of isolated branches of steady states and the text on the iterative linear systems solvers has been shortened.

Concerning the application of dynamical systems theory to the large-scale ocean circulation and El Niño in the chapters 5 to 7, many recent results on the dynamics of these flows are added. I have been tempted to include results on stochastic dynamics, on the use of dynamical systems theory in data analysis (e.g., attractor reconstruction techniques), and on ergodic theory. It would have been difficult, however, to keep the book self-contained. Therefore, I decided to restrict the text to results on bifurcation diagrams for deterministic models. Most of the changes have occurred in chapter 6, where I have omitted the material on flux-corrected models and shortened the section on the zonally averaged models. Instead, I have added recent work on the multiple equilibria of the thermohaline circulation in a hierarchy of three-dimensional models and on the multidecadal variability in the

North Atlantic. The introduction to El Niño in section 7.1 has been revised and results on the simulation of ENSO in coupled general circulation models have been added in section 7.9.

Each chapter now contains five exercises and two more computational oriented projects. In many of the exercises, additional material is introduced which could not be covered in the main text. For each exercise, references are provided to help with obtaining the answers to the problems. For the computational projects in each chapter, links and references to software needed to perform the computations is provided. In many of these exercises, the software package AUTO (which is freely available via FTP from the directory pub/doedel/auto at ftp.cs.concordia.ca) is used; AUTO runs on many UNIX platforms and PC's. If you have MATLAB available, I would recommend using the software package MATCONT (download from http://allserv.rug.ac.be/~ajdhooge/research.html) For alternative software, you may have a look at Gabriel Lord's website at http://www.ma.hw.ac.uk/~gabriel/auto/index.html.

Going over the final manuscript one more time, I think that the second edition of the book is in good shape to be used within a graduate course in physical oceanography, for example on 'nonlinear dynamics of the ocean circulation', combining chapters 1, 2, 3, 5 and 6. Another possibility is to use the book for a 'special topics' (capita selecta) course in climate dynamics. Combining chapters 1, 3 and 7 could provide material for a course on El Niño or on 'nonlinear tropical climate dynamics'. The computational aspects in chapter 4 are mainly of interest for those who intend to apply the techniques themselves to large-dimensional dynamical systems and the numerical methodology can be added to a graduate course, if desired.

The application of dynamical systems theory certainly has its limitations, but it provides an elegant framework for the interpretation of results from ocean and climate models. It is a pity that its powerful concepts and methods are still mostly ignored by researchers in physical oceanography and climate dynamics. I therefore sincerely hope that this second edition of the book will be used in future classes and that the material will find its way to the generation of future scientists in these fields.

JANUARY 2005, FORT COLLINS, USA

Acknowledgments

The preparation of this second edition would not have been possible without the support of others.

First of all, Will de Ruijter (IMAU, Utrecht) is thanked for his unfaltering support of my research activities. He and the rest of the faculty, staff, postdocs and students at IMAU, have been responsible for creating the environment in which most of the results presented in this book have been generated.

Joint work with Michael Ghil (LMD, Paris, France), David Neelin (UCLA, Los Angeles, USA) and Eric Simonnet (INLN, Nice, France) has been very important for the development of the material in this second edition. I thank David for support of visits to Los Angeles, Eric for the nice discussions on the wind-driven circulation, and Michael for 'spreading' many of the results around the (scientific) world.

Mark Wimbush (URI, Providence, USA) is thanked for his detailed comments on the text of the first version of the book. Sun Liang (LASG, Beijing, China) made very insightful remarks which have been incorporated in this second edition. Eric Chassignet (RSMAS, Miami, USA) is thanked for providing figures and references for chapter 5.

Joint work with the experts in linear system solvers and eigenvalue solvers at the Research Institute for Mathematics and Computing Science at the University of Groningen (Netherlands) has been essential to be able to apply continuation methods to ocean and climate models. In particular, I thank Fred Wubs, Arie de Niet, Auke van der Ploeg (now at MARIN) and Eugen Botta for always sharing their newest solvers with me and for their help in implementing these solvers in the ocean models.

My present (and former) students and postdocs at IMAU — Lianke te Raa, Janine Nauw-van der Vegt, Caroline Katsman, Wilbert Weijer, Ernst van der Avoird, Jeroen Gerrits, Jeroen Molemaker, Nathalie Rittemard, Maurice Schmeits, Paul van der Vaart, Arjen Terwisscha van Scheltinga, Anna von der Heydt, Anne Willem Omta and Hakan Öksüzoğlu — are thanked for all their contributions. In addition, I thank my colleagues at the Royal Dutch Meteorological Institute (KNMI),

Gerrit Burgers, Geert Jan van Oldenborgh, Sybren Drijfhout and Frank Selten for the nice collaboration over the years.

The generous support from the Netherlands Organization for Scientific Research (NWO) through a six year PIONIER (NWO-ALW) project: "Stability and variability of the climate system" over the years 1996-2002 and the NWO-E project: "Rapid Changes in Complex Flows" over the years 2002-2004 is much appreciated. This support has been essential to be able to carry out the research on the applications of dynamical systems theory in oceanography and climate dynamics. The support in supercomputing time on the machines at the Academic Computing Center in Amsterdam (SARA) through the projects SC-498, SC-029 and SC-122 from the National Computing Facilities Foundation (NCF) is also acknowledged.

The publisher at Kluwer Academic (which recently merged with Springer), Gert-Jan Geraeds, has been very supportive and patient in waiting for the final version of the manuscript. I thank Marie Johnson for going through the text and for providing useful comments on layout and style. Mieke van der Fluit and André Tournois are thanked for their comments on the layout and for processing the many last minute changes during the production stage of the manuscript.

Finally, the second edition of this book has been written in a relatively unhappy period in my life. I thank Julia and Joost, Bas and Anneloes for always being there; otherwise the book would not have been finished.

Chapter 1

INTRODUCTION

Chapter 1

INTRODUCTION

Patterns and their rhythms fill the spheres.
Evocation. Preludios Americanos I, A. Carlevaro

At the beginning of this century, the concept of 'climate' belonged to meteorologists and was considered to be a long term average state of temperature and precipitation. Later, other quantities were added to describing the average state of the atmosphere more accurately. At the moment, the atmosphere is considered to be only one of the components of a larger entity. The atmosphere (the world of air) together with the hydrosphere (the world of water), the cryosphere (the world of ice), the biosphere (the world of living beings) and the lithosphere (the world of the solid Earth) can be logically studied as one system: the climate system.

Earth's climate system displays variability on a multitude of time scales. Over long periods in Earth's history, large parts have been covered with ice, with warmer periods in between. On the very short time scale, the fluctuations of the weather on a day to day basis are experienced. In this first chapter, some motivating examples of climate variability are described. In particular, examples are chosen for which it is very plausible that changes in the ocean circulation are or have been involved.

In section 1.1, a short description of the history of Earth's climate sets the context for the discussion of the Younger Dryas event and the Little Ice Age period. Both phenomena illustrate that climate can undergo relatively rapid transitions which are not expected *a priori* from changes in the forcing conditions. In section 1.2, the present large-scale ocean circulation is introduced by sketching its forcing, the mean circulation patterns and the associated transport of heat and freshwater. Section 1.3 contains a brief description of two climate phenomena of current interest, the El Niño /Southern Oscillation in the Tropical Pacific and the Atlantic Multidecadal Oscillation. In both phenomena, there are significant changes in the sea-surface temperature and the ocean circulation. The central questions addressed in this book, and a motivation for the approach chosen towards possible answers, follows in section 1.4.

1.1. Past Climate Variability

Until fairly recently, the climates of the past had been described only qualitatively. At the moment, many techniques are available to construct climatic records from geological, biological and physical data (Bradley, 1999). Much information has been obtained through measurement of isotope content (such as oxygen and carbon isotopes) in material derived from ocean sediments and from ice cores. Accurate dating techniques are essential to interpret these measurements. For example, the carbonate in shells of marine organisms (e.g., foraminifera) and water in ice caps contain two isotopes of oxygen, ^{18}O and ^{16}O. The normalized isotope ratio $\delta^{18}O$ is calculated as a deviation from a reference sample as

$$\delta^{18}O = \frac{\left(\frac{^{18}O}{^{16}O}\right)_{sample} - \left(\frac{^{18}O}{^{16}O}\right)_{reference}}{\left(\frac{^{18}O}{^{16}O}\right)_{reference}}$$

where the reference sample is different for ice cores (i.e., standard mean ocean water) than for carbonate shells (i.e., a specific fossil Cretaceous species). The isotope ^{16}O is lighter than ^{18}O so that water containing ^{16}O is preferentially evaporated and a temperature-dependent fractionation occurs. Under cold conditions, less water containing ^{18}O is able to evaporate into the atmosphere.

Changes in $\delta^{18}O$ reflect the combined effect of changes in global ice volume and temperature at the time of deposition of the material. The two effects cannot be separated easily, and only recently an additional temperature indicator (Ca/Ma-thermometry) has been used to accomplish this (Lear *et al.*, 2000). During very cold conditions, global ice volume is relatively large and sea level is low, which enriches water in the ocean with ^{18}O. Because of the colder temperatures, also more ^{18}O remains in the ocean and less ^{18}O becomes locked in the ice. Hence, in ocean sediments the ratio $\delta^{18}O$ will increase under cold conditions, whereas in ice cores it will decrease.

When corrections for global ice volume (with respect to the reference sample) are made, $\delta^{18}O$ can be used as an indicator for the temperature at the time of deposition. In the next subsections, $\delta^{18}O$ records from the last 2.5 million years (My) will be shortly discussed. All these data (and many more) are available through NOAA's Paleoclimatology site (http://www.ngdc.noaa.gov/paleo/paleo.html). In addition to the information provided at this site, the books by Broecker (1995), Bradley (1999) and Ruddiman (2001) – where most of the references to the original studies can be found – are recommended sources of information.

1.1.1. The last 2.5 million years

Based on the oxygen isotope record of benthic foraminifera in ocean sediments and the reconstructed deep ocean temperature it is found that about 55 My ago (after the warm Cretaceous period), a gradual cooling started on Earth. Three major steps have occurred within this gradual cooling, one near 36 My, one near 14 My and the last near 3 My ago, each involving a temperature decrease of about 2-3°C, which induced an increase in $\delta^{18}O$ from near zero (note that the reference value was in the Cretaceous) to the current value of about 4.0. The $\delta^{18}O$ record from the last 2.5 My, as obtained from deep sea sediments (ODP site 677 in the equatorial Pacific at 1°N, 84°W), is plotted in Fig. 1.1; data are based on Raymo *et al.* (1990).

One observes the variations in climate superposed on a gradual cooling trend, with a change in pace about 0.7 My ago. From then on, a dominant period of about 100,000 year is found reflecting the frequency of major glaciations which occurred in the northern hemisphere. Termination of these glaciations seems to be rather abrupt and leads to warmer periods, called interglacials; at the moment, we live in the Holocene interglacial. The previous interglacial (the Eemian) can be seen as a peak of relatively small $\delta^{18}O$ at about 140,000 years ago. The transitions between glacials and interglacials are global in extent, since their signatures are found in available data all over the globe.

The oxygen isotope record of the last 110,000 years within the GRIP ice core from Greenland (Johnsen *et al.*, 1997) is plotted in Fig. 1.2. Note that, contrary to the values of ocean sediments, values of $\delta^{18}O$ are now negative and cool periods have smaller (larger negative) values than warm periods. No further smoothing was done on the 0.55 m averaged values (the total core depth is about 3 km); the

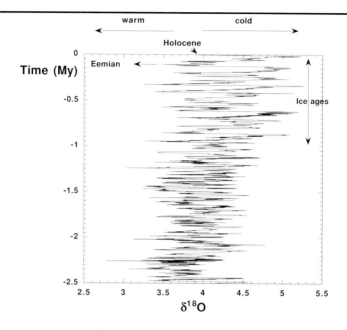

Figure 1.1. Isotope ratio $\delta^{18}O$ for benthic foraminifera at Ocean Drilling Program site 677 located in the eastern equatorial Pacific at $1°N$, $84°W$ (Raymo et al., 1990). A change in $\delta^{18}O$ of 0.23 units can be translated into a temperature change of about $1°C$ (Broecker, 1995).

time scale zero point is at 1950 AD. From the Eemian, the transition to the last glacial period has been in several stages, with again warmer periods alternating with cold intervals. These are the Heinrich (1988) events, with a near-periodicity of 6–7 kyr, and the Dansgaard-Oeschger cycles (Dansgaard *et al.*, 1989) with an average period around 1–2.5 kyr. Rapid changes in temperature, of up to one half of the amplitude of a typical glacial-interglacial temperature difference, occurred during Heinrich events and somewhat smaller ones over a Dansgaard-Oeschger cycle. Progressive cooling through several of these cycles followed by an abrupt warming defines a Bond cycle (Bond *et al.*, 1995). In North Atlantic sediment cores, the coldest part of each Bond cycle is marked by a so-called Heinrich layer that is rich in ice-rafted debris.

The Last Glacial Maximum (LGM) occurred at about 20,000 years ago, and the temperature difference between LGM and Eemian is about $10°C$. When the transition of the Last Glacial Maximum to the Holocene is considered in more detail, a rapid transition is observed near 12 kyr, where temperatures drop again about $5°C$. This transition, called the Younger Dryas, is also considered to be one of the Dansgaard-Oeschger events but it has been studied in much more detail.

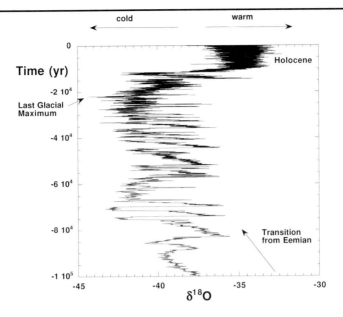

Figure 1.2. Oxygen isotope record of the last interglacial and the most recent glacial episode. A change in $\delta^{18}O$ of one unit corresponds to a temperature change of about $2°C$. Data are based on Johnsen et al. *(1997)*

1.1.2. The Younger Dryas

A magnification of the $\delta^{18}O$ record from the GRIP ice core is plotted in Fig. 1.3. The warming of the Earth from 20,000 years onward has been in several relatively distinctive stages. First, relatively fast transitions to the Bølling and Allerød interstadials occur, during which the temperature is relatively high. This is followed by a period of significant cooling between 12,500 and 11,500 years ago. The resulting stadial, during which the apparent warming trend was delayed for approximately 1,000 years, is referred to as the Younger Dryas. The period ends with a rapid shift to warmer temperatures into the beginning of the Holocene, with indications of a temperature rise of 1°C per decade (Broecker, 1995)!

Signatures of the Younger Dryas period are found at many locations on Earth using different indicators (Roberts, 1998; Bradley, 1999). In Scandinavia, reduced sedimentation and foraminiferal production is found in the north Norwegian Sea. Indications of changing vegetation have been found in southern Alaska, with expansion of tundra as a reaction to colder conditions. In the temperature record for northern Britain, which is reconstructed through Coleopteran (beetles) data (Lowe *et al.*, 1995), a very rapid change is observed at the beginning of the Younger Dryas (Fig. 1.4A). July temperatures dropped a few degrees during this period and at the end of the period a fast increase in temperature occurs. Sea-level reconstructions from drowned coral reefs at Bermuda (Fairbancks, 1990), shown

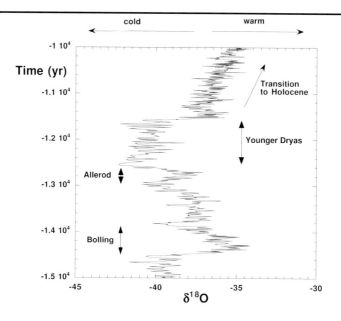

Figure 1.3. Oxygen isotope record from the GRIP ice core (shown in Fig. 1.2) over a smaller window of time, showing the Bølling and Allerød interstadials and the Younger Dryas stadial.

in Fig. 1.4D, indicate that during the Younger Dryas, the trend of increasing sea levels set by the deglacation from the Last Glacial Maximum has been retarded.
 The ice accumulation rate at Greenland (Fig. 1.4B) is small during the Younger Dryas and its dust content (Fig. 1.4C) is relatively high, which indicates that precipitation has been reduced at (northern) high latitudes.
 In the Southern Hemisphere, signatures of the Younger Dryas period are present as well, although not as clear as in the Northern Hemisphere. Evidence for a colder period has been found from pollen data of the southern part of South America and there are some indications that glaciers advanced during this period in New Zealand. On the other hand, analysis of a South Chinese Sea core shows a warming at 13,000 years ago, which continues uninterrupted into the Holocene. It therefore appears that the Younger Dryas is mainly a northern hemispheric, and in particular a North Atlantic, phenomenon and effects have propagated over the globe. A more detailed description of the climate changes during the Younger Dryas period can be found in Chapter 14 of Ruddiman (2001).

1.1.3. The Little Ice Age

The end of the Younger Dryas marks the beginning of the Holocene during which climate has been relatively stable (McManus *et al.*, 1994) with globally averaged temperature variations limited by an amplitude of about 2°C. Fossil

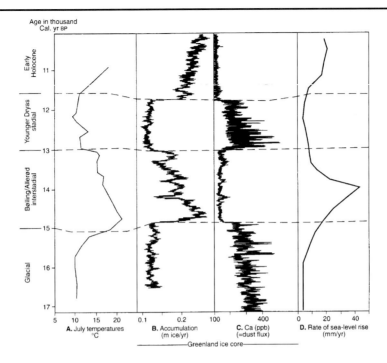

Figure 1.4.   Changes in proxy indicators during the Younger Dryas Period (Roberts, 1998). **A.** *Mean July temperatures in northern Britain based on insect (Coleopteran) data.* **B.** *Ice accumulation rate in the GISP2 ice core.* **C.** *Atmospheric dust in the GISP2 summit ice core.* **D.** *Rate of sea-level rise.*

records and lake level data indicate that during the climatic optimum, about 5,500 years ago, the temperature was about 1.5°C warmer than present and since then global temperatures have declined. Within the last two millennia, variability on century time scales is observed with an amplitude of about 1.0°C. For example, tree-ring data indicate that the Middle Ages were relatively warm and were followed by a colder period which is referred to as the Little Ice Age (Matthes, 1940; Grove, 1988). Although the conventional period of the Little Ice Age is between 1500 and 1850, a series of post-Medieval cool events started already in the fourteenth century. These events varied from region to region and it does not appear to have been uniformly colder in all regions over the whole period (Bradley, 1999). This can also be seen in a reconstruction of northern hemispheric temperature anomalies (with respect to the 1881-1960 mean) from tree-ring data, as plotted in Fig. 1.5.

These climate fluctuations had a significant impact on the lives of people in Europe. The years 1314 and 1319 saw harvests fail over large parts of Europe,

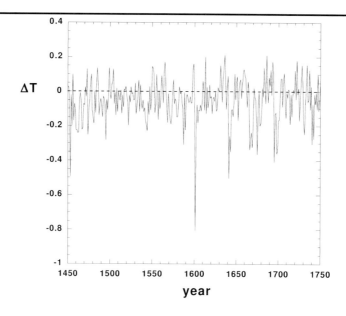

Figure 1.5.   Reconstruction from tree-ring data of the northern hemisphere temperature deviation ΔT over the period 1450-1750 from the mean temperature over the period 1881-1960; data from Briffa et al. (1998).

probably due to extremes in weather. Travel to Greenland became more difficult in these periods and finally people disappeared over there towards the end of
the fifteenth century. The river Thames was frozen over several winters during
the Little Ice Age, and one could skate on many Dutch rivers, as is beautifully
painted by the masters of that time. It was considerable more wet during this period and glaciers advanced, in correspondence with the decrease in temperatures
(Roberts, 1998). This also had immediate impact on the communities of people
during that time, in particular in the areas where conditions for growth of food are
critical. The timing of European glacier advances shows remarkable consistency
with those in other regions of the world, for example Canada, Alaska and Asia
(Grove, 1988). This may indicate that the Little Ice Age had global signatures,
but the issue is not yet settled. It has been well established that the cooling period
had ended by the mid-nineteenth century (Bradley, 1999).

1.1.4. Causes of past climate variability

In the previous subsections, typical examples of both fairly regular and dramatic climate change were shown. There is ample evidence that even such
changes as the Little Ice Age had major impacts on human beings. The records
force us to pose the question of how these climate changes were caused and not

only because of scientific curiosity: if these happened in the (far) past, they may happen again. Causes of climate change can be roughly divided into two types: external causes and internal causes.

1.1.4.1 External forcing: orbital and solar variations

Clear examples of external causes are variations in solar insolation (the amount of solar radiation arriving at the top of Earth's atmosphere), volcanic activity and continental drift. Obviously, the major forcing of the climate system is the radiation received on Earth from the Sun. However, this input of energy is not constant

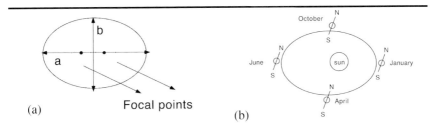

(a) Focal points   (b)

Figure 1.6.   (a) Sketch of the elliptical orbit of the Earth with the Sun in one of the focal points. The distance a is the length of half the major axis of the ellipse and likewise b is the length of half the minor axis; the eccentricity e is defined as $e^2 = 1 - (b/a)^2$. (b) The axis of the Earth makes an angle with respect to the normal of the surface which the orbit encloses; the angle is currently 23.5°. This angle is the origin of the seasons on Earth through its effect on the incidence of radiation. It is sketched here together with the current timing of the Earth-Sun distance which is minimal (maximal) in January (June).

at a particular point on Earth but varies as we know with season. This is mainly due to the tilt of Earth's spin axis — the obliquity — with respect to the normal of the plane of Earth's orbit around the Sun (Fig. 1.6). The Northern Hemisphere receives more heat during March - September while during the rest of the year preferential heating occurs over the Southern Hemisphere.

This seasonal contrast is modulated because the orbit of the Earth is not a perfect circle but rather an ellipse, with eccentricity $e^2 = 1 - (b/a)^2$; the Sun is located in one of its foci (Fig. 1.6a). The distance between the Earth and the Sun varies over the year and more energy is received when the Earth is closer to the Sun. At the moment, the Earth is closest to (farthest from) the Sun in January (June), such that the seasonal contrast is smaller in the Northern Hemisphere than in the Southern Hemisphere (Fig. 1.6b).

Over long time scales, the orbital characteristics of the Earth change and there are three types of motion relevant for the insolation on Earth. First, the spin axis of the Earth undergoes precession, which induces a shift of the seasons along the orbit (Fig. 1.7). About 12,000 years ago, the Earth was closest to the Sun in June and hence the seasonal contrast was larger in the northern hemisphere. The net effect of the precession of Earth's orbit is a fluctuation in solar insolation with a period of about 23,000 years. In addition, both the obliquity and the eccentric-

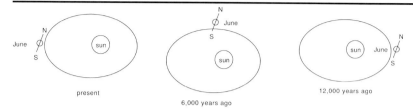

Figure 1.7.   Sketch of the effect of precession of Earth's spin axis on the phasing of the position of maximum incidence of radiation in the northern hemisphere, with (left) present situation, (middle) 6,000 year ago and (right) 12,000 year ago.

ity of Earth's orbit undergo periodic variations. The tilt angle changes in 41,000 years between $22°$ and $24°$ leading to variations in seasonal contrast, and the eccentricity varies from 0.0 (perfect circle) to about 0.05 with periodicities 100,000 and 450,000 years. Tilt changes are felt more strongly at high latitudes (if there was no tilt, the poles would receive no radiation) whereas the variations in eccentricity are felt over all latitudes. All these orbital variations do not cause any substantial change in the annual-mean solar insolation, but they induce changes in the seasonal contrast. Both precession and eccentricity have opposite effects in the Northern and Southern Hemisphere, but the effects of obliquity variations are similar in both hemispheres. This results in an equatorially-asymmetric effect of orbital changes on the insolation.

A plot of the June insolation at $60°$N for the last million years (Berger and Loutre, 1991) is shown in Fig. 1.8. For June, the amount of radiation decreases from north to south; in December this situation is reversed. The 100,000 year component of the signal is relatively weak, whereas the 23,000 and 41,000 components are quite strong. More details on the spectra of these time series can be found in Berger (1978) and in Bradley (1999). The different periods in the solar forcing variability are called Milankovitch cycles and the net changes in insolation the Milankovitch forcing. Because ice sheets in glacial times were located at northern high latitudes and since the mass balance of glaciers is sensitive to changes in summer heating, ice volume changes are most likely related to variations of insolation at $60°$N. Spectral analysis of the $\delta^{18}O$ record in ice cores has given fairly convincing evidence of the presence of the 23,000 and 41,000 year periods (Imbrie and Imbrie, 1980).

Apart from the orbital changes, the radiation from the Sun also changes due solar variability, in particular those associated with the 11 year Sunspot cycle. During this cycle, the number of Sunspots varies and the output of the Sun increases with increasing number of Sunspots. The peak-to-peak variations in the intensity are about 0.2% of the mean. During the Little Ice Age there was a minimum in Sunspots (the Maunder minimum) and hence a slightly smaller amount of insolation. Although it may have been a factor in the origin of the Little Ice Age, there is no obvious direct relation. The timing is far from perfect since the Maunder

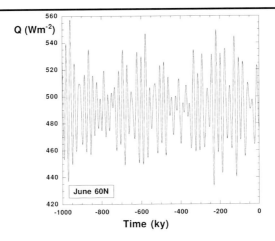

Figure 1.8. Variation of net insolation over the last million years for June at 60° N (data from Berger and Loutre (1991)).

minimum does not cover the whole period and also the variations in solar output are very small to be directly responsible for the temperature variations.

Clearly, variations in insolation have an effect on the climate system, leading to temperature variations on the same time scales as the forcing frequencies. One could view this as the deterministic linear response to the forcing and things would be dull if all climate variability could be explained in this way. Interestingly enough, however, it can not.

1.1.4.2 Internal variability

Looking at midlatitude weather maps, we discover a strong variability at time scales of 3-7 days associated with the development of high- and low-pressure systems. This variability is not related to any variation in the solar forcing and is known to develop through instabilities of the mean atmospheric flow. It is an example of so-called internal variability as the time scale is determined by nonlinear processes in the climate system itself (here the atmosphere). This variability would occur even when the forcing would be constant in time.

Internal variability is not restricted to the atmosphere, but also occurs in the other components of the climate system. In the ocean, internal processes cause the formation of ocean rings, ocean eddies and the meandering of the ocean currents such as the Gulf Stream. Going through the different components of the climate system, it is recognized that nonlinear processes are in every component and can lead to complex behavior on many different time scales. At the low-frequency end, there are the changes in land-ice distribution and at the high-frequency end, there are the variations found in the atmosphere. Interactions between components, for example in the ocean-ice system or the ocean-atmosphere system, may induce internal variability on other time scales than those present in the uncoupled systems.

The variability found on much longer time scales may also have internal causes. For example, the amplitude of the spectral component of the 100,000 years Milankovitch cycle is very weak and yet, there appears a strong signal of this period in the climate record over the last 700,000 years. Is the amplitude of this 100,000 year forcing signal simply amplified by linear processes in the climate system or are nonlinear processes involved (Ghil, 1994)? Also, the variations on suborbital time scales, for example the Dansgaard-Oeschger cycles, have no direct link to orbital variations in solar forcing and the climate system itself is likely to be responsible for this type of variability. Volcanic activity may have been an important factor, but is not expected to lead to the fairly regular cycles as observed in the climate record.

In summary, climate variations range from the large-amplitude climate excursions of the past millennia to smaller-amplitude fluctuations on shorter time scales. Several spectral peaks of variability are clearly related to forcing mechanisms; others simply can not. Processes internal to the climate system can also give rise to spectral peaks that are not related directly to the temporal variability of the forcing. Hence, even if the external forcing were constant in time — i.e., if no systematic changes in insolation or atmospheric composition (trace gases, aerosols) would occur — the climate system would display variability on many time scales. It is the interaction of this highly complex intrinsic variability with relatively small time-dependent variations in the forcing that is recorded in the proxy records and instrumental data.

The ocean, and in particular its circulation, takes a central position in this book as a starting point to understand the whole spectrum of processes which cause internal variability in the climate system. It is therefore time to have a closer look at the ocean circulation.

1.2. The Present Ocean Circulation

This section is devoted to an overview of the mean forcing fields of the ocean circulation (section 1.2.1), and the properties of the circulation and the water masses involved (section 1.2.2). The global ocean moderates climate through its large thermal inertia, its capacity to store enormous amounts of heat, and its poleward heat transport through ocean currents (section 1.2.3). The effects of the ocean circulation on the climate system will be addressed in the last section (section 1.2.4).

1.2.1. Surface forcing

On the large scale, the ocean circulation is driven by momentum fluxes (by the wind) and by fluxes of heat and freshwater at the ocean-atmosphere interface. The latter fluxes change the surface density of the ocean water and through mixing and advection, density differences are propagated horizontally and vertically. It is not straightforward to determine the annual-mean forcing fields and several methods have been used to compile these fields. Nearly all these data are available online (see, for example, http://ingrid.ldgo.columbia.edu/).

The annual-mean wind stress, as compiled in Trenberth *et al.* (1989), is shown in Fig. 1.9. Each arrow shows the direction of the wind stress and its length is an indication of its magnitude, with the arrow at the bottom having an amplitude of 0.25 Pa. Large amplitude easterly winds can be seen in equatorial latitudes, forming the trade wind system. Maximum wind stress values in this system are about 0.2 Pa and occur to the north of the equator in the eastern Pacific. At mid-latitudes, a belt of strong westerly winds is seen, with a maximum amplitude of about 0.3 Pa, which are particularly strong in the North Atlantic and over the Southern Ocean.

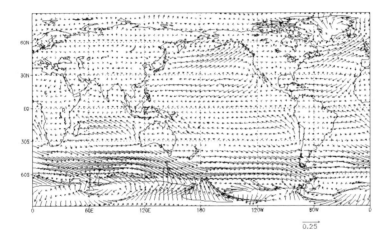

Figure 1.9. Wind-stress distribution (Pa) at the ocean - atmosphere interface as compiled by Trenberth et al. *(1989); the arrow shows a strength of 0.25 Pa.*

The total downward annual-mean heat flux into the ocean as compiled in Ober-huber (1988) is plotted in Fig. 1.10; the contour interval is 50 Wm^{-2}. There is net heat input near the equator and net heat loss at higher latitudes with a strong zonal asymmetry within the North Atlantic and North Pacific. In the areas near the eastern side of the continents, apparently the ocean is much warmer than the atmosphere with up to 150 Wm^{-2} transferring from the ocean to the atmosphere. The zonally averaged heat flux is fairly symmetric about the equator.

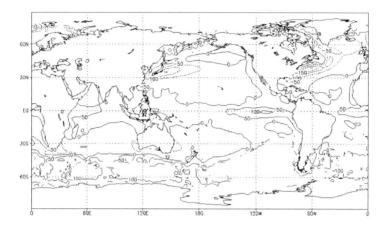

Figure 1.10.   Net downward annual-mean heat flux (in Wm^{-2}) into the ocean obtained from the Oberhuber (1988) data set.

The freshwater flux, compiled again by Oberhuber (1988), is plotted in Fig. 1.11 and shows a fairly zonally homogeneous pattern in both Atlantic and Pacific. The precipitation areas near the equator associated with the Intertropical Convergence Zone can be clearly distinguished. For example, in the eastern Tropical Pacific, values of 150 mm month^{-1} are found just north of the equator. In the subtropics, there are broad zones of net evaporation, for example in the North Atlantic and in the Southern Indian Ocean, with values of 100 mm month^{-1} excess evaporation over precipitation. The zonally averaged profile of the freshwater flux does not show any strong equatorial asymmetry, although the data from different sources show substantial variations (Zaucker *et al.*, 1994).   From the heat flux and freshwater flux, the surface buoyancy flux can be determined. The buoyancy flux and momentum flux largely determine the forcing of the ocean. There are other (more localized) sources which may be important, such as the heat released by deep ocean volcanic activity but these are not further considered here.

1.2.2. Ocean circulation patterns

A classical textbook picture of the surface ocean circulation (Peixoto and Oort, 1992) is shown in Fig. 1.12. This figure, although wrong in certain details, gives a good first impression of the current systems. The major current in the Southern

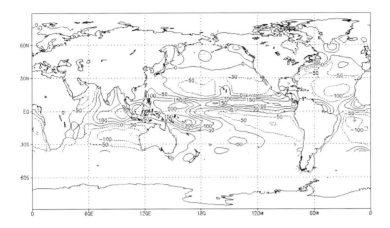

Figure 1.11.   Downward freshwater flux into the ocean as obtained from Oberhuber (1988). Units are in mm/month and contour levels are 50 mm/month.

Ocean is the Antarctic Circumpolar Current (ACC), which encircles the Antarctic continent from west to east. The ACC has an average volume transport of about 150 Sv (1 Sv = 10^6 m^{3}s^{-1}). In the southern hemisphere, the circulation between the ACC and the equatorial currents is dominated by cellular type circulations (called gyres), which rotate anti-clockwise. Although this is not so clear in the picture, the currents at the western side of each basin are the strongest, i.e., the East Australian Current in the South Pacific, the Brazil Current in the South Atlantic and the Agulhas Current in the South Indian Ocean. The major currents near the equator are westward, as one would expect from the direction of the trade winds. The eastward equatorial countercurrents just north of the equator in both the Atlantic and Pacific are peculiar features.

In the North Pacific and North Atlantic, cellular type motions are seen again, with a stronger clockwise rotating (subtropical) gyre and a weaker anti-clockwise (subpolar) gyre. Also here, currents at the western side of the basin are strongest, with the Gulf Stream in the Atlantic and the Kuroshio in the Pacific as major currents. The Gulf Stream can be viewed as an eastward jet being part of both the Atlantic subtropical and subpolar gyres. Typical horizontal velocities in the Gulf Stream are up to 1 ms^{-1}, whereas depth averaged velocities in the gyres

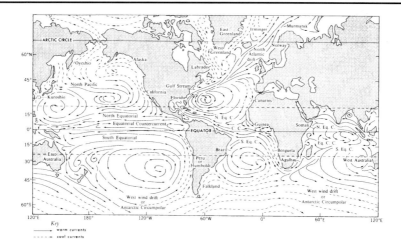

Figure 1.12.   Sketch of the global surface ocean circulation (Peixoto and Oort (1992)).

Some typical transports of the ocean currents		
Current	Location	Value
Agulhas[(a)]	31°S, Indian	70 Sv
Gulf Stream[(b)]	26°N, Atlantic	32 Sv
Gulf Stream[(c)]	38°N, Atlantic	88 Sv
Brazil Current[(d)]	28°S, Atlantic	22 Sv
Kuroshio[(e)]	25°N, Pacific	22 Sv
Kuroshio[(f)]	33°N, Pacific	57 Sv
East Australian [(g)]	30°S, Pacific	22 Sv
AAC[(h)]	150°E, Southern	147 Sv
ACC[(i)]	60°E, Southern	137 Sv

Table 1.1.   Some typical transports of the ocean currents as determined during WOCE. These transports can, for example, be used as checkpoints of ocean modeling output. (a): Bryden and Beal (2001); (b): Baringer and Larsen (2001); (c): Johns et al. (1995); (d): Mueller et al. (2000); (e): Johns et al. (2000); (f): Imawaki et al. (2001); (g): Mata et al. (2000); (h): Rintoul et al. (2001); (i): Cunningham et al. (2003).

are of order 0.01 ms^{-1}. Typical volume transports of the Gulf Stream and other western boundary currents are given in Table 1.1.

The circulation of heat and salt through the ocean basins is called the thermohaline circulation (Wunsch, 2002), usually abbreviated with THC. Since the transport of both salt and heat is quite advection dominated and both quantities are not mixed very well once deep below the surface, a particular amount of water can be traced back to its origin. Hence, such a volume of water can be characterized by its temperature and salinity at formation and is called a *water mass*. For example, in the North Atlantic, the relatively warm and saline water transported by the Gulf Stream is cooled on its way northward. In certain regions, i.e., the Greenland Sea and the Labrador Sea, the water column becomes unstably stratified and vigorous convection occurs. The net effect is the formation of a relatively dense water mass called the North Atlantic Deep Water (NADW), with a temperature of 2–4°C and a salinity of 34.9–35.0 psu. This water is transported southwards at mid-depth as a deep (western boundary) current, it crosses the equator and connects to the water masses of the Southern Ocean.

In the North-Pacific, no deep water is formed because the surface waters are too fresh and hence there is no equivalent of NADW. Deep water formation also occurs near the Antarctic continent. In the Pacific, this inflow of heavy deep water is compensated by a surface return flow which again connects with water masses in the Southern Ocean. The water mass entering the Atlantic from the south is an even denser water mass than the North Atlantic Deep Water, called the Antarctic Bottom Water (AABW), with a temperature of -0.5–0.0°C and a salinity of 34.6–34.7 psu. The outflow of NADW in the Atlantic is, apart from AABW, also compensated by surface inflow of water coming from the Indian Ocean and water coming through Drake Passage (Schmitz, 1995). For the Atlantic basin, the structure of these water masses shows up as layers, which gives a stepwise impression of the vertical stratification in the basin. An important additional water mass found between 20°N and 40°N, with a temperature of 10°C and a relatively high salinity of 35.5 psu, has its origin in the Mediterranean.

The three-dimensional flow of different water masses through the ocean basins has been termed (Gordon, 1986; Broecker, 1991) the 'Ocean Conveyor'. The simplest picture (Schmitz, 1995) arises when the vertical structure of the flow field is separated in a shallow and a deep flow (Fig. 1.13). In this figure, the dark (light) shaded flow is the surface (deep) water and the numbers indicate volume transports in Sv. Fig. 1.13 suggests a strong coherence of the flows in the basins. The three-dimensional ocean circulation, however, is a complex flow with different levels of coherence on different scales. Its properties, even at the very large scales are not well-determined yet because of a lack of observations over the whole globe. Analysis of the section data of the World Ocean Circulation Experiment (WOCE, see http://oceanic.cms.udel.edu/woce/), combined with inversion studies have lead to a more detailed estimates over the volume transports through the world oceans (Ganachaud and Wunsch, 2000). In Fig. 1.14, the zonally integrated mass transports over several sections are presented. The boundaries between water masses are taken as certain density surfaces (defined by a value of the quantity γ^n). In this way, the red arrows represent the surface transport, the blue arrows show the transport at intermediate depths and the green arrows indicate the transport in the deep ocean. Upwelling and downwelling are indicated by arrows and dots, respectively and their color indicates from which level the water is coming.

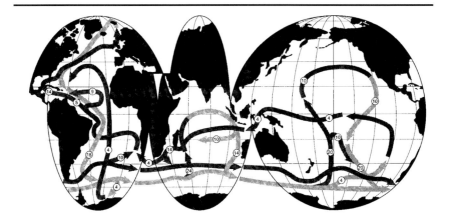

Figure 1.13.   Sketch of the three-dimensional ocean circulation as a two-layer flow (after Schmitz (1995), but the figure is taken from Bradley (1999)) with dark (light) shaded indicating flow of surface (deep) water.

1.2.3. Heat and freshwater transport

The oceans take care of about one third to a half of the total meridional heat transport of the combined ocean-atmosphere system. The total meridional heat transport due to the ocean circulation is difficult to measure directly and only a few estimates at certain locations have been obtained (Hall and Bryden, 1982; Wunsch *et al.*, 1983; Bryden *et al.*, 1991; Rintoul, 1991). Recent inversion studies of the WOCE-section data have lead to section estimates (Ganachaud and Wunsch, 2000) and a summary result is presented in Fig. 1.15. The meridional heat transport in the Atlantic is positive over the whole basin with a maximum of about 1.3 PW at 30°N. In the Pacific, the heat transport is at least a factor two smaller than in the Atlantic. The meridional heat transport in the Indian Ocean is mainly southward with a maximum of 1.8 PW near 20°S. Best estimates of the zonally averaged meridional heat transport are also presented in Ganachaud and Wunsch (2000) with a maximal northward heat transport of about 1.8 PW at 30°N.

Estimates of the freshwater transport through the oceans are also hard to obtain from direct observations. As can be seen from the surface freshwater flux in Fig. 1.11, there is net precipitation in the tropical, middle and high-latitude regions, and there is net evaporation in the subtropics. This leads to a net surface freshwater flux which is fairly equatorially symmetric. The ocean circulation must transport water into the evaporative zones and away from precipitation regions for compensation. Wijffels *et al.* (1992) present estimates of this freshwater transport (Fig. 1.16) and demonstrate the importance of the Bering Strait through-flow. The Pacific is a net precipitative basin with much of the gain occurring between 0-15 °N (the location of the Intertropical Convergence Zone), while the Atlantic and

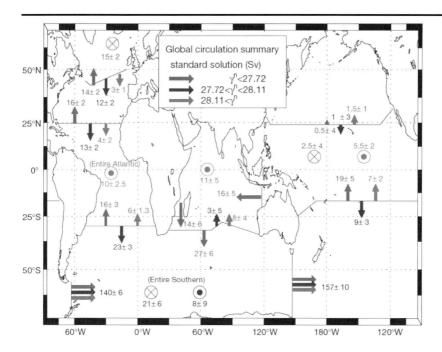

Figure 1.14. (in color on page 517). Estimated section integrated mass transports as determined in Ganachaud and Wunsch (2000) from the WOCE data.

Indian Ocean are evaporative basins. Over the whole North Atlantic Ocean, there is southward transport of freshwater with a maximum of about 0.95×10^{9} kgs^{-1} at 60°N.

1.2.4. Ocean circulation and past climate variability

With the limited information of the ocean circulation as provided in the previous section, already effects of changes in the ocean circulation on climate can be anticipated. Most of the heat transport in the Atlantic is determined by the overturning component of the circulation. Warm surface water moves northwards, sinks and the mass balance is closed by cold deep water moving southward, which effectively induces heat transport northward. In the gyre part of the circulation also warm surface water moves northwards, but the mass balance is closed by slightly cooler water moving southward. Since the THC is believed to be strongly influenced by the surface buoyancy forcing, changes in the buoyancy of the upper ocean can lead to substantial changes in poleward heat transport and hence to climate changes.

Are there any indications that these changes indeed happened, for example, during the Younger Dryas? For a full account of the evidence of these changes,

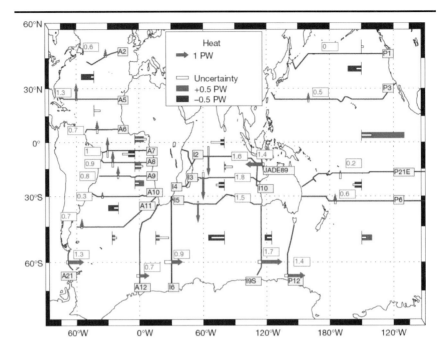

Figure 1.15.   (in color on page 517). Estimated section averaged heat transport over WOCE sections (Ganachaud and Wunsch, 2000); 1 PW = 10^{15} W.

sections 6.10 and 6.11 of Bradley (1999) are recommended. The interpretation of these changes is still under discussion, but several scenarios have been proposed. One of these is that meltwater pulses into the North Atlantic, due to the melting of the ice caps in the transition from the Last Glacial Maximum to the Holocene, reduced the surface salinity significantly. For example, indications for such an increased flux at about 14,500 years ago have been attributed to meltwater coming from the Fennoscandian ice sheet. The lower surface salinity in the North Atlantic may have shut down formation of North Atlantic Deep Water, which in turn interrupted the THC. As a consequence, less heat was transported northward and cooling occurred leading to the Younger Dryas period (Rooth, 1982; Broecker *et al.*, 1985; Broecker, 2000).

Alternative scenarios for the cause of the Younger Dryas are discussed in (Berger and Jansen, 1994); one of these involves the Bering Strait transport. During the Last Glacial Maximum, the global sea level was about 100 m lower than today and the Bering Strait was closed off by land. When sea level was rising, at some point (timed near the Younger Dryas onset) the Bering Strait was flooded and a current from the Pacific to the Atlantic developed. This flow discharged an enormous amount of sea ice into the North Atlantic, which again through melt-

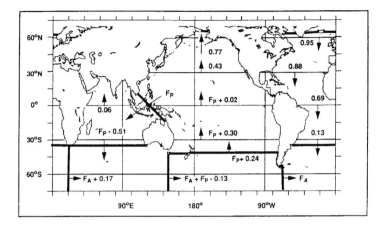

Figure 1.16. Meridional freshwater transport (in 10^9 kg/s) in the ocean, with the quantities F_P and F_A referring to the freshwater transport of the Pacific-Indian through flow and that of the Antarctic Circumpolar Current at Drake Passage, respectively (Wijffels et al., 1992).

water, reduced the strength of the THC. It is still unclear which scenario is most consistent with the present database of observations; both scenarios, however, involve changes in the THC.

1.3. Present Climate Variability

In this section two examples are given, which show the involvement of the ocean circulation in climate variability on interannual-to-multidecadal time scales. The first example is the El Niño /Southern Oscillation phenomenon in the equatorial Pacific, which will be discussed in chapter 7. The second example is the less known Atlantic Multidecadal Oscillation (AMO) in the North Atlantic, of which the physics is a main topic in chapter 6.

1.3.1. ENSO

Once about every four years, the sea-surface temperature (SST) in the eastern equatorial Pacific increases by a few degrees over a period of about one year. These events are called El Niño (literally: the little boy), referring to the Christ Child, since the maximum of the event is usually around December. This phenomenon arises through large-scale interaction between the Pacific Ocean and the overlying atmosphere and is associated with variations in the equatorial surface winds. Normally, the equatorial Pacific surface winds, the trade winds, are directed westward and are driven by a pressure difference between a high-pressure region in the east (e.g., at Tahiti) and a low-pressure region in the west (e.g., at Darwin). During an El Niño , the pressure is lower than normal in Tahiti and higher than normal in Darwin. These variations in atmospheric pressure are known as the Southern Oscillation. (Philander, 1990). The El Niño /Southern Oscillation (ENSO) phenomenon is the most prominent interannual signal in the climate system and has large effects on the weather, even far outside the Pacific basin. During the strong El Niño -event in 1997-1998, anomalies in the eastern Pacific SST exceeded 5°C locally, and the trade winds changed direction in the western equatorial Pacific. The associated shift in convection zones caused severe drought in Indonesia and torrential rains in Peru and Ecuador.

ENSO is one of the best studied climate phenomena, and from observations the relationship between the oceanic and atmospheric variables, the relevant spatial structures and their temporal development are well-known (Wallace *et al.*, 1998). The time-averaged SST is characterized by a cold tongue (Fig. 1.17a) of 24°C water in the Eastern Pacific and a warm pool of 30°C near the western boundary of the basin. The 1997/1998 El Niño had one of the largest amplitudes of this century; the SST anomaly pattern for December 1997 (with respect to the mean state of Fig. 1.17a) is plotted in Fig. 1.17b. El Niño is seen as a basin wide SST perturbation with a (monthly averaged) maximum amplitude near the South-American coast of about 3°C.

One of the measures of the temperature variations of the eastern equatorial Pacific is the NINO3 index, which gives the SST anomaly with respect to the mean state averaged over the box [150°W-90°W] × [5°S-5°N]; this index is positive during an El Niño . The thick curve in Fig. 1.18 shows the course of this index from 1900-2000. El Niño episodes occur once every three to seven years and last more than one year, with substantial variations in strength. The strongest El Niño 's were those of 1982/1983 and 1997/1998. The sea level pressure anomaly pattern of El Niño can be obtained through correlation with the NINO3 index. The pattern is more global than that of sea-surface temperature. An index that captures the amplitude of this sea level pressure pattern is the Southern Oscillation Index (SOI), which is the normalized difference of the pressure anomalies between Tahiti (18°S, 150°W) and Darwin (12°S, 131°E). The SOI is plotted as the thin curve in Fig. 1.18. When this index is negative, the westward trade winds are weak and vice versa.

Although the strong negative correlation between SOI and NINO3 in Fig. 1.18 is obvious, it took until 1969 (Bjerknes, 1969) before it was realized that the changes occurring in the atmosphere and the ocean are closely related. Warm water in the eastern Pacific causes a weakening of the trade winds, which in turn drives changes in the oceanic circulation which influence SST. The El Niño /Southern Oscillation is therefore a coupled ocean-atmosphere phenomenon. The warm phase of the oscillation coincides with El Niño (positive NINO3) in the ocean and with weak trade winds (negative SOI) in the atmosphere. The cold phase (also called La Niña meaning the little girl) coincides with strong trade winds (positive SOI) and lower than normal SST (negative NINO3) in the eastern part of the equatorial Pacific.

(a)

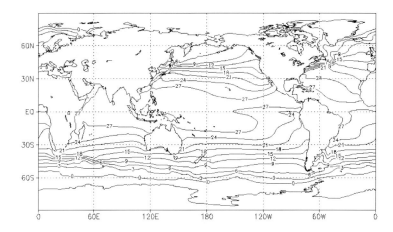

(b)

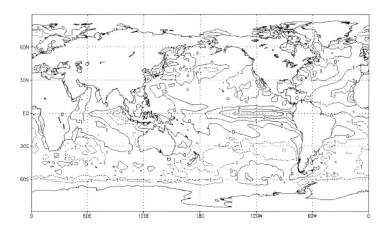

Figure 1.17. (a) Annual-mean SST in the Tropical Pacific from the Reynolds and Smith (1994) data set, with contour levels in °C (b) SST anomaly pattern for December 1997 with a maximum amplitude of 3°C.

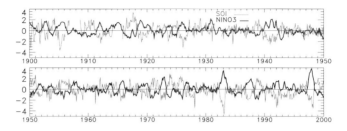

Figure 1.18.   Time series of the SST anomaly averaged over the box [150° W-90° W] × [5° S-5° N] (NINO3) and the difference in sea level pressure (SOI) between Tahiti (Eastern Pacific) and Darwin (Western Pacific). The figure is taken from Dijkstra and Burgers (2002).

1.3.2. The Atlantic Multidecadal Oscillation

In Fig. 1.19a, the famous Central England annual-mean temperature time series (available through the British Archive and Data Centre, http://www.badc.rl.ac.uk/) is plotted.   As can be seen by inspection, variability on interdecadal-to-multidecadal time scales seems likely, but there is also smaller time scale variability. Plaut *et al.* (1995) analyzed this data set in detail and found strong indications of interannual and interdecadal variability with a maximum amplitude of about 1°C (Fig. 1.19b). In general, it is difficult to extract the signal of variability (pattern and amplitude) on decadal and larger time scales because of the low signal-to-noise ratio.

The North Atlantic SST appears to have a distinct signal of multidecadal variability. Schlesinger and Ramankutty (1994) presented first indications of this variability from a singular spectrum analysis of four global-mean temperature records. Through objective analysis of the SST records in the North Atlantic over the years 1950-1990, Kushnir (1994) found a specific pattern characterizing the difference between the years 1950-1964 and 1970-1984. The dominant SST pattern found is basin scale, is strongest in winter and displays maximum amplitudes in the vicinity of Iceland and the Labrador Sea (Fig. 1.20). The North Atlantic was warmer in 1950-1964 than in 1970-1984.

By subsequent analysis of longer and better quality SST and sea level pressure fields, the pattern of multidecadal variability has been characterized more accurately (Deser and Blackmon, 1993; Latif, 1998; Moron *et al.*, 1998). Delworth and Mann (2000) extended the instrumental record with proxy data and demonstrated that there is a significant spectral peak in the 50-70 year frequency band. It was named the Atlantic Multidecadal Oscillation (AMO) by Kerr (2000) and an AMO index was defined by Enfield *et al.* (2001) as the 10-year running mean of the detrended SST anomalies north of the equator (Fig. 1.21).

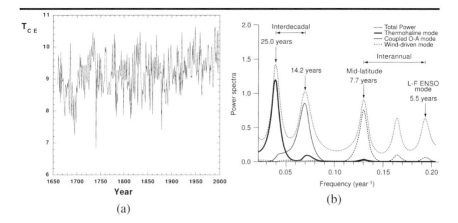

Figure 1.19. *(a) Plot of the annual-mean Central England Temperature from 1650 to present. The data were made available by the British Archive and Data Centre (http://www.badc.rl.ac.uk/). (b) Spectrum of this time series as determined in Plaut* et al. *(1995), showing energy at interannual, decadal and interdecadal time scales.*

Enfield *et al.* (2001) showed that there is a strong negative correlation of the AMO with US continental rainfall, with less (more) rain over most of the central US during a high (low) AMO-index period. The Mississippi outflow is about 5 % smaller than average during high-AMO periods. High-positive correlations

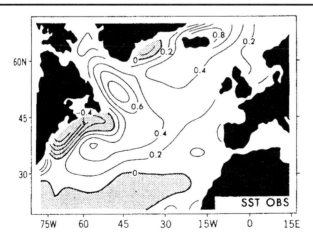

Figure 1.20. *Pattern of multidecadal SST anomalies obtained as the difference in SST of the relatively warm period 1950-1964 and the relatively cold period 1970-1984 (Kushnir, 1994). The figure is from Latif (1998).*

have also been found between the AMO and Sahel rainfall and between the AMO
and the hurricane intensity in the Atlantic (Gray *et al.*, 1997). During the positive AMO-index period 1950-1964, there were 47 intense (class 3-4-5) hurricanes
originating east of $60°$W, whereas in a same length low AMO-index period 1970-
1984, there were only 19.

1.4. Physics of Climate Variability

Many more examples of low-frequency climate variability have been found
in observations and although it would be desirable to have a text where most of
them were described, it is not the purpose of this book to cover that material.
Rather, focus will be on the more abstract problem related to the origin of the
variability. Can the physics of the variability be identified and the patterns of the
different fields and their temporal development be understood through elementary
mechanistic description?

1.4.1. The system view

On choosing a particular time scale of variability, it makes sense to decide how
to handle much smaller and much larger time scales of variability. In studying
the present ENSO variability, one obviously does not need to consider continental
drift and variations in the solar insolation due to orbital changes Hence, the physical processes of relatively large time scale changes do not have to be taken into
account and the situation they create can be assumed as a fixed boundary and/or
forcing condition. But what to do with the smaller time scale variations, for example, the variations of the weather in the Tropical Pacific on a day to day basis? In
ENSO studies, these rapid atmospheric fluctuations can be modelled by assuming
a random component in the response of the atmospheric winds to SST anomalies,
superposed on the deterministic response. In this way, when the space and time
scales of the phenomena under study are chosen, the evolution of the deterministic system at these scales is subjected to noise with given statistical properties,

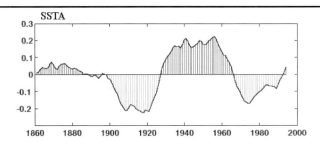

*Figure 1.21.   Detrended 10-year running mean of the Atlantic SST anomaly north of the equator
over the period 1860–1995 (Enfield et al., 2001).*

while all long time scale variations are considered fixed through the boundary and forcing conditions.

It is possible that high-frequency variability (noise) in one component of the system (e.g., the atmosphere) is able to create low-frequency variability when coupled to a slower (e.g., the ocean) component (Hasselmann, 1976). This is the 'default' explanation of interannual climate variability at midlatitudes (Frankignoul, 1999). However, it is unlikely that a 'noise' driven linear system can explain all of the low-frequency variability, because certain preferential spatial patterns and preferential time scales seem to be involved. If so, the nonlinear nature of physical processes within individual components of the system and interactions between these components is likely to be important for the selection of these patterns. The question then becomes which physical processes are responsible for the pronounced signals of climate variability in the past (Younger Dryas, Little Ice Age) as well as for the interannual-to-multidecadal variability (ENSO, AMO) in the instrumental record.

It is known that nonlinear systems can display very irregular behavior. Much of this complex behavior is related to instabilities which occur when a certain threshold is exceeded. Such a threshold indicates sensitivity of the system to particular perturbations having a certain spatial pattern. Physically, these phenomena are related to feedbacks which only become active when dissipative effects in a system can be overcome by active energy producing processes. In this way, relatively fast changes may occur and the behavior of the system may change unexpectedly. However, combined with noise, even stable states may show preferential scales, when the linearized operator on this state is non-normal (Trefethen *et al.*, 1993; Farrell and Ioannou, 1996). Rapid amplification of components may occur due to this non-normality which, combined with noise, leads to complicated behavior. In this way, noise can have unexpected effects when put on a rather passive slow system which is advecting quantities for example due to spatial resonance (Saravanan and McWilliams, 1997, 1998). When noise is added to strongly nonlinear systems, the realm of stochastic differential equations is entered (Oksendal, 1995). Initially, it was intended to cover part of this material also in this book, but it turned out to be too ambitious. Therefore, focus in this book will only be on deterministic nonlinear systems.

1.4.2. Central questions

Looking at the ocean circulation, it is likely that a three-dimensional complex and highly nonlinearly interacting stratified rotating flow is sensitive to perturbations on a large range of scales. Sensitivity can occur within one particular basin, causing changes in surface or overturning circulation. It may also occur on a global scale, involving a transition to different patterns within the 'Ocean Conveyor'. As changes in ocean currents can lead to substantial changes in heat and freshwater transports, these may influence the climate over large areas on the globe.

From this viewpoint, the main questions of low-frequency climate variability are the following:

(i) Can each of the characteristic frequencies of variability, such as the 50-year
     AMO period and the 4-year ENSO period be understood as internal oscilla-
     tions of the climate system? Can one trace the origin of these oscillations to
     instabilities of the circulation due to nonlinear interactions? Is it possible to
     identify the feedback mechanisms responsible for these instabilities?

(ii) How sensitive is the climate system to (not necessarily) small perturbations?
     For example, in view of future climate changes, it is important to know
     whether the present ocean circulation is close to undergoing a transition to
     a different mean state and to which perturbations it will be most vulnerable. In
     other words, how stable is the current climate system? Can rapid transitions,
     such as the Dansgaard-Oeschger oscillations, be explained through instabili-
     ties of the THC?

Many types of ocean models have been developed to tackle these problems and
identify the physics of both temporal variability and stability of the ocean circulation and its effect on climate. Ocean models contain many parameters, for
example, the strength of the wind-stress forcing and the mixing coefficients of
heat and salt. To understand the physics of particular phenomena in the ocean
circulation, solutions of these models are required at different values of these parameters. In this way, the influence of key physical processes on the phenomena
can be monitored and causal chains can be described.

Time-dependent solutions at values of parameters considered in the 'realistic'
range usually display a very complicated spatial-temporal behavior. Mean ocean
currents appear very sensitive to different types of perturbations. Through nonlinear interactions of these perturbations, the energy is distributed over many degrees
of freedom which, in general, leads to complex behavior. In addition, either wind
- or buoyancy forcing of the circulation (or some of the parameters) may have
a random component, which usually adds irregularity to the already inherently
irregular flow. How does one analyze the physical processes leading to this complexity?

1.4.3. Approach

In many studies, complex flows are analyzed with statistical techniques,
such as Empirical Orthogonal Function (EOF) analysis techniques and several
more sophisticated versions of these (Preisendorfer, 1988; Vautard *et al.*, 1992;
Von Storch, 1995; Von Storch *et al.*, 1995). The patterns found in this way are
associated with maximum variance of the flow in some norm. These techniques
have been extensively used in climate variability and physical oceanography but
the drawbacks are that additional modeling is needed to determine causal relationships between the responsible physical quantities. Only very recently have
techniques of nonlinear time series analysis (Kantz and Schreiber, 1997) found
application in climate research.

A second type of analysis of complex flows is to monitor integral quantities,
such as the volume averaged kinetic energy of the flow. These analyses have
been very useful to find a description of the interaction processes between mean

flows and perturbations (Pedlosky, 1987). In many ocean circulation problems, however, this type of analysis has not been performed routinely either because of the complexity involved or because integral balances may not be well satisfied in observations or ocean models.

A third way to proceed is to study how this complex behavior arises from simpler situations at different values of the parameters. For example, in a highly viscous ocean there is usually a unique sluggish flow. When the viscosity is decreased, the circulation becomes unstable through successive instabilities. Any instability introduces extra degrees of freedom which can take up energy within the flow. In general, more active degrees of freedom give more possibilities for irregularity in the flow. To understand the role of the nonlinear processes in the climate system is a tremendous task, knowing the trouble and effort it takes to understand the behavior of fluids in laboratory situations. In the latter field, for example, much understanding has been obtained by approaching problems using a hierarchy of models. For each of the models within this hierarchy, one can then follow the approach to complexity by changing parameters. The mathematical theory underlying these types of studies in nonlinear models is that of dynamical systems. The most outstanding advantage of this theory is its systematics, which allows classification of behavior.

It is this last approach which is pursued in this book and followed to understand (variability of) the ocean circulation and El Niño. A hierarchy of models is used and with techniques of the theory of dynamical systems, the solution structures of these models are analyzed. Central focus will be on equilibria of these models, which may be either steady states or periodic orbits. It is aimed to provide sufficient details of the models and parameter volumes investigated such that readers will be able to reproduce many of the results provided. In chapter 2, the origin of the different models is presented and the terminology used in the physics of the ocean circulation is introduced. The theory of dynamical systems is introduced in chapter 3 by using a simple example; this is followed by the more abstract theory. Chapter 4 provides details on the numerical techniques needed to obtain results for meaningful ocean and climate models. Application of the methodology starts in chapter 5 with the wind-driven ocean circulation and in chapter 6 these methods are applied to study the thermohaline circulation. In chapter 7, one extension into the coupled ocean/atmosphere climate system is described, focussing on the physics of El Niño in the equatorial Pacific.

1.5. Exercises on Chapter 1

(E1.1) *Radiation equilibrium temperature*

A black body with a temperature T emits an amount of radiation with a flux density $\mathcal{I}$ (in Wm^{-2}) according to

$$\mathcal{I} = \sigma T^4$$

where $\sigma = 5.67 \times 10^{-8}$ $\text{Wm}^{-2}\text{K}^{-4}$ is the Stefan-Boltzmann constant.

a. Assume that the Sun is a black body and that the average temperature of its photosphere is about 6000 K. Calculate the flux density at the photosphere.

The average distance of the photosphere to the center of the Sun is $r_p = 7.0 \times 10^8$ m. The solar constant is the flux density at a distance d from the Sun.

b. With r_e being the distance between the Earth and the Sun ($r_e = 1.5 \times 10^{11}$ m), calculate Earth's solar constant Σ_0.

c. Consider the radiation from the Sun absorbed by the Earth and assume that the Earth is a black body with a mean temperature T_e. Show that

$$\frac{\Sigma_0}{4}(1 - \alpha_p) = \sigma T_e^4$$

where α_p is Earth's mean planetary albedo. Calculate T_e for $\alpha_p = 0.3$. Why is T_e in reality much larger?

d. Variations in the solar constant Σ_0 due to the 11-year Sunspot cycle are about 0.2% (about 2 Wm^{-2}) of the mean value. Calculate the peak-to-peak changes in the equilibrium temperature T_e due to this solar variability.

Further reading: Hartmann (1994), chapter 2.

(E1.2) *Surface forcing*

Consider a layer of ocean water (with heat capacity $C_p = 4.281 \times 10^3$ $\text{Jkg}^{-1}\text{K}^{-1}$ and density $\rho = 1.027 \times 10^3$ kgm^{-3}) having a thickness of 50 m and a surface area of 10^4 m^2.

a. Calculate the amount of energy needed to warm this layer by 1 K.

Assume that the downward heat flux from the atmosphere into the ocean over the area is $Q_{oa} = 400$ Wm^{-2}.

b. How long does it take to raise the temperature of the layer by 1 K?

The mean wind speed U over the layer is 10 ms^{-1} and its direction is pure zonal. The zonal wind stress τ is calculated from a so-called 'bulkformula' as

$$\tau = C_D \rho_a U^2$$

where $C_D \approx 2 \times 10^{-3}$ is a semi-empirical constant and $\rho_a = 1.0$ kgm^{-3} is the density of air.

c. Calculate the magnitude of the wind stress (in Pa) over the layer.

d. Assume that the layer is initially motionless and that the wind stress calculated in c. accelerates the water in the layer. Calculate the zonal flow velocity in the layer after three hours.

Suppose finally that the initial salinity of the water is constant and equal to $\bar{S} = 35$ ppt. The net downward freshwater flux over the area is $F_S = 10^{-8}$ms^{-1}.

e. How long does it take to change the salinity of the layer by 1 ppt?

Further reading: Gill (1982), chapter 2.

(E1.3) *Gulf Stream*

Warm western-boundary currents such as the Gulf Stream transport heat from low-latitude regions towards polar regions and hence are important in the climate system. The Gulf Stream is about 100 km wide and 1000 m deep and has an average velocity of about 0.5 ms^{-1}.

a. Provide an estimate for the volume flux of the Gulf Stream (in Sv).

To estimate the heat transport, a characteristic horizontal temperature difference associated with the northward and southward flowing water is needed.

b. Provide an estimate for this characteristic temperature difference (consider, for example, Fig. 1.17a).

c. Provide an estimate for the meridional heat transport (the heat capacity of water is 4281 $JK^{-1}kg^{-1}$).

d. Why is this estimate different from that deduced from observations (as provided in Fig. 1.15)?

Further reading: Hartmann (1994), chapter 7.

(E1.4) *Heat transport*

In Fig. 1.22, the observed zonally averaged meridional heat transport in the three ocean basins and of the global ocean is plotted.

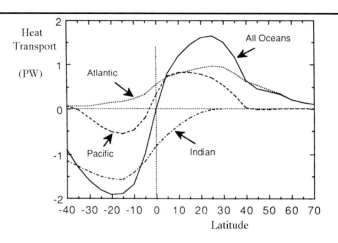

Figure 1.22.   Zonally averaged meridional heat transport in 10^{15} W over the different ocean basins and of all oceans together. The figure is slightly modified from the one in Hsiung (1985).

a. Using the fact that average deep temperatures in the North Atlantic are about 4°C, estimate the meridional heat transport of the zonally averaged North Atlantic circulation (see, e.g. Fig. 1.13).

b. Why is the meridional heat transport in the South Atlantic Ocean equatorward while it is poleward in the South Pacific?

c. Why is the meridional heat transport in the Indian Ocean poleward?

Further reading: Hsiung (1985) and Ganachaud and Wunsch (2000).

(E1.5) *Climate variability: stochastic models*

The different components of the climate system (ocean, atmosphere, cryosphere) have very different adjustment time scales to perturbations. The adjustment times in the atmosphere are typically a few weeks while those in the ocean can vary between a few weeks (upper ocean) to thousand of years (deep ocean). The relatively slow ocean component is hence forced by a rapidly varying atmosphere.

A scalar forcing field, say ζ, is called a white noise forcing if its mean $< \zeta >$ is zero and its autocorrelation $c(t) = \sigma^2 \delta(t)$ (where δ is the delta-distribution); such a field has a flat power spectrum $P(\omega)$. More explicitly,

$$< \zeta(t) > = \lim_{\tau \to \infty} \int_{-\tau/2}^{\tau/2} \zeta(t)\, dt = 0$$

$$< c(\eta) > = < \zeta(t)\zeta(t+\eta) > = \sigma^2 \delta(\eta)$$

$$P(\omega) = \int_{-\infty}^{\infty} c(t)e^{i\omega t} dt = \sigma^2$$

where τ is a typical averaging time, ω is the frequency and σ^2 is the variance of the forcing.

Suppose that we have a layer of water in which the temperature T is damped by ocean-atmosphere interaction (with damping factor α) and forced by a white noise heat flux ζ. To model the changes of the temperature T, a stochastic model (a Langevin equation) of the form

$$\frac{dT}{dt} = -\alpha T + \zeta$$

can be used (Hasselmann, 1976).

a. If $C(\eta)$ indicates the autocorrelation of T, then derive that

$$\frac{dC}{d\eta} = -\alpha C$$

and determine the solution $C(\eta)$ up to an integration constant C_0.

b. Show that the powerspectrum of T is given by

$$P(\omega) = \frac{2C_0\alpha}{\omega^2 + \alpha^2}$$

c. Sketch and interpret the spectrum in the limits (i) $\omega \ll \alpha$ and (ii) $\omega \gg \alpha$.

Further reading: Hasselmann (1976) and Gardiner (2002), chapter 4.

(P1.1) *Data visualization and analysis*

Nowadays there are many tools available to visualize and analyse observations. It is important to get an impression of the patterns and time scales of climate variability quickly using efficient software. There are three aspects to this: (i) finding the data, (ii) visualizing the data and (iii) obtaining statistical characteristics of the data.

For obtaining the data there are many sites available on the internet. The IRI/LDEO (http://ingrid.ldeo.columbia.edu/) is probably the best starting point for instrumental data as is the NOAA Paleoclimatogy site (http://www.ngdc.noaa.gov/paleo/paleo.html) for paleoclimatic data.   For visualization, there are many programs available. Most used is the package GRADS (http://grads.iges.org/grads/) because it is designed for plots of climatic data and it is freely available. For analysis of the data, GRADS provides also some tools, but for more in-depth statistical analysis, the SSA-MTM toolkit (http://www.atmos.ucla.edu/tcd/ssa/) is a freely available and recommended tool.

A website which integrates availability of data and analysis is the Climate Explorer at the Royal Dutch Meteorological Institute (http://climexp.knmi.nl/) designed by Geert Jan van Oldenborgh. In this exercise, some experience is gained in using this tool. So direct your browser to the URL of the Climate Explorer and login with your email address.

a. Select the NINO3 index (see section 3.1) from the Kaplan data set (starting at 1858) and plot the anomalies with respect to the seasonal cycle.

b. Plot the autocorrelation function $c(\tau)$ of the NINO3 index. At which lag does the first zero of $c(\tau)$ occur? How do you interpret this first zero crossing?

c. Plot the powerspectrum $P(\omega)$. At which period does the maximum power occur? How does this depend on the sampling of the time series?

d. Make a running mean of the original NINO3 time series using different windows. Can you discover any variability on lower frequencies than the typical El Niño frequency?

e. Repeat items a-d for the time series of the North Atlantic Oscillation (and any other you like).

Further reading: Chatfield (2004).

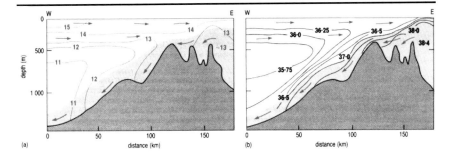

Figure 1.23. Sketch of the flow conditions in the Gibraltar Strait with both mean temperature (left) and salinity (right) profiles (the figure is taken from (OU-Staff, 1989)).

(P1.2) *Mediterranean Outflow and Climate*

In the present ocean, the densest water is produced in high-latitude marginal seas and the deep ocean is filled with cold, polar and subpolar waters. There are two important northern outflows (from the Norwegian-Greenland Sea and the Labrador Sea) and there is at least one important southern outflow (from the Weddell Sea). These northern and southern sources produce deep waters that have different chemical characteristics and leave a distinctive imprint in deep-sea sediments.

Another important source of deepwater is that produced in the Mediterranean Sea. From here, warm and salty water that is very dense enters the Atlantic. The density of the outflow water is such that it enters at about 1000 m depth. The aim of this exercise is to learn about the influence of the Mediterranean outflow on the Atlantic THC and consequently on the climate system.

Consider a flow over a sill (such as Gibraltar Strait) as sketched in Fig. 1.23. Let the volume flux from the Mediterranean basin into the Atlantic be indicated by Q (in Sv, 1 Sv = 10^6 m^{3}s^{-1}) The water west of the Strait has a salinity S_a and east of the Strait (in the Mediterranean basin), salinity is indicated by S_b.

The difference between evaporation E en precipitation P is about 1 m year^{-1} for the Mediterranean basin; the latter has a surface area of about $A = 2.5 \times 10^{12}$ m^2.

a. Show that

$$Q(S_a - S_b) = (E - P)AS_0$$

where $S_0 = 35$ is a reference salinity used to convert the freshwater flux to a salinity flux.

b. Estimate S_a and S_b from Fig. 1.23 and determine Q.

In a controversial article, R.G. Johnson (Johnsen, 1997) suggested to build a dam in the Gibraltar Strait to prevent catastrophic climate changes due to human activities.   You can download the paper from http://www.agu.org/sci_soc/eosrjohnson.html.   It nicely illustrates the complexity of global warming, and therefore the difficulty of estimating its effects.

c. Read this paper and formulate the precise arguments why Johnson thinks that building a dam in the Gibraltar Strait can prevent a next Ice Age.

Next   read   the   article   by   S.   Rahmstorf   (Rahmstorf,   1998), which   you   can   download   from   http://www.pik-potsdam.de/   stefan/Publications/Journals/gibraltar.html.

d. Describe the influence of Mediterranean outflow on the northern North Atlantic ocean surface temperatures as deduced from the model study. Do you think that building a dam in the Gibraltar Strait would have the effect that Johnson suggested?

Further reading: Tomczak and Godfrey (1994), chapter 16.

Chapter 2

BACKGROUND MATERIAL

The beginning of a framework: mastering the language.
Etude No 5., H. Villa-Lobos

It is assumed that the reader has an elementary knowledge of vector analysis, differential equations and (geophysical) fluid dynamics. To make reading through the chapters 5-7 more easy, some background material is included in this chapter. The general equations of motion are presented in section 2.1; this serves also to introduce the notation used in the book. There are many textbooks available where these equations are derived and discussed (Batchelor, 1974; Pedlosky, 1987; Cushman-Roisin, 1994). In geophysical fluid dynamics and dynamical oceanography, many results are interpreted in terms of vorticity transport within the flow. In the sections 2.2 and 2.3, the mechanisms of vorticity transport and the concept of potential vorticity are illustrated by using simple examples. These examples serve as a reference for the terminology used in later chapters. The last piece of background material is elementary hydrodynamic stability theory. In section 2.4, Joseph (1976) is followed in a general discussion on stability bounds. Some more mathematical issues are placed in technical boxes and can be skipped on first reading.

2.1. Basic Equations

Standard notation as used in the field of geophysical fluid dynamics (such as in Pedlosky (1987)) is adopted. All dimensional dependent variables have a * subscript. This is useful to distinguish dimensionless and dimensional equations in later sections. The inner product is just indicated with a dot (.) and for the vector product, the $\wedge$ notation is used.

2.1.1. Coordinate free

Consider a flow of water within a bounded region $\mathcal{V}$ on Earth's sphere, an example shown in Fig. 2.1. The region rotates with the movement of the Earth, having rotation vector $\mathbf{\Omega}$ and angular frequency $\Omega = |\mathbf{\Omega}|$. The equations of motion described from a reference frame moving along with the earth are (Pedlosky, 1987)

$$\rho_* \left[\frac{D\mathbf{v}_*}{dt_*} + 2\mathbf{\Omega} \wedge \mathbf{v}_* \right] = -\nabla p_* + \rho_* \nabla \Phi + \rho_* \mathcal{F}_{I*} \qquad (2.1a)$$

$$\frac{D\rho_*}{dt_*} + \rho_* \nabla.\mathbf{v}_* = 0 \qquad (2.1b)$$

Here, $D/dt_* = \partial/\partial t_* + \mathbf{v}_*.\nabla$ is the material derivative. The vector $\mathbf{v}_*$ is the velocity field of the flow, p_* is the pressure field, and ρ_* is the density of the water. The quantity Φ is the geopotential, where the dominant term is given by the gravitational acceleration. In spherical coordinates $-\nabla \Phi = g\mathbf{i}_r$, with g the acceleration due to gravity and $\mathbf{i}_r$ the unit vector in radial-direction. The vector $\mathcal{F}_{I*}$ $[ms^{-2}]$ represents the accelerations due to random motions (mixing) and its form will be discussed in section 2.1.3.

Although this set-up is general, an approximation which is made in nearly all modelling studies is the Boussinesq approximation. In this approximation, only the effect of density differences is considered in the volume (e.g., gravity) force,

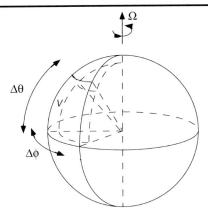

Figure 2.1.   A sector flow domain $\mathcal{V}$ on Earth's sphere, which is rotating with angular frequency $\Omega = |\mathbf{\Omega}|$.

whereas the effects of density variations are neglected in the continuity equation, momentum equations and temperature and salinity equations. This can be justified rigorously in many cases and the reader is referred, for example, to Batchelor (1974). Under this approximation, the flow is (to a good approximation) incompressible and the density ρ_* in all equations above, except in the vertical momentum equation, can be changed to a constant reference density ρ_0. Under the Boussinesq approximation, the equations (2.1) become

$$\rho_0 \left[\frac{D\mathbf{v}_*}{dt_*} + 2\mathbf{\Omega} \wedge \mathbf{v}_* \right] = -\nabla p_* + \rho_* \nabla \Phi + \rho_0 \mathcal{F}_{I*} \qquad (2.2a)$$

$$\nabla . \mathbf{v}_* = 0 \qquad (2.2b)$$

Local conservation of heat and salt is formulated as

$$\rho_0 C_p \frac{DT_*}{dt_*} = \mathcal{F}_{T*} + Q_{T*} \qquad (2.3a)$$

$$\rho_0 \frac{DS_*}{dt_*} = \mathcal{F}_{S*} + Q_{S*} \qquad (2.3b)$$

where T_* and S_* indicate temperature and salinity of the water. The scalars $\mathcal{F}_{T*}$ $[Wm^{-3}]$ and $\mathcal{F}_{S*}$ $[kg\ m^{-3}\ s^{-1}]$ represent the effect of random motion (diffusion and mixing) on the local changes of heat and salt. The scalar quantities Q_{T*} $[Wm^{-3}]$ and Q_{S*} $[kg\ m^{-3}\ s^{-1}]$ represent the internal sources and sinks of heat and salt. The quantity C_p is the heat capacity of the liquid. The unknowns in these equations are the three components of the velocity field $\mathbf{v}_*$, the pressure p_*, the density ρ_*, the temperature T_* and the salinity S_*, in total 7. Since there are only six equations (2.2) and (2.3), an additional relation is necessary. This is the equation of state, which for liquids such as salt water is given by

$$\rho_* = \rho_0 \rho(T_*, S_*, p_*) \tag{2.4}$$

For ocean water, the standard equation of state can be found in Gill (1982). At every (reasonable) temperature in the ocean, the density is nearly linear with the salinity S_* and only varies with pressure for large depth changes. In Fig. 2.2, the density is plotted as a function of temperature for three depths and constant salinity $S_* = 35$. The dependence of the density on salinity $\partial \rho_* / \partial S_* \approx 0.77 \, [kg \, m^{-3}]$ for each depth and is slightly larger for colder temperatures and slightly smaller under higher pressure.

2.1.2. Spherical coordinates

It is natural to use spherical coordinates on a sphere. In meteorology and physical oceanography positions in the flow domain are described in coordinates longitude, latitude and depth (Fig. 2.3a). Note that in the usual spherical coordinates the polar angle $\vartheta = \pi/2 - \theta$ is used. When converting the equations of motion from many textbooks on fluid mechanics (Batchelor, 1974) to those below, this has to be taken into account along with a velocity vector $\mathbf{v}_* = (u_*, v_*, w_*)$ for the velocity field in zonal, meridional and vertical (radial) direction.

In the coordinate system so defined, the Coriolis acceleration term is written (see Fig. 2.3b), as

$$2\mathbf{\Omega} \wedge \mathbf{v}_* = \begin{pmatrix} 2\Omega(w_* \cos\theta - v_* \sin\theta) \\ 2\Omega u_* \sin\theta \\ -2\Omega u_* \cos\theta \end{pmatrix} \tag{2.5}$$

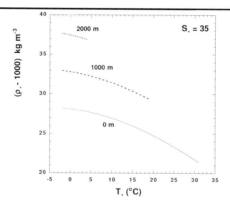

Figure 2.2.   Density $\rho_* - 1000 \, [kgm^{-3}]$ as a function of temperature T_* (in $°C$) for constant salinity $S_* = 35$ for three different depths (0, 1000 and 2000 m). Note that the range of temperatures decreases with depth.

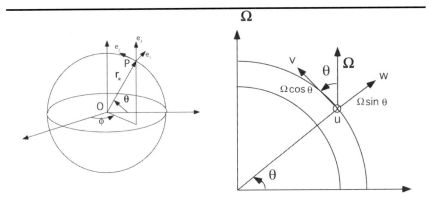

Figure 2.3. (a) *Coordinate system used on the sphere with radius* r_*, *longitude* ϕ *and latitude* θ; *the point P is an arbitrary point on the sphere.* (b) *Illustration to help to determine the components of the Coriolis acceleration.*

and the equations (2.1), (2.3) and (2.4) in coordinates (ϕ, θ, r_*) become

$$\frac{Du_*}{dt_*} + \frac{u_* w_*}{r_*} \quad - \quad \frac{u_* v_*}{r_*} \tan \theta - 2\Omega \left(v_* \sin \theta - w_* \cos \theta \right) =$$
$$- \frac{1}{\rho_* r_* \cos \theta} \frac{\partial p_*}{\partial \phi} + \mathcal{F}_{I*}^{\phi} \tag{2.6a}$$

$$\frac{Dv_*}{dt_*} + \frac{w_* v_*}{r_*} \quad + \quad \frac{u_*^2}{r_*} \tan \theta + 2\Omega u_* \sin \theta =$$
$$- \frac{1}{\rho_0 r_*} \frac{\partial p_*}{\partial \theta} + \mathcal{F}_{I*}^{\theta} \tag{2.6b}$$

$$\frac{Dw_*}{dt_*} - \frac{u_*^2 + v_*^2}{r_*} \quad - \quad 2\Omega u_* \cos \theta =$$
$$- \frac{1}{\rho_0} \frac{\partial p_*}{\partial r_*} - g + \mathcal{F}_{I*}^{r} \tag{2.6c}$$

$$\frac{\partial w_*}{\partial r_*} + \frac{2w_*}{r_*} \quad + \quad \frac{1}{r_* \cos \theta} \left(\frac{\partial (v_* \cos \theta)}{\partial \theta} + \frac{\partial u_*}{\partial \phi} \right) = 0 \tag{2.6d}$$

$$\rho_0 C_p \frac{DT_*}{dt_*} \quad = \quad \mathcal{F}_{T*} + Q_{T*} \tag{2.6e}$$

$$\rho_0 \frac{DS_*}{dt_*} \quad = \quad \mathcal{F}_{S*} + Q_{S*} \tag{2.6f}$$

$$\rho_* \quad = \quad \rho_0 \rho(T_*, S_*, p_*) \tag{2.6g}$$

with the material derivative written out as

$$\frac{D}{dt_*} = \frac{\partial}{\partial t_*} + \frac{u_*}{r_* \cos \theta} \frac{\partial}{\partial \phi} + \frac{v_*}{r_*} \frac{\partial}{\partial \theta} + w_* \frac{\partial}{\partial r_*} \tag{2.7}$$

As the ocean depths are much smaller than the radius of the Earth, it is convenient to define a reference radius r_0 from the center of the Earth to the average height of the ocean-atmosphere surface, and then define a depth coordinate

$$z_* = r_* - r_0 \qquad (2.8)$$

To close the system of equations, the mixing terms ($\mathcal{F}$) have to be expressed in terms of the velocity field ($\mathcal{F}_{\mathcal{I}*}$), the temperature ($\mathcal{F}_{\mathcal{T}*}$) and the salinity ($\mathcal{F}_{\mathcal{S}*}$).

2.1.3. Dissipative processes

The representation of the dissipative processes in models of the large-scale ocean circulation and climate is one of the stumbling blocks for substantial progress. Anyone who has travelled at sea will have noticed the vigorous processes at the sea surface, where waves are breaking on a highly turbulent surface flow. Indeed, the ocean circulation is a turbulent flow and although the degree of turbulence is certainly higher at the surface than in the deeper ocean, one *a priori* cannot model the circulation as if it were a laminar flow.

This is also directly concluded from the values of parameters in a typical large-scale ocean flow when using the expression of viscous (molecular) momentum transfer of water (a Newtonian liquid),

$$\mathcal{F}_{I*} = \nu \nabla^2 \mathbf{v}_* \qquad (2.9)$$

where ν is the kinematic viscosity of water ($\nu = 10^{-6}$ $[m^2 s^{-1}]$) and ∇^2 is the Laplace operator. If U and L denote typical velocity and length scales of the flow, then the ratio of inertial accelerations $\mathbf{v}_*.\nabla \mathbf{v}_*$ and frictional ones $\nu \nabla^2 \mathbf{v}_*$ is given by the Reynolds number

$$Re = \frac{U^2/L}{\nu U/L^2} = \frac{UL}{\nu} \qquad (2.10)$$

For flows on a moderate horizontal scale $L = 100$ $[km]$ and $U = 10^{-2}$ $[ms^{-1}]$, it is found that $Re \approx 10^9$. On a larger scale, or for larger velocities, this value is even larger. In nearly all laboratory situations, flows under the conditions represented by these parameter values are turbulent.

In turbulent flows, the energy is spread over an enormous range of spatial scales, many of which cannot be resolved in large-scale ocean models. Hence, for each model, a pragmatic approach is needed to represent the effects of unresolved scales on the large-scale flow. The most simple of these approaches is to model the effects as downgradient diffusion, which is also referred to as a first order turbulence closure. The flux $\mathbf{\Phi}_*$ of any such scalar quantity ϕ_* is then assumed proportional to the gradient $\nabla \phi_*$, with proportionality constant K,

$$\mathbf{\Phi}_* = -K\nabla \phi_* \qquad (2.11)$$

Mixing coefficients for momentum are indicated by A, while those for heat and salt will be indicated by K^T and K^S. This way of representing mixing of momentum and tracers in ocean models is usually referred to as Laplacian mixing, since the divergence of the flux $\mathbf{\Phi}_*$ enters the equations.

Many other approaches have been suggested and an overview is given in Large *et al.* (1994). In many high-resolution ocean models (see section 5.8), a biharmonic operator representation of the divergence of $\mathbf{\Phi}_*$ is used for the representation of horizontal mixing, because it has specific scale selective properties (McWilliams and Chow, 1981; McWilliams, 1996). There have been many efforts to represent the effect of unresolved scales of motion on the transport of the tracers, such as temperature and salinity. For example, the effect of eddies on tracer transport in low resolution (non-eddy resolving) models can be parameterized by additional advection and diffusion (Gent and McWilliams, 1990; Gent *et al.*, 1995). The latter parameterization provides mixing of tracers more along constant density surfaces (isopycnals) than the simple representation (2.11).

One can imagine that with such a crude representation of mixing processes as in (2.11), the values of mixing coefficients are not well known. In fact, values used in the literature vary by orders of magnitude. Rough numbers can be obtained through dimensional considerations, using the fact that K can be seen as a product of a length scale and a velocity scale, but the problem is that in many cases these relevant scales are not easily recognized. Because of the enormous length scale differences between the horizontal and vertical directions, usually different mixing coefficients (e.g., A_H and A_V for momentum) are taken (Large *et al.*, 2001).

In this way, the accelerations through dissipative processes are represented by

$$\mathcal{F}_{I*} = \nabla . \mathcal{T}_* \tag{2.12}$$

where $\mathcal{T}_*$ is the part of the stress tensor representing shear forces. In the coordinates (ϕ, θ, z_*), $\mathcal{T}_*$ can be written as

$$\mathcal{T}_* = A_H(\nabla_H \otimes \mathbf{v}_* + (\nabla_H \otimes \mathbf{v}_*)^T) + A_V(\nabla_z \otimes \mathbf{v}_* + (\nabla_z \otimes \mathbf{v}_*)^T) \tag{2.13}$$

where ∇_H is the horizontal gradient operator, $\nabla_z = (0, 0, \partial/\partial z_*)$ and the superscript T indicates transpose. The notation $\otimes$ is the dyadic product of the vectors that is defined by

$$\mathbf{a} \otimes \mathbf{b} = \begin{pmatrix} a_1 b_1 & a_1 b_2 & a_1 b_3 \\ a_2 b_1 & a_2 b_2 & a_2 b_3 \\ a_3 b_1 & a_3 b_2 & a_3 b_3 \end{pmatrix} \tag{2.14}$$

Similarly, the terms representing turbulent 'diffusive' transport of heat and salt are written as

$$\mathcal{F}_{T*} = \rho_0 C_p (\nabla_H . (K_H^T \nabla_H T_*) + \frac{\partial}{\partial z_*}(K_V^T \frac{\partial T_*}{\partial z_*})) \tag{2.15a}$$

$$\mathcal{F}_{S*} = \rho_0 (\nabla_H . (K_H^S \nabla_H S_*) + \frac{\partial}{\partial z_*}(K_V^S \frac{\partial S_*}{\partial z_*})) \tag{2.15b}$$

2.1.4. Boundary conditions

The flow domain $\mathcal{V}$ is horizontally bounded by continents, vertically bounded by an ocean floor and an ocean-atmosphere interface (Fig. 2.4). If the average depth of the water is given by d, then the specified bottom topography can be specified as a function $z_* = -d + h_b(\phi, \theta)$. On this lower boundary, the velocity is zero and there is no transport of heat and salt. The boundary conditions then become

$$z_* = -d + h_b(\phi, \theta) : \quad \mathbf{n} . \mathbf{v}_* \quad = \quad 0 \tag{2.16a}$$
$$\mathbf{t}_i . \mathbf{v}_* \quad = \quad 0, \ i = 1, 2 \tag{2.16b}$$
$$\mathbf{n} . \nabla T_* \quad = \quad 0 \tag{2.16c}$$
$$\mathbf{n} . \nabla S_* \quad = \quad 0 \tag{2.16d}$$

where $\mathbf{n}$ is the outward normal at the bottom and $\mathbf{t}_1$ and $\mathbf{t}_2$ are the two tangent vectors. The ocean-atmosphere interface is written as

$$z_* = \eta_*(\phi, \theta, t_*) \tag{2.17}$$

with the average position being at $z_* = 0$. The pressure and shear stress are continuous over the interface; the latter is modelled as a material surface.

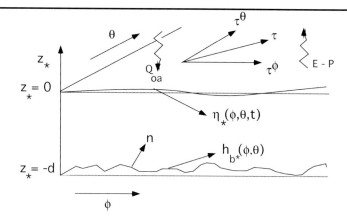

Figure 2.4. Sketch of the vertical structure of the flow domain. The downward heat flux is indicated by Q_{oa}, the wind stress by τ, and the freshwater flux by $E - P$.

If the atmospheric pressure just at the ocean-atmosphere interface is indicated by $p_a(\phi, \theta, t_*)$, then the boundary conditions at $z_* = \eta_*(\phi, \theta, t_*)$ can be formulated as

$$\frac{D}{dt_*}(z_* - \eta_*(\phi, \theta, t_*)) \quad = \quad 0 \tag{2.18a}$$
$$\rho_0 A_V r_* \frac{\partial}{\partial r_*}\left(\frac{u_*}{r_*}\right) + \frac{\rho_0 A_H}{r_* \cos \theta} \frac{\partial w_*}{\partial \phi} \quad = \quad \tau^\phi \tag{2.18b}$$
$$\rho_0 A_V r_* \frac{\partial}{\partial r_*}\left(\frac{v_*}{r_*}\right) - \frac{\rho_0 A_H}{r_*} \frac{\partial w_*}{\partial \theta} \quad = \quad \tau^\theta \tag{2.18c}$$
$$p_* \quad = \quad p_a(\phi, \theta, t_*) \tag{2.18d}$$

where τ^ϕ $[N/m^2]$ and τ^θ $[N/m^2]$ are the zonal and meridional components of the shear stress generated by the wind. The mixing coefficients of momentum (A_V and A_H) now turn up in the representation of the shear stress in the liquid, according to (2.13). Note that there is one condition less at the bottom boundary than at the surface, because h_{b*} is prescribed, whereas η_* is a dependent quantity. The boundary conditions for temperature and salinity at $z_* = \eta_*(\phi, \theta, t_*)$ are given by

$$\rho_0 C_p K_V^T \frac{\partial T_*}{\partial z_*} = Q_{oa} \qquad (2.19a)$$

$$K_V^S \frac{\partial S_*}{\partial z_*} = (E - P)S_0 \qquad (2.19b)$$

where E and P represent evaporation and precipitation, both in $[ms^{-1}]$, and S_0 is a reference salinity needed to convert the freshwater flux into an equivalent salt flux (Huang, 1993). The quantity Q_{oa} is the downward heat flux in $[Wm^{-2}]$ at the ocean-atmosphere surface.

On the continental boundaries of the domain, conditions of zero velocity (no-slip) and no heat flux and no salt flux are usually prescribed but their formulation will be presented for each particular application in later chapters.

2.1.5. Integral constraints

In the general model formulation above, several integral conditions can be derived. Satisfying these conditions is important in numerical models to be discussed later on. Two obvious constraints arise through the notions of conservation of total heat and salt. When the temperature equation is integrated over a fixed flow domain $\mathcal{V}$, one obtains

$$\int_{\mathcal{V}} \rho_0 C_p \left(\frac{\partial T_*}{\partial t_*} + \mathbf{v}_* . \nabla T_* \right) d^3 x_* =$$

$$\rho_0 C_p \int_{\mathcal{V}} \left(\nabla_H . (K_H^T \nabla_H T_*) + \frac{\partial}{\partial z_*} (K_V^T \frac{\partial T_*}{\partial z_*}) \right) d^3 x_* + \int_{\mathcal{V}} Q_{T*} d^3 x_*$$

where $d^3 x_* = r_0^2 \cos \theta \, d\phi \, d\theta \, dz_*$ in spherical coordinates. Under the boundary conditions of no-normal heat flux at the bottom and at the continental boundaries, and zero normal velocities at all boundaries (which is a kinematic constraint in confined flows), then using incompressibility, one can write $\mathbf{v}_* . \nabla T_* = \mathbf{v}_* . \nabla T_* + T_* \nabla . \mathbf{v}_* = \nabla . (\mathbf{v}_* T_*)$. Hence,

$$\int_{\mathcal{V}} \nabla . (\mathbf{v}_* T_*) d^3 x_* = \int_{\mathcal{S}} \mathbf{n} . \mathbf{v}_* \, T_* \, d^2 x_* = 0$$

$$\int_{\mathcal{V}} \left[\nabla_H . (K_H^T \nabla_H T_*) + \frac{\partial}{\partial z_*} (K_V^T \frac{\partial T_*}{\partial z_*}) \right] d^3 x_* = \int_{\mathcal{S}_{oa}} \frac{Q_{oa}}{\rho_0 C_p} d^2 x_*$$

where $\mathcal{S}$ is the total surface enclosing $\mathcal{V}$, the subscript oa indicates the ocean-atmosphere interface and $d^2 x_* = r_0^2 \cos \theta \, d\phi \, d\theta$. This leads to the integral constraint

$$\int_{\mathcal{V}}(\frac{\partial T_*}{\partial t_*} - \frac{Q_{T*}}{\rho_0 C_p})\, d^3x_* = \int_{S_{oa}} \frac{Q_{oa}}{\rho_0 C_p} d^2x_* \qquad (2.20)$$

which expresses total conservation of heat. In the special case of a steady state and zero internal sources of heat, a consequence is that the surface integral of the downward heat flux must be zero.

For the salt equation, a similar derivation leads to the integral constraint representing total salt conservation as

$$\int_{\mathcal{V}}(\frac{\partial S_*}{\partial t_*} - \frac{Q_{S*}}{\rho_0})\, d^3x_* = \int_{S_{oa}}(E - P)S_0\, d^2x_* \qquad (2.21)$$

Again in steady state and without internal sources of salt, the surface integral over the net freshwater flux has to be zero. In this situation, the constraint is essential, since the freshwater flux does not depend on salinity itself and hence, the salinity is determined up to an additive constant. The constraint (2.21) provides the only reasonable regularization.

Other integral constraints may be important. One of them, the mechanical energy balance, will for example be used in chapter 6. This balance is derived by multiplying the momentum equations (2.2b) by $\mathbf{v}_*$ and integrating the equations over the domain, to give

$$\int_{\mathcal{V}} \rho_0(\mathbf{v}_*\cdot\frac{D\mathbf{v}_*}{dt_*} + \mathbf{v}_*.(2\mathbf{\Omega} \wedge \mathbf{v}_*))\, d^3x_* =$$
$$\int_{\mathcal{V}} [-\mathbf{v}_*.\nabla p_* - w_*\rho_* g + \rho_0 \mathbf{v}_*.\nabla \mathcal{T}_*]\, d^3x_*$$

Using kinematic boundary conditions and the vector identity $\mathbf{v}_*.\nabla \mathcal{T}_* = \nabla.(\mathcal{T}_*\mathbf{v}_*) - \nabla \otimes \mathbf{v}_* : \mathcal{T}_*$, where : is the tensor direct product defined by

$$\mathcal{S} : \mathcal{T} = \sum_{i=1,3}\sum_{j=1,3} \mathcal{S}_{i,j}\mathcal{T}_{i,j} \qquad (2.22)$$

one can eventually derive the integral balance

$$\int_{\mathcal{V}} \frac{\rho_0}{2}\frac{\partial(\mathbf{v}_*.\mathbf{v}_*)}{\partial t_*} d^3x_* = -\int_{\mathcal{V}} w_*\rho_* g\, d^3x_* +$$
$$+\rho_0(\int_{S} \mathbf{v}_*.(\mathcal{T}_*\mathbf{n})\, d^2x_* - \int_{\mathcal{V}} \nabla \otimes \mathbf{v}_* : \mathcal{T}_*\, d^3x_*) \qquad (2.23)$$

The left hand side is the change in volume averaged kinetic energy. This quantity is balanced by the buoyancy production (first term in the right-hand side), the work of shear stress (e.g. wind) at the boundaries of the domain (second term in the right-hand side) and the dissipation (third term in the right-hand side). Several other constraints can be derived, for example those related to entropy production in the system (Ozawa *et al.*, 2003), but are not considered here.

2.2. Vorticity transport

Vorticity is an important concept in dynamical oceanography. Because the frame of reference from which flows are described is itself rotating, a distinction is made between the planetary vorticity $2\boldsymbol{\Omega}$ and the relative vorticity $\boldsymbol{\omega}_* = \nabla \wedge \mathbf{v}_*$. The absolute vorticity $\boldsymbol{\omega}_{a*}$ is defined as $\boldsymbol{\omega}_{a*} = \boldsymbol{\omega}_* + 2\boldsymbol{\Omega}$.

The vorticity vector leads to two important concepts: the vortex line and the vortex tube. At a fixed time t_0, a curve in the fluid is a vortex line if, for each point on the curve, the tangent vector is the vorticity vector $\boldsymbol{\omega}_{a*}$. If a vortex line is parametrized by a curve $\boldsymbol{\sigma} : [a, b] \in \mathbb{R} \to \mathbb{R}^3$ (Fig. 2.5), then at $t_* = t_0$,

$$\boldsymbol{\sigma}'(s) = \boldsymbol{\omega}_{a*}(t_0, \boldsymbol{\sigma}(s)) \qquad (2.24)$$

where $s \in [a, b]$. A vortex tube is formed by vortex lines that go through a closed curve (Fig. 2.5). If we take two closed curves C_1 and C_2 enclosing a vortex tube as in Fig. 2.5 then it is easy to prove the Helmholtz theorem (see Technical box 2.1)

$$\Gamma_1 = \int_{C_1} \mathbf{v}_*.\mathbf{ds} = \int_{C_2} \mathbf{v}_*.\mathbf{ds} = \Gamma_2 \qquad (2.25)$$

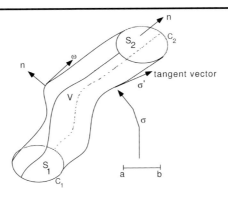

Figure 2.5.   Sketch of a vortex tube consisting of vortex lines, which pass through a closed curve C_1.

The quantity Γ is called the circulation of the velocity field around a closed curve C and the integral is taken in an anti-clockwise direction. Through Stokes' theorem, the circulation is directly related to the local vorticity through

$$\Gamma = \int_C \mathbf{v}_*.\mathbf{ds} = \int_S \boldsymbol{\omega}_* . \mathbf{n} \, dS \qquad (2.26)$$

where $\mathbf{n}$ is the outward normal to the surface S enclosed by the curve C. Hence, if a vortex tube compresses (for example, through stretching), the area of a typical surface S decreases and with constant circulation, the vorticity must increase.

Similarly, if a vortex tube is expanded the area enclosed becomes larger and the vorticity decreases. Although this shows that the vortex tube concept is useful, it does not show the processes by which vorticity is transported through the flow.

Technical box 2.1:
Helmholtz theorem

Starting point of the proof of the theorem is that $\nabla.\boldsymbol{\omega} = \nabla.\nabla \wedge \mathbf{v} = 0$. If we integrate this function over the volume V enclosed by the vortex tube (Fig. 2.5), then we get

$$0 = \int_V \nabla.\boldsymbol{\omega} \, dV = \int_S \boldsymbol{\omega}.\mathbf{n} \, dS$$

where S is the surface of the vortex tube and $\mathbf{n}$ the outward normal to the surface. By definition of the vortex tube, the vorticity vector is tangential to the surface of the vortex tube and hence $\boldsymbol{\omega}.\mathbf{n} = 0$. The contributions to the integral come from the surfaces enclosed by the curves C_1 and C_2 and with help of Stokes' theorem the final result follows as

$$0 = \int_S \boldsymbol{\omega}.\mathbf{n} \, dS = \int_{S_1} \boldsymbol{\omega}.\mathbf{n} \, dS + \int_{S_2} \boldsymbol{\omega}.\mathbf{n} \, dS = \int_{C_1} \mathbf{v}.\mathbf{ds} - \int_{C_2} \mathbf{v}.\mathbf{ds}$$

where the minus sign arises through the different orientation of the normals.

To study vorticity transport, the local vorticity balance is needed and it can be obtained by taking the curl of the momentum balance. Using the identity $\mathbf{v}.\nabla\mathbf{v} = \frac{1}{2}\nabla\mathbf{v}.\mathbf{v} + \boldsymbol{\omega} \wedge \mathbf{v}$, the coordinate free form of the vorticity equation becomes

$$\frac{\partial \boldsymbol{\omega}_{a*}}{\partial t_*} + \nabla \wedge ((2\boldsymbol{\Omega} + \boldsymbol{\omega}_*) \wedge \mathbf{v}_*) = \rho_*^{-2}\nabla\rho_* \wedge \nabla p_* + \nabla \wedge \mathcal{F}_{I*} \qquad (2.27)$$

This equation can be reduced with help of the identities

$$\nabla.\boldsymbol{\omega} = 0$$

and

$$\nabla \wedge (\mathbf{v} \wedge \boldsymbol{\omega}) = \boldsymbol{\omega}.\nabla\mathbf{v} - \boldsymbol{\omega}\nabla.\mathbf{v} - \mathbf{v}.\nabla\boldsymbol{\omega}$$

resulting in

$$\frac{\partial \boldsymbol{\omega}_{a*}}{\partial t_*} = -\mathbf{v}_*.\nabla\boldsymbol{\omega}_{a*} + \boldsymbol{\omega}_{a*}.\nabla\mathbf{v}_* - \boldsymbol{\omega}_{a*}\nabla.\mathbf{v}_* + \rho_*^{-2}\nabla\rho_* \wedge \nabla p_* + \nabla \wedge \mathcal{F}_{I*} \quad (2.28)$$

First term in the right hand side is the change in vorticity due to advection. The second and third terms are associated with changes in vorticity due to vortex stretching and tilting, the fourth term is associated with baroclinic vorticity changes and the last term gives the dissipation of vorticity due to frictional processes. From these processes, advection and diffusion (dissipation) do not need no further explanation. However, the other vorticity changing processes are considered in more detail below by looking at specific examples.

2.2.1. Vortex stretching and vortex tilting

Consider in a local Cartesian coordinate system the situation where the absolute vorticity vector $\boldsymbol{\omega}_a$ is parallel to the z-as, i.e. $\boldsymbol{\omega}_a = \bar{\omega}\,(0,0,1)^T$, with $\bar{\omega} > 0$. If the velocity components are indicated by (u, v, w) then direct computation gives

$$\boldsymbol{\omega}_a.\nabla \mathbf{v} - \boldsymbol{\omega}_a \nabla.\mathbf{v} = (\bar{\omega}\frac{\partial u}{\partial z},\; \bar{\omega}\frac{\partial v}{\partial z},\; -\bar{\omega}(\frac{\partial u}{\partial x} + \frac{\partial v}{\partial y}))^T \qquad (2.29)$$

If all other vorticity changing effects are absent, we see from (2.28) that the tendency of the z-component of $\boldsymbol{\omega}_a$ is proportional to the horizontal divergence of the velocity field in the plane orthogonal to the z-axis. If $\partial u/\partial x + \partial v/\partial y < 0$, then there is a local convergence of mass. Consider a (local) vortex tube parallel to the z-axis (Fig. 2.6a). This vortex tube is compressed and as a consequence of the Helmholtz theorem (Technical Box 2.1), the vorticity in the z-direction increases, which follows also directly from (2.29). This process of vorticity production is called vortex stretching.

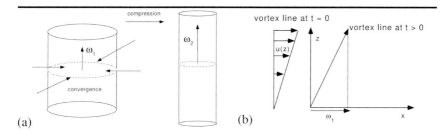

Figure 2.6.   *(a) Sketch of the mechanism of vortex stretching. A vortex tube is compressed through convergences in the flow. (b) Sketch to illustrate the mechanism of vortex tilting. A vortex line which is parallel to the z-axis at $t = 0$ is deformed through vertical shear of the background flow. As a consequence, it is tilted and generates a vorticity component in the x-direction.*

From the first component of (2.29), it follows that the tendency of the x-component of $\boldsymbol{\omega}_a$ is proportional to $\bar{\omega}\,\partial u/\partial z$. Consider a vortex line which is initially parallel to the z-axis in a flow for which $\partial u/\partial z > 0$. Because of the vertical shear, the vortex line will tilt and gives a contribution to the tendency of the vorticity component in the x-direction (Fig. 2.6b). This mechanism can also produce vorticity in the y-direction when $\partial v/\partial z \neq 0$, and is called vortex tilting.

2.2.2. Baroclinic vorticity production

The vector $\nabla\rho \wedge \nabla p$ in (2.28) is called the baroclinic vector. According to (2.28), vorticity is produced if this vector is nonzero. When surfaces of constant pressure (isobars) are also surfaces of constant density (isopycnals), then the pressure p is a unique function of the density ρ, i.e. $p = p(\rho)$. In this so-called barotropic case, the baroclinic vector is zero and there is no baroclinic vorticity production.

In order to show how vorticity is produced when $\nabla\rho \wedge \nabla p \neq 0$, consider the following example. For $z \in [-1, 0]$, let the pressure be given by $p(z) = -z$ and for $x \in [0, 1]$, the density be given by $\rho = \rho_0 - \delta z - \gamma x$, with $\delta > 0, \gamma > 0$. The surfaces of constant pressure (isobars) and constant density (isopycnals) are sketched in Fig. 2.7. Let's focus on two fluid elements which are at the same height, but at different lateral positions x, say x_1 and x_2 $(> x_1)$. The fluid element

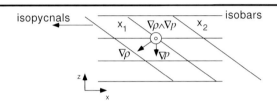

Figure 2.7.   Sketch of baroclinic vorticity production. Two liquid elements at x_1 and x_2 have different density, but are situated at the same isobar. Since density decreases downwards, the fluid element at x_1 will move downward and that at x_2 will move upwards, inducing an anti-clockwise rotation.

at x_1 has a larger density than that at x_2, whereas the same pressure holds at both positions. The fluid element at x_1 will therefore move downwards with respect to that at x_2. It introduces a vorticity component in the negative y-direction (anti-clockwise), i.e., out of the plane of the paper. This also follows through direct calculation from

$$\nabla\rho \wedge \nabla p = \begin{pmatrix} 0 \\ -\gamma \\ 0 \end{pmatrix} \qquad (2.30)$$

2.3. Potential Vorticity (PV)

Having considered the elementary concepts of vorticity change in a rotating (geophysical) flow, one may ask whether there are a priori constraints on the change in vorticity. In non-rotating ideal (i.e., inviscid and incompressible) barotropic flows, an example of such a constraint is the Kelvin theorem (Batchelor, 1974) which deals with the changes of the circulation within the flow. Although the Kelvin theorem can be generalized for rotating flows, it appears that there is a concept in rotating stratified flows, which gives stronger constraints on the flow, because it is a restriction on the evolution of a scalar property. This property is called potential vorticity, which is introduced next in its most general form.

2.3.1. The Ertel theorem

Consider any scalar quantity λ_*, which satisfies

$$\frac{D\lambda_*}{dt_*} = \mathcal{F}_{\lambda_*} \tag{2.31}$$

then the potential vorticity Π_{λ_*} is defined as

$$\Pi_{\lambda_*} = \frac{\boldsymbol{\omega}_* + 2\boldsymbol{\Omega}}{\rho_*}.\nabla\lambda_* \tag{2.32}$$

For every λ_*, there is a corresponding potential vorticity Π_{λ_*} and hence a context is needed for this quantity.

Technical box 2.2:
Potential vorticity equation

Within this technical box all quantities are dimensional, but for clarity of notation, no $*$ subscript will be used. The ith component of $D(\nabla\lambda)/dt$ can be written as

$$(\frac{D}{dt}\nabla\lambda)_i = \left[\frac{\partial}{\partial t} + \sum_j v_j \frac{\partial}{\partial x_j}\right]\frac{\partial\lambda}{\partial x_i}$$

If the inner product of this vector is taken with $\boldsymbol{\omega}_a/\rho$, one obtains (with the notation $\omega_i = (\boldsymbol{\omega}_a)_i$),

$$\frac{\boldsymbol{\omega}_a}{\rho}.\frac{D(\nabla\lambda)}{dt} = \sum_i \frac{\omega_i}{\rho}\left[\frac{\partial}{\partial t} + \sum_j v_j \frac{\partial}{\partial x_j}\right]\frac{\partial\lambda}{\partial x_i} =$$

$$= \sum_i \frac{\omega_i}{\rho}\frac{\partial}{\partial x_i}\left[\frac{\partial}{\partial t} + \sum_j v_j \frac{\partial}{\partial x_j}\right]\lambda - \sum_i \frac{\omega_i}{\rho}\sum_j \frac{\partial\lambda}{\partial x_j}\frac{\partial v_j}{\partial x_i} =$$

$$= \sum_i \frac{\omega_i}{\rho}\frac{\partial}{\partial x_i}\left[\frac{\partial}{\partial t} + \sum_j v_j \frac{\partial}{\partial x_j}\right]\lambda - \sum_j \frac{\partial\lambda}{\partial x_j}\sum_i \frac{\omega_i}{\rho}\frac{\partial v_j}{\partial x_i} =$$

$$= \frac{\boldsymbol{\omega}_a}{\rho}.\nabla\frac{D\lambda}{dt} - \nabla\lambda.\frac{\boldsymbol{\omega}_a}{\rho}.\nabla\mathbf{v}$$

which is identity A. Next, the vorticity equation (2.28) is written as

$$\frac{D\boldsymbol{\omega}_a}{dt} = \boldsymbol{\omega}_a.\nabla\mathbf{v} - \boldsymbol{\omega}_a\nabla.\mathbf{v} + \frac{\nabla\rho \wedge \nabla p}{\rho^2} + \nabla \wedge \mathcal{F}_I$$

Writing $\nabla.\mathbf{v} = -\rho^{-1}D\rho/dt$ (using the general continuity equation (2.1b)), dividing this expression by ρ, substituting it in the equation above and taking the inner product of the result and $\nabla\lambda$, one obtains

$$\nabla\lambda.(\rho^{-1}\frac{D}{dt}\boldsymbol{\omega}_a - \frac{\boldsymbol{\omega}_a}{\rho^2}\frac{D\rho}{dt}) = \nabla\lambda.\frac{D}{dt}\frac{\boldsymbol{\omega}_a}{\rho} =$$

$$\nabla\lambda. \left(\frac{\boldsymbol{\omega}_a}{\rho}.\nabla\mathbf{v} + \frac{\nabla\rho \wedge \nabla p}{\rho^3} + \frac{\nabla \wedge \mathcal{F}_I}{\rho}\right)$$

which is identity B. Adding both identities A and B and using (2.31) it is found that

$$\frac{D\Pi_\lambda}{dt} = \frac{\boldsymbol{\omega}_a}{\rho}.\nabla\mathcal{F}_\lambda + \nabla\lambda. \left(\frac{\nabla\rho \wedge \nabla p}{\rho^3} + \frac{\nabla \wedge \mathcal{F}_I}{\rho}\right)$$

In Technical box 2.2, the equation for the evolution of the potential vorticity $\Pi_{\lambda*}$ is derived from the vorticity equation with the result

$$\frac{D\Pi_{\lambda*}}{dt_*} = \frac{\boldsymbol{\omega}_{a*}.\nabla\mathcal{F}_{\lambda*}}{\rho_*} + \nabla\lambda_*. \left(\frac{\nabla\rho_* \wedge \nabla p_*}{\rho_*^3} + \frac{\nabla \wedge \mathcal{F}_{I*}}{\rho_*}\right) \qquad (2.33)$$

Strong constraints on the flow appear in the following theorem due to Ertel (Ertel, 1942). If the following conditions are all satisfied,

(i) The quantity λ_* is a conserved quantity, i.e., $\mathcal{F}_{\lambda*} = 0$,

(ii) Diffusion of momentum can be neglected, i.e., $\mathcal{F}_{I*} = 0$,

(iii) The quantity $\lambda_* = \lambda_*(\rho_*, p_*)$ is a function of only density and pressure,

then it follows from (2.33) that

$$\frac{D\Pi_{\lambda*}}{dt} = 0 \qquad (2.34)$$

and hence the potential vorticity $\Pi_{\lambda*}$ is a conserved quantity. The example below provides an application of the use of potential vorticity (or shortly PV) conservation.

2.3.2. PV conservation

Consider an isothermal, inviscid, and constant density flow with vertical length scale d and horizontal length scale L in a horizontally unbounded rotating channel ($\boldsymbol{\Omega} = (0, 0, f/2)$) bounded by a free surface $z_* = h_*(x, y, t)$ and bottom topography $z_* = -d + h_b(x, y)$. It can be shown that under the condition $d/L \ll 1$, the relative height of a fluid element in the liquid, i.e.,

$$\lambda_* = \frac{z_* - h_b}{H_*} \; ; \; H_* = h_* + d - h_b$$

is a conserved quantity (see for example Pedlosky (1987), section 3.4). Hence, in these types of flows all three conditions above are satisfied, and the potential vorticity associated with λ_* is

$$\Pi_{\lambda*} = \frac{\boldsymbol{\omega}_* + 2\boldsymbol{\Omega}}{\rho_0} \cdot \nabla \frac{z_* - h_b}{H_*} \approx \frac{\zeta_* + f}{\rho_0 H_*}$$

where ζ_* is the vertical component of the vorticity vector. The other two components of the vorticity vector are small, since both the vertical velocity as well as the vertical shear are small in these flows. Consider the flow over the bottom

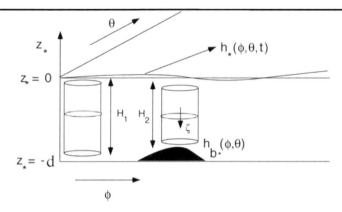

Figure 2.8. *Example of how the constraint of conservation of potential vorticity leads to determination of the relative vorticity of the flow. A water column of constant density moves from left to right over bottom topography.*

topography as sketched in Fig. 2.8 and assume that the flow is parallel upstream of the topography and hence, $\zeta_{1*} = 0$. The upstream potential vorticity is thus given by $\Pi_{1*} = f/(\rho_* H_1)$. On passing over the topography, the liquid depth decreases, since $H_2 < H_1$. As potential vorticity is conserved, it follows that

$$\frac{\zeta_* + f}{\rho_0 H_2} = \frac{f}{\rho_0 H_1} \Rightarrow \zeta_* = -f(1 - \frac{H_2}{H_1}) < 0$$

and hence the column is starting to rotate in a clockwise direction when it moves over the topography.

2.4. Stability

Let a solution $(\bar{\mathbf{v}}, \bar{p}, \bar{T}, \bar{S}, \bar{\rho})$ of the governing equations (2.18) be determined for a certain choice of parameters. A central question is then how sensitive this solution is to perturbations. Perturbations may occur in the initial conditions but also perturbations can be considered in the governing equations, the parameters and boundary conditions. For example, we may pose the question whether the solution changes essentially when a different representation of a particular process is incorporated in the governing equations. The above sentence already indicates that the notions of sensitivity and 'essential change' have to be defined more accurately. This is done through the concept of stability.

Intuitively, for a certain fixed set of governing equations, parameters and boundary conditions, a flow is stable when initial perturbations to the flow decay to zero in time. In the stability theory of fluid motions, this is made more precise through the concept of asymptotic stability in the mean (Joseph, 1976). For simplicity, consider a constant density steady flow with velocity field $\bar{\mathbf{v}}$ in a fixed flow domain $\mathcal{V}$. For a velocity perturbation $\tilde{\mathbf{v}}$ on $\bar{\mathbf{v}}$, consider the evolution of the volume averaged kinetic energy E of the perturbation, given by

$$E(t) = \int_{\mathcal{V}} \frac{1}{2} \tilde{\mathbf{v}}^2 \, d^3 x \qquad (2.35)$$

Note that the mechanical energy balance, as described in section 2.1.5 can be used to derive the equations for the evolution of the perturbations.

The solution $\bar{\mathbf{v}}$ is said to be asymptotically stable in the mean if

$$\lim_{t \to \infty} \frac{E(t)}{E(0)} = 0 \qquad (2.36)$$

where $E(0)$ is the initial volume averaged kinetic energy of the perturbation. If there exists a positive value δ such that (2.36) holds only when $E(0) < \delta$, then the basic state is said to be conditionally stable. If $\delta \to \infty$, then the basic state is globally stable and if (2.36) is satisfied and $dE(t)/dt < 0$ holds for all $t > 0$, then the basic state is said to be monotonically stable. Note that this definition of stability does not imply that the perturbations should be small a priori.

Let one of the parameters in a particular model be indicated by R. According to the notions above, a picture (Fig. 2.9) can be drawn of the different possibilities. In region I, the basic state is monotonically stable; all perturbations, whatever their amplitude, have a monotonically decaying kinetic energy. In region II, there may be perturbations which initially grow (not necessarily exponentially), but the energy eventually decays to zero for all initial amplitudes of the perturbations. Region III is a region of conditional instability. If the initial amplitude of the perturbations is small enough ($E(0) < \delta(R)$) the perturbation energy decays to zero, whereas if it larger than some particular value $\delta(R)$, the perturbation energy will increase. In the latter case, the perturbed state will evolve to a different state than $\bar{\mathbf{v}}$ and the state $\bar{\mathbf{v}}$ is said to be (nonlinearly) unstable to finite amplitude perturbations.

From Fig. 2.9, stability boundaries can be defined according to the evolution of the perturbation kinetic energy E. When $\mathbf{R} < \mathbf{R}_G$ then the basic state is globally stable and every perturbation decays to zero in time; $\mathbf{R}_G$ is the global stability limit and provides *sufficient conditions for stability*. If $\mathbf{R} < \mathbf{R}_E$, the basic state is monotonically stable; $\mathbf{R}_E$ is called the energy stability limit. If $\mathbf{R}_G < \mathbf{R} < \mathbf{R}_L$, then the basic state is conditionally stable: small amplitude disturbances decay whereas too large perturbations grow. Beyond the linear stability boundary

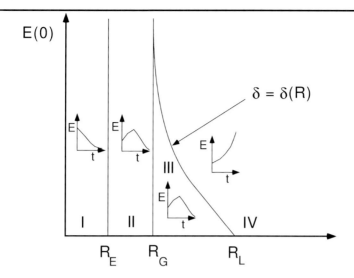

Figure 2.9. Plot of the different stability regimes, with I: monotonic stability, II: global stability, III: conditional stability and IV: instability. The control parameter is R and the values of R_E, R_G and R_L are the energy, global and linear stability boundaries, respectively. The curve $\delta(R)$ bounds the region of conditional stability. Typical trajectories of the volume averaged kinetic energy are sketched to illustrate the different behavior in each domain.

$\mathbf{R}_L$, infinitesimally small perturbations will grow and this stability bound provides *sufficient conditions for instability*. In summary, there are two cases of instability

(i) Subcritical instability: $\mathbf{R}_G < \mathbf{R} < \mathbf{R}_L$, i.e., not globally stable.

(ii) Supercritical instability: $\mathbf{R} > \mathbf{R}_L$, i.e., not linearly stable

Determination of the global and energy stability boundaries has to be done with the full nonlinear equations and exercise (P2.1) is provided to obtain experience with these calculations. Use is made of variational principles and many examples are provided in Joseph (1976) and Straughan (2004). The linear stability boundary is obtained by linearizing the evolution equations for the perturbations in their amplitude, which is infinitesimally small. This linear stability problem leads to an eigenvalue problem (see problem P2.2) of which several examples are shown in later chapters and which, in general, also has to be solved numerically.

2.5. Exercises on Chapter 2

(E2.1) *Inertial motion*

Consider a situation where the wind stress accelerates a horizontally un-bounded layer of ocean water (having a constant density ρ_0) up to a time t_0 and then decreases instantaneously to zero. Our aim is to describe the motion for $t > t_0$ – when the layer is unforced – in a rectangular coordinate system. We assume a linear friction formulation in the water, i.e., $\mathcal{F}_I^x = -ru$ and $\mathcal{F}_I^y = -ru$ and consider solutions $(u(t), v(t), w(t), p(t))$ which are spatially independent.

a. Show that the horizontal momentum equations (2.1a) reduce to

$$\frac{du}{dt} = f_0 v - ru$$
$$\frac{dv}{dt} = -f_0 u - rv$$

where f_0 is the local Coriolis parameter.

b. Determine the solutions $(u(t), v(t))$.

c. Sketch (plot) and describe the motion of the water over time.

Further reading: Cushman-Roisin (1994), chapter 6.

(E2.2) *Taylor column*

Consider a flow with velocity field $\mathbf{v} = (u, v, w)$ in a horizontally unbounded layer of liquid with constant density ρ and thickness D. The layer rotates with constant angular velocity Ω around the z-as.

a. What is the vorticity equation for this flow?

Consider variations in the flow on a timescale τ, with a characteristic velocity U and on a horizontal length scale L.

b. Estimate the terms in the vorticity equation under a.

Let $f = 2\Omega$. A special kind of flow appears in the limit $\tau >> 1/f$ and $\epsilon = U/(fL) << 1$.

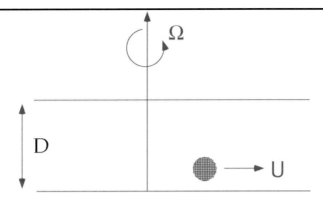

Figure 2.10.   *Sketch of the flow configuration in problem E2.2*

c. What type of special flow develops?

d. A sphere with radius R, with $R < L$, is drawn with velocity U over the bottom of the layer (see Fig. 2.10). After a while, a stationary flow appears with $\epsilon \ll 1$. Provide a sketch of the flow field.

Further reading: Batchelor (2000), chapter 7.

(E2.3) *Stress tensor*

The shear part of the stress tensor $\mathcal{T}_*$ is given in coordinate free form in equation (2.13).

a. Determine $\mathcal{T}_*$ for a Cartesian coordinate system.

b. To what expression does $\mathcal{T}_*$ reduce in a Cartesian formulation in case $A_H = A_V$?

(E2.4) *Application of the Helmholz theorem*

Consider a tornado as a thin vortex tube. Assume that the relative vorticity is constant over the cross section of the tube.

a. Show that the vorticity increases when the cross section of the tube decreases.

b. At about 10 m from the center of a certain tornado, there are wind speeds of 200 km/hour. What are the pressure variations when such a tornado passes by?

(E2.5) *Topographic Steering*

For large-scale flows in the deep ocean and far away from the equator, the relative vorticity can be neglected with respect to the planetary vorticity.

a. Show that for a constant density ocean, the potential vorticity of these flows is given by

$$\Pi = \frac{f}{H}$$

where H is the depth of the liquid column and $f = 2\Omega \sin\theta$ is the Coriolis parameter.

Consider a flow with upstream zonal velocity field U over a seamount with scale h_0 and horizontal length scale L (see Fig. 2.11).

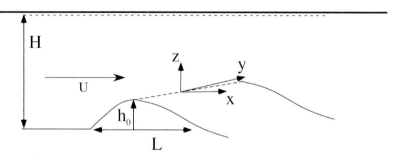

Figure 2.11.   Sketch of the flow configuration for problem E2.5

b. Describe the flow over the seamount in case of (i) $L/h_0 \ll 1$ and (ii) $h_0/H \ll 1$.

Further reading: Pedlosky (1996), chapter 3.

(P2.1) *Energy Stability Boundary*

One of the important stability bounds is the energy stability boundary as defined in section 2.4. To illustrate the computation of this stability boundary, consider the dimensional Navier-Stokes equations for a nonrotating, incompressible liquid in a bounded domain V. These are

$$\frac{\partial \mathbf{u}}{\partial t} + \mathbf{u}.\nabla \mathbf{u} = -\frac{1}{\rho}\nabla p + \nu \nabla^2 \mathbf{u}$$

$$\nabla.\mathbf{v} = 0$$

on a domain with no-slip conditions at the boundary of V. Here, ρ and ν are the constant density and kinematic viscosity of the liquid.

Assume that a steady state $(\mathbf{U}, P)$ exists and that it is subjected to a perturbation $(\tilde{\mathbf{u}}, \tilde{p})$.

a. Determine the nonlinear evolution equations of this perturbation.

Define the perturbation kinetic energy $\mathcal{E}$ through

$$\mathcal{E} = \frac{1}{2} < \tilde{\mathbf{u}}.\tilde{\mathbf{u}} >= \frac{1}{2}\int_V \tilde{\mathbf{u}}.\tilde{\mathbf{u}}\ d^3x$$

b. Multiply the momentum equations under a. with $\tilde{\mathbf{u}}$ and integrate over V. Show that the energy balance can be written as

$$\frac{d\mathcal{E}}{dt} = \mathcal{I} - \mathcal{D}$$

where $\mathcal{I}$ is the energy production through the Reynolds' stresses and $\mathcal{D}$ is the dissipation. Give the expressions for $\mathcal{I}$ and $\mathcal{D}$.

Let the space of kinematically allowed velocity vectorfunctions $\tilde{\mathbf{u}}$ be indicated by $\mathcal{H}$. This space consist of vectorfunctions that are divergence free, and which satisfy the kinematic boundary conditions. Consider now the problem

$$\max_{\mathcal{H}} \frac{1}{\mathcal{E}}\frac{d\mathcal{E}}{dt} = \max_{\mathcal{H}} \frac{\mathcal{I} - \mathcal{D}}{\mathcal{E}}$$

and assume that this maximum exists and is equal to a scalar λ_E.

c. Show that $\mathcal{E}$ is monotonically decreasing for $\lambda_E < 0$.

The determination of the energy stability boundary hence boils down to the solution of the variational problem in the space $\mathcal{H}$. The divergence free constraint is taken into account through a Lagrange multiplier and the functional J is defined as

$$J(\mathbf{v}) = \mathcal{I} - \mathcal{D} - p\nabla.\mathbf{v}$$

where $\mathbf{v} \in \mathcal{H}$ and p is a Lagrange multiplier.

d. Formulate the Euler-Lagrange equations for this functional and derive the problem to determine λ_E.

Further reading: Straughan (2004), chapter 2.

(P2.2) *Linear stability analysis*

Consider again the Navier-Stokes equations as in (P2.1) now describing the two-dimensional constant density flow between two horizontally unbounded parallel plates. The plates are a distance H apart and the flow is driven by a constant horizontal pressure gradient α.

a. Show that the two-dimensional Navier-Stokes equations for the horizontal velocity u, the vertical velocity w and the pressure p can be written as

$$\frac{\partial u}{\partial t} + u\frac{\partial u}{\partial x} + w\frac{\partial u}{\partial z} = -\frac{1}{\rho}\frac{\partial p}{\partial x} + \nu\nabla^2 u$$

$$\frac{\partial w}{\partial t} + u\frac{\partial w}{\partial x} + w\frac{\partial w}{\partial z} = -\frac{1}{\rho}\frac{\partial p}{\partial z} + \nu\nabla^2 w$$

$$\frac{\partial u}{\partial x} + \frac{\partial w}{\partial z} = 0$$

b. With no-slip boundary conditions at the walls, show that the steady parallel flow solution $\bar{u} = \bar{u}(z), \bar{w} = 0$ is given by

$$\bar{u}(z) = \frac{\alpha}{\rho\nu}(z^2 - zH)$$

Now assume perturbations $u = \bar{u} + \tilde{u}, w = \tilde{w}, p = \bar{p} + \tilde{p}$.

c. Substitute these into the equations above and determine the evolution equations for the perturbation fields.

Next, we linearize these equations assuming that the amplitude of the perturbations is infinitesimally small such that quadratic terms in these amplitudes can be neglected. Make the resulting equations dimensionless with scales H for length, $V = \bar{u}(H/2)$ for velocity, $\rho\nu V/H$ for pressure and H/V for time. Next, eliminate the pressure and introduce the streamfunction ψ through

$$\tilde{u} = \frac{\partial\psi}{\partial z} \ , \ \tilde{v} = -\frac{\partial\psi}{\partial x}$$

d. Show that the resulting equation governing ψ can be written as

$$\frac{\partial\nabla^2\psi}{\partial t} + \bar{U}\frac{\partial\nabla^2\psi}{\partial x} - \bar{U}''\frac{\partial\psi}{\partial x} = \mathbf{R}\nabla^2\nabla^2\psi$$

where $\bar{U} = \bar{u}/V$ and $\mathbf{R} = HV/\nu$ is the Reynolds number.

e. Formulate the boundary conditions for ψ.

Now consider the limit $\mathbf{R} \to \infty$ and expand ψ into normal modes

$$\psi(x, z, t) = \phi(z)e^{i(kx-ct)}$$

where k is the wavenumber and $c = c_r + ic_i$ is the complex growth factor.

f. Show that the problem for ϕ becomes

$$\phi'' - k^2\phi - \frac{\bar{U}''}{\bar{U} - c}\phi = 0$$
$$\phi(0) = \phi(1) = 0$$

Finally, multiply the equation for ϕ by its complex conjugate ϕ^*, integrate over the layer, use the boundary conditions and show that

$$c_i \int_0^1 \frac{\bar{U}''}{|\bar{U} - c|^2} |\phi|^2 \, dz = 0$$

g. Show that the steady state $\bar{U}$ is linearly stable.

Further reading: Drazin and Reid (2004), chapter 4.

Chapter 3

A DYNAMICAL SYSTEMS POINT OF VIEW

Capturing all flavors of development.
Valse, M. Ponce

Transition phenomena which occur in model solutions as parameters are varied can be analyzed in detail using concepts and techniques of the theory of dynamical systems. Being a qualitative theory, its concepts are powerful and they provide a strong link to the physics associated with the transition behavior. It is through the latter that complexity in irregular flows can be understood. In section 3.1, the qualitative theory is introduced through an elementary but oceanographically relevant problem, dealing with different density-driven flows under similar forcing conditions. Although there are many textbooks available on dynamical systems theory, many of them are too mathematical to be readily accessible to oceanographers and other geoscientists. In the remainder of this chapter, an overview of the more abstract qualitative theory is given with such a reader in mind. Hence, quite elementary mathematics is used and where possible, links with the example in section 3.1 are given.

3.1. An Elementary Problem

As was discussed in chapter 1, there is heat input at low latitudes and heat loss at higher latitudes in the North Atlantic Ocean. This causes a density driven surface flow from the equator to the poles because of sinking of the colder water in the north (Fig. 3.1, left panel). On the other hand, there is substantial evaporation at low latitudes which increases the salinity of the low-latitude water and hence its density. If there was no meridional temperature gradient, an equatorward density-driven surface flow would result, because water would sink near the equator (Fig. 3.1, right panel). The fact that the surface freshwater flux and heat flux have opposing effects on the large scale ocean circulation raises the interesting question: What happens if the circulation is driven by both surface fluxes?

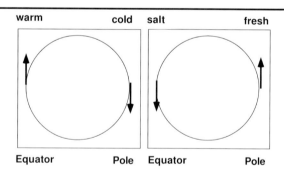

Figure 3.1.   Sketch of temperature (left panel) and salinity (right panel) driven (ocean) circulation patterns in a single equator-to-pole basin.

3.1.1. The Stommel two-box model

In the model proposed by Stommel (1961), this problem is studied in its most essential form, using two boxes having volumes V_p and V_e. These contain well-mixed water of temperature and salinity (T_{e*}, S_{e*}) and (T_{p*}, S_{p*}), the subscripts 'e' and 'p' indicating the equatorial and polar box, respectively. The boxes are connected at the surface by an overflow region and at the bottom by a capillary tube. The flow rate Ψ_* is directed from high to low pressure and is assumed to be linearly related to the density difference of the liquid between the boxes, i.e.

$$\Psi_* = \gamma \frac{\rho_{p*} - \rho_{e*}}{\rho_0} \qquad (3.1)$$

where ρ_0 is a reference density and γ [m^{3}s^{-1}] a hydraulic constant. The flow rate is taken positive if the liquid is heavier in the polar box. The exchange of properties does not depend on the sign of Ψ_*, because it only matters that properties from one box are transported to the other box. Because mass is conserved, the pathway (either through the overflow, or through the capillary) is unimportant. A linear equation of state is assumed of the form

$$\rho_* = \rho_0(1 - \alpha_T(T_* - T_0) + \alpha_S(S_* - S_0)) \qquad (3.2)$$

where T_0 and S_0 are reference values. The quantities α_T and α_S are the thermal expansion and haline contraction coefficients, respectively.

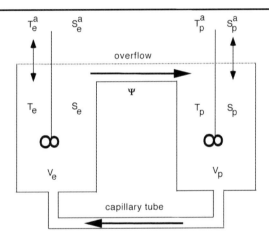

Figure 3.2. Sketch of the two-box model set-up as in Stommel (1961). Two reservoirs contain well-mixed water and are connected through an overflow and a capillary tube. The circulation is driven by density gradients between the water in both boxes, which are caused by the exchange of buoyancy at the surface. The sign of Ψ is positive when there is sinking in the polar box.

Exchange of heat and salt in each box due the surface forcing is modelled through a relaxation to a prescribed surface temperature and salinity (T^a, S^a) with relaxation coefficients C^T and C^S. These coefficients are different for each box and for each quantity considered (heat or salt). In this way, the balances of heat and salt in each box are given by (Stommel, 1961; Thual and McWilliams, 1992)

$$V_p \frac{dT_{p*}}{dt_*} = C_p^T (T_p^a - T_{p*}) + |\Psi_*| (T_{e*} - T_{p*}) \tag{3.3a}$$

$$V_e \frac{dT_{e*}}{dt_*} = C_e^T (T_e^a - T_{e*}) + |\Psi_*| (T_{p*} - T_{e*}) \tag{3.3b}$$

$$V_p \frac{dS_{p*}}{dt_*} = C_p^S (S_p^a - S_{p*}) + |\Psi_*| (S_{e*} - S_{p*}) \tag{3.3c}$$

$$V_e \frac{dS_{e*}}{dt_*} = C_e^S (S_e^a - S_{e*}) + |\Psi_*| (S_{p*} - S_{e*}) \tag{3.3d}$$

In the following we will restrict to the case of realistic forcing, for which $T_e^a - T_p^a > 0$ and $S_e^a - S_p^a > 0$. For simplicity, it is assumed that the relaxation times for temperature to the surface forcing in both boxes is proportional to their volume and hence $C_p^T/V_p = C_e^T/V_e \equiv R_T$ is constant. The same simplification is made for salinity with $R_S = C_p^S/V_p = C_e^S/V_e$. When time, temperature, salinity and flow rate are scaled with $1/R_T$, $V_e V_p R_T/(\gamma \alpha_T (V_e + V_p))$, $V_e V_p R_T/(\gamma \alpha_S (V_e + V_p))$ and $V_e V_p R_T/((V_e + V_p))$, the dimensionless equations become

$$\frac{dT}{dt} = \eta_1 - T(1 + |T - S|) \tag{3.4a}$$

$$\frac{dS}{dt} = \eta_2 - S(\eta_3 + |T - S|) \tag{3.4b}$$

where $T = T_e - T_p$, $S = S_e - S_p$ and $\Psi = T - S$ is the dimensionless flow rate. Three parameters appear in the equations (3.4) which are given by

$$\eta_1 = \frac{(T_e^a - T_p^a)\, \gamma \alpha_T (V_e + V_p)}{V_e V_p R_T}$$

$$\eta_2 = \frac{R_S}{R_T} \frac{(S_e^a - S_p^a)\, \gamma \alpha_S (V_e + V_p)}{V_e V_p R_T} \tag{3.5}$$

$$\eta_3 = \frac{R_S}{R_T}$$

The model is a two-dimensional system of ordinary differential equations describing the evolution of the temperature and salinity difference between the boxes; it contains three independent parameters η_i, $i = 1, 2, 3$. Here η_1 is a measure of the thermal forcing, η_2 of the saline forcing and η_3 is a ratio of adjustment time scales to heat and salt perturbations at the surface.

Clearly, the equations (3.4) form a relatively simple mathematical model which immediately attracts to proceed with analytical methods. However, imagine that we had many more, say ten or more, of these boxes all coupled through exchanges of heat and salt. This would model the horizontal and vertical structure of the exchanges of these properties in more and more detail. A typical way to proceed would then be to choose parameter values as 'realistic' as possible and compute the time evolution of the temperature and salinity in the boxes, starting from some initial state. Such a time dependent solution of the model is called a *trajectory*. Starting from the initial state $(0,0)$ $(T = 0, S = 0)$, a trajectory is shown in Fig. 3.3a for the parameter values $\eta_1 = 3.0$, $\eta_2 = 0.5$ and $\eta_3 = 0.3$. In this case, the freshwater forcing is relatively small and the flow evolves to a steady state with sinking in the north, since $\Psi = T - S > 0$. It turns out that for these parameter values, whatever initial condition one takes, this same steady state is always reached.

However, one usually likes to know sensitivity to changes in parameters. So we fix $\eta_1 = 3.0$ and $\eta_3 = 0.3$ and plot three trajectories in Fig. 3.3b for the case $\eta_2 = 1.0$. The trajectories starting at the initial conditions $(0,0)$ and $(2.5, 2.5)$ approach a steady state with sinking in the north similar as in the case $\eta_2 = 0.5$. However, the trajectory from the initial condition $(3.0, 3.0)$ approaches a steady state with sinking in the south, since $\Psi = T - S < 0$. Apparently, there are multiple steady states under the same forcing conditions in (3.4) if η_2 is large enough. But what is the limiting value of η_2, where these multiple equilibria just appear (we can guess already that it is somewhere between 0.5 and 1.0)? This question motivates us to look at the steady equations and find their solutions directly as functions of parameters.

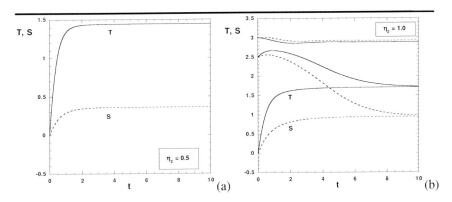

Figure 3.3. (a) *Trajectory starting from the zero solution* ($T = S = 0$) *for the model (3.3) with* $\eta_3 = 0.3$, $\eta_1 = 3.0$ *and* $\eta_2 = 0.5$. (b) *For* $\eta_3 = 0.3$, $\eta_1 = 3.0$ *and* $\eta_2 = 1.0$, *trajectories from three different initial conditions lead to two different steady states. In the figure, the values of* T *are drawn whereas the values of* S *are dashed.*

3.1.2. Equilibrium solutions

For steady states, the time derivatives in (3.4) are zero which gives the solutions

$$T = \frac{\eta_1}{1 + |\Psi|} \ ; \ S = \frac{\eta_2}{\eta_3 + |\Psi|} \tag{3.6}$$

where Ψ has to be solved from the implicit equation

$$\Psi = \frac{\eta_1}{1+ \mid \Psi \mid} - \frac{\eta_2}{\eta_3 + \mid \Psi \mid} \qquad (3.7)$$

When $\eta_2 = 0$, then $S = 0$ and hence $\Psi = T > 0$. The solution for Ψ follows from a quadratic equation and its positive root gives the solution

$$T = -\frac{1}{2} + \sqrt{\frac{1}{4} + \eta_1} \; ; \; S = 0 \qquad (3.8)$$

which is referred to as a TH state. The flow, with sinking in the northern box is driven by the temperature difference between equator and pole with warm water flowing through the overflow and cold water going through the tube (Fig. 3.2).

When $\eta_1 = 0$, there is no heat forcing and hence $T = 0$. It follows that $\Psi = -S < 0$ and hence a flow driven by the salinity gradient is obtained. The solution is

$$T = 0 \; ; \; S = -\frac{1}{2}\eta_3 + \sqrt{\frac{1}{4}\eta_3^2 + \eta_2}$$

which is referred to as a SA state.

Although the structure of the equilibrium solutions can be explicitly solved (Thual and McWilliams, 1992), it is more illustrative to show some typical results. With fixed $\eta_3 = 0.3$, a plot of steady solutions T and S versus η_2 are shown in Fig. 3.4a for $\eta_1 = 0.25$. This diagram, where a property of the solution (here T or S) is plotted versus a parameter, is our first example of a *bifurcation diagram*. There is a unique solution which is thermally driven for small η_2 (in this case, $\Psi = T - S > 0$), it is motionless at $\eta_2 = 0.1$ (at the intersection of the T and S curves) and becomes saline driven at larger η_2. Hence, with increasing η_2 the solution changes from TH type to SA type.

The same diagram is shown for $\eta_1 = 3.0$ in Fig. 3.4b. As the time-dependent results for $\eta_1 = 3.0$ in Fig. 3.3b already indicated, there are multiple steady solutions over a certain interval in η_2. Up to the point L_1 in Fig. 3.4b, the solution is of TH type and unique. Between the points L_1 and L_2, both TH and SA solutions exist and for values of η_2 beyond L_2 only the SA solution exists. The points L_1 and L_2 exactly bound the region of multiple equilibria. When the positions of these points are determined for additional values of η_1, the area in the (η_1, η_2) parameter plane where both TH and SA solutions occur is bounded by two curves (Fig. 3.4c). To the right of the curve of points L_1, there is a unique northern sinking (TH) solution, whereas to the left of the curve of points L_2, there is a unique equatorial sinking (SA) solution. At the point Q both curves of saddle-node bifurcations intersect. The diagram in Fig. 3.4c, where the boundaries of the different solution regimes are plotted in a two-parameter plane, is a first example of a *regime diagram*.

In summary, the most important result from the analysis of the model (3.4) is that for some of the parameter values different steady states, with opposite circulation directions, exist. Two points on the steady solution branches, indicated by

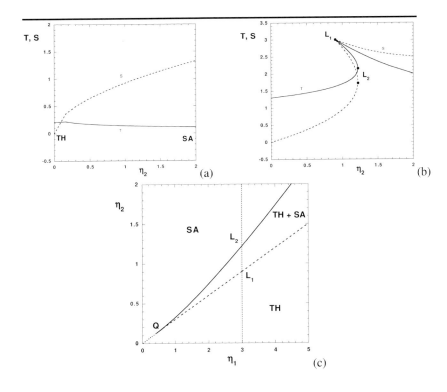

Figure 3.4. (a) Plot of values of temperature (T) and salinity (S) for steady states of the model (3.3) for different η_2 with fixed $\eta_1 = 0.25$ and $\eta_3 = 0.3$. (b) Same as in (a) but for $\eta_1 = 3.0$. (c) Path of the points L_1 and L_2 in (b) in the (η_1, η_2) parameter plane. These paths intersect at the point Q and connect to the curve $\eta_2 = \eta_1 \eta_3$ which defines the motionless flow, $\Psi = 0$, and extends from Q to the origin.

L_1 and L_2 in Fig. 3.4b, play a central role since they bound the multiple equilibria regime in parameter space. The regime diagram in Fig. 3.4c provides a complete overview of where in the two-parameter plane these multiple equilibria occur.

3.1.3. Stability of steady solutions

If a particular steady state from the previous section is indicated by $(\overline{T}, \overline{S})$, a next step is to consider the evolution of perturbations $(\tilde{T}, \tilde{S})$ on this steady state,

$$T = \overline{T} + \tilde{T} \tag{3.9a}$$
$$S = \overline{S} + \tilde{S} \tag{3.9b}$$

For the box model (3.4), we will now use the notation $\mathcal{M}$ for a smoothed version of the modulus function,

$$\mathcal{M}(\Psi) = [\mathcal{H}(\Psi) - \mathcal{H}(-\Psi)]\,\Psi$$

where $\mathcal{H}$ is a smoothed version of the Heaviside function. One expression which is used frequently is, with $\epsilon \ll 1$,

$$\mathcal{H}(\Psi) = \frac{1}{2}(1 + \tanh\frac{\Psi}{\epsilon}) \qquad (3.10)$$

such that derivatives of $\mathcal{M}$ exist. For the linear stability boundary, quadratic interactions in the perturbations are neglected and using the expansion

$$\mathcal{M}(\overline{\Psi} + \tilde{\Psi}) = \mathcal{M}(\overline{\Psi}) + \mathcal{M}'(\overline{\Psi})\tilde{\Psi} + \cdots$$

leads to the evolution equations

$$\frac{d\tilde{T}}{dt} = -\left[(1 + \mathcal{M}(\overline{\Psi}))\,\tilde{T} + \mathcal{M}'(\overline{\Psi})\overline{T}\,(\tilde{T} - \tilde{S})\right] \qquad (3.11a)$$

$$\frac{d\tilde{S}}{dt} = -\left[(\eta_3 + \mathcal{M}(\overline{\Psi}))\,\tilde{S} + \mathcal{M}'(\overline{\Psi})\overline{S}\,(\tilde{T} - \tilde{S})\right] \qquad (3.11b)$$

with $\tilde{\Psi} = \tilde{T} - \tilde{S}$. These equations admit solutions of the form

$$\tilde{T} = \hat{T}\,e^{\sigma t}\,;\; \tilde{S} = \hat{S}\,e^{\sigma t} \qquad (3.12)$$

where $\sigma = \sigma_r + i\sigma_i$ is the complex growth factor. The real part σ_r monitors the (exponential) growth rate of the perturbations. When $\sigma_r < 0$ for a particular perturbation $(\hat{T}, \hat{S})$, this perturbation damps and when $\sigma_r > 0$ it will grow, leading to instability of the steady state. Substituting these expressions into the equations (3.11) gives an eigenvalue problem

$$\begin{pmatrix} -(1 + \mathcal{M}(\overline{\Psi}) + \mathcal{M}'(\overline{\Psi})\overline{T}) & \mathcal{M}'(\overline{\Psi})\overline{T} \\ -\mathcal{M}'(\overline{\Psi})\overline{S} & -(\eta_3 + \mathcal{M}(\overline{\Psi}) - \mathcal{M}'(\overline{\Psi}))\overline{S} \end{pmatrix} \begin{pmatrix} \hat{T} \\ \hat{S} \end{pmatrix} = \sigma \begin{pmatrix} \hat{T} \\ \hat{S} \end{pmatrix}$$
$$(3.13)$$

The matrix in the left hand side of (3.13) is called the Jacobian matrix and will be indicated by J.

In Fig. 3.5a, the solutions in Fig. 3.4b are replotted, now with $\overline{\Psi}$ on the vertical axis. Along the branches, the signs ($\pm$) of both real eigenvalues σ of (3.13) are shown. For values of η_2 up to the point L_2, the TH state is stable and similarly for values beyond L_1, the SA state is stable. On the branch of solutions connecting the solutions at L_1 and L_2, one of the eigenvalues is positive. According to (3.12), small perturbations will grow on this steady state and hence it is unstable. This is demonstrated by computing the time evolution of the temperature and salinity fields starting exactly (within computer precision) at a steady state on this branch (point A, $T = 2.80, S = 2.74$) as plotted in Fig. 3.5b. The time-dependent state diverges away from the unstable steady state and eventually the steady TH state at point B is reached.

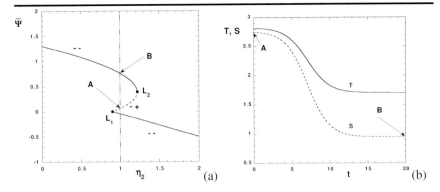

Figure 3.5.   (a) Plot of steady values of the flow $\bar{\Psi}$ for the model (3.3) for different η_2 with fixed $\eta_1 = 3.0$ and $\eta_3 = 0.3$. The $\pm$ signs indicate the sign of the (real) eigenvalues of the Jacobian matrix of each steady solution; drawn (dashed) curves indicate stable (unstable) steady states. (b) Evolution of the temperature and salinity fields for $\eta_1 = 3.0$, $\eta_2 = 1.0$ and $\eta_3 = 0.3$ starting at the steady state at point A ($T = 2.80, S = 2.74$).

With the analysis of the steady states and their linear stability in parameter space, the trajectories computed for $\eta_2 = 1.0$ in Fig. 3.3 can also be understood. For $\eta_2 = 0.5$, the system is in the unique stable TH regime according to Fig.3.4c. For $\eta_2 = 1.0$, the system is in the regime of overlapping stable TH states and SA states and hence trajectories with two different initial conditions can approach different steady states.

3.1.4. The presence of symmetry

An extension of the Stommel (1961) box model to include a southern polar box (Fig. 3.6) has been studied by Welander (1986) and Thual and McWilliams (1992). Denoting temperature and salinity in the northern and southern box as (T_{n*}, S_{n*}) and (T_{s*}, S_{s*}), respectively, the equations are a direct extension of (3.3) and given by

$$V_s \frac{dT_{s*}}{dt_*} = C_s^T(T_s^a - T_{s*}) + \mid \Psi_{s*} \mid (T_{e*} - T_{s*}) \tag{3.14a}$$

$$V_e \frac{dT_{e*}}{dt_*} = C_e^T(T_e^a - T_{e*}) + \mid \Psi_{s*} \mid (T_{s*} - T_{e*})$$
$$+ \mid \Psi_{n*} \mid (T_{n*} - T_{e*}) \tag{3.14b}$$

$$V_n \frac{dT_{n*}}{dt_*} = C_n^T(T_n^a - T_{n*}) + \mid \Psi_{n*} \mid (T_{e*} - T_{n*}) \tag{3.14c}$$

$$V_s \frac{dS_{s*}}{dt_*} = C_s^S(S_s^a - S_{s*}) + \mid \Psi_{s*} \mid (S_{e*} - S_{s*}) \tag{3.14d}$$

$$V_e \frac{dS_{e*}}{dt_*} = C_e^S(S_e^a - S_{e*}) + \mid \Psi_{s*} \mid (S_{s*} - S_{e*})$$

$$+ \; | \, \Psi_{n*} \, | \, (S_{n*} - S_{e*}) \tag{3.14e}$$

$$V_n \frac{dS_{n*}}{dt_*} = C_n^S(S_n^a - S_{n*}) + | \, \Psi_{n*} \, | \, (S_{e*} - S_{n*}) \tag{3.14f}$$

with the flow rates

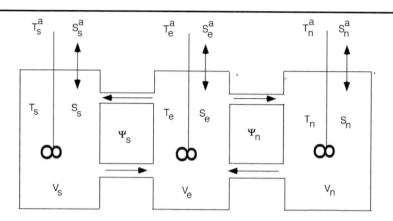

Figure 3.6.   Sketch of the three-box set-up as in Welander (1986) and Thual and McWilliams (1992).

$$\Psi_{s*} = \gamma(\alpha_T(T_{e*} - T_{s*}) - \alpha_S(S_{e*} - S_{s*}))$$
$$\Psi_{n*} = \gamma(\alpha_T(T_{e*} - T_{n*}) - \alpha_S(S_{e*} - S_{n*}))$$

The sign of each Ψ_* is positive if the surface circulation is from equator to pole. Again with

$$C_s^T/V_s = C_e^T/V_e = C_n^T/V_n = R_T$$
$$C_s^S/V_s = C_e^S/V_e = C_n^S/V_n = R_S$$

and new variables

$$\Theta_{s*} = T_{e*} - T_{s*} \, , \; \Theta_{n*} = T_{e*} - T_{n*} \, , \; \Sigma_{s*} = S_{e*} - S_{s*} \, , \; \Sigma_{n*} = S_{e*} - S_{n*}$$

the dimensionless form of the equations is derived by scaling Θ_*, Σ_* and time with $V_s R_T/(2\gamma\alpha_T)$, $2V_s R_T/(2\gamma\alpha_S)$ and R_T^{-1}, respectively. For $V_s = V_n = V_e/2$, one then obtains (Thual and McWilliams, 1992)

$$\frac{d\Theta_s}{dt} = \alpha_s - \Theta_s(1 + \frac{3}{4} | \, \Psi_s \, |) - \frac{1}{4} | \, \Psi_n \, | \, \Theta_n \tag{3.16a}$$

$$\frac{d\Theta_n}{dt} = \alpha_n - \Theta_n(1 + \frac{3}{4} | \, \Psi_n \, |) - \frac{1}{4} | \, \Psi_s \, | \, \Theta_s \tag{3.16b}$$

$$\frac{d\Sigma_s}{dt} = \beta_s - \Sigma_s(\eta_3 + \frac{3}{4} | \, \Psi_s \, |) - \frac{1}{4} | \, \Psi_n \, | \, \Sigma_n \tag{3.16c}$$

$$\frac{d\Sigma_n}{dt} = \beta_n - \Sigma_n(\eta_3 + \frac{3}{4} | \, \Psi_n \, |) - \frac{1}{4} | \, \Psi_s \, | \, \Sigma_s \tag{3.16d}$$

with $\eta_3 = R_S/R_T$ and

$$\Psi_s = \Theta_s - \Sigma_s \quad ; \quad \Psi_n = \Theta_n - \Sigma_n$$

$$\alpha_s = \frac{2\gamma\alpha_T}{V_s R_T}(T_e^a - T_s^a) \quad ; \quad \alpha_n = \frac{2\gamma\alpha_T}{V_s R_T}(T_e^a - T_n^a)$$

$$\beta_s = \frac{R_S}{R_T}\frac{2\gamma\alpha_S}{V_s R_T}(S_e^a - S_s^a) \quad ; \quad \beta_n = \frac{R_S}{R_T}\frac{2\gamma\alpha_S}{V_s R_T}(S_e^a - S_n^a)$$

The forcing is symmetric with respect to the equator if $\alpha_n = \alpha_s$ and $\beta_n = \beta_s$. In this case, the equations possess a reflection symmetry meaning that north and south are indistinguishable in the model. The symmetry can be represented by interchanging Θ_n and Θ_s and simultaneously Σ_s and Σ_n in the equations (3.16); this leaves them invariant.

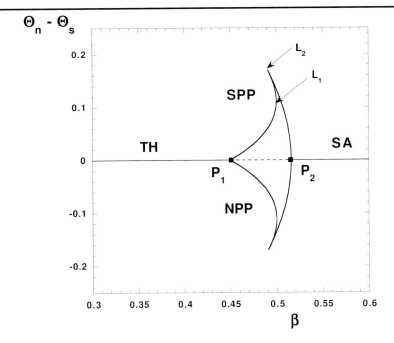

Figure 3.7. Branches of steady solutions for the 3-box model (3.16) with symmetric surface forcing $\alpha_s = \alpha_n = \alpha = 1.5$, $\beta_s = \beta_n = \beta$ and $\eta_3 = 0.3$. In the diagram, β is used as a control parameter and along the vertical axis, $\Theta_n - \Theta_s$ is plotted (which is proportional to the temperature difference between the southern and northern box). It is zero when the solutions are symmetric with respect to the equator.

The presence of the equatorial symmetry has a striking influence on the structure of the steady solutions and their stability. In Fig. 3.7, the bifurcation diagram is plotted for fixed $\alpha_s = \alpha_n = \alpha = 1.5$ using $\beta_s = \beta_n = \beta$ as a control parameter. The difference $\Theta_n - \Theta_s$ was chosen as a property of the solution because it is is zero for equatorially-symmetric solutions. The stability of the solutions is again shown through the linestyle with drawn (dashed) branches indicating stable (unstable) steady solutions; the short branch between L_1 and L_2 contains unstable solutions.

For small β, there is only one symmetric solution with upwelling at the equator and sinking at the poles (Fig. 3.8a) and hence it is of TH type. At the point labelled P_1, the TH solution becomes unstable and two asymmetric solutions – labelled NPP and SPP – appear. The equatorial symmetry is apparently spontaneously broken and asymmetric solutions appear under symmetric forcing conditions. For these NPP and SPP solutions, shown in Fig. 3.8c and Fig. 3.8d, there is no longer equatorial upwelling or downwelling but there is only downwelling at one of the poles. At large β, only the SA solution exists with downwelling at the equator (Fig. 3.8b). This solution becomes unstable for values of β below those at the point P_2 leading also to the asymmetric NPP and SPP solutions. The branches coming from both points P_1 and P_2 are connected through the two points L_1 and L_2.

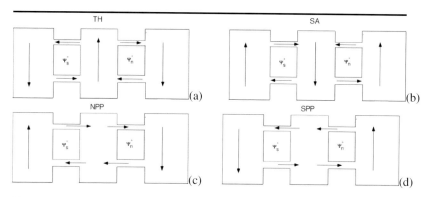

Figure 3.8.   Sketch of the four different solutions within the 3-box model. (a) TH solution; (b) SA solution; (c) NPP solution; (d) SPP solution.

3.1.5. Imperfections

In the three-box model (3.14), the equatorial symmetry is no longer present as soon as $\alpha_n \neq \alpha_s$ or $\beta_n \neq \beta_s$ (or both). Let us consider the case $\beta_s = \beta_n = \beta$ with $\alpha_n = \alpha_s(1 + \epsilon)$ for $\epsilon > 0$. Physically, this means that the equator-to-pole temperature difference is slightly larger in the northern hemisphere than in the southern hemisphere. Imagining only a thermally driven flow, this would induce a preference for northern sinking. When $\epsilon = 0.01$ the points P_1 and P_2, present for $\epsilon = 0$ in Fig. 3.7, no longer exist (Fig. 3.9a). The SPP branch becomes an isolated branch and separates from the connected branch. The latter branch consists of a connection between the symmetric branch of solutions and the NPP branch. With increasing ϵ, the isolated branch shrinks to one point and it has disappeared for $\epsilon = 0.1$ (Fig. 3.9b).

When Fig. 3.7a and Fig. 3.9 are compared, one sees that although the connection between the branches is broken, one can still understand the origin of the solution branches in the asymmetric case from the equatorially-symmetric case. The latter hence serves as a reference case, called the perfect case, because of its higher symmetry. The equatorial asymmetry hence introduces *imperfections*[1] which leads to isolated branches of steady states. The imposed asymmetry (in this case mimicking $T_n^a < T_s^a$) introduces a preference for the northern sinking (NPP) state.

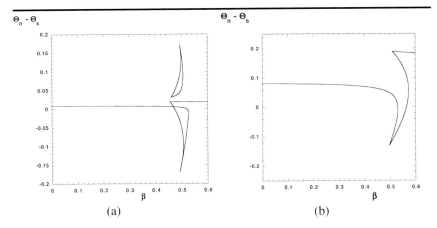

Figure 3.9. (a) Branches of steady solutions for $\epsilon = 0.01$ of the three-box model (3.14) with $\alpha_n = \alpha_s(1 + \epsilon)$ and $\beta_s = \beta_n = \beta$ and $\eta_3 = 0.3$. (b) Same as (a), but for $\epsilon = 0.1$. The stability of the branches is not indicated.

3.2. Dynamical Systems: Fixed Points

The equations governing the evolution of salinity and temperature of the two-box model (3.4) in section 3.1 are a special case of a general system of ordinary differential equations (ODEs), which can be written as

$$\frac{d\mathbf{x}}{dt} = \mathbf{f}(\mathbf{x}, \lambda, t) \qquad (3.18)$$

where $\mathbf{x}$ is the state vector in the state space $\mathbb{R}^n$, $\mathbf{f}$ is a smooth (sufficiently differentiable) vector field, λ is a parameter in the control or parameter space $\mathbb{R}^p$ and t denotes time. This equations (3.18) define a continuous time dynamical system,

---

[1]From a computational point of view (see chapter 4) it is very advantageous if branches of solutions are connected because it is quite difficult to find isolated branches. Hence, from this simple example we see already that it is advisable to start computations from the most symmetric model context, such that all branches are connected. Once these are determined, the imperfections can be studied systematically.

which is called autonomous if the vector field $\mathbf{f}$ does not depend explicitly on time; otherwise it is called non-autonomous. A trajectory of the dynamical system, starting for example at $\mathbf{x}_0$, is a curve in the state space $\mathbf{x}(t)$ satisfying (3.18), i.e., at each point of the curve the vector field is tangent to it.

Many introductory textbooks on bifurcation theory and its applications are available (Guckenheimer and Holmes, 1990; Wiggins, 1990; Kuznetsov, 1995; Nayfeh and Balachandran, 1995), but are quite mathematically oriented. In Strogatz (1994), a nice introduction to bifurcation theory is given, with focus on applications in physics, biology and engineering. In the next sections, the basic ingredients of the more abstract approach are sketched with much reference to the example in section 3.1.

3.2.1. Elementary concepts

A solution $\bar{\mathbf{x}}$ of an autonomous dynamical system at a parameter value $\bar{\lambda}$ is a fixed point if

$$\mathbf{f}(\bar{\mathbf{x}}, \bar{\lambda}) = 0 \tag{3.19}$$

and hence any trajectory with initial conditions at the fixed point will remain there forever.

▶

Example 3.1: The Stommel two-box model

The Stommel two-box model (3.4) can be written, with $x_1 = T, x_2 = S$, as the two-dimensional system of ODEs,

$$\frac{dx_1}{dt} = \eta_1 - x_1(1 + \mathcal{M}(x_1 - x_2))$$

$$\frac{dx_2}{dt} = \eta_2 - x_2(\eta_3 + \mathcal{M}(x_1 - x_2))$$

The state space is two-dimensional ($n = 2$), the parameter space is three-dimensional ($p = 3$) and the dynamical system is autonomous because the right hand side does not depend explicitly on t. Examples of trajectories were plotted in Fig. 3.3b, one of them starting at $\mathbf{x}_0 = (0, 0)$. Examples of fixed points were given in the previous section, i.e., for $\eta_2 = 0$ a fixed point is given by $\bar{x}_1 = -1/2 + \sqrt{1/4 + \eta_1}, \bar{x}_2 = 0$.

◀

In the analysis of the linear stability of a particular fixed point $\bar{\mathbf{x}}$ at $\bar{\lambda}$, small perturbations $\mathbf{y}$ are assumed to be present according to

$$\mathbf{x} = \bar{\mathbf{x}} + \mathbf{y} \tag{3.21}$$

and linearization of (3.18) around $\bar{\mathbf{x}}$ gives

$$\frac{d\mathbf{y}}{dt} = J(\bar{\mathbf{x}}, \bar{\lambda})\mathbf{y} \tag{3.22}$$

where J is the Jacobian matrix given by

$$J = \begin{bmatrix} \frac{\partial f_1}{\partial x_1} & \cdots & \frac{\partial f_1}{\partial x_n} \\ \cdots & \cdots & \cdots \\ \frac{\partial f_n}{\partial x_1} & \cdots & \frac{\partial f_n}{\partial x_n} \end{bmatrix} \qquad (3.23)$$

For example, for the Stommel two-box model the Jacobian matrix was given in (3.13). The set of equations (3.22) has solutions $\mathbf{y}(t) = e^{\sigma t}\hat{\mathbf{y}}$ which, when substituted, leads to an eigenvalue problem for the complex growth factor $\sigma = \sigma_r + i\sigma_i$,

$$\sigma\hat{\mathbf{y}} = J(\bar{\mathbf{x}}, \bar{\lambda})\hat{\mathbf{y}} \qquad (3.24)$$

A special role have those fixed points, for which $\sigma_r \neq 0$ for all eigenvalues σ, and these are called hyperbolic fixed points. In Technical box 3.1, the local behavior near fixed points in a two-dimensional system is given. The eigenvalues of the Jacobian totally determine the local behavior of the trajectories near the fixed points, except in nonhyperbolic cases (for example, at a center).

 ### Technical box 3.1: Stability of fixed points in $\mathbb{R}^2$

Consider the two-dimensional linear system $\dot{\mathbf{x}} = A\mathbf{x}$ or written out as

$$\frac{dx}{dt} = ax + by \qquad (3.25a)$$

$$\frac{dy}{dt} = cx + dy \qquad (3.25b)$$

and assume that $\det A = ad - bc \neq 0$. The only fixed point of these equations is $x = y = 0$. Its stability is determined by the eigenvalues of A; indicate these with σ and τ. These can both be real or they can form a complex conjugate pair; in the latter case $\sigma = \alpha + i\beta$, $\tau = \alpha - i\beta$. From linear algebra, we use the fact that there exists a nonsingular matrix T such that with the transformation $\tilde{\mathbf{x}} = T\mathbf{x}$, the dynamical system is written as

$$\dot{\tilde{\mathbf{x}}} = TAT^{-1}\mathbf{x} = B\tilde{\mathbf{x}}$$

where B has the same eigenvalues as A.

The different cases are related to the different forms of the matrix B and can be conveniently visualized by plotting the local trajectories (also called *phase portraits*) near the fixed point (Fig. 3.10). The matrix B has one of the following expressions:

(i) Improper node I

$$B = \begin{pmatrix} \sigma & 0 \\ 0 & \tau \end{pmatrix}, \ \sigma < \tau < 0 \text{ or } 0 < \sigma < \tau$$

In this case, we have solutions $x(t) = C_1 exp(\sigma t), y(t) = C_2 exp(\tau t)$. For $\sigma < \tau < 0$, the fixed point is stable and for $0 < \sigma < \tau$ it is unstable. The phase portrait for an example is plotted in Fig. 3.10a.

(ii) Improper node II

$$B = \begin{pmatrix} \sigma & 0 \\ 1 & \sigma \end{pmatrix}, \ \sigma \neq 0$$

In this case, we have solutions $x(t) = C_1 exp(\sigma t), y(t) = (C_2 + t) exp(\sigma t)$ and hence the fixed point is unstable if $\sigma > 0$ and stable when $\sigma < 0$. The phase portrait for an example is plotted in Fig. 3.10b.

(iii) Proper node

$$B = \begin{pmatrix} \sigma & 0 \\ 0 & \sigma \end{pmatrix}, \ \sigma \neq 0$$

In this case, we have solutions $x(t) = C_1 exp(\sigma t), y(t) = C_2 exp(\sigma t)$ and hence $y = kx$, where $k = C_2/C_1$; the fixed point is unstable if $\sigma > 0$ and stable when $\sigma < 0$. The phase portrait for an example is plotted in Fig. 3.10c.

(iv) Focus

$$B = \begin{pmatrix} \alpha & \beta \\ -\beta & \alpha \end{pmatrix}, \ \alpha \neq 0, \beta \neq 0$$

In this case, if we transform to polar coordinates $x = r \cos \theta, y = r \sin \theta$, we have solutions $r(t) = C_1 exp(\alpha t), \theta(t) = -\beta t + C_2$. When $\alpha > 0$, the fixed point is unstable while for $\alpha < 0$, it is stable. The phase portrait for an example is plotted in Fig. 3.10d.

(v) Center

$$B = \begin{pmatrix} 0 & \beta \\ -\beta & 0 \end{pmatrix}, \ \beta \neq 0$$

In this case, if we again transform to polar coordinates $x = r \cos \theta, y = r \sin \theta$, we have solutions $r(t) = C_1, \theta(t) = -\beta t + C_2$. The fixed point is neither stable nor unstable, and it is called neutrally stable. The phase portrait for an example is plotted in Fig. 3.10e.

(vi) Saddle

$$B = \begin{pmatrix} \sigma & 0 \\ 0 & \tau \end{pmatrix}, \ \sigma < 0 < \tau$$

In this case, we have solutions $x(t) = C_1 exp(\sigma t), y(t) = C_2 exp(\tau t)$ and hence the fixed point is always unstable. The phase portrait for an example is plotted in Fig. 3.10f.

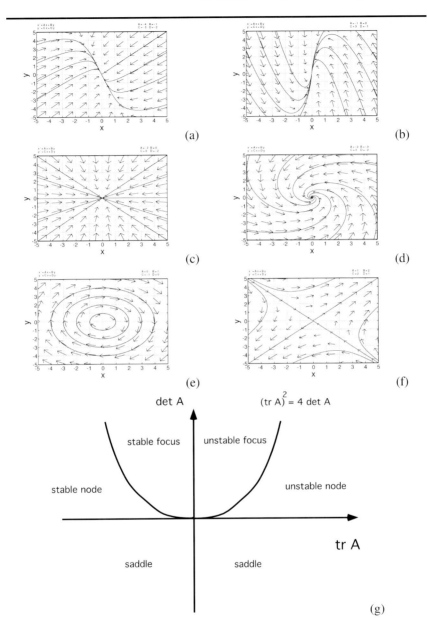

Figure 3.10. Overview of the different local phase portraits for a two-dimensional dy-namical system. Here, a, b, c and d are entries of the matrix A and σ, τ are its eigenvalues. The figures below were made with the Matlab package pplane6; see also http://www.math.iupui.edu/~mtc/Chaos/phase.htm (a) stable improper node I; a = -4, b = -1, c = -3, d = -2; $\sigma = -1, \tau = -5$. (b) stable improper node II; a = -1, b = 0, c = 3, d = -1; $\sigma = -1$, $\tau = -1$. (c) stable proper node; a = -2, b = 0, c = 0, d = -2; $\sigma = -2, \tau = -2$. (d) stable focus; a = -2, b = -3, c = 3, d = -2; $\sigma = -2 + 3i, \tau = -2 - 3i$. (e) center; a = 0, b = 1, c = -1, d = 0; $\sigma = i, \tau = -i$. (f) saddle; a = 1, b = 2, c = 2, d = 1; $\sigma = -1, \tau = 3$. (g) Different behavior of local trajectories in the (det A, tr A) plane.

The different cases can be conveniently ordered in a diagram (Fig. 3.10g) of the determinant of A versus the trace of A, where

$$det A \ = \ ad - bc = \sigma\tau \qquad (3.26a)$$
$$tr A \ = \ a + c = \sigma + \tau \qquad (3.26b)$$

The curve $(tr A)^2 = 4\, det A$ separates the nodes from the focus phase portraits.

Near a hyperbolic fixed point, the local solution structure of the linearized system is the same as that of the nonlinear system. This is a consequence of the so-called Hartman-Grobman theorem (Guckenheimer and Holmes, 1990). When qualitative changes occur in the fixed point solutions of the dynamical system, such as the changes in type or number of solutions, the dynamical system is said to have undergone a bifurcation. This can only occur at nonhyperbolic points. For example in Fig. 3.7, the number of solutions changes when the value of β crosses the points P; the same happened in Fig. 3.4b at the points L. In the state-parameter space formed by $(\mathbf{x}, \lambda)$, locations at which bifurcations occur are called bifurcation points. The points L in Fig. 3.4b and P in Fig. 3.7 are examples of such bifurcation points. A bifurcation diagram is a graph in which the variation of the solutions of a particular problem is displayed in the state-control parameter space and Fig. 3.4b and Fig. 3.7 are notable examples. A bifurcation that needs at least m parameters to occur is called a codimension-m bifurcation. As both the bifurcation points L and the bifurcation points P occur as one parameter (η_2 and β, respectively) is varied, these are examples of codimension-1 bifurcations.

3.2.2. Codimension-1 bifurcations

Transition behavior in the structure of fixed points will occur near bifurcation points. How this will happen depends on how eigenvalues of the Jacobian matrix J will cross the imaginary axis as parameters are changed. When J has real coefficients, either real eigenvalues or complex conjugate pairs of eigenvalues cross this axis. Hence, when a single parameter is involved, two different situations can occur: a single real eigenvalue or a complex conjugate pair of eigenvalues can cross the imaginary axis.

3.2.2.1 A single zero eigenvalue

The simplest case is a crossing of one real eigenvalue at a non-hyperbolic point $(\bar{\mathbf{x}} = 0, \bar{\lambda} = 0)$ through variation of only one parameter. This, for example, happens in Fig. 3.5 at the point L_1, where one eigenvalue crosses the imaginary axis. At L_1, one eigenvalue in (3.11) is exactly zero ($\sigma = 0$); let the eigenvector be indicated by $\mathbf{v}$. General methods exists (see Technical box 3.2) which relate the solution structure of the full equations to a single scalar equation (near this bifurcation point) of the form

$$g(x, \lambda) = 0 \qquad (3.27)$$

where $\mathbf{v} = x\, \mathbf{v}_0$ for some fixed vector $\mathbf{v}_0$. The equation (3.27), a single equation for a scalar variable x, is called the *reduced* or *bifurcation* equation. The point

$x = 0$ is a fixed point of g and, since it is a bifurcation point, its Jacobian matrix must have a zero eigenvalue and hence $g_x = 0$ where the subscript indicates differentiation to x.

 Technical box 3.2:
Reduction methods

A typical reduction technique is the Lyapunov-Schmidt method which is described in detail in Golubitsky *et al.* (1988), page 25-35. One can decompose the space $\mathbb{R}^n$ into a vector along the nullspace of J, say $ker J$, and a remainder space M, i.e. $\mathbb{R}^n = ker(J) \oplus M$, where $\oplus$ indicates the direct product of subspaces. Application of the implicit function theorem (Guckenheimer and Holmes, 1990) on the projection of the equations on the space M, gives that for $\mathbf{v} \in ker(J), \mathbf{w} \in M$, one can write $\mathbf{w} = \mathbf{W}(\mathbf{v}, \lambda)$. Under very mild conditions on the vector field $\mathbf{f}$ it can be shown that there exists a mapping ϕ such that

$$\mathbf{f}(\mathbf{v} + \mathbf{W}(\mathbf{v}, \lambda), \lambda) = 0 \Leftrightarrow \phi(\mathbf{v}, \lambda) = 0$$

If $\mathbf{v}_0$ is a vector in $ker\ J$, then one can write $\mathbf{v} = x\mathbf{v}_0$, with $x \in \mathbb{R}$ and the zeroes of $\phi(\mathbf{v}, \lambda) = 0$ correspond to those of

$$g(x, \lambda) = <\mathbf{v}_0, \phi(x\mathbf{v}_0, \lambda)>$$

where $<,>$ is the standard inner product on $\mathbb{R}^n$. In this way, bifurcation equations can be derived for very general dynamical systems.

Within the theory, it can furthermore be proven that (Golubitsky *et al.*, 1988):

(i) A finite number of coefficients in the Taylor series of g in (3.27) near the point $(x = 0, \lambda = 0)$ fully determines the solution structure when λ is locally varied. Note that these coefficients are directly related to derivatives of g to x and λ near the origin, since

$$g(x, \lambda) = g(0, 0) + g_x(0, 0)x + g_\lambda(0, 0)\lambda + g_{xx}(0, 0)\frac{x^2}{2} + \dots$$

(ii) All dynamical systems having a bifurcation equation with the same conditions on the derivatives of g as a certain prototype system, called the *normal form*, may be transformed (through a coordinate transformation) into this normal form. These systems locally have thus the same solution structure as that of the normal form. Techniques how to derive these normal forms are well described in Guckenheimer and Holmes (1990).

The above remarks indicate that normal forms play a central role in the classi-fication of the changes in solution structure when parameters are changed. For the one-parameter case in which non-hyperbolicity is induced by one single zero eigenvalue, there are three important normal forms which arise in applications.

The first normal form is that of the fold or *saddle-node* bifurcation (also called limit point or turning point). For $\delta \in \{-1, 1\}$, it is given by

$$g(x, \lambda) = \lambda + \delta x^2 \qquad (3.28)$$

Indeed $g(0,0) = g_x(0,0) = 0$, but the derivatives $g_\lambda(0,0)$ and $g_{xx}(0,0)$ are nonzero. For $\delta = -1$, steady solutions $\bar{x} = \pm\sqrt{\lambda}$ only exist when $\lambda > 0$ (Fig. 3.11a). The stability of these states is determined by the sign of the 'Ja-cobian' $g_x = -2x$ and hence the solution $\bar{x} = \sqrt{\lambda}$ is stable and the solution $\bar{x} = -\sqrt{\lambda}$ is unstable. For $\delta = 1$, solutions $\bar{x} = \pm\sqrt{-\lambda}$ occur only for nega-tive λ; the branch with positive (negative) $\bar{x}$ is unstable (stable) (Fig. 3.11b). The points L_1 and L_2 in Fig. 3.4b are examples of saddle-node bifurcations.

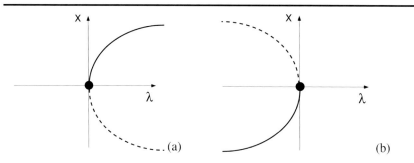

Figure 3.11.   Bifurcation diagram of the saddle-node bifurcation (3.28) for (a) $\delta = -1$ and (b) $\delta = 1$. A drawn (dashed) curve indicates stable (unstable) steady solutions.

The second normal form is that of the *transcritical* bifurcation which, again for $\delta \in \{-1, 1\}$, is given by

$$g(x, \lambda) = \lambda x + \delta x^2 \qquad (3.29)$$

Again $g(0,0) = g_x(0,0) = 0$, but now also $g_\lambda(0,0) = 0$. The first nonzero derivative is the second order derivate $g_{xx}(0,0)$. For $\delta = -1$, steady solutions are given by $\bar{x} = 0$ and $\bar{x} = \lambda$ whereas for $\delta = 1$, the steady solutions are $\bar{x} = 0$ and $\bar{x} = -\lambda$. With Jacobian $g_x = 2\delta x + \lambda$, the stability of each steady state is easily determined and the solutions are shown in Fig. 3.12a and Fig. 3.12b for $\delta = -1$ and $\delta = 1$, respectively. In both cases, there is a simple exchange of stability properties between two different steady states. Transcritical bifurcations did not appear in the Stommel two-box model.

The third normal form has one additional zero derivative, i.e., $g_{xx}(0,0) = 0$ and is that of the pitchfork bifurcation. The bifurcation equation is, for $\delta \in \{-1, 1\}$,

$$g(x, \lambda) = \lambda x - \delta x^3 \qquad (3.30)$$

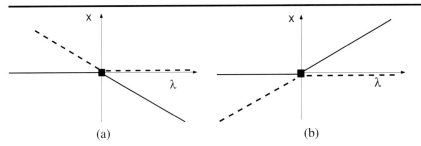

Figure 3.12. Bifurcation diagram of the transcritical bifurcation for (a) $\delta = -1$ and for (b) $\delta = 1$. A drawn (dashed) curve indicates stable (unstable) steady solutions.

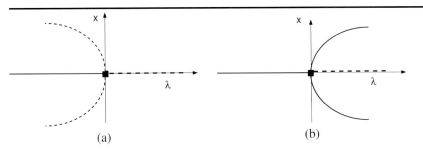

Figure 3.13. Bifurcation diagram of the pitchfork bifurcation for (a) $\delta = -1$ and (b) $\delta = 1$. A drawn (dashed) curve indicates stable (unstable) steady solutions.

Consider the case $\delta = 1$. For $\lambda < 0$, there is only one steady solution (or fixed point) $\bar{x} = 0$, but for $\lambda > 0$, three fixed points exist, i.e. $\bar{x} = 0$ and $\bar{x} = \pm\sqrt{\lambda}$. Hence, the number of fixed points changes as λ crosses zero. To determine the stability of the fixed points, the eigenvalues σ of (3.24) with the Jacobian matrix $g_x = \lambda - 3x^2$ must be considered at the fixed points. For $\bar{x} = 0$, it follows that $\sigma = \lambda$, indicating that $\bar{x} = 0$ is stable for $\lambda < 0$ but unstable for $\lambda > 0$. At both additional fixed points existing for $\lambda > 0$, it follows that $\sigma = -2\lambda$, showing that these are both stable (Fig. 3.13b). This bifurcation is called a supercritical pitchfork bifurcation since the symmetrically related nontrivial solutions exists for $\lambda > 0$. The case $\delta = -1$, a subcritical pitchfork bifurcation, is shown in Fig. 3.13a. The points P in Fig. 3.7 are examples of pitchfork bifurcations.

3.2.2.2 A single complex conjugate pair of eigenvalues

Whereas in the previous bifurcations, the number of fixed points changed as a parameter was varied, it is also possible that the character of the solution changes from steady to oscillatory. The normal form of this type of bifurcation, called

Poincaré-Andronov-Hopf bifurcation, or simply *Hopf* bifurcation is given by

$$\frac{dx}{dt} = \lambda x - \omega y - \delta x(x^2 + y^2) \tag{3.31a}$$

$$\frac{dy}{dt} = \lambda y + \omega x - \delta y(x^2 + y^2) \tag{3.31b}$$

It can be easily checked that at $\lambda = 0$, the Jacobian matrix J of the zero state has a pair of complex conjugate eigenvalues $\sigma = \pm i\omega$. Using the transformation $x = r \cos \theta$, $y = r \sin \theta$, (3.31) is transformed into

$$\frac{dr}{dt} = \lambda r - \delta r^3 \tag{3.32a}$$

$$\frac{d\theta}{dt} = \omega \tag{3.32b}$$

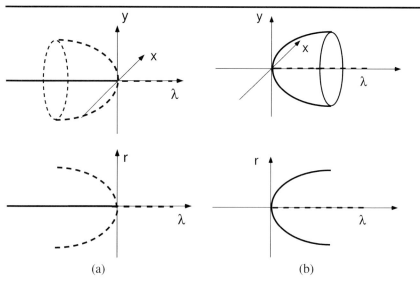

(a)   (b)

Figure 3.14.   Bifurcation diagrams of the Hopf bifurcation for (a) the subcritical case $\delta = -1$ and (b) the supercritical case $\delta = 1$. In the (r, λ) diagrams, a drawn (dashed) curve with $r = 0$ indicates a stable (unstable) steady state. A drawn (dashed) curve with $r \neq 0$ indicates a stable (unstable) periodic orbit. The periodic orbits and steady states are also drawn in the (x, y, λ) diagrams.

Comparing (3.32a) with (3.30), it can be seen that a pitchfork bifurcation occurs at $\lambda = 0$ in the (r, λ) plane; consider again $\delta = 1$. For $\lambda < 0$, only one stable fixed point exists, which corresponds to a steady solution of the original equations (3.31). However, for $\lambda > 0$ the stable nontrivial fixed point is now a closed trajectory with a period $2\pi/\omega$ and this is a periodic solution of the original equations (3.31). Hence, a transition of steady behavior to periodic behavior occurs as λ crosses zero; a Hopf bifurcation occurs. The bifurcation diagram is shown in Fig. 3.14a for the subcritical case ($\delta = -1$); the supercritical case ($\delta = 1$) is shown in Fig. 3.14b. Hopf bifurcations did not occur in the Stommel two-box model discussed in section 3.1.

The emergence of the periodic orbit can be seen explicitly by computing trajectories of the equations (3.31) with $\delta = \omega = 1$ for a value of λ below ($\lambda = -0.1$) the Hopf bifurcation and one above ($\lambda = 0.1$) the Hopf bifurcation. As initial condition, the point $(x, y) = (0, 2)$ is chosen. For $\lambda = -0.1$ (Fig. 3.15a), the trajectory spirals in and finally ends up at the stable fixed point $(0, 0)$. However, for $\lambda = 0.1$ (Fig. 3.15b), it spirals to a periodic orbit with radius $r = 1/\sqrt{10}$ which arises through the Hopf bifurcation. A trajectory starting at $(0, 0.1)$ is also plotted as the dashed curve in Fig. 3.15b and demonstrates that the origin $(x, y) = (0, 0)$ has become an unstable fixed point; also this trajectory spirals to the periodic orbit.

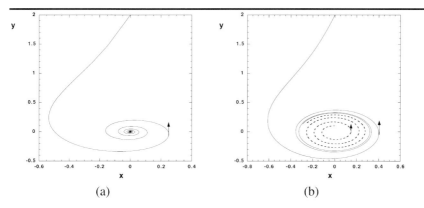

(a)   (b)

Figure 3.15.   Trajectories of the equations (3.31) with $\omega = \delta = 1$ and (a) $\lambda = -0.1$ and (b) $\lambda = 0.1$. Initial condition for both drawn trajectories in (a) and (b) is the point $(0, 2)$; in (b) also a trajectory starting at $(0, 0.1)$ is shown as the dashed curve. Arrows indicate the direction of time along the curves.

3.2.3. Imperfection theory

Normal forms play a central role in the theory of bifurcations and are the most elementary equations exhibiting some type of bifurcation behavior. In the codimension-1 case here, a normal form is expressed as a single bifurcation equation $g(x, \lambda) = 0$. One may ask what happens when g is slightly perturbed, for example when small changes occur in the model under investigation. Does the solution structure change and if yes, can one determine *a priori* how? The changes in solution structure under perturbations of the vector field is usually referred to as imperfection theory. An illustrative example is the imperfection of the pitchfork bifurcation, with normal form (3.30). The latter normal form exhibits a reflection symmetry with respect to $x = 0$, since $g(-x, \lambda) = -g(x, \lambda)$. This symmetry property of the reduced equation is inherited from a symmetric model set-up, due to geometrical symmetry or symmetry in the forcing conditions. For example, the three-box model (3.16) has a reflection symmetry under equatorially-symmetric forcing conditions.

Slight perturbations from symmetry in a particular model will also destroy the symmetry in the reduced equation. For the three-box model (3.16), an example was shown when the forcing was considered slightly asymmetric ($\alpha_n > \alpha_s$). This asymmetry immediately shows up in the reduced equation which becomes

$$G(x, \lambda) = \epsilon + \lambda x - x^3 \qquad (3.33)$$

for some (small) ϵ. When $\epsilon = 0$, a pitchfork bifurcation occurs at $(x = 0, \lambda = 0)$, but for $\epsilon \neq 0$, the pitchfork bifurcation is no longer present because the reflection symmetry no longer exists. Bifurcation diagrams are sketched for both negative and positive ϵ in Fig. 3.16 and show different reconnections of the branches. In both cases, there is only one solution for negative λ (approximately given by $x = -\epsilon/\lambda < 0$ for small ϵ) and three solutions exist for large positive ϵ.

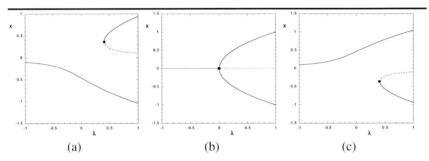

(a)   (b)   (c)

Figure 3.16.   Bifurcation diagrams for the reduced equation $G(x, \lambda) = \epsilon + \lambda x - x^3$ for different values of ϵ with (a) $\epsilon = -0.1$, (b) $\epsilon = 0$ and (c) $\epsilon = 0.1$. A drawn (dashed) curve indicates stable (unstable) steady solutions.

Another example is the imperfection of the transcritical bifurcation given by

$$G(x, \lambda) = \epsilon + \lambda x + x^2 \qquad (3.34)$$

For $\epsilon = 0$, a transcritical bifurcation occurs at $(x = 0, \lambda = 0)$, but for $\epsilon \neq 0$, this bifurcation is no longer present. The different reconnections for this case are shown in Fig. 3.17 both for positive and negative ϵ. For $\epsilon < 0$, the stable and unstable branches connect up, whereas for $\epsilon > 0$ an unstable and stable branch connect to give two saddle-node bifurcations. In the latter case, a window in λ opens where no steady solutions exist.

More general theory for imperfections of bifurcations exists (Golubitsky *et al.*, 1988). In short, the concept needed is that of a $k-$parameter unfolding of g (already indicated by G above) which can be considered as a perturbation on g, i.e.,

$$\begin{aligned} G(x, \lambda, \alpha_1, ..., \alpha_k) &= 0 \\ G(x, \lambda, 0, ..., 0) &= g(x, \lambda) \end{aligned}$$

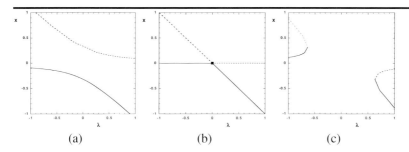

Figure 3.17. Bifurcation diagrams for the reduced equation $G(x, \lambda) = \epsilon + \lambda x + x^2$ for different values of ϵ with (a) $\epsilon = -0.1$, (b) $\epsilon = 0$ and (c) $\epsilon = 0.1$. A drawn (dashed) curve indicates stable (unstable) steady solutions.

where the α_k are additional parameters. For example in (3.33), $k = 1$, $\alpha_1 = \epsilon$ and $g(x, \lambda) = \lambda x - x^3$. The theory provides a classification of bifurcation behavior through the following results

(i) For each normal form, there is one unfolding – the universal unfolding – which captures every solution structure for every possible perturbation of g.

(ii) This universal unfolding requires a minimum number of parameters, say indicated by k_{min}.

For the saddle-node bifurcation, the universal unfolding is the normal form itself and $k_{min} = 0$. This indicates that slight changes in the vector field will have no effect on the local solution structure of the saddle-node bifurcation. For the transcritical bifurcation, the universal unfolding is given by

$$G(x, \lambda, \epsilon) = \delta x^2 + \lambda x + \epsilon \tag{3.36}$$

with $k_{min} = 1$. This defines a one-parameter family of bifurcation diagrams with parameter ϵ as shown for $\delta = 1$ in Fig. 3.17.

For the pitchfork bifurcation, the universal unfolding is

$$G(x, \lambda, \alpha, \beta) = \lambda x - \delta x^3 + \alpha x + \beta \tag{3.37}$$

with $k_{min} = 2$, defining a two-parameter family of bifurcation diagrams. Whereas in Fig. 3.16, the special case $\delta = 1$, $\alpha = 0$ and $\beta = \epsilon$ was presented, the more general picture of different solutions in the (α, β)-plane is shown in Fig. 3.18 for $\delta = 1$. For $\alpha = 0$, indeed the changes occur from regime (1) to (2) when β is varied, which corresponds to Fig. 3.16, but for $\alpha \neq 0$ additional saddle-node bifurcations may occur on the disconnected branches (regimes (3) and (4) in Fig. 3.18). Although quite abstract, the existence of a classification scheme of bifurcation behavior is a beautiful and powerful result.

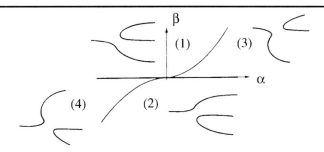

Figure 3.18. A two-parameter family of bifurcation diagrams of the unfolding $G(x, \lambda, \alpha, \beta) = \lambda x - \delta x^3 + \alpha x + \beta$ of the pitchfork bifurcation for $\delta = 1$ (Golubitsky et al., 1988).

3.2.4. Codimension-2 bifurcations

As we have seen, multiple stable steady states in a particular model can occur due to a pitchfork, a transcritical or a saddle-node bifurcation, whereas a Hopf bifurcation may introduce temporal variability through the appearance of a periodic orbit. These bifurcations occur when changing only one parameter in the system and are associated with the crossing of only one real eigenvalue (or a pair of complex conjugated eigenvalues) through the imaginary axis. There are several ways in which more complicated bifurcations may arise. Only two examples are given below for which also alternative representations of the dynamics are presented.

3.2.4.1 The cusp bifurcation

The first example is the *cusp* bifurcation, where higher-order derivatives in the normal form become zero through variation of a second parameter. The normal form of the cusp is given by

$$g(x, \lambda, \alpha) = \lambda - \alpha x - x^3 \qquad (3.38)$$

for which $g_{xx}(0, 0) = 0$, but $g_{xxx}(0, 0) \neq 0$. The different diagrams in the (α, λ) plane are sketched in Fig. 3.19. Instead of sketching the different bifurcation diagrams as in Fig. 3.18, now pictures of the function g are drawn as dotted curves. Intersections with $g = 0$ provide steady states and their stability is indicated by the arrows; if one of the arrows points away from the steady state, it is unstable. Hence, the arrows indicate the directions of the one-dimensional trajectories.

In the region within the two curves C_1 and C_2, there are three equilibria of which two are stable and one is unstable. On the curves C_1 and C_2, the function $g(x)$ just becomes tangent to the x-axis and two equilibria, instead of three, appear. For $\lambda = 0$, three solutions exist if $\alpha < 0$ and only one if $\alpha > 0$. For $\alpha = \lambda = 0$, the dotted curve shows the function $g(x) = -x^3$, which gives one fixed point, the cusp point, of which the stability can only be decided after further analysis. The point Q in Fig. 3.4c is an example of a cusp bifurcation.

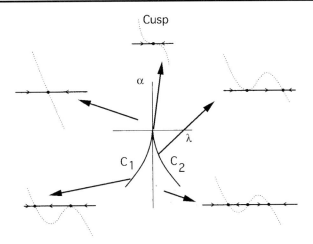

Figure 3.19.   Sketch of the solution structure in the (λ, α) plane of the reduced equation $g(x, \lambda) = \lambda - \alpha x - x^3$. Within the region bounded by C_1 and C_2, three solutions exist and outside this region there is one solution; the origin is a cusp point (Kuznetsov, 1995).

3.2.4.2 The Bogdanov-Takens bifurcation

Another possibility is that additional eigenvalues of the Jacobian matrix cross the imaginary axis. For example, when two parameters are varied within a dynamical system, paths of different pitchfork bifurcations may intersect. An example is the *Bogdanov-Takens* bifurcation whose normal form is given by

$$g_1(x, y) = y \tag{3.39a}$$
$$g_2(x, y) = \mu_1 + \mu_2 y + x^2 + x\,y \tag{3.39b}$$

In this case, two real eigenvalues of the Jacobian matrix

$$J = \begin{pmatrix} 0 & 1 \\ 2\bar{x} + \bar{y} & \bar{x} + \mu_2 \end{pmatrix} \tag{3.40}$$

monitoring the stability of the trivial solution ($\bar{x} = 0, \bar{y} = 0$) simultaneously cross the imaginary axis when μ_2 crosses zero. The solution structure can be analyzed in quite detail, but this would go too far within this text (see Guckenheimer and Holmes (1990)). The change in solution structure is plotted as a diagram (Fig. 3.20) of the trajectories in the two-parameter plane (μ_1, μ_2). Intersections of trajectories, or points where trajectories originate or disappear, indicate steady states while closed curves indicate periodic orbits. In the region where $\mu_1 < 0$ and $\mu_2 < 0$, there are two steady states of which one is stable (a focus, see Technical Box 3.1) and one is unstable (a saddle). Along the line $\mu_1 = 0$, both equilibria disappear through a saddle-node bifurcation and no steady states exist when

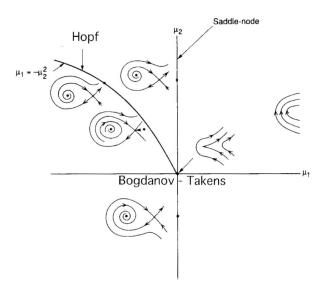

Figure 3.20.   Solution structure near a Bogdanov-Takens bifurcation (Guckenheimer and Holmes, 1990). This codimension-2 bifurcation is located at the origin and phase space trajectories (the curves containing an arrow) are plotted to show the solution structure at nearby values of the parameters. Closed curves are periodic orbits. At the line $\mu_1 = 0$, saddle-node bifurcations occur, while along the curve $\mu_1 = -\mu_2^2$, Hopf bifurcations occur.

$\mu_1 > 0$. Hopf bifurcations occur along the curve described by $\mu_1 = -\mu_2^2$ leading to periodic orbits left of this curve.

3.3.   Periodic Solutions and their Stability

A solution of a general dynamical system is a periodic solution with period $\mathcal{P}$ if it satisfies

$$\frac{d\mathbf{x}}{dt} = \mathbf{f}(\mathbf{x}, \lambda, t) \qquad (3.41a)$$

$$\mathbf{x}(t + \mathcal{P}) = \mathbf{x}(t) \qquad (3.41b)$$

such that $\mathbf{x}(t + \tau) \neq \mathbf{x}(t)$ for $0 \leq \tau \leq \mathcal{P}$. It will be useful to relate periodic solutions to fixed points of a mapping; this is the subject of section 3.3.1. Next, the stability of periodic orbits is considered (in section 3.3.2) through Floquet analysis.

3.3.1.   Poincaré section and Poincaré map

To define a Poincaré map, a hypersurface Σ in the state space $\mathbb{R}^n$, for example a line segment in two-dimensional state space or a plane in three-dimensional state space, is chosen such that the periodic orbit is not tangent to it for all time t, i.e., when

$$\mathbf{n}.\mathbf{f} \neq 0 \qquad (3.42)$$

Here, $\mathbf{n}$ is the normal to the hypersurface (Fig. 3.21) and $\mathbf{f}$ the right hand side (the vector field) of (3.41a). This hypersurface is called a Poincaré section and it

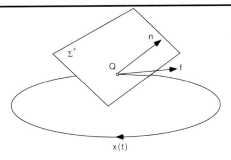

Figure 3.21.   Sketch of a one-sided Poincaré section. A periodic orbit is sketched which intersects the Poincaré section at the point Q. The vector $\mathbf{f}$ is the tangent to the trajectory at Q and $\mathbf{n}$ is the outward normal of Σ^+ at Q.

is one-sided (notation Σ^+) when the sign of $\mathbf{n}.\mathbf{f}$ at the intersection of trajectories and Σ are the same each time the surface is crossed.

In non-autonomous systems, the period $\mathcal{P}$ of the periodic orbit is usually explicitly known. In applications, this may be the period of the seasonal cycle, appearing for example, in a climate model with seasonal forcing. Remember that for such a periodic solution, the vector field $\mathbf{f}$ is periodic with period $\mathcal{P}$. This property can be explicitly used to construct a Poincaré section, as shown in Fig. 3.22, by stroboscopically measuring the state variables at intervals $\mathcal{P}$.

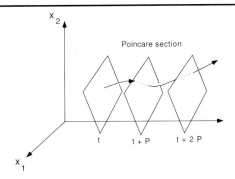

Figure 3.22.   Sketch of a one-sided Poincaré section for a non-autonomous system.

Let a periodic orbit intersect a one-sided Poincaré section Σ^+ at successive intersections indicated by $\{\mathbf{x}_1, \mathbf{x}_2, \mathbf{x}_3, \cdots\}$ then the Poincaré map $P : \Sigma^+ \rightarrow \Sigma^+$ is defined as

$$\mathbf{x}_{k+1} = P\mathbf{x}_k \qquad (3.43)$$

It immediately follows that if the Poincaré map has a fixed point, i.e. $P\mathbf{x}_k = \mathbf{x}_k$, then the dynamical system from which the map is derived has a periodic orbit. Hence, once a Poincaré section is found, periodic orbits can be directly related to fixed points of the Poincaré map on this section.

▶

Example 3.2: Poincaré section

Consider the system of equations (3.31) for $\delta = 1$ and $\omega \neq 0$, in particular its formulation in polar coordinates (3.32),

$$\frac{dr}{dt} = \lambda r - r^3$$
$$\frac{d\theta}{dt} = \omega$$

At $\lambda = 0$, a Hopf bifurcation occurs and for $\lambda > 0$, a periodic orbit having a period $\mathcal{P} = 2\pi/\omega$ exists and is described by

$$r = \sqrt{\lambda} \,;\, \theta = \omega t$$

A Poincaré section can be chosen (for certain θ_0) as

$$\Sigma^+ = \{(r, \theta) \in \mathbb{R} \times [0, 2\pi) \mid \theta = \theta_0\}$$

In this case, the normal in polar coordinates is given by $\mathbf{n} = (0, 1)$ and (on the periodic orbit) $\mathbf{f} = (0, \omega)$ such that Σ^+ is a one-sided Poincaré section if $\omega \neq 0$. If we choose $\theta_0 = \pi/2$, then in Cartesian coordinates, the Poincaré section is parallel to the $y-$axis. For example, in Fig. 3.15 one could take the interval $y \in (0, 1]$ as a Poincaré section.

There are very few examples where a Poincaré map can be explicitly computed but in this example it is possible because explicit solutions exists of the trajectories for all initial conditions (r_0, θ_0) at $t = t_0$. Using the indefinite integral,

$$\int \frac{dx}{\alpha_1 x^3 + \alpha_2 x} = \frac{1}{2\alpha_2} \ln \mid \frac{x^2}{\alpha_1 x^2 + \alpha_2} \mid$$

the closed form solution $(r(t), \theta(t))$ is

$$r(t; t_0) = \left[(\frac{1}{r_0^2} - \frac{1}{\lambda}) e^{-2\lambda(t-t_0)} + \frac{1}{\lambda} \right]^{-\frac{1}{2}}$$
$$\theta(t; t_0) = \omega(t - t_0) + \theta_0$$

A trajectory with initial conditions at (r_0, θ_0) intersects Σ^+ at times $t_k = t_0 + 2k\pi/\omega$. As the time difference between subsequent intersections (needed for the Poincaré map) is $2\pi/\omega$, this gives

$$r_{k+1} = P(r_k) = \left[(\frac{1}{r_k^2} - \frac{1}{\lambda}) e^{\frac{-4\pi\lambda}{\omega}} + \frac{1}{\lambda} \right]^{-\frac{1}{2}}$$

Fixed points of the Poincaré map $P(r)$ are defined by $P(r^*) = r^*$ and a short calculation gives indeed $r^* = \sqrt{\lambda}$. This Poincaré map is plotted as the dotted

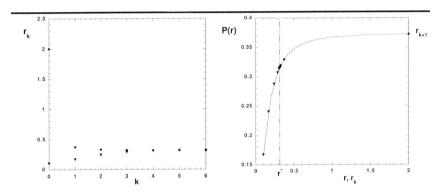

Figure 3.23. (a) Intersections r_k of the two trajectories in Fig. 3.15b (for $\lambda = 0.1$ and $\omega = 1.0$) with the one-sided Poincaré section $r > 0, \theta_0 = \pi/2$. Subsequent intersections are labelled with k; the diamonds and triangles represent the intersections of the trajectories starting in (0,2) and (0,0.1), respectively. (b) Plot of the Poincaré map $P(r)$ as in (3.22) together with the intersections r_k as in (a), but replotted as r_{k+1} versus r_k. The fixed point of the Poincaré map $P(r)$ is at $r^* = \sqrt{\lambda} = 0.3162$.

curve in Fig. 3.23b for the case $\lambda = 0.1$, $\omega = 1$ for which trajectories were plotted in Fig. 3.15b. The intersections r_k of the two trajectories with the Poincaré section defined by $r > 0, \theta_0 = \pi/2$ are plotted versus k — which monitors the subsequent intersections — in Fig. 3.23a. They are replotted in Fig. 3.23b as r_{k+1} versus r_k and indeed move along the Poincaré map. With increasing k, the fixed point r^* is reached.

◀

3.3.2. Floquet theory

To determine the stability of a periodic solution $\mathbf{X}(t)$ at $\lambda = \lambda_0$, which satisfies

$$\frac{d\mathbf{X}}{dt} = \mathbf{f}(\mathbf{X}, \lambda_0, t) \tag{3.46}$$

and has period $\mathcal{P}$, we consider infinitesimally perturbations $\mathbf{y}$ such that

$$\mathbf{x}(t) = \mathbf{X}(t) + \mathbf{y}(t) \tag{3.47}$$

Substituting (3.47) into (3.41) and neglecting quadratic interactions in $\mathbf{y}$, the evolution equations for the perturbations are

$$\frac{d\mathbf{y}}{dt} = J(\mathbf{X}(t), \lambda_0, t)\mathbf{y} \tag{3.48}$$

where J is the Jacobian matrix, which now contains time-periodic coefficients.

Consider first the case of an autonomous system. The linear system of equations (3.48) has n independent solutions $\mathbf{y}_j$, $j = 1, \cdots, n$ which can be written in matrix form as $Y(t) = [\mathbf{y}_1, ..., \mathbf{y}_n]$, i.e.

$$\frac{dY}{dt} = J(\mathbf{X}(t), \lambda_0)Y \tag{3.49}$$

The matrix Y is called the fundamental solution. If we change variables $\tau = t + \mathcal{P}$ then $Y(\tau)$ is also a solution of (3.49) since $J(\mathbf{X}(\tau - \mathcal{P}), \lambda_0) = J(\mathbf{X}(\tau), \lambda_0)$. Because the $\mathbf{y}_j$ are linearly independent, $Y(t + \mathcal{P})$ can be expressed into $Y(t)$ through

$$Y(t + \mathcal{P}) = \Phi Y(t) \tag{3.50}$$

where Φ is a constant matrix, that is called the *monodromy* matrix. If one chooses $Y(0) = I$, where I is the identity matrix, then $\Phi = Y(\mathcal{P})$.

The eigenvalues ρ_j of Φ are called Floquet multipliers. In the simplest case, these are all different and, if Z is the matrix of eigenvectors, the eigenvalue problem can be written as

$$\Phi Z = ZR \tag{3.51}$$

where R is a diagonal matrix with the Floquet multipliers as entries. If we multiply (3.50) and (3.51) from the left with Z^{-1} and introduce new variables $V = Z^{-1}Y$, we find from (3.50) that $V(t + \mathcal{P}) = RV(t)$. Written out in components, this gives

$$\begin{aligned} \mathbf{v}_j(t + \mathcal{P}) &= \rho_j \mathbf{v}_j(t) \\ \mathbf{v}_j(t + 2\mathcal{P}) &= \rho_j \mathbf{v}_j(t + \mathcal{P}) = \rho_j^2 \mathbf{v}_j(t) \end{aligned} \tag{3.52}$$

and more general

$$\mathbf{v}_j(t + N\mathcal{P}) = \rho_j^N \mathbf{v}_j(t) \tag{3.53}$$

The last equation shows that, since the $\mathbf{y}_j$ and hence the $\mathbf{v}_j$ provide the direction of the disturbances from the periodic orbit, the periodic orbit is stable when all Floquet multipliers ρ_j are within the unit circle, i.e., $|\rho_j| < 1$ for all $j = 1, \cdots, n$.

If a trajectory starts exactly at the periodic orbit it will stay on the orbit. However, when a trajectory starts near the orbit, it may diverge from or converge to the orbit which defines the stability of the orbit. The degree of divergence/convergence is measured by the Floquet multipliers. Already from this

view, it is clear that perturbations along the orbit do neither diverge nor converge
from the periodic orbit; hence for autonomous systems one of the Floquet multipliers is unity. This can be easily proved, since when $\mathbf{X}(t)$ is a periodic solution
of the autonomous system, then for every τ also $\mathbf{X}(t + \tau)$ is a periodic solution.
For small τ, define $\mathbf{y}(t) = \mathbf{X}(t + \tau) - \mathbf{X}(t)$, then $\mathbf{y}(0)$ is an initial disturbance
along the orbit of the periodic solution $\mathbf{X}$. For every integer N, then

$$\mathbf{y}(t + N\mathcal{P}) = \mathbf{X}(t + N\mathcal{P} + \tau) - \mathbf{X}(t + N\mathcal{P}) = \mathbf{X}(t + \tau) - \mathbf{X}(t) = \mathbf{y}(t) \quad (3.54)$$

and hence the Floquet multiplier for this disturbance is unity.

For non-autonomous systems, the stability of periodic orbits is determined in
the same way as for autonomous systems, by considering the Floquet multipliers,
through the construction of the monodromy matrix. The only difference with the
autonomous case is that a unit Floquet multiplier does not have to be present.

Technical box 3.3:
Unstable directions of a periodic
orbit

In many applications, one wants to determine the directions into which trajectories diverge from/converge to a periodic orbit of period $\mathcal{P}$. Multiplying (3.53)
for $N = 1$ by $e^{-\sigma_j(t+\mathcal{P})}$, where σ_j is a complex number, we obtain

$$e^{-\sigma_j(t+\mathcal{P})}\mathbf{v}_j(t + \mathcal{P}) = \rho_j e^{-\sigma_j(t+\mathcal{P})}\mathbf{v}_j(t)$$

If we define the σ_j (through the ρ_j) as $\rho_j = e^{\sigma_j \mathcal{P}}$, this becomes

$$e^{-\sigma_j(t+\mathcal{P})}\mathbf{v}_j(t + \mathcal{P}) = e^{-\sigma_j t}\mathbf{v}_j(t)$$

It follows that the vector $e^{-\sigma_j t}\mathbf{v}_j(t)$ is periodic with period $\mathcal{P}$. The coefficients σ_j
are called the Floquet exponents and are determined from the Floquet multipliers
ρ_j (up to a multiple of $2k\pi i$) through

$$\sigma_j = \frac{\ln|\rho_j| + 2k\pi i}{\mathcal{P}} \; ; \; k = 0, 1, 2, 3, \ldots$$

To determine the $\mathbf{y}_j$ once the Floquet multipliers (and the corresponding eigenvectors in the matrix Z) are calculated, we substitute $\mathbf{y}(t) = e^{-\sigma t}\mathbf{v}_j(t)$ into (3.48)
and have to solve $\mathbf{v}_j$ from

$$\frac{d\mathbf{v}_j}{dt} = J(\mathbf{X}(t), \lambda)\mathbf{v}_j + \sigma_j \mathbf{v}_j$$

with $\mathbf{v}_j(0) = Z^{-1}\mathbf{e}_j$ as initial condition; once this has been done, the $\mathbf{y}_j = Z\mathbf{v}_j$
are calculated.

3.4. Bifurcations of Periodic Orbits

The bifurcation behavior of periodic orbits is rich and only a few interesting and relevant examples will be given. Periodic orbits become unstable when Floquet multipliers ρ cross the unit circle. In the most simple situation where there is only one control parameter λ, there are three possible cases.

3.4.1. The cyclic-fold bifurcation

A case where a real Floquet multiplier $\rho(\lambda) > 1$ may occur in several situations which look similar to the codimension-one bifurcations of fixed points. In a *cyclic-fold* bifurcation, the periodic orbit exists for $\lambda < \lambda_0$, but it does not exist for $\lambda > \lambda_0$. This is similar to the structure of fixed point solutions near a saddle-node bifurcation. The bifurcation diagram together with some trajectories are plotted in Fig. 3.24. For $\lambda_1 < \lambda_0$, two periodic orbits (Fig. 3.24a) exist of which one is

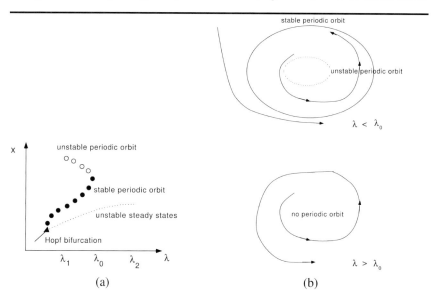

Figure 3.24.   (a) Typical bifurcation diagram of a cyclic-fold bifurcation, as seen as a limit point on a branch of periodic solutions. (b) Sketch of orbits and trajectories for values of the control parameter left and right of the cyclic-fold bifurcation.

stable (closed circles) and one is unstable (open circles). Trajectories are attracted to the stable orbit and diverge from the unstable one (Fig. 3.24b). For $\lambda_2 > \lambda_0$, in Fig. 3.24a, periodic orbits do not exist anymore.

3.4.2. The period-doubling bifurcation

Consider the case where a real Floquet multiplier $\rho(\lambda) < -1$. Once a Floquet multiplier becomes -1, it follows from (3.53) that $v_j(t + N\mathcal{P}) = (-1)^N v_j(t)$. Hence, after two periods of oscillation, the perturbation is returning to its original value indicating that this bifurcation introduces a subharmonic frequency (with twice the period of the original orbit) into the behavior of the dynamical system. This bifurcation is therefore called a *period-doubling* or flip bifurcation. The bi-

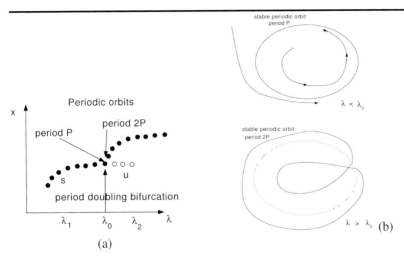

(a)

(b)

Figure 3.25.   *(a) Typical bifurcation diagram of a period-doubling bifurcation at which a new branch of periodic solutions arises. (b) Sketch of orbits and trajectories for values of the control parameter left and right of the period-doubling bifurcation.*

furcation diagram and some trajectories are sketched in Fig. 3.25. For values $\lambda < \lambda_0$, there is a stable periodic orbit of period $\mathcal{P}$, whereas for $\lambda > \lambda_0$ another stable periodic orbit of period $2\mathcal{P}$ arises, whereas the original orbit becomes unstable.

Technical box 3.4:
Stability of fixed points of maps

The Poincaré map is an example of a general class of maps defined by

$$\mathbf{x}_{k+1} = \mathbf{F}(\mathbf{x}_k, \lambda) \tag{3.55}$$

where λ is again a parameter and $\mathbf{x}$ a n-dimensional vector. A solution $\bar{\mathbf{x}}$ is a fixed point of the map $\mathbf{F}$ if

$$\bar{\mathbf{x}} = \mathbf{F}(\bar{\mathbf{x}}, \lambda) \tag{3.56}$$

To investigate the stability of this fixed point, we again look at perturbations $\mathbf{x}_{k+1} = \bar{\mathbf{x}} + \mathbf{y}_{k+1}$. Substitution in (3.55) and linearization around $\bar{\mathbf{x}}$ leads to

$$\mathbf{y}_{k+1} = J\mathbf{y}_k \tag{3.57}$$

where J is the Jacobian matrix of $\mathbf{F}$. Suppose the eigenvalues are real and can be ordered as $\sigma_1 > \sigma_2 > \ldots > \sigma_n$, then there exists a matrix T such that with $\mathbf{z}_k = T\mathbf{y}_k$, the linearized system (3.57) transforms into

$$\mathbf{z}_{k+1} = \begin{pmatrix} \sigma_1 & \ldots & 0 \\ 0 & \sigma_i & 0 \\ 0 & \ldots & \sigma_n \end{pmatrix} \mathbf{z}_k \qquad (3.58)$$

Hence, we have $z_k^j = \sigma_j z_{k-1}^j = \sigma_j^2 z_{k-2}^j = \ldots = \sigma_j^k z_0^j$ and the fixed point is stable if all eigenvalues σ_j satisfy $\mid \sigma_j \mid < 1$.

3.4.3. The Naimark-Sacker bifurcation

In this case, a complex conjugate pair of Floquet multipliers, say $\rho = e^{\pm i\tilde{\omega}}$, crosses the unit circle. The imaginary part of the pair of Floquet multipliers corresponds to a new frequency. In general, this frequency $\tilde{\omega}$ is unrelated to the frequency ω of the original periodic orbit. The frequencies are said to be incommensurate if the equation

$$p\omega + q\tilde{\omega} = 0 \qquad (3.59)$$

has no solutions for integer values of p and q. In this case, the trajectory moves on a torus, where one frequency of the trajectory corresponds to the frequency of the original periodic orbit. The second frequency originates from the destabilizing perturbation. This bifurcation is called a *Naimark-Sacker*, or torus, bifurcation.

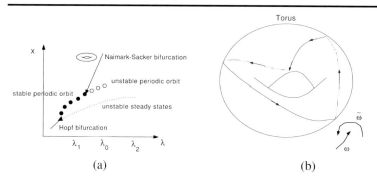

(a)   (b)

Figure 3.26.   (a) Typical bifurcation diagram of a Naimark-Sacker bifurcation, which destabilizes a branch of periodic solutions. (b) Sketch of a trajectory on a torus, arising when the two frequencies $\tilde{\omega}$ and ω are incommensurate.

The bifurcation diagram and the motion on the torus are sketched in Fig. 3.26. In a typical situation, a Naimark-Sacker bifurcation arises as a secondary bifurcation (Fig. 3.26a), after a fixed point has become unstable at a Hopf bifurcation. In the sketch of the trajectory in Fig. 3.26b, the motion associated with $\tilde{\omega}$ is associated with motion normal to the axis of the torus, while the motion along its axis has the original frequency ω.

▶

Example 3.3: Floquet multipliers

Consider again the periodic orbit as in Example 3.2, which is now written as

$$x(t) = \sqrt{\lambda}\,\cos\omega t$$
$$y(t) = \sqrt{\lambda}\,\sin\omega t$$

for $\lambda > 0$ and $\omega \neq 0$. The Jacobian matrix at the periodic orbit can be obtained from (3.31) and, for $\delta = 1$, is given by

$$J(\mathbf{X}(t), \lambda) = \begin{pmatrix} \lambda - 3x^2(t) - y^2(t) & -\omega - 2x(t)y(t) \\ \omega - 2x(t)y(t) & \lambda - 3y^2(t) - x^2(t) \end{pmatrix}$$

Next, to determine the monodromy matrix, the system $d\mathbf{y}_j/dt = J(\mathbf{X}(t), \lambda)\mathbf{y}_j$ has to be solved for $j = 1, 2$ with $y_1(0) = (1, 0)$ and $y_2(0) = (0, 1)$ as initial conditions. This has to be done numerically and the monodromy matrix Φ is found from

$$\Phi = (y_1(\frac{2\pi}{\omega}), y_2(\frac{2\pi}{\omega}))$$

The Floquet multipliers are determined as the two eigenvalues of the matrix Φ. The first one ($\rho_1 = 1$) is unity and the second one determines the stability of the periodic orbit, and in this case it is within the unit circle ($\mid \rho_2 \mid < 1$) for $\lambda > 0$. The periodic orbit in Fig. 3.15b is therefore stable, as was already indicated by the trajectories.

For this particular case there is also a quicker way to determine the stability of the periodic orbits. This is done by looking at the stability of the fixed point of the Poincaré map (see Technical Box 3.4), which was explicitly computed in Example 3.2. The Jacobian of the Poincaré map will indicate whether intersections of trajectories drift away (if positive) or are attracted to (if negative) the fixed point of the Poincaré map. For the periodic orbit in Example 3.2, the stability can be determined from

$$\frac{dP}{dr}\mid_{r=\sqrt{\lambda}} = \frac{d}{dr}\left[(\frac{1}{r^2} - \frac{1}{\lambda})e^{\frac{-4\pi\lambda}{\omega}} + \frac{1}{\lambda}\right]^{-\frac{1}{2}}\mid_{r=\sqrt{\lambda}} = e^{\frac{-4\pi\lambda}{\omega}} \tag{3.60}$$

The norm of the right hand side of this equation is smaller than unity for $\lambda > 0$ and hence the periodic orbit is stable.

◀

3.5. Global bifurcations

So far, only the changes in the local structure of the solutions was considered. By varying a parameter, however, also changes in the global structure of the solutions can occur. These changes may, for example, involve more than one fixed point and/or periodic orbit and are associated with so-called global bifurcations. Below we present only one type of global bifurcation, the homoclinic bifurcation, and give an example of a simple dynamical system (the famous Lorenz model) where it occurs.

3.5.1. Homoclinic orbits

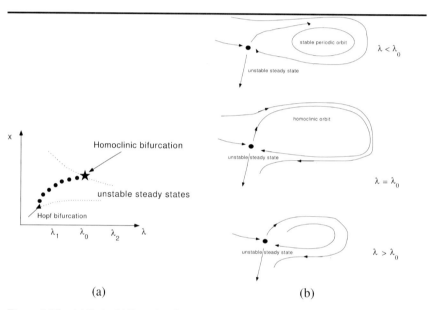

(a)   (b)

Figure 3.27.   (a) Typical bifurcation diagram showing the origin of a homoclinic orbit. Along a branch of stable periodic solutions, the period increases through the presence of a nearby unstable steady state. At $\lambda = \lambda_0$, a homoclinic bifurcation occurs and the periodic orbit ceases to exist. (b) Sketch of the trajectories near a homoclinic bifurcation, showing the presence of a homoclinic orbit at $\lambda = \lambda_0$.

As an example of a global bifurcation, the interaction between an unstable fixed point and a periodic orbit is considered (Fig. 3.27). Note that this behavior can only occur in three- or higher dimensional dynamical systems. For a parameter λ_1, both fixed point and orbit are well separated but as the parameter λ is increased, both they interact and have influence on the trajectories. The period of the orbit becomes longer when λ approaches λ_0 and at λ_0 the period becomes infinity (Fig. 3.27a). The trajectory along the unstable direction of the steady state approaches the same steady state again along a stable direction in infinite time; this trajectory is called a homoclinic orbit. For values of $\lambda > \lambda_0$, the periodic orbit has disappeared and the unstable steady state remains (Fig. 3.27b). The presence of homoclinic orbits can lead to very complicated, and even often chaotic, behavior.

In Wiggins (1990), it is shown that the linear stability properties of the unstable fixed point, together with the symmetry properties of the equations, are crucial to the type of behavior near the homoclinic bifurcation. Considering the eigenvalues σ of the Jacobian matrix at the unstable fixed point, there are basically two cases: (i) the first three eigenvalues σ of the linear stability problem are real, the first one is positive ($\sigma^U > 0$) and in absolute value larger than the second and third (σ^S) eigenvalue ($\sigma^U > -\sigma^S$); (ii) the first two eigenvalues form a complex conjugate pair with negative real part (σ_r^S) < 0) and the third eigenvalue is real and positive ($\sigma^U > 0$).

In both cases, the fixed point is unstable to only one real mode (and hence one direction in state space), but stable to all others. If the unstable steady state is pictured within a plane in Fig. 3.28, then the stable directions can be sketched as occurring in that plane and the unstable direction as perpendicular to it. In that case, the homoclinic orbit connects the unstable direction with one of the stable directions. The different cases are distinguished by how the attraction in the stable directions occur. In the first case, the behavior of the system is akin to the Lorenz (1963) system (Fig. 3.28a) and in the second case (Fig. 3.28b), it displays so-called Shilnikov-type phenomena (Shilnikov, 1965).

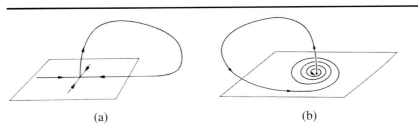

(a)   (b)

Figure 3.28.   Sketch of the two cases of homoclinic bifurcation. In panel (a), there are two stable attracting directions in the plane and one expanding direction perpendicular to it. Both expansion and attraction are exponential, characteristic of the Lorenz case. In panel (b), the attraction is oscillatory whereas the expansion is exponential, characteristic of the Shilnikov case.

3.5.2. The Lorenz (1963) dynamical system

The occurrence of homoclinic orbits is not limited to the case above and can even occur in the absence of a periodic orbit. A notable example of the latter is in the well-known Lorenz model (Lorenz, 1963), described by the three-dimensional system of equations

$$\frac{dx}{dt} = -\alpha(x - y) \tag{3.61a}$$

$$\frac{dy}{dt} = \rho x - y - xz \tag{3.61b}$$

$$\frac{dz}{dt} = xy - \beta z \tag{3.61c}$$

with fixed $\alpha = 10$, $\beta = 8/3$ and using ρ as a control parameter. The Jacobian matrix is easily determined as

$$J = \begin{pmatrix} -\alpha & \alpha & 0 \\ \rho - z & -1 & -x \\ y & x & -\beta \end{pmatrix} \tag{3.62}$$

For $0 < \rho < 1$ there is a unique steady state given by $x = y = z = 0$, which is stable since all eigenvalues of J are real and have negative real parts (indicated by the minus signs in Fig. 3.29a). At $\rho = 1$, the trivial state loses stability at a pitchfork bifurcation and two new steady states (C_1 and C_2) given by

$$x = y = \sqrt{\beta(\rho - 1)}, \ z = \rho - 1 \tag{3.63a}$$

$$x = y = -\sqrt{\beta(\rho - 1)}, \ z = \rho - 1 \tag{3.63b}$$

appear, which are stable. At about $\rho = 1.35$, two of the three real eigenvalues monitoring the stability of the nontrivial steady states coalesce and form a complex conjugate pair (indicated by the signs within an ellipse in Fig. 3.29a). As a result, the trajectories spiral into these steady states, while there is one direction (associated with the real positive eigenvalue) in which trajectories diverge from the trivial steady state as is shown in Fig. 3.29b. With increasing ρ, the trajectories from the unstable direction of the trivial steady state make wider turns and finally a homoclinic orbit appears near $\rho \approx 15$, as such a trajectory connects back to the trivial state along an attracting direction (Fig. 3.29b).

3.6. Resonance phenomena: frequency locking

An important phenomenon which often occurs in periodically forced systems is frequency locking. This phenomenon can be described as the existence of periodic orbits with a frequency, which is a rational multiple of the externally imposed frequency. Hence, response frequency and forcing frequency are commensurate. Remember from section 3.2.1 that for non-autonomous systems, a Poincaré section is easily constructed and hence the periodic orbits can be studied through a Poincaré map.

A typical example of a map where frequency locking occurs is the circle map given by

$$x_{n+1} = P(x_n) = x_n + \Omega - \frac{K}{2\pi} \sin 2\pi x_n \ (mod \ 1) \tag{3.64}$$

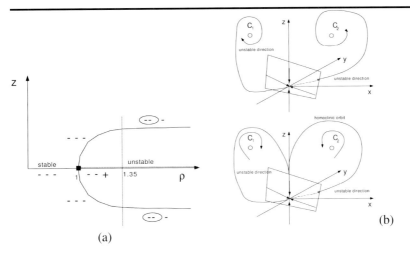

Figure 3.29. *(a) Bifurcation diagram of the Lorenz system up to the point, where a homoclinic orbit exists. (b) Sketch of trajectories in state space illustrating the origin of the homoclinic orbit.*

on the interval $[0, 1)$. Here, mod 1 indicates that if x_n becomes larger than 1 or smaller than 0, its value is adjusted by adding or subtracting unity to get a value within the unit interval. $\Omega \in [0, 1)$ can be considered the forcing frequency and the control parameter K is a measure of the nonlinearity of the system. When $K = 0$, the iterates move over a distance Ω in every step.

The rotation number W measures the average movement of iterates over the interval and is defined as

$$W = \lim_{n \to \infty} \frac{x_n - x_0}{n} \qquad (3.65)$$

In this definition, the value of x_n is considered without the mod 1 function. Hence, for $K = 0$ it follows that $W = \Omega$. Examples of the iterates are shown in Fig. 3.30 for $K = 0.1$ and several values of Ω ($\Omega = 0.05$, $\Omega = 1/3$ and $\Omega = 0.5$) with the rotation numbers W provided in the caption. If the rotation number is irrational, the orbit just fills up the whole interval and there is no periodic motion related to Ω. This can be seen in Fig. 3.30a, where the first 500 iterates are plotted for $\Omega = 0.05$. However, for some values of Ω, periodic orbits occur. For example, a period-3 orbit appears for $\Omega = 1/3$ (Fig. 3.30b) and a period-2 orbit appears for $\Omega = 1/2$ (Fig. 3.30c). What is remarkable is that these periodic orbits do occur also for irrational values of Ω, for example, for $K = 0.95, \Omega = 1/2 - 1/(10\pi)$ (Fig. 3.30d). Apparently, these iterates are frequency locked to the period-2 orbits.

To look for frequency locked regions, one varies Ω for fixed K and searches for rational rotation numbers W. The result for the circle map is shown in Fig. 3.31 and shows tongues, so called Arnold' tongues, in which the rotation number W is rational. Note that all the results in Fig. 3.30 are consistent with this picture. To

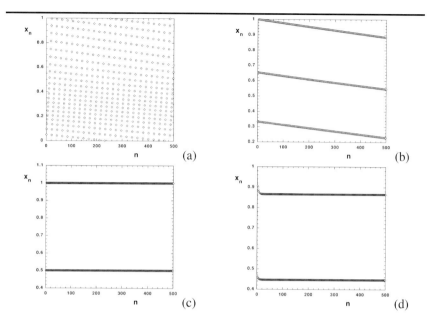

*Figure 3.30.   Iterates x_n versus n for the circle map and several values of Ω for fixed $K = 0.1$.
(a) $\Omega = 0.05, W = 4.72 \times 10^{-2}$, (b) $\Omega = 1/3, W = 1/3$, (c) $\Omega = 1/2, W = 1/2$. In panel
(d) the iterates for $K = 0.95, \Omega = 1/2 - 1/(10\pi)$ are plotted. For this case, the rotation number
$W = 1/2$.*

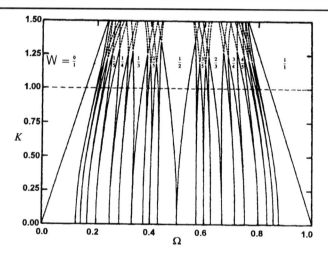

*Figure 3.31.   Plot of regions of rational rotation number in the $\Omega - K$ plane displaying the
Arnold tongues (Bak et al. (1985)) for the circle map. Note that the value of Ω at the wedge starting
at $\Omega = 0$, $K = 0$ ends up at $\Omega = K/(2\pi) = 0.239$ for $K = 1.5$.*

see how these Arnold' tongues arise, consider the interval in Ω for which $W = 0$. From the definition (3.65), one obtains $x_n = x_0$, $\forall n > 1$ and there is a fixed point of the mapping P in (3.64). Hence,

$$x_1 = x_0 + \Omega - \frac{K}{2\pi} \sin 2\pi x_0 = x_0 \tag{3.66}$$

By taking different x_0, for example $x_0 = 0.5$ and $x_0 = 1.0$, it is observed that this fixed point only occurs when $-K/(2\pi) < \Omega < K/(2\pi)$. Hence, within this wedge the rotation number $W = 0$ and this is exactly the wedge near $\Omega = 0$ in Fig. 3.31. Similar calculations can be made for the other tongues. If one plots

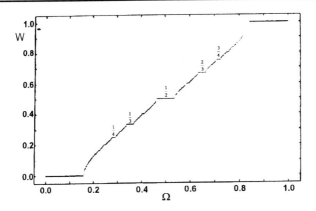

Figure 3.32. Plot of the regions of rational rotation number for $K = 1$ showing the Devil's staircase (Bak et al. (1985)).

the intervals of Ω against the rational values of the rotation number for fixed K, a peculiar structure arises of a staircase where over certain intervals, W is constant. Such a staircase, which is the famous Devil's staircase (Bak *et al.*, 1985), is plotted for $K = 1$ in Fig. 3.32. In the interval Ω for which $W = 1/2$ periodic motion occurs with a period which is commensurate with the forcing frequency Ω. Nonlinear resonances between both forcing and internal frequency hence cause the synchronization of the resulting signal to multiples of the forcing frequency. The reader is referred to the beautiful book of Pikovsky *et al.* (2001) for an extensive discussion of these synchronization phenomena.

The type of dynamical behavior discussed in this chapter forms only the surface of the deep ocean of rich dynamics displayed by maps and differential equations. While much more is known, the different behavior sketched above give sufficient background to understand the phenomena as presented in later chapters. In the last section of this chapter, the question is addressed what we learn from this bifurcation behavior with respect to the physics of the phenomena under study.

3.7. Physics of Bifurcation Behavior

There is quite a nice connection between the physical understanding of the time-evolution of a system and the mathematical understanding of the trajectories in the associated dynamical system. The systematics of the theory of dynamical systems is therefore very helpful to understand physical mechanisms, since a classification of qualitative type of behavior is provided.

Note that in models of geophysical flows, one does not easily escape from the notion of bifurcation behavior. At very large viscosity (or friction), these flows are usually very sluggish and, in most cases, unique steady flows exist. When friction or diffusion is decreased in these models eventually the flow in the 'realistic' parameter regime is very complex. In this case, bifurcation behavior will always have occurred in the interval over which the friction or diffusion parameter has been varied.

3.7.1. Physical constraints

Inspection of the model equations which represent dominant balances of momentum, heat, freshwater and maybe other properties in the physical system may already a priori indicate what type of bifurcations can be expected. Usually, these models contain a number of parameters and a modeler will pick a few of them as control parameters, the choice being related to the question which is aimed to be answered.

There are a few issues which are important during first inspection.

(i) *Symmetry.* When symmetry is present in the model equations, a restriction will be put on the type of codimension-1 bifurcations which can occur. For example, consider the presence of a reflection symmetry in the physical model, such as that in the three-box model (3.14). Through the reduction process as sketched in Technical box 3.2, such a symmetry will be inherited by the reduced equation and for the normal form of any bifurcation, there will be a requirement

$$g(-x, \lambda) = -g(x, \lambda)$$

Hence, when a bifurcation occurs as a parameter λ is varied, one expects pitchfork bifurcations rather than transcritical bifurcations, because the normal form of the latter does not satisfy the requirement above. This is only a very elementary example of the constraints put on bifurcation diagrams through symmetry. An enormous amount of literature exists on this subject and interested readers are referred to Golubitsky *et al.* (1988). For systems of equations which have no symmetry, transcritical, Hopf and saddle-node bifurcations are expected to occur.

(ii) *Special solutions.* Of particular importance are solutions which remain a solution for all values of a particular control parameter λ. Sometimes this occurs in a subtle way; examples will be shown in later chapters of so-called flux-corrected ocean/atmosphere models where such a solution is constructed. Note that when bifurcations occur from these solutions, the requirement on the reduced equation will be

$$g(0, \lambda) = 0$$

for every value of λ. This excludes the occurrence of saddle-node bifurcations on this branch of solutions, as inspection of (3.28) confirms.

Many more of these examples, which demonstrate that *a priori* information on the solution structure can be derive, exist

3.7.2. Qualitative versus quantitative sensitivity

From the very brief overview of the theory of the behavior of dynamical systems as described in the previous sections, it has become clear that bifurcations are the most interesting points from the physical point of view, because these mark boundary of intervals of different qualitative behavior of trajectories. For example, the picture where trajectories are attracted to a single steady state may change at a Hopf bifurcation into a picture where trajectories are attracted to a periodic orbit. When changing a parameter, the state of the physical system suddenly becomes quite different. There is qualitative sensitivity near a bifurcation point and the system is in a different regime on either side of the bifurcation.

When a system has more than one control parameter, say μ, the boundaries between different regimes are marked by paths of bifurcation points in the (λ, μ) plane. For example, a path of a supercritical Hopf bifurcations in a two-parameter plane separates the regimes of steady and periodic behavior in that plane. Another example of such a regime diagram is Fig. 3.4c, where the different regimes (regimes of unique steady solutions versus a regime of multiple equilibria) are bounded by paths of saddle-node bifurcations.

Knowing regime diagrams and bifurcation diagrams in several control parameters leads to an overview of possible behavior of trajectories in the model and the sensitivity to changes in parameters. The regime diagrams are particularly important in models of physical systems for which parameters are poorly known. Ocean and climate models are an example of these type of models since they, for example, lack an adequate description of mixing processes. This introduces substantial uncertainty into the representation of subgrid-scale processes, reflected in uncertain values of mixing coefficients.

Basically, there are two types of sensitivities of model behavior to parameters: qualitative sensitivity and quantitative sensitivity. These concepts are explained with help of Fig. 3.33. In Fig. 3.33a, there are two back-to-back saddle-node bifurcations on a single branch of solutions. Suppose first that the parameter λ is slowly varied from zero to more positive values. Between the changes in λ, enough time is provided to let the system equilibrate to steady state. In this way, the state will remain close to that of the branch C_1 when λ is increased up to $\lambda = \lambda_1$. For larger values of λ, there will a sudden transition to a state on the branch C_2, since this is the only steady state and the state will remain near C_2 for larger values of λ. When the same procedure is followed with decreasing λ, the state of the system will remain close to C_2 until the value of λ_2, where a sudden jump occurs to a state on C_1. Hence, transitions will occur at different values of the control parameter depending on its direction of change and this is characteristic of hysteresis behavior. Apart from this hysteresis, one also knows that the solutions on the stable branches, within the region of multiple equilibria ($\lambda_2 < \lambda < \lambda_1$) are sensitive to finite amplitude perturbations. For example, a perturbation exists such that there is a transition from a stable state on C_1 to one on C_2.

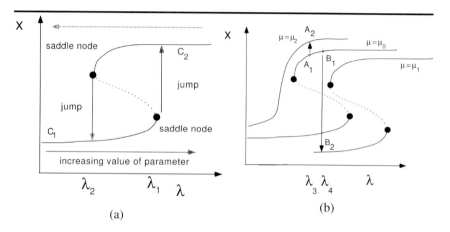

Figure 3.33. (a) A sketch of a typical bifurcation diagram, similar to that in Fig. 3.4a, displaying hysteresis behavior. (b) Sketch of possible changes in the bifurcation diagram as a second parameter μ is changed with $\mu_2 < \mu_0 < \mu_1$.

In many ocean modeling studies, sensitivity of solutions is studied to parameters, for example parameters in parameterizations of mixing and effects of topography, to name a few. In many cases, a parameter value is changed by a specific amount, the model is rerun and the changes in behavior are monitored. An example of the changes in bifurcation diagram with the secondary parameter μ is shown in Fig. 3.33b. Suppose, the original bifurcation diagram corresponds to that of $\mu = \mu_0$ and the two other diagrams are for μ_1 and μ_2, with $\mu_2 < \mu_0 < \mu_1$. The qualitative features of the bifurcation diagram are unchanged in case $\mu = \mu_1$, but these features are totally different for $\mu = \mu_2$. In the latter case, no multiple equilibria occur anymore over the whole range of λ. Hence, there is qualitative sensitivity to a decrease of μ, since a different regime is entered. However, the system is not qualitatively sensitive to an increase in μ, since the same transition behavior with λ is still possible at $\mu = \mu_1$. Only the values of λ at which these transitions occur are quantitatively changed (shifted) and consequently there is only quantitative sensitivity.

It is difficult to obtain a correct overview of this sensitivity by computation of only a few trajectories of a model. Imagine that the value of λ is chosen such that for $\mu = \mu_0$, multiple equilibria occur (λ_3 in Fig. 3.33b). In the first trajectory, the model has been integrated to the steady solution marked A_1. When the model is then run for $\mu = \mu_2$, one will monitor only modest changes in model behavior when it evolves towards the state A_2. Hence, one easily will conclude that the effect represented by μ is not important, although a big qualitative change in the bifurcation diagram has occurred. Whereas the solution A_2 is globally stable (section 2.3) to finite amplitude perturbations, the state A_1 is only conditionally stable.

As a second case, consider the value λ_4, which is also in the region of multiple equilibria for $\mu = \mu_0$ and the state B_1 is obtained. When the model is now rerun at $\mu = \mu_1$, one will find a totally different state, because λ_4 is not longer in the region of multiple equilibria. Hence, one will easily attribute an enormous importance to the change in the physical effect representing an increase in μ, whereas no qualitative changes have occurred in the bifurcation diagram. Only computation of full bifurcation diagrams will give correct information on the sensitivity of a particular physical process on the solutions of a particular system.

3.7.3. Instability mechanisms

At bifurcation points, something special is happening and this has a meaning in physical terms. Of course, precise interpretation depends on the model under study, but a general approach can be followed as explained in the next subsections for the codimension-1 bifurcations.

3.7.3.1 Saddle-node bifurcation

Consider a fixed point depending on the value of a particular parameter λ, for example as in the Stommel two-box model in section 3.1 and assume that this steady solution is linearly stable. This means that a trajectory nearby the steady state will be attracted towards this state and that for $t \to \infty$, the exact steady physical balances of the model are satisfied. As λ is increased, a saddle-node bifurcation is encountered, which is, for example, found through numerical computation. Could we have argued from the physics that this bifurcation would be located at a specific of λ? This question turns out to be difficult to answer in general. However, in many applications, the governing equations are representations of balances (of mass, momentum or energy) of physical quantities. Hence, we may look at an alternative problem: What happens to these balances, when a saddle-node bifurcation appears? When a saddle-node bifurcation is encountered at λ_0, this implies that the steady balances in the model cannot be satisfied for val-

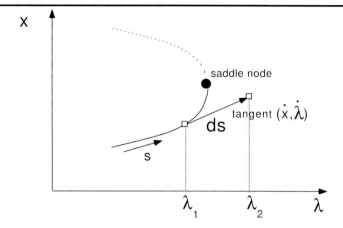

Figure 3.34. Sketch of a situation of a saddle-node bifurcation, illustrating the approach to determine its physical meaning.

ues of λ beyond λ_0. One can look at which balances fail to be satisfied (most) by computing (for a solution $(\mathbf{x}_1, \lambda_1)$) the tangent $(\dot{\mathbf{x}}, \dot{\lambda})$ on the branch of solutions, say parameterized by s, and determine the state $(\mathbf{x}_2, \lambda_2)$ by (Fig. 3.34)

$$\mathbf{x}_2 = \mathbf{x}_1 + ds\, \dot{\mathbf{x}} \tag{3.67a}$$

$$\lambda_2 = \lambda_1 + ds\, \dot{\lambda} \tag{3.67b}$$

for a small value of ds (Fig. 3.34). The state $(\mathbf{x}_2, \lambda_2)$ is not a steady solution and one can compute the non-zero residual

$$\frac{d\mathbf{x}}{dt} = \mathbf{f}(\mathbf{x}_2, \lambda_2) \neq \mathbf{0} \tag{3.68}$$

when substituting this solution back into the equations. The tendencies in the left-hand side now indicate which components of the solution vector will drift away from the state $(\mathbf{x}_2, \lambda_2)$. This can give useful information on the balances which are difficult to satisfy in steady state for $\lambda > \lambda_2$. Balances may be difficult to satisfy near the boundaries or integral conditions can no longer be satisfied. This can guide physical arguments explaining the existence of the saddle-node bifurcation.

3.7.3.2 Transcricital and pitchfork bifurcation

If a transcritical bifurcation occurs at $\lambda = \lambda_0$, this means that a stable solution exchanges stability with some other solution. This may occur, for example, when there is no internal symmetry in the particular model under study and a solution for all values of λ exists. Hence, there is a smooth modification of solutions when a parameter is varied and the question is why there is a difference in preference of both solutions just at $\lambda = \lambda_0$. The same question applies to the supercritical pitchfork bifurcation, where, in general, a stable symmetric solution is replaced by asymmetric, but symmetry related, stable solutions.

The method to analyze this problem is to look at the eigenvectors of the linear stability problem just at bifurcation. Left of λ_0, all eigenvalues of the trivial state are in the left half of the complex plane. At λ_0 one of them moves through the imaginary axis and the spatial pattern of the corresponding eigenvector is important. It means that the steady solution becomes unstable to that particular pattern. One can either start a time integration with an initial condition into this direction or alternatively look at the perturbation equations for integral quantities, such as the mechanical energy balances. However, a more mechanistic description of the instability mechanism can be attempted in the following way.

To show that a steady state becomes unstable to a particular disturbance, one must describe the causal chain of how this disturbance is amplified through its interaction with the steady state. A mechanistic understanding of the physics of this process can be obtained by dividing the instability process into two virtual stages, an initiation stage and a growth stage. During the initiation stage, a perturbation is assumed to be present in the system and the causal chain is described of how the perturbation changes the steady state. During the growth stage, a description is provided how the changes in the steady state have feedback on the original perturbation leading to its amplification (Fig. 3.35). Although it is in general not

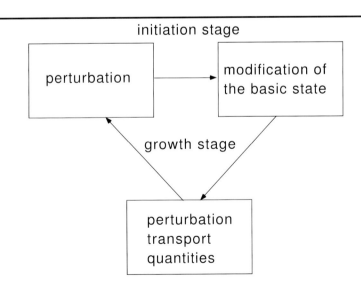

Figure 3.35.   Sketch of the initiation and growth stages of an instability.

straightforward to obtain such a description, several examples will be shown in
later chapters.

3.7.3.3 Hopf bifurcation

Finally at a Hopf bifurcation, a steady solution becomes unstable to an oscillatory perturbation. The pattern of the most unstable oscillatory mode can be
determined from the solution of the linear stability problem (3.24) which leads to
an eigenvalue problem with eigenvector(s) $\hat{\mathbf{y}}$ and complex growth factor σ. The
corresponding complex eigenfunction $\hat{\mathbf{y}} = \hat{\mathbf{y}}_R + i\hat{\mathbf{y}}_I$ provides the disturbance
structure $\mathbf{\Phi}(t)$ with angular frequency σ_i and growth rate σ_r to which the steady
state may become unstable, i.e.

$$\mathbf{\Phi}(t) = e^{\sigma_r t} \left[\hat{\mathbf{y}}_R \ \cos(\sigma_i t) - \hat{\mathbf{y}}_I \ \sin(\sigma_i t) \right] \tag{3.69}$$

The evolution of this perturbation can be followed by looking, near the Hopf bi-
furcation, at $\mathbf{\Phi}(\frac{-\pi}{2\sigma_i}) = \hat{\mathbf{y}}_I$ and then at $\mathbf{\Phi}(0) = \hat{\mathbf{y}}_R$. An attempt can be made to
understand the instability mechanism using these different phases of the perturba-
tions along similar lines as in Fig. 3.35. In most cases, the propagation mechanism of the oscillation can be determined, but it is more difficult to determine the
growth of the perturbations over one period of the oscillation.

As a bridge to the next chapter, we can pose the question here whether knowing
what happens at particular bifurcation points is useful to understand the physics
of the behavior where one is interested in. When bifurcations occur at parameter
values far from those values thought as being 'realistic' to represent the physical
system, what do physical explanations near bifurcation points mean? The answer
to this question will depend strongly on the application at hand and on the extent to
which one is able to compute the different bifurcation diagrams. But first, one has
to be able to compute bifurcation diagrams on meaningful models of the ocean,
atmosphere and climate. This problem is addressed in the next chapter.

3.8. Exercises on Chapter 3

(E3.1) *The Rooth three-box model*

One of the shortcomings of the Stommel two-box model (section 3.1) is that it only describes flow in the northern hemisphere. In a box model designed by Rooth (1982), the flow is driven by a north-south density difference, caused by freshwater fluxes H_S and H_N only.

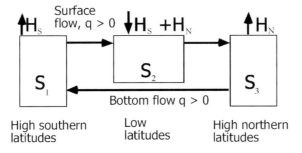

Figure 3.36.   Overview of the three-box model suggested by Rooth (1982)

Assume that the temperature is constant and only salinity varies in the boxes. It follows then that

$$q = k(S_3 - S_1)$$

where k is a transport coefficient.

a. Formulate the equations describing the evolution of the salinity in the boxes.

b. Determine the steady states of the model.

c. Show that the strength of the overturning circulation only depends on the freshwater flux H_N. What is the physical explanation for this?

Further reading: Rooth (1982) and Marotzke (2000).

(E3.2) *Intersection of saddle-node bifurcations*

The point Q in Fig. 3.4c is located at the boundary of the multiple equilibria regime and the unique regime in the (η_1, η_2) plane.

a. Show that Q has to be on the curve of motionless solutions (*Hint: consider the sign of Ψ at each saddle-node bifurcation L*).

b. Show that the coordinates of Q in the (η_1, η_2) are given by

$$Q = \left(\frac{\eta_3}{1 - \eta_3}, \frac{\eta_3^2}{1 - \eta_3} \right)$$

c. What type of codimension-2 bifurcation occurs at Q?

Further reading: Thual and McWilliams (1992).

(E3.3) *Imperfection of the transcritical bifurcation*

The bifurcation equation for the imperfect transcritical bifurcation is given by

$$g(x, \lambda) = \epsilon + \lambda x + x^2$$

a. Determine all solutions of this equation for both $\epsilon > 0$ and $\epsilon < 0$.

b. Determine the values of λ at the saddle-node bifurcations for $\epsilon > 0$.

Further reading: Iooss and Joseph (1997), chapter 2.

(E3.4) *Floquet multipliers*

The equation describing the motion of a pendulum with a periodically excited support may be written as a nonlinear Mathieu equation

$$\theta'' + (\alpha^2 + \beta \cos t) \sin \theta = 0$$

where θ is the angle with the vertical and $\beta \geq 0$. Note that for $\beta = 0$, the nonlinear pendulum equations are obtained.

a. Write the Mathieu equation as a three-dimensional first order system of equations for θ, θ', ϕ, with $\phi' = 1$.

b. Show that for all values of β, there exist two periodic orbits, $(0, 0, \phi(t))$ and $(\pi, 0, \phi(t))$.

Let $\alpha \neq n/2, n = 0, 1, 2, ...$

c. Show with Floquet theory that for $\beta = 0$, the orbit $(0, 0, \phi(t))$ is stable, while the orbit $(\pi, 0, \phi(t))$ is unstable.

Further reading: Guckenheimer and Holmes (1990), chapter 3.

(E3.5) *Circle map*

For the circle map as in section 3.6

$$P(x_n) = x_{n+1} + \Omega - \frac{K}{2\pi} \sin 2\pi x_n \ (mod1)$$

consider the interval in Ω for which $W = 1/2$ (see Fig. 3.31).

a. From the definition of W, show that in this interval

$$x_2 = x_0 + 1$$

Let $\Omega = 1/2 + \Delta\Omega$.

b. Show that by applying P twice that the $\Delta\Omega$ interval for which $W = 1/2$ is determined by

$$\Delta\Omega = \pm\frac{\pi K^2}{2}$$

c. Check these values of $\Delta\Omega$ in Fig. 3.31 for $K = 0.1$.

Further reading: Pikovsky et al. (2001), chapter 2.

(P3.1) *The Van der Pol equation*

In this exercise, we study the Van der Pol equation

$$\theta'' + (\delta + \epsilon(\theta^2 - 1))\theta' + \theta = 0$$

a. With $x_1 = \theta$ and $x_2 = \theta'$, write this equation as a first order system.

b. Show that the equations are invariant under the transformation

$$x_1 \to -x_1, x_2 \to -x_2$$

c. Compute the steady state of this equation and, for fixed δ, determine the value of ϵ at the first Hopf bifurcation. Determine the period $\mathcal{P}$ of the oscillation at the Hopf bifurcation.

Choose $\delta = 0$. An approximation of the periodic orbit near the Hopf bifurcation can be obtained by the so-called *Method of Averaging*. We seek solutions of the form

$$
\begin{aligned}
x_1(t) &= a(t)\cos(t + \psi(t)) \\
x_2(t) &= -a(t)\sin(t + \psi(t))
\end{aligned}
$$

This type of solution is motivated by the fact that near the Hopf bifurcation, the solution has this form with $a(t)$ and $\psi(t)$ being constants.

d. If we write $F(x_1, x_2) = (1 - x_1^2)x_2$, show that

$$\frac{da}{dt} = -\epsilon \sin(t + \psi)F(a\cos(t + \psi), -a\sin(t + \psi))$$

$$\frac{d\psi}{dt} = -\frac{\epsilon}{a}\cos(t + \psi)F(a\cos(t + \psi), -a\sin(t + \psi))$$

With $\epsilon \ll 1$, we introduce a 'slow' time scale $\tau = \epsilon t$ and expand the solution as

$$a(t; \tau) = \bar{a}(\tau) + \epsilon a_1(t; \tau) + \mathcal{O}(\epsilon^2)$$
$$\psi(t; \tau) = \bar{\psi}(\tau) + \epsilon \psi_1(t; \tau) + \mathcal{O}(\epsilon^2)$$

e. Substitute these expansions into the equations for $a(t)$ and $\psi(t)$, integrate over the period of the oscillation $\mathcal{P}$ and show that

$$\frac{d\bar{a}}{d\tau} = \epsilon \frac{\bar{a}}{8}(4 - \bar{a}^2)$$

$$\frac{d\bar{\psi}}{d\tau} = 0$$

f. Show that the periodic orbit is stable for $\epsilon > 0$.

Further reading: Sanders and Verhulst (1985).

(P3.2) *Center manifold*

Consider the Lorenz (1963) system of equations,

$$\frac{dx}{dt} = -\alpha(x - y)$$

$$\frac{dy}{dt} = \rho x - y - xz$$

$$\frac{dz}{dt} = xy - \beta z$$

In this exercise, we are going to investigate the behavior of the system near the pitchfork bifurcation at $\rho = 1$ in more detail.

a. Determine the eigenvalues and eigenvectors of the Jacobian matrix for $\rho = 1$.

Write the eigenvectors q_1, q_2 and q_3 as $Q = [q_1, q_2, q_3]$.

b. Transform coordinates through $(u, v, w)^T = Q(x, y, z)^T$ and derive the nonlinear dynamical system in (u, v, w).

In the u-direction, the linearized equations show no attraction to or repulsion from the steady state and in the v- and $w-$ direction, trajectories are attracted to the steady state. There exists a so-called 'center manifold' of the form

$$(v(u), w(u))^T = (h_1(u), h_2(u))^T$$

on which the essential dynamics takes place. This manifold is tangent to the attracting directions at the bifurcation and hence $h_1(0) = h_2(0) = h'_1(0) = h'_2(0) = 0$.

c. Expand the functions h_1 and h_2 into polynomials in u, and substitute these into the equations for (u, v, w). Derive that in the center manifold, the dynamics is governed by the reduced equation

$$\frac{du}{dt} = -\gamma u^3$$

and determine γ.

Further reading: Guckenheimer and Holmes (1990), chapter 3.

Chapter 4

NUMERICAL TECHNIQUES

Andante

For progress, certain skills are necessary.
Recuerdos d'Alhambra, F. Tarrega

If one wants to use the systematic methodology of chapter 3 on meaningful ocean and climate models, one must be able to determine at least the codimension-1 bifurcation points. As will become clear in this chapter, the detail of the dynamical behavior which can be analyzed depends on the dimension, say N, of the dynamical system. For systems of ordinary differential equations of small dimension ($N < 10$), the origin of very complex spatial and temporal dynamics can be investigated. For example, codimension-2 bifurcations can be determined numerically by software packages such as CONTENT (Kuznetsov, 1995) (available from http://ftp.cwi.nl/CONTENT/), DSTOOL (Guckenheimer and Kim, 1991) (available from http://www.cam.cornell.edu/guckenheimer/dstool.html) and MATCONT (available from http://allserv.rug.ac.be/∼ajdhooge/research.html)

For somewhat larger dimensional models, with dimensions up to $N = 100$, also software is available to perform analysis of the bifurcation behavior of the model, but the detail of analysis becomes already less. One highly recommended code is AUTO (Doedel, 1980) which is available from http://indy.cs.concordia.ca/auto/. In the very clear manual, the many capabilities of this program are described. For news on these packages, see for example http://www.amsta.leeds.ac.uk/Applied/news.dir/bifurcation.html.

For systems of partial differential equations, such as arising from ocean models (typically $N = 10^5$), two public domain packages is available. First package is the code PDECONT, described in Lust *et al.* (1998) and Lust and Roose (2000), which can be obtained through http://www.cs.kuleuven.ac.be/∼kurt/r_PDEcont.html. The package LOCA is developed by Andy Salinger at Sandia National Laboratories and is available through http://www.cs.sandia.gov/loca/.

In this chapter, numerical techniques to apply bifurcation analysis to large-dimensional dynamical systems will be described. Doedel and Tuckermann (2000) provide an overview of the many techniques around. The aim here is to sketch the methods available in the code (BOOM) which has been developed in Utrecht over the years and will be further developed in Fort Collins (see http://fractal.atmos.colostate.edu). It simultaneously provides a background on the computational approaches used in subsequent chapters and hopefully a good entrance to the literature for readers interested to pursue this subject further.

The starting point is a given set of partial differential equations which can be written in operator form as

$$\mathcal{M}\frac{\partial \mathbf{u}}{\partial t} + \mathcal{L}\mathbf{u} + \mathcal{N}(\mathbf{u}) = \mathbf{F} \qquad (4.1)$$

where $\mathcal{L}$, $\mathcal{M}$ are linear operators, $\mathcal{N}$ is a nonlinear operator, $\mathbf{u}$ is the vector of dependent quantities and $\mathbf{F}$ contains the forcing of the system. To get a well-posed problem, appropriate boundary conditions have to be added to this set of equations. A typical problem will be given in section 4.1 and this problem also serves as a testcase for illustrating the methods.

The computational approach can be divided into three separate parts. First a discrete representation of the model equations has to be obtained through some kind of discretization procedure (Fig. 4.1). An overview of discretization methods is not provided here. Instead, a specific example will be considered using only one finite difference technique (section 4.2). Having a discrete form of the

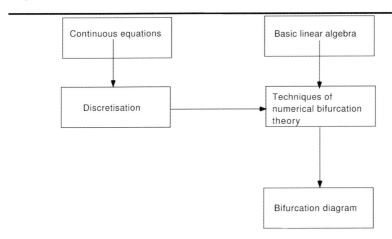

Figure 4.1. Sketch of the scheme of the computational work involved to compute bifurcation diagrams.

equations, in many cases the model can now be treated as a set of ordinary differential equations with or without algebraic constraints. The second part of the computational work is to apply specific techniques of numerical bifurcation theory (Fig. 4.1). These are the same techniques which are used in the packages for the smaller dimensional systems, such as AUTO, MATCONT and CONTENT.

For the example problem, the scheme below will be followed in subsequent sections. Note that this is a similar scheme as was used to analyze the Stommel (1961) two-box model in section 3.1.

(i) Determine the fixed points $\bar{u}$ of the system of equations when parameters are changed, i.e., solve the problem

$$\mathcal{L}\bar{u} + \mathcal{N}(\bar{u}) = \mathbf{F} \qquad (4.2)$$

This will be done using continuation methods which are presented in section 4.2. Their description follows closely texts in Seydel (1994) and Nayfeh and Balachandran (1995).

(ii) When one is able to compute a branch of steady solutions in a control parameter, one wants to know whether a bifurcation point has been crossed, whether other branches exist and if yes, how they can be reached. Practical techniques to do so are provided in section 4.3.

(iii) If a steady state is computed, one wants to assess its linear stability. With $\mathbf{u} = \bar{\mathbf{u}} + \tilde{\mathbf{u}}$, linearizing (4.1) around $\bar{\mathbf{u}}$ and separating $\tilde{\mathbf{u}} = \hat{\mathbf{u}} \, e^{\sigma t}$ gives an eigenvalue problem of the form

$$(\mathcal{L} + \mathcal{N}_u(\mathbf{u}))\hat{\mathbf{u}} = -\sigma \mathcal{M} \hat{\mathbf{u}} \qquad (4.3)$$

where $\mathcal{N}_u$ is the derivative of the operator $\mathcal{N}$ with respect to $\mathbf{u}$. The solution of these eigenvalue problems is discussed in section 4.4

(iv) Finally, one wants to compute trajectories of the model under investigation, either in the regime where bifurcation behavior is known to occur, or to compute periodic orbits. As a spin-off of the methodology above, the use of implicit time-dependent methods will be discussed in section 4.5

As it will turn out, an important part of the computational work is the solution of large linear systems of equations. The success of the latter methods mainly determines the dimension of the dynamical system which can be handled. Whereas for small dimensional dynamical systems robust so-called direct techniques (section 4.6) can be used, for giant dimensional systems one must turn to sophisticated (and non-robust) iterative solvers. Section 4.7 provides an introduction into these techniques. In section 4.8, the whole scheme will be applied to the example problem described below.

4.1. An Example Problem

The Rayleigh-Bénard problem discussed in this section has been used as a test-problem during a workshop on "Application of Continuation Methods in Fluid Mechanics" in 1998 (Henry and Bergeon, 2000). It is a relatively simple problem, and hence techniques can be easily illustrated. The physics of the problem is also very transparent, making it a nice prototype system to use here. A third advantage of the problem is that it introduces the fluid mechanics of buoyancy driven flows which are the central topic within chapter 6.

4.1.1. Introduction

The Rayleigh-Bénard problem is one of the 'classics' in fluid dynamics and one in the area of cellular convection. It is motivated by results from a (conceptually) simple experiment (Fig. 4.2). A rectangular container is filled with a viscous liquid such as silicone oil. Air is situated above the upper surface of the liquid and the temperature far from the air-liquid interface is nearly constant. When the initially motionless liquid is heated from below, the liquid remains motionless below a critical value of the vertical temperature gradient. In this case, the heat transfer through the layer is only by heat conduction. When the temperature gradient slightly exceeds the critical value, the liquid is set into motion and after a while the flow organizes itself into cellular patterns (Fig. 4.3).

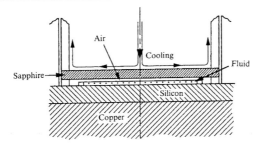

Figure 4.2.   Sketch of the experimental set-up; the liquid is situated on the (heated) silicon block and separated from the (cooled) sapphire block by a small air gap (Koschmieder and Switzer, 1992).

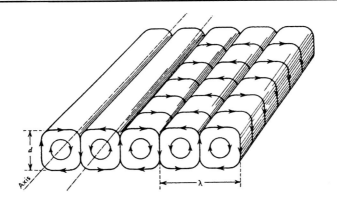

Figure 4.3.   Example of a flow pattern consisting of cellular rolls (also called roll cells) arising in a liquid layer heated from below (Koschmieder, 1993).

The motion of the liquid can also be detected by measuring the horizontally averaged vertical heat flux. A measure for the increase of heat transport due to convection is the Nusselt number Nu. This dimensionless scalar is the ratio of the heat transfer due to combined conduction and convection and the heat transfer due to conduction only; hence $Nu = 1$ in case there is no convection. In Fig. 4.4, Nu is plotted as a function of the vertical temperature difference over the layer. The onset of convection in the liquid is shown by the increase of Nu above unity. From the experimental data, one can guess that some bifurcation is involved where the steady motionless state becomes unstable and new cellular type of solutions stabilize. From the symmetry properties of the flow — one can imagine to rotate the container over $180°$ and get the same experimental results — a pitchfork bifurcation is anticipated. One of the relevant problems with respect to the experiment is to determine the vertical temperature gradient associated with this bifurcation point.

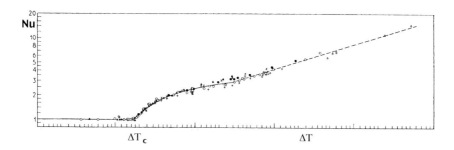

Figure 4.4.   Plot of the Nusselt number Nu as a function of the vertical temperature difference ΔT; $Nu = 1$ if the heat transport is by conduction only and when $Nu > 1$ there is convection in the liquid; ΔT_c is the critical temperature gradient (Chandrasekhar, 1961).

4.1.2. Model

The equations governing the flow follow from the equations (2.2), for $\mathbf{\Omega} = 0$, and are

$$\rho_0 \left[\frac{\partial \mathbf{v}_*}{\partial t_*} + \mathbf{v}_* . \nabla \mathbf{v}_* \right] = -\nabla p_* + \mu \nabla^2 \mathbf{v}_* - \rho_* g \mathbf{e}_3 \qquad (4.4a)$$

$$\nabla \cdot \mathbf{v}_* = 0 \qquad (4.4b)$$

$$\rho_0 C_p \left[\frac{\partial T_*}{\partial t_*} + \mathbf{v}_* . \nabla T_* \right] = \lambda_T \nabla^2 T_* \qquad (4.4c)$$

In these equations, (x_*, y_*, z_*) are the Cartesian coordinates of a point in the liquid layer, t_* denotes time, $\mathbf{v}_* = (u_*, v_*, w_*)$ is the velocity vector, p_* denotes pressure, $\mathbf{e}_3$ the unit vector in z-direction and T_* is the temperature. The quantities ρ_0, g, C_p, μ and λ_T are the reference density, the acceleration due to gravity, the heat capacity, the dynamic viscosity and the thermal conductivity, respectively. The thermal diffusivity κ and kinematic viscosity ν are given by $\nu = \mu/\rho_0$, $\kappa = \lambda_T/(\rho_0 C_p)$ and all these quantities will be assumed constant. A linear equation of state

$$\rho_* = \rho_0 (1 - \alpha_T (T_* - T_0)) \qquad (4.5)$$

is assumed, where α_T is the thermal compressibility coefficient and T_0 a reference temperature. The lower boundary of the liquid is considered to be a very good conducting boundary on which the temperature is constant T_B, and no-slip conditions apply. On the lateral walls (at $x_* = 0, L_x$ and $y_* = 0, L_y$) no-flux and

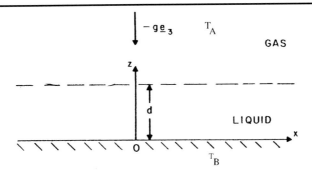

Figure 4.5.   Sketch of the model set-up and boundary conditions of the Rayleigh-Bénard problem.

no-slip conditions are prescribed. Let the non-deforming gas-liquid interface be located at $z_* = d$, then the boundary conditions become (Fig. 4.5)

$$z_* = d \quad : \quad \frac{\partial u_*}{\partial z_*} = \frac{\partial v_*}{\partial z_*} = w_* = 0 \; ; \; \lambda_T \frac{\partial T_*}{\partial z_*} = h(T_A - T_*) \quad (4.6a)$$

$$z_* = 0 \quad : \quad T_* = T_B \; ; \; u_* = v_* = w_* = 0 \quad\quad\quad (4.6b)$$

$$x_* = 0, L_x \quad : \quad u_* = v_* = w_* = \frac{\partial T_*}{\partial x_*} = 0 \quad\quad\quad (4.6c)$$

$$y_* = 0, L_y \quad : \quad u_* = v_* = w_* = \frac{\partial T_*}{\partial y_*} = 0 \quad\quad\quad (4.6d)$$

where h is an interfacial heat transfer coefficient and T_A is the temperature of the gas far from the interface.

4.1.3. Motionless solution

For $\bar{\mathbf{v}}_* = 0$, there is a steady state given by

$$\bar{T}_*(z_*) = T_B - \beta z_* \; ; \; \beta = \frac{h(T_B - T_A)}{\lambda_T + hd} \quad\quad\quad (4.7)$$

The quantity β is the vertical temperature gradient over the layer. The corresponding pressure distribution is readily determined from (4.4a) and if one chooses $T_0 = T_A$, this gives

$$\bar{p}_*(z_*) = p_0 + \rho_0 g([\alpha_T(T_B - T_A) - 1]z_* - \frac{\alpha_T \beta}{2} z_*^2) \quad\quad\quad (4.8)$$

This motionless solution is characterized by only conductive heat transfer and is easily realized in laboratory experiments. Note that such a motionless solution exists for all values of the vertical temperature difference $\Delta T = \beta d$. Hence, according to theory presented in section 3.7, we would not expect saddle-node bifurcations to occur on the branch of motionless solutions.

4.1.4. Dimensionless equations

The equations and boundary conditions are next non-dimensionalized using scales κ/d for velocity, d^2/κ for time and d for length. Moreover, a dimensionless temperature T is introduced through $T_* = (T_B - T_A)T + T_A$ and a dimensionless pressure p through $p_* = p_0 + p(\mu\kappa/d^2)$. This leads to the non-dimensional problem

$$Pr^{-1}\left[\frac{\partial \mathbf{v}}{\partial t} + \mathbf{v}.\nabla\mathbf{v}\right] = -\nabla p + \nabla^2\mathbf{v} + Ra\, T\, \mathbf{e}_3 \tag{4.9a}$$

$$\nabla \cdot \mathbf{v} = 0 \tag{4.9b}$$

$$\frac{\partial T}{\partial t} + \mathbf{v}.\nabla T = \nabla^2 T \tag{4.9c}$$

with boundary conditions

$$z = 1 \quad : \quad \frac{\partial u}{\partial z} = \frac{\partial v}{\partial z} = w = 0; \frac{\partial T}{\partial z} = -Bi\, T \tag{4.10a}$$

$$z = 0 \quad : \quad T = 1\, ; \, u = v = w = 0 \tag{4.10b}$$

$$x = 0, A_x \quad : \quad u = v = w = \frac{\partial T}{\partial x} = 0 \tag{4.10c}$$

$$y = 0, A_y \quad : \quad u = v = w = \frac{\partial T}{\partial y} = 0 \tag{4.10d}$$

In the equations (4.9)-(4.10), the dimensionless parameters Pr (Prandtl), Ra (Rayleigh), A_x, A_y (Aspect ratios) and Bi (Biot) appear which are defined as

$$Ra = \frac{\alpha_T g(T_B - T_A)d^3}{\nu\kappa}; \, Pr = \frac{\nu}{\kappa}; \, Bi = \frac{hd}{\lambda_T}$$

$$A_x = L_x/d; \, A_y = L_y/d \tag{4.11}$$

and hence there are five parameters in this system of equations. This number reduces to four in the two-dimensional case since one of the aspect ratios disappears.

The dimensionless motionless solution is given by

$$\bar{u} = \bar{v} = \bar{w} = 0 \quad ; \quad \bar{T}(z) = 1 - z\frac{Bi}{Bi + 1} \tag{4.12a}$$

$$\bar{p}(z) = Ra\left[z - \frac{Bi}{(1 + Bi)}\frac{z^2}{2}\right] \tag{4.12b}$$

and this is a solution for all values of Ra and Bi which makes it an ideal starting point for the computations below.

4.2. Computation of Steady Solutions

In this section, the methods to determine steady state solutions in parameter space are presented. The example problem from the previous section is used in section 4.2.1 to illustrate the discretization methods. In section 4.2.2 the so-called continuation methods, used to follow branches of steady states, are described.

4.2.1. Discretization

For the example problem, many type of discretization methods have been used, e.g., finite differences, finite elements and spectral methods. To illustrate the use of finite differences, consider the two-dimensional case in the example problem above, i.e., restricting to solutions $\mathbf{v} = (u, 0, w)$, and p and T, which are independent of y. A staggered grid is used, with u, w at boundaries and p, T at center points of the grid cells (see Fig. 4.6a). The horizontal momentum equation

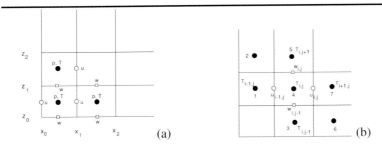

Figure 4.6.   (a) Sketch of the staggered grid, with points $i = 0, ..., I$ and $j = 0, ..., J$ in the x, z-direction, respectively. (b) Local stencil around the point $T_{i,j}$.

is enforced at $u-$points, the vertical momentum equation at $w-$points and the continuity and temperature equation at center (p, T) points.

For the discretization, it is efficient to define the discrete operators on a local stencil and subsequently assemble the operators over the whole domain. This is particularly useful when the nonlinear interactions in a model are at most quadratic, such as in the Navier-Stokes equations. In the latter case, the nonlinear operator $\mathcal{N}$ in (4.1) can be written as $\mathcal{N}(\mathbf{u})\mathbf{u}$. For each variable, a local stencil is defined such as in Fig. 4.6b for the temperature point $T_{i,j}$. As an example, consider the discretization of the horizontal diffusion operator, which is linear, using central differences. In this case, at point (i, j)

$$\frac{\partial^2 T}{\partial x^2} \approx \frac{T_{i+1,j} + T_{i-1,j} - 2T_{i,j}}{\Delta x^2} \tag{4.13}$$

According to the stensil (Fig. 4.6b), one now defines local operators $L_{i,j}^{TT}[1, \cdots, 7]$. The first superscript in L^{TT} indicates which equation is handled (in this case the temperature equation). The second superscript indicates by which unknown the coefficient has to be multiplied to get the right equations (in this case again the temperature). The index $[*]$ refers to the stencil points and hence

$$L_{i,j}^{TT}[1] = \frac{1}{\Delta x^2} \tag{4.14a}$$

$$L_{i,j}^{TT}[7] = \frac{1}{\Delta x^2} \qquad (4.14b)$$

$$L_{i,j}^{TT}[4] = -\frac{2}{\Delta x^2} \qquad (4.14c)$$

with all other $L_{i,j}^{TT}[*]$ being zero. The local operator is then built up as

$$\sum_{l=1}^{7} L_{i,j}^{TT}[l] T[l] \qquad (4.15)$$

where $T[l]$ refers to the stensil location around (i,j) with $T[1] = T_{i-1,j}$ and $T[5] = T_{i,j+1}$.

Next consider the nonlinear horizontal advection term for heat, which is discretized at $T-$ points as

$$\frac{\partial(uT)}{\partial x} = u_{i,j} \frac{T_{i+1,j} + T_{i,j}}{2\Delta x} - u_{i-1,j} \frac{T_{i,j} + T_{i-1,j}}{2\Delta x} \qquad (4.16)$$

This term is a part of the nonlinear operator in the $T-$ equation associated with the operator $\mathcal{N}$ in (4.1). One defines a local nonlinear operator $N_{i,j}^{TT}[1,\cdots,7]$ as

$$N_{i,j}^{TT}[1] = -\frac{u_{i-1,j}}{2\Delta x} \qquad (4.17a)$$

$$N_{i,j}^{TT}[7] = \frac{u_{i,j}}{2\Delta x} \qquad (4.17b)$$

$$N_{i,j}^{TT}[4] = \frac{u_{i,j} - u_{i-1,j}}{2\Delta x} \qquad (4.17c)$$

with other stensil coefficients zero. The discretized equations of the local nonlinear operator are then build up according to

$$\sum_{l=1}^{7} N_{i,j}^{TT}[l] T[l] \qquad (4.18)$$

In this way, it is relatively easy to include boundary conditions. For example, imagine the implementation of a no-flux condition ($\partial T/\partial x = 0$) for the temperature at $x = 0$. Using central differences, this becomes

$$\frac{\partial T}{\partial x} = 0 \Rightarrow T_{0,j} = T_{1,j}$$

If the total stensil coefficient is indicated by $A^{TT} = L^{TT} + N^{TT}$, then the boundary condition can be accounted for by correcting the stensil coefficient $A_{1,j}^{TT}[4]$ as

$$\tilde{A}_{1,j}^{TT}[4] = A_{1,j}^{TT}[4] + A_{1,j}^{TT}[1]$$

and thereafter setting $A_{1,j}^{TT}[1] = 0$.

The boundary condition $T = 1$ for the temperature at $z = 0$ is discretized as

$$\frac{1}{2}(T_{i,0} + T_{i,1}) = 1 \Rightarrow T_{i,0} = 2 - T_{i,1}$$

This can be accounted for by correcting the stensil coefficient $A_{i,1}^{TT}[4]$ as

$$\tilde{A}_{i,1}^{TT}[4] = A_{i,1}^{TT}[4] - A_{i,1}^{TT}[3]$$

by including a forcing term, $F_{i,1}^T = 2\, A_{i,1}^{TT}[3]$, and by setting $A_{i,1}^{TT}[3] = 0$ thereafter. Assembly of the total operators can be accomplished by one big loop over the grid points and the stensil points.

To determine the linear stability of a steady state, we see from (4.3) that not only the discretized operator $\mathcal{N}$ is needed, but also its derivative $\mathcal{N}_u$ around a certain solution $(\bar{u}, \bar{w}, \bar{p}, \bar{T})$. For the horizontal advection operator in (4.16), this derivative becomes

$$\frac{\partial(\bar{u}T)}{\partial x} + \frac{\partial(u\bar{T})}{\partial x} \tag{4.19}$$

When discretized with central differences, the coefficients for the first term are similar to those in the operator $N_{i,j}^{TT}$ in (4.17) but with u substituted by $\bar{u}$. For the second term, an additional operator $N_{i,j}^{TU}$ is needed, which is defined by

$$N_{i,j}^{TU}[1] = -\frac{\bar{T}_{i-1,j} + \bar{T}_{i,j}}{2\Delta x} \tag{4.20a}$$

$$N_{i,j}^{TU}[4] = \frac{\bar{T}_{i+1,j} + \bar{T}_{i,j}}{2\Delta x} \tag{4.20b}$$

such that the term from this operator in the Jacobian matrix is built up as

$$\sum_{l=1}^{7} \left[N_{i,j}^{TT}[l]T[l] + N_{i,j}^{TU}[l]u[l] \right] \tag{4.21}$$

where again $T[l]$ and $u[l]$ refer to stensil point values, i.e., $u[4] = u_{i,j}$. Corrections due to boundary conditions and assembly of the matrices can be accomplished in the same way for the other operators in (4.22).

The discretized equations can thus be written as a nonlinear system of ordinary differential equations with algebraic constraints which has the form

$$\mathcal{M}_N \frac{\partial \mathbf{x}}{\partial t} + \mathcal{L}_N \mathbf{x} + \mathcal{N}_N(\mathbf{x}) = \mathcal{F}_N \tag{4.22}$$

where $\mathbf{x}$ indicates the total N-dimensional vector of unknowns. The operators depend on parameters and their subscript N indicates that they are discrete equivalents of the continuous operators. In the two-dimensional example problem, $\mathbf{x}$ is given by

$$\mathbf{x} = (u_{0,0}, w_{0,0}, p_{0,0}, T_{0,0}, u_{1,0}, ..., T_{I-1,J}, u_{I,J}, w_{I,J}, p_{I,J}, T_{I,J})^T \tag{4.23}$$

and $N = 4\,(I+1)\,(J+1)$.

4.2.2. Pseudo-arclength continuation

To determine steady solutions of (4.22), we need to solve the set of nonlinear algebraic equations

$$\Phi(\mathbf{x}, \lambda) = \mathcal{L}_N \mathbf{x} + \mathcal{N}_N(\mathbf{x}) - \mathcal{F}_N = 0 \qquad (4.24)$$

where λ indicates a control parameter which appears in the operators $\mathcal{L}_N$ and/or $\mathcal{N}_N$ and/or $\mathcal{F}_N$. For example, in the example problem this is the parameter Ra, since this parameter represents the vertical temperature gradient.

For reasons which will be made clear below, it is advantageous to parametrize branches of solutions with an arclength parameter s as sketched in Fig. 4.7. A branch γ of steady solutions $(\mathbf{x}(s), \lambda(s))$, $s \in [s_a, s_b]$ is a smooth one-parameter family of solutions of (4.24). Since an extra degree of freedom is introduced by the arclength s, a normalization condition of the form

$$\Sigma(\mathbf{x}(s), \lambda(s), s) = 0 \qquad (4.25)$$

is needed to close the system of equations. We thus end up to solve a system of nonlinear algebraic equations of dimension $N + 1$ for the $N + 1$ unknowns $(\mathbf{x}(s), \lambda(s))$. But where to start and how to choose the normalization?

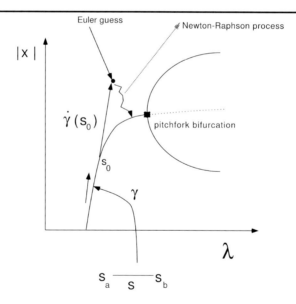

Figure 4.7.   Sketch of the parametrization of branches of steady solutions by an arclength parameter s and the tangent $\dot{\gamma}$ along the branch in a typical bifurcation diagram.

In the more specialized literature (Seydel, 1994), several alternatives are described. In the example problem, and in many others, we know already a steady solution of the system (the motionless solution). In many applications, some trivial state can always be found, for example, under zero forcing conditions. With respect to the normalization issue, we consider the geometry of the problem. Aim is to determine the range of a curve $\gamma : I \subseteq \mathbb{R} \to \mathbb{R}^{N+1}$, with $\gamma(s) = (\mathbf{x}(s), \lambda(s))$ such that (4.24) is satisfied. Assuming that we now know, at some point s_0, a solution $(\mathbf{x}_0, \lambda_0)$, then the tangent space of the curve at $s = s_0$ is spanned by the vector $\dot{\gamma}(s_0) = (\dot{\mathbf{x}}(s_0), \dot{\lambda}(s_0))^T$ (Fig. 4.7). It is advantageous (the reason being the solution method in the next section) to take a normalization of the length of the tangent,

$$\dot{\mathbf{x}}_0^T \dot{\mathbf{x}}_0 + \dot{\lambda}_0^2 = 1 \tag{4.26}$$

In some applications, the initial tangent is analytically available. For example, in the example problem we know that the motionless solution is a solution for all values of Ra and hence with $\lambda = Ra$, we find $\dot{\mathbf{x}}_0 = 0$ and $\dot{\lambda}_0 = 1$.

A more general way of computing the tangent is the following. By differentiating $\Phi(\gamma(s)) = 0$ to s we find

$$[\Phi_x \ \Phi_\lambda]\dot{\gamma}(s) = \begin{pmatrix} \frac{\partial \Phi_1}{\partial x_1} & \cdots & \frac{\partial \Phi_1}{\partial x_N} & \frac{\partial \Phi_1}{\partial \lambda} \\ \frac{\partial \Phi_1}{\partial x_N} & \cdots & \frac{\partial \Phi_N}{\partial x_N} & \frac{\partial \Phi_N}{\partial \lambda} \end{pmatrix} \dot{\gamma}(s) = 0 \tag{4.27}$$

If $(\mathbf{x}_0, \lambda_0)$ is not a bifurcation point, then $\dim(\ker([\Phi_x \ \Phi_\lambda])) = 1$ and therefore $[\Phi_x \ \Phi_\lambda]$ has rank N. Hence, we can determine $\dot{\gamma}(s_0)$ as the null space of the $N(N+1)$ matrix $[\Phi_x \ \Phi_\lambda]$.

First, the matrix $[\Phi_x \ \Phi_\lambda]$ is triangulated into the form

$$\begin{pmatrix} * & * & * & * \\ 0 & * & * & * \\ 0 & 0 & * & * \end{pmatrix} \tag{4.28}$$

where this matrix (a $*$ indicates a possible nonzero element) is shown for $N = 3$. The last row cannot be entirely zero, and therefore the (permuted) tangent vector $\mathbf{v} = (\dot{\mathbf{x}}_0, \dot{\lambda}_0)$ can be computed by solving

$$\begin{pmatrix} * & * & * & * \\ 0 & * & * & * \\ 0 & 0 & * & * \\ 0 & 0 & 0 & 1 \end{pmatrix} \mathbf{v} = \begin{pmatrix} 0 \\ 0 \\ 0 \\ 1 \end{pmatrix} \tag{4.29}$$

and its length is normalized as in (4.26).

Once $\mathbf{x}_0, \lambda_0, \dot{\mathbf{x}}_0$ and $\dot{\lambda}_0$ are determined, a further point on the same branch can be calculated by taking

$$\Sigma(\mathbf{x}, \lambda, s) = \dot{\mathbf{x}}_0^T(\mathbf{x} - \mathbf{x}_0) + \dot{\lambda}_0(\lambda - \lambda_0) - (s - s_0) \tag{4.30}$$

and solve the total system of equations (4.24) and (4.30) given a prescribed step length $\Delta s = s - s_0$. In this form, the continuation method is called a pseudo-arclength method (Keller, 1977). The name derives from the fact that (4.30) is an approximation to (4.26). The advantage of this method is that the Jacobian of the extended system (4.24) and (4.30) is non-singular at saddle-node bifurcations, whereas the Jacobian Φ_x is singular. Hence, one can easily follow a branch around a saddle-node bifurcation (Keller, 1977).

4.2.3. The Euler-Newton method

To solve the equations (4.24) and (4.30), an Euler predictor/Newton corrector algorithm is applied. The Newton-Raphson technique is a robust technique to solve for zeroes of nonlinear equations. It is best illustrated through its one-dimensional form, aiming to compute a zero of the scalar equation (with a function $f : \mathbb{R} \to \mathbb{R}$)

$$f(x) = 0 \tag{4.31}$$

Here, one starts from an initial guess x^0 and computes the tangent of the function at the point $f(x^0)$, i.e., this line is given by $y = f'(x^0)(x - x^0) - f(x^0)$. A next

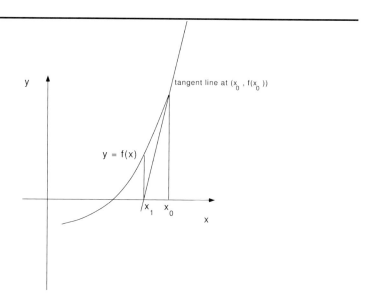

Figure 4.8.   Sketch of the Newton-Raphson technique to determine a zero of a function $f : \mathbb{R} \to \mathbb{R}$ using iterates $x_0, x_1, \cdots$ according to (4.32).

iterate is computed by the intersection of this tangent and the x-axis (Fig. 4.8), which gives $x^1 = x^0 - f(x^0)/f'(x^0)$. The general iterative technique becomes

$$x^{k+1} = x^k - \frac{f(x^k)}{f'(x^k)} \Rightarrow f'(x^k)(x^{k+1} - x^k) = -f(x^k) \qquad (4.32)$$

where k is the Newton-Raphson iteration index. If at some k, $|\ x_{k+1} - x_k\ |$ becomes smaller than some tolerance then, under fairly mild conditions, it can be shown that a zero of $f(x)$ has been found (Atkinson, 1976)).

The Newton-Raphson technique can be easily generalized to nonlinear systems of equations and hence to solve (4.24) and (4.30). Let the steady state which is already known be indicated by $\mathbf{x}^0$, then a good guess for the next steady state is the Euler predictor given by

$$\mathbf{x}^1 = \mathbf{x}_0 + \Delta s\ \dot{\mathbf{x}}_0 \qquad (4.33a)$$
$$\lambda^1 = \lambda_0 + \Delta s\ \dot{\lambda}_0 \qquad (4.33b)$$

where again the dot indicates differentiation to s. Once $\mathbf{x}^k$, $\dot{\mathbf{x}}^k$, λ^k and $\dot{\lambda}^k$ are calculated the equations (4.24) and (4.30) are linearized around this solution, i.e.,

$$\mathbf{x}^{k+1} = \mathbf{x}^k + \Delta \mathbf{x}^{k+1} \qquad (4.34a)$$
$$\lambda^{k+1} = \lambda^k + \Delta \lambda^{k+1} \qquad (4.34b)$$

In every Newton-Raphson iteration, the solutions $(\Delta \mathbf{x}^{k+1}, \Delta \lambda^{k+1})$ are determined by solving the linear system

$$\begin{pmatrix} \Phi_x(\mathbf{x}^k, \lambda^k) & \Phi_\lambda(\mathbf{x}^k, \lambda^k) \\ \dot{\mathbf{x}}_0^T & \dot{\lambda}_0 \end{pmatrix} \begin{pmatrix} \Delta \mathbf{x}^{k+1} \\ \Delta \lambda^{k+1} \end{pmatrix} =$$
$$= \begin{pmatrix} -\Phi(\mathbf{x}^k, \lambda^k) \\ \Delta s - \dot{\mathbf{x}}_0^T(\mathbf{x}^k - \mathbf{x}_0) - \dot{\lambda}_0(\lambda^k - \lambda_0) \end{pmatrix} \qquad (4.35)$$

which is just the generalization of (4.32) to N–dimensions. Hence, within each iteration, a linear system of equations has to be solved. If the Newton-Raphson process has converged up to a desired accuracy, a new steady state has been found.

One can split the solution of (4.35) into two steps in which only linear systems with Φ_x are solved. Let $\mathbf{r} = -\Phi(\mathbf{x}^k, \lambda^k)$ and $r_{N+1} = \Delta s - \dot{\mathbf{x}}_0^T(\mathbf{x}^k - \mathbf{x}_0) - \dot{\lambda}_0(\lambda^k - \lambda_0)$, then if $\mathbf{z}_1$ and $\mathbf{z}_2$ are solved from

$$\Phi_x(\mathbf{x}^k, \lambda^k)\mathbf{z}_1 = \mathbf{r} \qquad (4.36a)$$
$$\Phi_x(\mathbf{x}^k, \lambda^k)\mathbf{z}_2 = \Phi_\lambda(\mathbf{x}^k, \lambda^k) \qquad (4.36b)$$

then the solution $(\Delta \mathbf{x}^{k+1}, \Delta \lambda^{k+1})$ is found from

$$\Delta \lambda^{k+1} = \frac{r_{N+1} - \dot{\mathbf{x}}_0^T \mathbf{z}_1}{\dot{\lambda}_0 - \dot{\mathbf{x}}_0^T \mathbf{z}_2} \qquad (4.37a)$$
$$\Delta \mathbf{x}^{k+1} = \mathbf{z}_1 - \Delta \lambda^{k+1} \mathbf{z}_2 \qquad (4.37b)$$

One of the problems involved is the determination of the Jacobian matrix Φ_x and the derivative vector Φ_λ. One can do this in, at least, four ways: (i) 'by hand', (ii) symbolically using Mathematica or Maple, (iii) use automatic differentiation software which provides the code for the Jacobian matrix Φ_x from that of the right hand side Φ — an example of such a program is ADIFOR (http://www-unix.mcs.anl.gov/autodiff/ADIFOR/) — or (iv) compute it numerically by finite differences through

$$\frac{\partial \Phi_k}{\partial x_l} \approx \frac{\Phi_k(x_l + \epsilon) - \Phi_k(x_l)}{\epsilon} \tag{4.38}$$

for $k = 1, .., N; l = 1, ..., N$ and small ϵ. My experience is that it is usually faster to do it 'by hand', but if one is really handy with the symbolic manipulation programs or automatic differentiation codes, ... , one should do this.

4.3. Detection and Switching

In the previous section, a method has been described to perform steady state continuation in a single parameter. Suppose, we have computed the points on a branch of steady solutions as indicated in Fig. 4.9 by varying a parameter λ. In this case, the method would just pass the pitchfork bifurcation point P (Fig. 4.9). How do we determine that this bifurcation has occurred?

One way to do this is to solve the eigenvalue problem associated with the stability of the steady state at each point. We know that for a pitchfork bifurcation, a single real eigenvalue must cross the imaginary axis and by monitoring this eigenvalue, the pitchfork bifurcation can be detected. In many applications, however, the solution of the eigenvalue problem is computationally expensive. Hence, simpler and cheaper indicator functions may be desired and some of these are described below.

4.3.1. Detection of bifurcations

To determine simple codimension-1 bifurcation points (transcritical, pitchfork and saddle-node bifurcations), the determinant of the Jacobian matrix (det Φ_x) can be monitored. For many large dimensional problems this determinant is expensive to compute and other alternatives must be considered. In Seydel (1994), a family of test functions τ_{pq} is obtained as follows: let Φ_x^{pq} be the Jacobian matrix Φ_x in which the pth row is replaced by the qth unit vector. If we solve the linear system

$$\Phi_x^{pq} \mathbf{v} = \mathbf{e}_p \tag{4.39}$$

for $\mathbf{v}$, where $\mathbf{e}_p$ is the pth unit vector, then it can be shown (Seydel, 1994) that

$$\tau_{pq} = \mathbf{e}_p^T \Phi_x \mathbf{v} = \mathbf{e}_p^T (\Phi_x^{pq})^{-1} \mathbf{e}_p \tag{4.40}$$

changes sign when Φ_x is singular. In principle, the choices of q and p are arbitrary as long as Φ_x^{pq} is nonsingular. Of course, for any solution method, it is

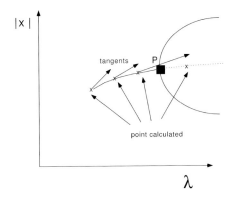

Figure 4.9.   Example of computations of steady states versus a parameter λ on a branch passing through a pitchfork bifurcation P.

advantageous that Φ_x^{pq} and Φ_x have the same matrix structure. However, in specific problems, not all values of q and p can be chosen and it is advisable to make a choice based on the knowledge of the properties of the solutions (e.g., symmetry) of the particular problem.

Saddle-node bifurcations can be easily detected by following $\dot{\lambda}$ along a branch, where the dot indicates differentiation to the arclength parameter s. For Hopf bifurcation points, also more sophisticated methods exist (Kuznetsov, 1995), but usually these points are determined by solving the linear stability problem which is discussed in the next section. In this case, a complex conjugate pair of eigenvalues $\sigma = \sigma_r + i\,\sigma_i$ crosses the imaginary axis and a zero of the function $\sigma_r(\lambda)$ has to be calculated to obtain the location of the Hopf bifurcation.

Once a change in sign is found in one of these scalar quantities (e.g., $\dot{\lambda}$, det Φ_x, τ_{pq} or $\sigma_r(\lambda)$) between two points along a branch, say s_a and s_b, a secant process can be used to locate the zero of each function exactly. In more detail, let either function be indicated by $f(s)$ then a zero of $f(s)$ is determined by

$$s_{l+1} = s_l - f(s_l)\frac{s_l - s_{l-1}}{f(s_l) - f(s_{l-1})} \tag{4.41a}$$

$$s_0 = s_a \quad ; \quad s_1 = s_b \tag{4.41b}$$

When $s_a \neq 0$, the stopping criterion on the iteration can be chosen as

$$\frac{|\,s_{l+1} - s_l\,|}{s_a} < \varepsilon$$

where ε must be chosen according to the desired accuracy. In some cases, a larger ε must be taken because the matrix Φ_x may become nearly singular during the iteration. It is recommended to check *a postiori* that the value of $f(s)$ is substantially smaller than the value of this function at both s_a and s_b.

4.3.2. Branch switching

If, for example, $\det \Phi_x$ changes sign but $\dot{\lambda}$ does not, a simple bifurcation point (transcritical or pitchfork) is detected. Subsequently, a branch switch process can be started to locate solutions on the nearby branch. In Fig. 4.10, this situation is sketched near a pitchfork bifurcation. Let $\hat{\Phi}_x$ be the Jacobian matrix at the bifurcation point $(\mathbf{x}_*, \lambda_*)$ just after the secant iteration (see the previous section) has converged. Furthermore, let the tangent along the already known branch in $s = s_a$ be indicated by $(\dot{\mathbf{x}}_0, \dot{\lambda}_0)$.

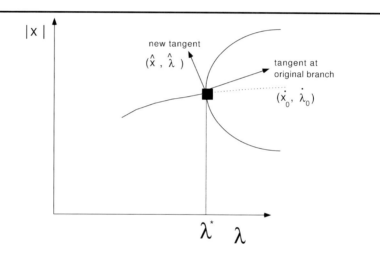

Figure 4.10.   Example of branch switching near a pitchfork bifurcation.

First, the null vector ϕ of $\hat{\Phi}_x$ is calculated, for example by inverse iteration (Atkinson, 1976); the latter method is described in Technical box 4.1. Next, a vector $(\hat{\mathbf{x}}, \hat{\lambda})$ is constructed which is orthogonal to $(\dot{\mathbf{x}}_0, \dot{\lambda}_0)$ by solving

$$\begin{pmatrix} \hat{\Phi}_x & \hat{\Phi}_\lambda \\ \dot{\mathbf{x}}_0^T & \dot{\lambda}_0 \end{pmatrix} \begin{pmatrix} \hat{\mathbf{x}} \\ \hat{\lambda} \end{pmatrix} = \begin{pmatrix} \mathbf{0} \\ 0 \end{pmatrix} \qquad (4.42)$$

The solution of this problem is easily determined to be

$$\hat{\lambda} = \frac{-\dot{\mathbf{x}}_0^T \phi}{\dot{\lambda}_0 - \dot{\mathbf{x}}_0^T \mathbf{z}} \; ; \; \hat{\mathbf{x}} = \phi - \hat{\lambda} \mathbf{z}$$

where $\mathbf{z}$ is the solution of $\hat{\Phi}_x \mathbf{z} = \hat{\Phi}_\lambda$. To determine a point on the new branch (Fig. 4.10), the Newton process is started with Euler - predictor

$$\mathbf{x}^1 = \mathbf{x}_* \pm \Delta s\, \hat{\mathbf{x}}\; ;\; \lambda^1 = \lambda_* + \Delta s\, \hat{\lambda} \qquad (4.43)$$

The $\pm$ indicates that points can be found on either side of the known branch. When a point on a new branch is found, the pseudo-arclength procedure is again used to compute additional points on this branch.

Technical box 4.1:
Inverse iteration

One of most powerful tools to determine eigenvectors of a matrix once an eigenvalue is approximately known is inverse iteration. Let A be an $n \times n$ matrix with eigenvalues $\lambda_1, ..., \lambda_n$ and assume λ is an approximation to one of the eigenvalues λ_k. The method starts with an initial guess $\mathbf{z}^0$ and successively the vectors $\mathbf{z}^{m+1}$ and $\mathbf{w}^{m+1}$ are calculated from

$$(A - \lambda I)\mathbf{w}^{m+1} = \mathbf{z}^m\; ;\; \mathbf{z}^{m+1} = \frac{\mathbf{w}^{m+1}}{\|\mathbf{w}^{m+1}\|}$$

for $m = 0, 1, \cdots$, where I is the identity matrix. For this method, one can show that $\mathbf{z}^m \to \mathbf{x}_k$, where $A\mathbf{x}_k = \lambda_k \mathbf{x}_k$ (see Atkinson (1976), p548). There are optimal choices for the starting vector in some situations, but starting from $\mathbf{z}^0 = (1, ..., 1)^T$ works well in most cases.

If one already anticipates a pitchfork bifurcation, one can also determine the other branch by a technique which makes use of the imperfections of the pitchfork as described in section 3.2.3. Suppose, two points A and B on a branch are computed where the stability is different (Fig. 4.11a) or where some τ_{pq} from (4.40) changes sign. Now one knows that there is an internal symmetry of the system associated with a pitchfork bifurcation. By introducing an additional parameter p_s which breaks the symmetry (for example, introducing some asymmetric component in the forcing), the pitchfork no longer exists for small p_s (see section 3.2.3). One continues a few steps into this parameter from point A up to $p_s = \varepsilon$. Then, a point C on the bifurcation diagram as in Fig. 4.11b is obtained. Next, the parameter λ is increased up to the value of λ at point B; in this way point D is reached (Fig. 4.11b). As a last step, p_s is continued back to zero and point E is obtained (Fig. 4.11c). By following the branch back (from point E) in λ, the pitchfork bifurcation is easily found as the point where λ changes sign.

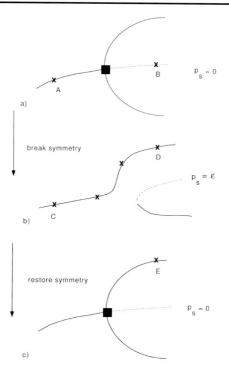

Figure 4.11.   Example of how knowledge of imperfections can be used to locate bifurcation points. The control parameter in the horizontal is λ. *(a) Symmetric situation with computed points A and B, where a sign switch in one of the indicator functions has been detected. (b) The imperfect pitchfork bifurcation is created by adding artificial asymmetry into the set of equations using a parameter* p_s. *Point A is followed up to point C in* p_s. *As a next step, one continues from A to D for a value of* λ *approximately up to the value at B. (c) Finally, symmetry is restored, point D is followed up to E and the pitchfork can be found as the point where* $\dot{\lambda}$ *changes sign.*

4.3.3. Finding isolated branches

In many applications, there exist branches of steady state solutions that are disconnected from the branch containing a trivial starting solution; these branches are the so-called isolated branches. An example of such an isolated branch can be seen in Fig. 3.9a, where it arises through an imperfection of the pitchfork bifurcations in the three-box model. One can already anticipate that in a dynamical system in which there is no symmetry, it is likely that isolated branches are present.

There are at least four methods to compute these isolated branches but it is never guaranteed that one will find all branches with either of these methods. Two of those are more or less 'trial and error' while in the latter two, a more systematic approach is followed.

(i) Transient integration.

In this approach, a set of initial conditions is chosen and a transient computation is started, for example by using an implicit method as described in section 4.5. If one is lucky, a trajectory starting at one of the initial conditions is attracted to a stable steady state on the isolated branch. Once found (Fig. 4.12a), one can continue tracing this branch using the pseudo-arclength continuation method.

(ii) Isolated Newton-Raphson search.

One can also start a Newton-Raphson process uncoupled from the pseudo-arclength continuation from several chosen starting points. Since the convergence of the Newton-Raphson process is only good when one is near the steady state, this method may not work very well, but again, if one is very lucky an isolated branch might be found (Fig. 4.12b).

(iii) Two-parameter continuation.

In many cases, a second parameter can be varied such that the isolated branch targeted connects to an already known branch. An important example is the case in which the dynamical system has a reflection symmetry at one particular value of a second parameter and a pitchfork bifurcation occurs. Once such a connection is present, the isolated branch can be computed by restoring the second parameter to its original value (Fig. 4.12c).

(iv) Residue continuation.

This method is a special case of a two-parameter continuation where one starts with a guess of the solution on the isolated branch, say indicated by x_G, at some value of a parameter λ. Because this is no steady solution, it follows that

$$\mathbf{f}(\mathbf{x}_G, \lambda) = \mathbf{r}_G \neq 0 \qquad (4.44)$$

where $\mathbf{r}_G$ is the nonzero residue. One now defines a second (so-called 'homotopy') parameter α and considers the problem

$$\mathbf{f}(\mathbf{x}, \lambda) - (1 - \alpha)\mathbf{r}_G = 0 \qquad (4.45)$$

For $\alpha = 0$, the solution is given by x_G (by construction) and hence this is the starting point of the pseudo-arclength continuation. By tracing the steady solution branch from $\alpha = 0$ to $\alpha = 1$, we may eventually find an isolated branch (Fig. 4.12d).

4.4. Linear Stability Problem

Suppose a stationary solution $\bar{\mathbf{x}}$ at a certain value of λ has been determined. Then its linear stability is investigated by considering perturbations $\mathbf{x} = \bar{\mathbf{x}} + \tilde{\mathbf{x}}$. When substituted into the general equations (4.22), and omitting quadratic terms in the perturbations quantities, one gets

$$\mathcal{M}\frac{\partial \tilde{\mathbf{x}}}{\partial t} + \mathcal{L}\tilde{\mathbf{x}} + \mathcal{N}_x(\bar{\mathbf{x}})\tilde{\mathbf{x}} = 0 \qquad (4.46)$$

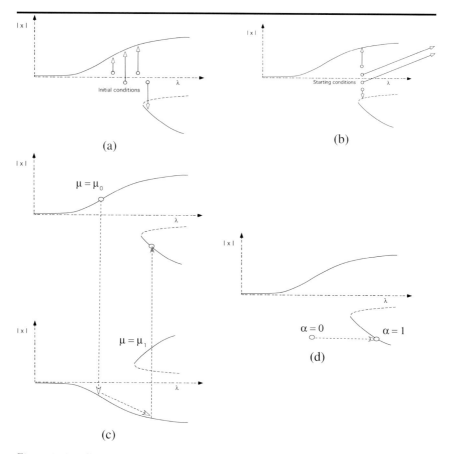

Figure 4.12.   Illustrations of the computation of isolated branches using four different methods. (a) Transient integration; the open circles indicate the initial conditions and the arrows the direction of the trajectories. Note that only stable steady states can be reached. (b) Isolated Newton-Raphson search; the open circles indicate the starting points.   The two large arrows indicate a possible divergence of the Newton-Raphson process. (c) Two-parameter continuation; a pitchfork occurs for $\mu_0 < \mu < \mu_1$. (d) Residue continuation, where α is the 'homotopy' parameter.

where the subscript N in (4.22) has been omitted for clarity and $\mathcal{N}_x(\bar{\mathbf{x}})$ is the Jacobian matrix at $\bar{\mathbf{x}}$. These equations admit solutions of the form $\tilde{\mathbf{x}} = \hat{\mathbf{x}}\, e^{\sigma t}$. The linear stability problem of a particular steady state leads, after discretization, to a generalized matrix eigenvalue problem of the form

$$Ax = \sigma B\mathbf{x} \qquad\qquad (4.47)$$

with $A = \mathcal{L} + \mathcal{N}_x(\bar{\mathbf{x}})$ and $B = -\mathcal{M}$. The matrix B may be singular; in the example problem, time derivatives are absent in the continuity equation and hence zeroes on the diagonal of B appear. The pair (A, B) is called a matrix pensil and some properties of the spectrum of matrix pensils are given in Technical box 4.2.

Technical box 4.2:
Eigenvalues of (matrix) pensils

Let A and B be two $n \times n$ matrices, then the set of matrices of the form $A - \sigma B$ is called a pensil. The eigenvalues of the pensil belong to the set $\sigma(A, B)$ defined by

$$\sigma(A, B) = \{z \in \mathbb{C} \mid \det(A - zB) = 0\}$$

When B is non-singular, the generalized eigenvalue problem is equivalent to an ordinary eigenvalue problem

$$B^{-1}A\mathbf{x} = \sigma\mathbf{x}$$

When B is singular, which occurs in many applications, complications arise. The set $\sigma(A, B)$ may consists of a finite number of eigenvalues, no eigenvalue may exist or infinitely many may occur. In Golub and Van Loan (1983), two examples of the latter situations are given.

$$A = \begin{pmatrix} 1 & 2 \\ 0 & 3 \end{pmatrix} \; ; B = \begin{pmatrix} 0 & 1 \\ 0 & 0 \end{pmatrix} \Rightarrow \sigma(A, B) = \emptyset$$

$$A = \begin{pmatrix} 1 & 2 \\ 0 & 0 \end{pmatrix} \; ; B = \begin{pmatrix} 1 & 0 \\ 0 & 0 \end{pmatrix} \Rightarrow \sigma(A, B) = \mathbb{C}$$

The traditional (robust) method to solve these generalized eigenvalue problems by which all eigenvalues and, if desired, all eigenvectors can be computed is the QZ-method. Details of this method are described in section 7.7 of Golub and Van Loan (1983) and its understanding requires quite some background in numerical linear algebra. The code of this method is available in standard numerical libraries such as NAG (http://www.nag.com/) and IMSL (http://www.imsl.com/). In studies of bifurcation behavior, as seen in chapter 3, the first few bifurcation points only involve a small number of eigenvalues and one is only interested to compute the eigenvectors with eigenvalues closest to the imaginary axis (the so-called 'most dangerous' modes). Fortunately, special methods are available to perform these type of computations and two of them are discussed below.

4.4.1. The simultaneous iteration method

This method belongs to the class of generalized power methods. As a preparation to apply the method, a complex mapping of the form (Fig. 4.13)

$$\sigma = b + a\frac{\mu - 1}{\mu + 1} \tag{4.49}$$

is applied, with $b \in \mathbb{C}, a \in \mathbb{R}^+$. This transforms the eigenvalue problem (4.47) into

$$Ex = D^{-1}Cx = \mu x \qquad (4.50)$$

where $C = -(A + (a - b)B)$ and $D = A - (a + b)B$.

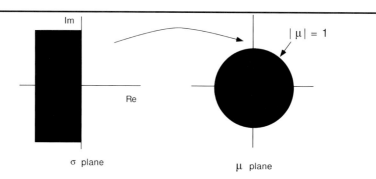

Figure 4.13.   Sketch of the complex mapping (4.49) for $a = 1$ and $b = 0$.

By the mapping (4.49), the most dangerous modes (those closest to the imaginary axis) of the original problem (4.47) are mapped onto the most dominant modes (those with largest norm) of the problem (4.50). This can be seen by writing $\sigma - b = x + iy$ and $\mu = p + iq$ which gives the relations

$$p = \frac{a^2 - x^2 - y^2}{(x - a)^2 + y^2} \; ; \; q = \frac{2ay}{(x - a)^2 + y^2} \qquad (4.51)$$

On the problem (4.50), the Simultaneous Iteration Techniques (SIT) described by Steward and Jennings (1981) is applied. With this method a few, say m, eigenvalues with largest norm are determined. The method proceeds in two steps:

(i) **The filtering step**

First step is to start with m initial vectors $\mathbf{u}_j, j = 1, \ldots, m$; the notation

$$U^0 = [\mathbf{u}_1, \ldots, \mathbf{u}_m] \qquad (4.52)$$

is used for the $N \times m$ matrix U^0. During the filtering stage of the SIT, for $l = 1, \cdots, L$ the product

$$U^l = EU^{l-1} \qquad (4.53)$$

is computed by solving the linear systems (note that $E = D^{-1}C$)

$$DU^l = CU^{l-1} \qquad (4.54)$$

Let Λ denote the diagonal matrix with the eigenvalues ordered according to their norm and Q the corresponding eigenvectors, i.e.,

$$\Lambda = \begin{pmatrix} \mu_1 & 0 & 0 \\ 0 & \dots & 0 \\ 0 & 0 & \mu_N \end{pmatrix} ; Q = [\mathbf{q}_1, \dots, \mathbf{q}_N] \qquad (4.55)$$

with $E\mathbf{q}_i = \mu_i \mathbf{q}_i$. Using the $m \times m$ matrix C_a and the $(N-m) \times m$ matrix C_b, one can decompose

$$U = U^0 = Q_a C_a + Q_b C_b \qquad (4.56)$$

where the $N \times m$ matrix Q_a contains the first m columns of Q and Q_b the remainder of the columns of Q. After one step (4.54) we obtain

$$VU^1 = E(Q_a C_a + Q_b C_b) = Q_a \Lambda_a C_a + Q_b \Lambda_b C_b \qquad (4.57)$$

where Λ_a is the $N \times m$ matrix of the first m columns of Λ and Λ_b consists of the remaining columns. Hence, after each filtering step the components of the starting vectors in the direction of the dominant eigenvectors increase most in amplitude. As already mentioned, infinite (numerically very large) eigenvalues can occur in generalized eigenvalue problems. Hence, during the filtering stage directions into these (undesired) eigenvectors gain in amplitude. One way to avoid this problem is to apply inverse iteration (see Technical box 4.1) on the matrix E^{-1} since singular vectors of E^{-1} correspond to infinite eigenvalues of E.

After a certain number of filtering steps L, usually 5 to 10, a reorientation step is performed to obtain a better approximation to the eigensolution of (4.50).

(ii) **The reorientation step**

First, the $m \times m$ matrices

$$F = U^T U \; ; \; G = U^T V \qquad (4.58)$$

where $U = U^{L-1}$ and $V = EU$ are computed. Subsequently, the matrix H is solved from

$$FH = G \rightarrow H = F^{-1}G \qquad (4.59)$$

For simplicity, consider the case $L = 1$. The eigenvalues of the matrix H are an approximation of the eigenvalues of E, because

$$\begin{aligned} FH &= U^T U H = U^T[Q_a C_a + Q_b C_b]H \\ G &= U^T V = U^T(Q_a \Lambda_a C_a + Q_b \Lambda_b C_b) \end{aligned} \qquad (4.60)$$

As the matrix $U^T Q_a$ is nonsingular, we obtain from (4.59), when we omit the part associated with Λ_b from (4.60), a low dimensional $m \times m$ eigenvalue problem from which approximations to the first m eigenvalues can be computed, i.e.,

$$C_a H = \Lambda_a C_a \qquad (4.61)$$

In applications, the eigenvalue problem

$$HP = P\Lambda_a \rightarrow P = C_a^{-1} \tag{4.62}$$

is solved and a next approximation to the eigenvectors is then found from the first m columns of the matrix W given by

$$W = VP = Q_a\Lambda_a + Q_b\Lambda_b C_b C_a^{-1} \tag{4.63}$$

The accuracy of the approximation of the eigensolution after reorientation k is determined by substituting the eigenvalues and eigenvectors back into the original eigenvalue problem (4.47), computing the residue and dividing it by the L_2 norm of the corresponding eigenvector. The stopping criterion is that this residue is smaller than some prescribed accuracy. In most cases, this process converges within a few (about 10) reorientations. When converged, the eigenvalues of the original problem are computed by the complex mapping (4.49).

4.4.2. The Jacobi-Davidson QZ-method

With the Jacobi-Davidson QZ-method (JDQZ) method, one can compute several, say m, eigenvalues and (optionally) eigenvectors of the generalized eigenvalue problem

$$\beta A\mathbf{q} = \alpha B\mathbf{q} \tag{4.64}$$

near a specified target $\tau \in \mathbb{C}$. Here A, B can be matrices with complex entries and α and β are complex numbers and the pair (α, β) is called an eigenvalue with corresponding eigenvector $\mathbf{q}$. The method is described in detail elsewhere (Sleijpen and Van der Vorst, 1996; Van Dorsselaer, 1997) but in a very general way, which makes it hard to understand because of the technicalities. Instead, here an attempt is made to describe the method in a simpler way, by going in detail through the first two steps of the algorithm.

 Technical box 4.3:
Schur decompositions of a
matrix

The Schur decomposition of a matrix is one of the many decompositions known in linear algebra. Let A be an $n \times n$ real matrix, then there exists an orthogonal $n \times n$ matrix Q (with $Q^T Q = QQ^T = I$) such that

$$Q^T AQ = R = \begin{pmatrix} R_{11} & R_{12} & \cdot & R_{1m} \\ 0 & R_{22} & \cdot & R_{2m} \\ \vdots & \vdots & \ddots & \vdots \\ 0 & 0 & \cdot & R_{mm} \end{pmatrix}$$

where each R_{ii} is either a 1×1 matrix or a 2×2 matrix, the latter having complex conjugate eigenvalues. This result shows that any real matrix is orthogonally similar to an upper quasi-triangular matrix. There is also a complex form of the Schur decomposition (see Golub and Van Loan (1983), p192). A partial generalized Schur decomposition of a matrix pensil (A, B) is given by

$$AQ = ZS \; ; \; BQ = ZT$$

where Q and Z are $N \times m$ orthogonal matrices and S and T are $m \times m$ upper triangular matrices.

Suppose we want to compute only specific eigenvalues near the imaginary axis, and hence we set the target $\tau = 0$; in the general method, τ can be chosen as desired. The JDQZ method then proceeds in two steps:

(i) **The Jacobi-Davidson step**

As a first step in the JDQZ method, a decomposition of the matrix pensil (A, B) into a partial generalized Schur form (see Technical box 4.3) is computed. These computations are outlined below and the end result is

$$AQ_1 = Z_1 S_1 \; ; \; BQ_1 = Z_1 T_1 \tag{4.65}$$

where Q_1 and Z_1 are normalized $N \times 1$ matrices (in this case vectors) and S_1 and T_1 are 1×1 matrices (in this case scalars). We can therefore write $Q_1 = \mathbf{q}_1$ and $Z_1 = \mathbf{z}_1$. The vector $\mathbf{q}_1$ is an eigenvector of the problem (4.64) with eigenvalue $\alpha/\beta = S_1/T_1$, since

$$\beta A \mathbf{q}_1 = \beta \mathbf{z}_1 S_1 = \mathbf{z}_1 \alpha T_1 = \alpha B \mathbf{q}_1 \tag{4.66}$$

The vector $\mathbf{q}_1$ is called the first generalized Schur vector and it is computed with the so-called Jacobi-Davidson method, as follows.

Let $\mathbf{v}_1$ be an initial guess for the vector $\mathbf{q}_1$, then we compute $\mathbf{w}_1 = A\mathbf{v}_1$ and scale both $\mathbf{v}_1$ and $\mathbf{w}_1$ such that their norm is unity. These vectors define the initial search space $V = [\mathbf{v}_1]$ and the initial test space $W = [\mathbf{w}_1]$. A new approximation of the eigenvector $\tilde{\mathbf{q}}_1 = u\mathbf{v}_1$ and of the eigenvalue $(\tilde{\alpha}, \tilde{\beta})$ are found through the projected eigenvalue problem

$$\tilde{\beta} \mathbf{w}_1^* A \mathbf{v}_1 u = \tilde{\alpha} \mathbf{w}_1^* B \mathbf{v}_1 u \tag{4.67}$$

where $\mathbf{w}_1^*$ is the adjoint (transposed and complex conjugated) vector of $\mathbf{w}_1$. This is a scalar eigenvalue problem which is easily solved. The eigenvalue $(\tilde{\alpha}, \tilde{\beta})$ closest to the target $\tau = 0$ is selected (with 'eigenvector' u) and $\tilde{\mathbf{q}}_1 = u\mathbf{v}_1$ is the first approximate to the eigenvector $\mathbf{q}_1$. The residual $\mathbf{r} = \tilde{\beta} A \tilde{\mathbf{q}}_1 - \tilde{\alpha} B \tilde{\mathbf{q}}_1$ is computed next.

In general, $\|\mathbf{r}\|$ will be larger than a given tolerance and the spaces V and W have to be extended to get sufficiently accurate approximations to $\mathbf{q}_1$. Thereto, first the vector $\tilde{\mathbf{z}}_1 = A\tilde{\mathbf{q}}_1$ is computed and normalized. Next, the vector $\mathbf{v}_2$ is chosen such that

$$(I - \tilde{\mathbf{z}}_1\tilde{\mathbf{z}}_1^*)(\tilde{\beta}A - \tilde{\alpha}B)(I - \tilde{\mathbf{q}}_1\tilde{\mathbf{q}}_1^*)\mathbf{v}_2 = -\mathbf{r} \tag{4.68}$$

where I is the $N \times N$ identity matrix. Note that always $(I - \mathbf{x}\mathbf{x}^*)\mathbf{x} = 0$ for every normalized vector $\mathbf{x}$ and hence $\mathbf{v}_2$ is orthogonal to $\tilde{\mathbf{q}}_1$. This is a linear system of equations which has to be solved for the vector $\mathbf{v}_2$. Once done, this vector is orthogonalized with respect to $\mathbf{v}_1$ and added to the basis V. Next, the space W is extended with the vector $\mathbf{w}_2 = A\mathbf{v}_2$, orthogonalized with respect to $\mathbf{w}_1$.

With the new basis $V = [\mathbf{v}_1, \mathbf{v}_2]$ and $W = [\mathbf{w}_1, \mathbf{w}_2]$ the projected eigenvalue problem now becomes

$$\tilde{\beta}W^*AV\mathbf{u} = \tilde{\alpha}W^*BV\mathbf{u} \tag{4.69}$$

with $\mathbf{u}$ now being a 2×1 vector. The traditional QZ method (Golub and Van Loan, 1983) can be used to solve this eigenvalue problem and again the eigensolution with the eigenvalue $(\tilde{\alpha}, \tilde{\beta})$ closest to the target τ is chosen and a next approximation $\tilde{\mathbf{q}}_1 = V\mathbf{u}$ follows. One can now imagine how the basis is extended each step, the projected eigenvalue problem is solved until $\|\mathbf{r}\|$ is small enough and the final eigenvector $\mathbf{q}_1$ is determined. Note that when $Q_1 = \mathbf{q}_1$ is determined, also $Z_1 = \mathbf{z}_1$ can be directly solved from (4.65).

(ii) **The extension step**

Once the first Schur vector has been obtained, the next problem is to compute the extended partial generalized Schur form

$$AQ_2 = Z_2S_2 \, ; \; BQ_2 = Z_2T_2 \tag{4.70}$$

where $Q_2 = [\mathbf{q}_1, \mathbf{q}_2]$ and $Z_2 = [\mathbf{z}_1, \mathbf{z}_2]$ are normalized $N \times 2$ matrices and S_2, T_2 are yet unknown 2×2 matrices. When the small eigenvalue problem $\beta S_2\mathbf{x} = \alpha T_2\mathbf{x}$ is considered and the (two) eigenvalues (α, β) are determined, then the (first two) eigenvectors of (4.64) are found from $Q_2\mathbf{x}$ since

$$\beta AQ_2\mathbf{x} = \beta Z_2S_2\mathbf{x} = Z_2\alpha T_2\mathbf{x} = \beta BQ_2\mathbf{x} \tag{4.71}$$

As one eigenvector $Q_1 = \mathbf{q}_1$ is already known, the next step is to determine the vector $\mathbf{q}_2$. Write (4.70) as

$$A(Q_1 \quad \mathbf{q}_2) = (Z_1 \quad \mathbf{z}_2)\begin{pmatrix} S_1 & s \\ 0 & \alpha \end{pmatrix} \tag{4.72a}$$

$$B(Q_1 \quad \mathbf{q}_2) = (Z_1 \quad \mathbf{z}_2)\begin{pmatrix} T_1 & t \\ 0 & \beta \end{pmatrix} \tag{4.72b}$$

Eliminating Q_1 leads to

$$\beta A\mathbf{q}_2 - \alpha B\mathbf{q}_2 = Z_1(\beta s - \alpha t)$$

Applying the operator $(I - Z_1 Z_1^*)$ to both sides, it can be shown that the vector $\mathbf{q}_2$, which is made orthogonal to $\mathbf{q}_1$, can be obtained from the generalized eigenvalue problem

$$(I - Z_1 Z_1^*)(\beta A - \alpha B)(I - Q_1 Q_1^*)\mathbf{q}_2 = 0 \qquad (4.73)$$

Again, this is a similar problem as before (for $\mathbf{q}_1$) and it can again be solved by the Jacobi-Davidson method. In this way, two eigenvalues (one related to $\mathbf{q}_1$ and the other to $\mathbf{q}_2$) near the specified target are found. Hence, in the method, the matrices T_2 and S_2 are actually not explicitly computed. One can now imagine how the next partial Schur decomposition is constructed and how finally m eigenvalues close to the specified target τ are computed.

The software for the JDQZ method is available through *http://www.math.uu.nl/~people/bomhof/jd.html*. A short (but clear) manual is available with the code. This manual is needed, because there are several technical details involved and several choices have to be made for certain parameters, such as the target τ and the dimensions of the search and test spaces.

4.5. Implicit Time Integration

In many ocean models, there is an explicit time marching procedure, which can be represented by, using (4.22),

$$\mathcal{M}_N \mathbf{x}^{n+1} = \mathcal{M}_N \mathbf{x}^n + \Delta t \, \mathcal{G}(\mathbf{x}^n) \qquad (4.74)$$

where $\mathcal{G}_N = \mathcal{F}_N - (\mathcal{L}_N + \mathcal{N}_N)$. Explicit schemes allow relatively easy implementation of all kinds of physical processes and details of boundary conditions, but suffer from a substantial drawback. The time step is limited because of numerical amplification of truncation errors (through well-known stability criteria) rather than by the changes in the actual solution (Roache, 1976). This limitation is even more restrictive as the spatial resolution increases. These properties are extremely undesirable, for example, in model studies of changes in the thermohaline circulation where integration times of at least a few thousand years are desired.

A nice spin-off of continuation methods is the immediate availability of implicit time integration schemes. Using a time step Δt, and a time index n, this scheme becomes for $\omega \in [0, 1]$,

$$\mathcal{M}_N \frac{\mathbf{x}^{n+1} - \mathbf{x}^n}{\Delta t} + (1 - \omega)\mathcal{F}(\mathbf{x}^n) + \omega\mathcal{F}(\mathbf{x}^{n+1}) = 0 \qquad (4.75)$$

For $\omega = 1/2$ and $\omega = 1$, these are the Crank-Nicholson method and backward Euler method, respectively (Atkinson, 1976).

The equations for $\mathbf{x}^{n+1}$ are solved by the Newton-Raphson technique and lead to the same type of numerical problems as that for the steady state computation. It is well-known that the second-order Crank-Nicholson scheme is unconditionally stable for linear equations. This does not mean that one can take any time step, as this quantity is still limited by two factors. One of these factors is accuracy:

although the scheme is second-order accurate in time, large discretization errors occur when too large time steps are used. A second limitation on the time step is the convergence domain of the Newton-Raphson process, which does not necessarily converge for every chosen time step. For many applications, however, much larger time steps (up to a factor 100 or 1000) can be taken than in similar explicit models.

4.6. Linear System Solvers: Direct Methods

As has become clear from the previous sections, the basic problem to overcome in the computation of the steady states for large-dimensional dynamical systems — as well as in the linear stability problem — is the solution of large linear systems of equations. Depending on the sparsity of the matrices involved, different methods are available to solve this basic problem of linear algebra efficiently. In this and the next section, such an $N \times N$ matrix is indicated by A and the right hand side of the algebraic system of equations is indicated by $\mathbf{b}$. Hence, the problem is to solve the vector $\mathbf{x}$ from the equations

$$A\mathbf{x} = \mathbf{b} \qquad (4.76)$$

Two types of methods can be distinguished: direct methods and iterative methods. If most of the coefficients of the matrix A are nonzero, the matrix is referred to as a dense matrix. In this case, memory constraints will soon limit the dimension N of the system of equations which can be solved, since the whole matrix has to be stored. For sparse matrices, where most of the matrix elements are zero, systems of much higher dimension may be solved and several methods are available to perform this efficiently. The most common direct methods will be described in this section while the iterative methods are presented in section 4.7.

4.6.1. Basic principle

The basic method which probably everyone has learned in high school is based on successive elimination of unknowns and is called Gaussian elimination. The idea is to reduce the system $A\mathbf{x} = \mathbf{b}$ to an equivalent system $U\mathbf{x} = \mathbf{g}$, where U is an upper triangular matrix. The algorithm consists of the following steps; it is illustrated by the 3×3 problem

$$\begin{pmatrix} a_{11} & a_{12} & a_{13} \\ a_{21} & a_{22} & a_{23} \\ a_{31} & a_{32} & a_{33} \end{pmatrix} \begin{pmatrix} x_1 \\ x_2 \\ x_3 \end{pmatrix} = \begin{pmatrix} b_1 \\ b_2 \\ b_3 \end{pmatrix} \qquad (4.77)$$

which is indicated by by $A^{(1)}\mathbf{x} = \mathbf{b}^{(1)}$.

Assume $a_{11}^{(1)} \neq 0$ and define the row multipliers

$$m_{21} = \frac{a_{21}^{(1)}}{a_{11}^{(1)}} \; ; \; m_{31} = \frac{a_{31}^{(1)}}{a_{11}^{(1)}} \qquad (4.78)$$

These multipliers are used to eliminate the x_1 coefficients in the rows 2 and 3, by defining

$$a_{ij}^{(2)} = a_{ij}^{(1)} - m_{i1}a_{1j}^{(1)}, \quad i = 2, 3 \; ; \; j = 1, 2, 3$$
$$b_i^{(2)} = b_i^{(1)} - m_{i1}b_1^{(1)}, \quad i = 2, 3$$

At the end of this step, the following system results

$$
\begin{pmatrix}
a_{11}^{(1)} & a_{12}^{(1)} & a_{13}^{(1)} \\
0 & a_{22}^{(2)} & a_{23}^{(2)} \\
0 & a_{32}^{(2)} & a_{33}^{(2)}
\end{pmatrix}
\begin{pmatrix}
x_1 \\
x_2 \\
x_3
\end{pmatrix}
=
\begin{pmatrix}
b_1^{(1)} \\
b_2^{(2)} \\
b_3^{(2)}
\end{pmatrix}
\tag{4.80}
$$

This step is repeated once more. Assume $a_{22}^{(2)} \neq 0$, define the multiplier m_{32} as below and eliminate the coefficient of x_2 in the third row through

$$m_{32} = \frac{a_{32}^{(2)}}{a_{22}^{(2)}}$$
$$a_{ij}^{(3)} = a_{ij}^{(2)} - m_{32}a_{2,j}^{(2)}, \quad i = 3 \; ; \; j = 2, 3 \tag{4.81}$$
$$b_3^{(3)} = b_3^{(2)} - m_{32}b_2^{(2)}$$

to give the problem

$$
\begin{pmatrix}
a_{11}^{(1)} & a_{12}^{(1)} & a_{13}^{(1)} \\
0 & a_{22}^{(2)} & a_{23}^{(2)} \\
0 & 0 & a_{33}^{(3)}
\end{pmatrix}
\begin{pmatrix}
x_1 \\
x_2 \\
x_3
\end{pmatrix}
=
\begin{pmatrix}
b_1^{(1)} \\
b_2^{(2)} \\
b_3^{(3)}
\end{pmatrix}
\tag{4.82}
$$

which is finally of the form $U\mathbf{x} = \mathbf{g}$, where U is upper triangular. The latter system is easy to solve by back substitution, since

$$x_3 = \frac{b_3^{(3)}}{a_{33}^{(3)}}$$

$$x_2 = \frac{b_2^{(2)} - a_{23}^{(2)}x_3}{a_{22}^{(2)}}$$

$$x_1 = \frac{b_1^{(1)} - a_{12}^{(1)}x_2 - a_{13}^{(1)}x_3}{a_{11}^{(1)}}$$

which completes the Gaussian elimination process.

If we collect the multipliers $m_{i,j}$ in a lower triangular matrix L, then one can easily verify that

$$
L\,U =
\begin{pmatrix}
1 & 0 & 0 \\
m_{21} & 1 & 0 \\
m_{31} & m_{32} & 1
\end{pmatrix}
\begin{pmatrix}
a_{11}^{(1)} & a_{12}^{(1)} & a_{13}^{(1)} \\
0 & a_{22}^{(2)} & a_{23}^{(2)} \\
0 & 0 & a_{33}^{(3)}
\end{pmatrix}
= A
\tag{4.84}
$$

and hence the elimination process has introduced a decomposition of the original matrix A. From this decomposition, the determinant of the matrix A, $\det(A)$, can be computed as

$$\det(A) = \det(L)\det(U) = \det(U) = a_{11}^{(1)}a_{22}^{(2)}a_{33}^{(3)} \tag{4.85}$$

At each stage of the elimination process, it is assumed that the diagonal element $a_{kk}^{(k)} \neq 0$. In practice, such an element may not be zero exactly but it may be very small. In these cases, problems may arise because of amplification of round-off errors and the accuracy of the final solution is deteriorated.

4.6.2. Pivoting

The solution to the problem mentioned at the end of the previous section is a pivoting strategy, which can be either partial (only rows are interchanged) or complete (rows and columns are interchanged). The partial pivoting strategy is illustrated with the example in the previous section. When $a_{22}^{(2)} = 0$, m_{32} in (4.81) cannot be defined. In the second column, one searches now for the largest element in the remaining rows to be eliminated (in this case only row 3). Assume that $a_{32}^{(2)} \neq 0$, then row 2 and row 3 are interchanged. This gives the system

$$\begin{pmatrix} a_{11}^{(1)} & a_{12}^{(1)} & a_{13}^{(1)} \\ 0 & a_{32}^{(2)} & a_{33}^{(2)} \\ 0 & a_{22}^{(2)} & a_{23}^{(2)} \end{pmatrix} \begin{pmatrix} x_1 \\ x_3 \\ x_2 \end{pmatrix} = \begin{pmatrix} b_1^{(1)} \\ b_3^{(2)} \\ b_2^{(2)} \end{pmatrix} \tag{4.86}$$

On this system, the last elimination step can be performed in the same way as above and the system can be solved. The decomposition is now changed into $LU = PA$, where P is a permutation matrix monitoring the row interchanges, i.e., in the example above

$$P = \begin{pmatrix} 1 & 0 & 0 \\ 0 & 0 & 1 \\ 0 & 1 & 0 \end{pmatrix} \tag{4.87}$$

In this elementary form the algorithm is, for example, presented in Atkinson (1976).

With complete pivoting, one searches also the other columns for largest elements and interchanges rows and columns to put that element on the diagonal. In practise, only partial pivoting is used, because it appears just as good as complete pivoting and it is cheaper. Many variants of the code are available as FORTRAN90 routines (for example in the NAG and/or IMSL libraries).

The classical Gaussian elimination technique is quite expensive and its memory usage and CPU time increase with N^3, with N being the dimension of the matrix. There are special cases where more efficient methods are possible. When the matrix has a band structure, such as often arises from discretization of partial differential equations, the computational costs and the amount of memory needed can be decreased substantially. For an overview of alternative direct methods, for example frontal methods, see Duff *et al.* (1986).

4.7. Linear System Solvers: Iterative Methods

Iterative methods are techniques which allow the solution of very large systems of sparse linear equations, such as those arising from the discretization of partial differential equations. The disadvantage of these techniques is that they are not robust and their performance is very application dependent. In this section, the basics of these iterative techniques is given which will give the reader an introduction into the literature on the subject.

4.7.1. Relaxation methods

The class of most simple methods are those for which iteratively one or more components are modified with the aim to annihilate components of the residual vector $\mathbf{r} = \mathbf{b} - A\mathbf{x}$. Nearly all these methods are based on a decomposition of the matrix of the form

$$A = D - E - F \tag{4.88}$$

where D is the diagonal of A, $-E$ its strict lower part and $-F$ its strict upper part. All methods start with an initial vector $\mathbf{x}^0$ and differ in the rules for annihilation of components of the residual vector. If we indicate the different iterates of the solution vector by $\mathbf{x}^k$, then the most important methods are given by

▪ Jacobi

$$D\mathbf{x}^{k+1} = (E + F)\mathbf{x}^k + \mathbf{b} \tag{4.89}$$

▪ Gauss-Seidel

$$(D - E)\mathbf{x}^{k+1} = F\mathbf{x}^k + \mathbf{b} \tag{4.90}$$

▪ Successive overrelaxation

$$(D - \omega E)\mathbf{x}^{k+1} = (\omega F + (1 - \omega))\mathbf{x}^k + \omega\mathbf{b} \tag{4.91}$$

In this procedure, ω is called a relaxation factor.

All these methods have a nice geometrical representation, which is best illustrated for the case $N = 2$. In this case, the intersection point of two lines in the plane[1] is sought as sketched in Fig. 4.14. An illustrative example are the equations

$$3x + y = 4 \tag{4.92a}$$
$$x + 2y = 3 \tag{4.92b}$$

where the iterative process is started from the initial condition $(3, 4)$. When starting from an initial point (x^0, y^0), the Gauss-Seidel method moves the iterate

---

[1]When $N = 3$, the coefficients in the set of equations represents the normals of three planes of which the intersection point is sought.

(x^k, y^k) into directions parallel to the coordinate axes. This is done by determining the intersections with lines parallel to coordinate axes and the actual lines defined by the set of equations. In Fig. 4.14a and Fig. 4.14b, the same system of equations is solved with matrices

$$A_1 = \begin{pmatrix} 3 & 1 \\ 1 & 2 \end{pmatrix} \text{ and } A_2 = \begin{pmatrix} 1 & 2 \\ 3 & 1 \end{pmatrix}$$

in the Gauss-Seidel method, respectively. The way in which the equations are ordered is important for the Gauss-Seidel method since for A_1 the method is convergent, while for A_2 the iterates diverge.

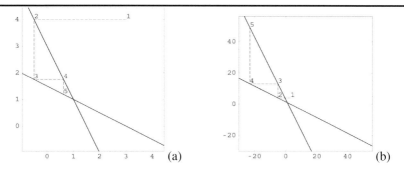

Figure 4.14.   Sketch of the (a) convergence and (b) divergence of the Gauss-Seidel method for a simple two-dimensional system of equations (4.92). Iterates are numbered and in (a), the matrix A_1 is used, whereas in (b) the matrix A_2 is used.

As is seen from the list of methods (4.89), (4.90) and (4.91), they all are of the form

$$\mathbf{x}^{k+1} = G\mathbf{x}^k + f \tag{4.93}$$

with different iteration matrices G. The performance of these methods depends on the properties of G and in particular on the spectral radius $\rho(G)$, which is the eigenvalue of G with maximum norm. Hence, if the set of eigenvalues of G is indicated by $\sigma(G)$, then

$$\rho(G) = \max_{\lambda \in \sigma(G)} |\lambda| \tag{4.94}$$

and a method is guaranteed to converge when $\rho(G) < 1$. For many problems this is the case, but convergence may be very slow because $\rho(G) \approx 1$.

4.7.2. Projection techniques

The general idea behind the projection methods is to obtain an approximation of the solution $\mathbf{x}$ of the system of n linear equations $A\mathbf{x} = \mathbf{b}$ in a relatively small dimensional subspace $K \subset \mathbb{R}^n$. If m is the dimension of K, then m constraints (defining equations) are necessary to generate this space. The residual is thereto required to be orthogonal to m linear independent vectors; the latter vectors define also a m-dimensional subspace L. With a starting vector $\mathbf{x}^0$ and hence an initial residue $\mathbf{r}^0 = \mathbf{b} - A\mathbf{x}^0$, the problem is formulated as (Fig. 4.15): find $\mathbf{x} = \mathbf{x}^0 + \mathbf{c}$, $\mathbf{c} \in K$ such that

$$< \mathbf{b} - A\mathbf{x}, \mathbf{w} > = < \mathbf{r}^0 - A\mathbf{c}, \mathbf{w} > = 0 \; , \; \forall \, \mathbf{w} \in L \qquad (4.95)$$

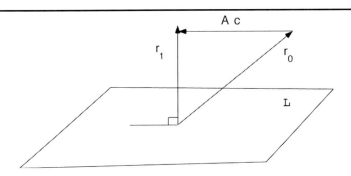

Figure 4.15.   Sketch of the geometry of a general projection method; here $\mathbf{r}_1 = \mathbf{r}^0 - A\mathbf{c}$.

Once the spaces K and L are chosen, the approximation $\mathbf{c}$ can be found in the following way (for simplicity, the case $m = 2$ is described). Let $V = [\mathbf{v}_1, \mathbf{v}_2]$ be a basis of K and $W = [\mathbf{w}_1, \mathbf{w}_2]$ a basis of L, where V and W are $N \times 2$ matrices. Then with $\mathbf{y} = (y_1, y_2) \in \mathbb{R}^2$, the representation of $\mathbf{c}$ is given by

$$\mathbf{c} = y_1 \mathbf{v}_1 + y_2 \mathbf{v}_2 = V\mathbf{y} \qquad (4.96)$$

and the orthogonality constraints (4.95) become

$$< \mathbf{r}^0 - y_1 A\mathbf{v}_1 - y_2 A\mathbf{v}_2, \mathbf{w}_1 > = 0 \qquad (4.97a)$$
$$< \mathbf{r}^0 - y_1 A\mathbf{v}_1 - y_2 A\mathbf{v}_2, \mathbf{w}_2 > = 0 \qquad (4.97b)$$

which can be rewritten in matrix form as

$$W^T A V \, \mathbf{y} = W^T \, \mathbf{r}^0 \qquad (4.98)$$

where W^T is the $2 \times N$ transpose matrix of W. The matrix $W^T A V$ is only 2×2 and can be easily inverted when it is nonsingular. Hence the solution $\mathbf{c}$ is found as

$$\mathbf{y} = (W^T A V)^{-1} W^T \, \mathbf{r}^0 \; ; \; \mathbf{c} = V\mathbf{y} \qquad (4.99)$$

In the above projection method, the subspaces still have an arbitrary form. Efficient techniques are obtained if one chooses a Krylov subspace $K_m(A, v)$ for the m-dimensional space K. Such a space is of the form

$$K_m(A, \mathbf{v}) = [\mathbf{v}, A\mathbf{v}, A^2\mathbf{v}, \cdots, A^{m-1}\mathbf{v}] \qquad (4.100)$$

Particular projection methods arise by the choice $K = L = K_m(A, \mathbf{r}^0)$ in which simultaneously the Krylov space is built up during the iteration process. Although there is a multitude of these methods (Saad, 1996), we will focus on only two of these methods, i.e., GMRES and BICGSTAB.

4.7.2.1 The GMRES technique

In this method, one successively builds a basis of the Krylov subspace $K_m(A, \mathbf{r}^0)$ and simultaneously solves a small system of equations to find an approximation of the solution $\mathbf{x}$ within this subspace. The mathematics of this method will be described below, again for a small value of $m = 2$.

(i) **Subspace generation**

Using the starting vector $\mathbf{x}^0$, the residual vector $\mathbf{r}^0$ is computed and its L_2-norm is indicated by $\beta = \|\mathbf{r}^0\|_2$. The basic vector of the Krylov subspace is

$$\mathbf{v}_1 = \frac{\mathbf{r}^0}{\beta}$$

which has unit L_2 norm. The next vector $\mathbf{v}_2$ of the subspace is determined through the so-called Arnoldi procedure as follows

$$
\begin{aligned}
h_{11} &= <\mathbf{v}_1, A\mathbf{v}_1> \\
\mathbf{w}_1 &= A\mathbf{v}_1 - h_{11}\mathbf{v}_1 \quad\quad\quad (4.101\text{a}) \\
h_{21} &= \|\mathbf{w}_1\|_2 \\
\mathbf{v}_2 &= \mathbf{w}_1/h_{21} \quad\quad\quad\quad\quad (4.101\text{b})
\end{aligned}
$$

Indeed, the vector $\mathbf{v}_2$ increases the span of the subspace in the direction of the vector $A\mathbf{v}_1$ and moreover the vectors $\mathbf{v}_1$ and $\mathbf{v}_2$ are orthogonal, which is easily seen through

$$<\mathbf{v}_1, \mathbf{v}_2> = <\mathbf{v}_1, \frac{1}{h_{21}}(A\mathbf{v}_1 - h_{11}\mathbf{v}_1)> = 0 \quad\quad (4.102)$$

It is illustrative to go through another Arnoldi step to construct $\mathbf{v}_3$

$$
\begin{aligned}
h_{12} &= <\mathbf{v}_1, A\mathbf{v}_2> \quad ; \quad h_{22} = <\mathbf{v}_2, A\mathbf{v}_2> \\
\mathbf{w}_2 &= A\mathbf{v}_2 - h_{12}\mathbf{v}_1 - h_{22}\mathbf{v}_2 \quad (4.103\text{a}) \\
h_{32} &= \|\mathbf{w}_2\|_2 \\
\mathbf{v}_3 &= \frac{\mathbf{w}_2}{h_{32}} \quad\quad\quad\quad\quad\quad (4.103\text{b})
\end{aligned}
$$

which again increases the span of the subspace in the direction of $A^2\mathbf{v}_1$ and $\mathbf{v}_3$ is orthogonal to both $\mathbf{v}_1$ and $\mathbf{v}_2$ and has unit norm. It is now clear how the following vector $\mathbf{v}_4$ of the Krylov subspace is constructed.

(ii) **The projected problem**

The equations (4.101) and (4.103) can be rewritten as

$$A\mathbf{v}_1 = h_{11}\mathbf{v}_1 + \mathbf{w}_1 = h_{11}\mathbf{v}_1 + h_{21}\mathbf{v}_2 \qquad (4.104a)$$
$$A\mathbf{v}_2 = h_{12}\mathbf{v}_1 + h_{22}\mathbf{v}_2 + h_{32}\mathbf{v}_3 \qquad (4.104b)$$

If H_2 denotes the 3×2 matrix of the coefficients h_{ij} (with $h_{31} = 0$) then the matrix form of the equations above is

$$AV_2 = (A\mathbf{v}_1 \quad A\mathbf{v}_2) = (\mathbf{v}_1 \quad \mathbf{v}_2 \quad \mathbf{v}_3)\, H_2 = V_3 H_2 \qquad (4.105)$$

In GMRES, the L_2−norm of the residue $\|\mathbf{b} - A\mathbf{x}\|_2$ is minimized. With $\mathbf{x} = \mathbf{x}^0 + V_2\mathbf{y}$ this can be written as

$$\begin{aligned}
\|\mathbf{b} - A\mathbf{x}\|_2 &= \|\mathbf{r}^0 - AV_2\mathbf{y}\|_2 = \\
\|\beta\mathbf{v}_1 - V_3 H_2\mathbf{y}\|_2 &= \|V_3(\beta\mathbf{e}_1 - H_2\mathbf{y}\|_2 = \\
\|\beta\mathbf{e}_1 &- H_2\mathbf{y}\|_2
\end{aligned} \qquad (4.106)$$

where the results above are used together with the fact that the matrix V_3 is orthogonal (its norm is unity) and $\mathbf{e}_1 = (1, 0, 0)$. The minimization of the norm (4.106) can be easily solved for $\mathbf{y}$ by least squares techniques since it is small dimensional.

One can now imagine, how the subspace is further extended and how successive approximations can be found through projection and solution of a system of equations of small dimension.

4.7.2.2 The BICGSTAB technique

In the previous method, an orthogonal basis of a Krylov subspace $K_m(A, \mathbf{v})$ was constructed. Other methods have been developed which rely on the construction on a bi-orthogonal basis of two subspaces $K_m(A, \mathbf{v}_1)$ and $K_m(A^T, \mathbf{w}_1)$. An approximation of the solution $\mathbf{x}$ is searched in the first subspace $K = K_m(A, \mathbf{v}_1)$, while being orthogonal to the space $L = K_m(A^T, \mathbf{w}_1)$. One of those techniques is the BICSTAB method, which is now briefly described (see Saad (1996) for details).

(i) **Construction of the subspaces**

Using a starting vector $\mathbf{x}^0$, the residual vector $\mathbf{r}^0$ is computed and again its L_2-norm is indicated by β. The basic vector of the Krylov subspace is the normalized vector

$$\mathbf{v}_1 = \frac{\mathbf{r}^0}{\beta}$$

For the testspace, one chooses $\mathbf{w}_1$ such that $< \mathbf{v}_1, \mathbf{w}_1 >= 1$. The bases of both spaces are extended by construction of vectors $\mathbf{v}_2$ and $\mathbf{w}_2$ through the so-called Lanczos algorithm as follows

$$\alpha_1 \;\; = \;\; < \mathbf{v}_1, A\mathbf{v}_1 >$$
$$\hat{\mathbf{v}}_2 = A\mathbf{v}_1 - \alpha_1 \mathbf{v}_1 \quad ; \quad \hat{\mathbf{w}}_2 = A^T \mathbf{w}_1 - \alpha_1 \mathbf{w}_1$$
$$\delta_2 = \| < \hat{\mathbf{v}}_2, \hat{\mathbf{w}}_2 > \|^{\frac{1}{2}} \quad ; \quad \beta_2 = \frac{< \hat{\mathbf{v}}_2, \hat{\mathbf{w}}_2 >}{\delta_2} \qquad (4.107)$$
$$\mathbf{w}_2 = \frac{\hat{\mathbf{w}}_2}{\beta_2} \quad ; \quad \mathbf{v}_2 = \frac{\hat{\mathbf{v}}_2}{\delta_2}$$

In this way, the vectors $\mathbf{v}_i$ and $\mathbf{w}_i$, $i = 1, 2$, are bi-orthogonal as can be easily checked from the expressions above, i.e.,

$$< \mathbf{v}_1, \mathbf{w}_1 >= 1, \;\; < \mathbf{v}_2, \mathbf{w}_1 >= 0, \;\; < \mathbf{v}_1, \mathbf{w}_2 >= 0, \;\; < \mathbf{v}_2, \mathbf{w}_2 >= 1$$
$$(4.108)$$

The next step proceeds in the same way: for example, with $\alpha_2 =< \mathbf{v}_2, A\mathbf{v}_2 >$, the new vector $\hat{\mathbf{v}}_3$ is defined as $\hat{\mathbf{v}}_3 = A\mathbf{v}_2 - \alpha_2 \mathbf{v}_2 - \beta_2 \mathbf{v}_1$.

(ii) **The projected problem**

In the same way as in the GMRES method, we write the equations for the basis vectors in matrix notation and find

$$W_2^T A V_2 = \begin{pmatrix} < \mathbf{w}_1, A\mathbf{v}_1 > & < \mathbf{w}_1, A\mathbf{v}_2 > \\ < \mathbf{w}_2, A\mathbf{v}_1 > & < \mathbf{w}_2, A\mathbf{v}_2 > \end{pmatrix} = \begin{pmatrix} \alpha_1 & \beta_2 \\ \delta_2 & \alpha_2 \end{pmatrix} = T_2$$
$$(4.109)$$

where for example the identities for $< \mathbf{w}_1, A\mathbf{v}_2 >$ and $< \mathbf{w}_2, A\mathbf{v}_1 >$ follow from

$$< \mathbf{w}_1, A\mathbf{v}_2 > \;\; = \;\; < \mathbf{w}_1, \hat{\mathbf{v}}_3 + \alpha_2 \mathbf{v}_2 + \beta_2 \mathbf{v}_1 >= \beta_2 \qquad (4.110a)$$
$$< \mathbf{w}_2, A\mathbf{v}_1 > \;\; = \;\; < \mathbf{w}_2, \hat{\mathbf{v}}_2 + \alpha_1 \mathbf{v}_1 >= \delta_2 \qquad (4.110b)$$

because of the bi-orthogonality of the bases. The projected problem now follows directly from the orthogonality constraints (4.95) which can be written, again with $\mathbf{x} = \mathbf{x}^0 + V_2 \mathbf{y}$ and $\mathbf{r}^0 = \beta \mathbf{v}_1$ as

$$W_2^T A V_2 \, \mathbf{y} = W^T \mathbf{r}^0 \;\; \Rightarrow \;\; T_2 \mathbf{y} = \beta \mathbf{e}_1 \qquad (4.111)$$

which is a small problem to solve. When an LU decomposition of T_2 is made, special advantages can be taken of the structure of the upper and lower triangular matrices to obtain an efficient algorithm; implementation details can be found in Saad (1996).

The software package SPARSKIT, which contains many of the iterative solvers in FORTRAN, can be obtained from http://www-users.cs.umn.edu/~saad/software/SPARSKIT/sparskit.html.   A   package containing C versions of the iterative methods is available through http://www.tu-dresden.de/mwism/skalicky/laspack/laspack.html. The iterative methods will not perform well on all systems of linear equations and are even not guaranteed to converge. This is contrary to direct methods, which always give an answer if the matrix is nonsingular; it may, however, be very (in most cases prohibitively!) expensive to get it. This has lead to the development of so-called preconditioning techniques (see Technical Box 4.4) for iterative solvers. Building an efficient preconditioner is usually the key to being able to solve giant dimensional sparse linear systems.

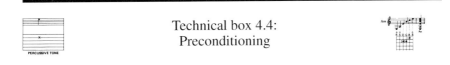

Technical box 4.4:
Preconditioning

The idea of preconditioning is that if $A\mathbf{x} = \mathbf{b}$ is difficult to solve — e.g., when A is approximately singular — then one may find a matrix M for which the system $M^{-1}A\mathbf{x} = M^{-1}\mathbf{b} = \mathbf{c}$ is more easy to solve. Geometrically, preconditioning has also a nice interpretation, best again illustrated in case $N = 2$.   If a matrix is

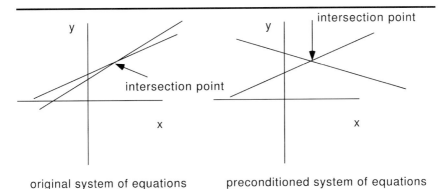

original system of equations   preconditioned system of equations

Figure 4.16.   Sketch of the effect of preconditioning in the case of $N = 2$.

nearly singular, this means that the two lines of which one wants to compute the intersection are nearly parallel. In the preconditioned system of equations, the two lines are more orthogonal (Fig. 4.16). There is a wealth of literature on these preconditioning techniques, which we cannot discuss here (but see Saad (1996)); we only sketch one approach.

For sparse matrices, a very effective way of obtaining a proper preconditioning matrix is to proceed with an LU decomposition, such as that constructed with a direct solver (section 4.6.1). To preserve the sparsity in the factors $\tilde{L}$ and $\hat{U}$ some or all elements causing fill-in additional to that of A are ignored. The matrix M is then chosen equal to $\tilde{L}\tilde{U}$. In this way, one obtains a splitting of A in which $R = A - \tilde{L}\tilde{U}$ is the so-called residual matrix. An iterative method, such as GMRES, is then applied to the system of equations

$$\tilde{U}^{-1}\tilde{L}^{-1}A\mathbf{x} = \tilde{U}^{-1}\tilde{L}^{-1}\mathbf{b} \qquad (4.112)$$

The way in which elements in the incomplete decomposition are ignored leads to many variants with names as ILU, NGILU and MRILU and more details of these methods are given in Van der Ploeg (1992), Saad (1996) and Botta and Wubs (1999). In the ILU technique, for example, an incomplete LU decomposition is used in which the sparsity pattern of $\tilde{L}$ and $\tilde{U}$ is based on a drop tolerance ε.

Also for the preconditioning techniques much software is (in most cases freely) available. For ILU-type methods, codes are available in SPARSKIT (http://www-users.cs.umn.edu/~saad/software/SPARSKIT/sparskit.html) and LASPACK (http://www.tu-dresden.de/mwism/skalicky/laspack/laspack.html). The MRILU method is available from http://www.math.rug.nl/~wubs.

4.8. Application to the Example Problem

In this section, a typical application of the methods above is presented for the Rayleigh-Bénard problem as described in section 4.2. All results below were computed with a version of the code BOOM, which has been developed in my group over the years. The BOOM (Dutch for 'tree' and abbreviation for Bifurcation Analysis ('Onderzoek' in Dutch) of Ocean Models) code combines the continuation method with a choice of eigenvalue solvers and iterative linear systems solvers. The user has to supply the discretized operators $\mathcal{L}_N$, $\mathcal{N}_N$, $\mathcal{M}_N$, $\mathcal{F}_N$ and the Jacobian matrix.

A starting point $(\mathbf{x}_0, \lambda_0)$ has to be prescribed and the number of eigenvalues m_e to compute within the linear stability analysis has to be chosen. The sequence of computations is the following:

1 Compute the tangent vector $(\dot{\mathbf{x}}_0, \dot{\lambda}_0)$, if necessary; sometimes it can be analytically determined.

2 Compute the Euler guess with chosen step length Δs,

$$\mathbf{x} = \mathbf{x}_0 + \Delta s\, \dot{\mathbf{x}}_0$$
$$\lambda = \lambda_0 + \Delta s\, \dot{\lambda}_0$$

3 Solve the system of nonlinear algebraic equations (4.24) and (4.30) using the Newton-Raphson method. Within each Newton iteration, one (or two) systems of linear equations have to be solved with a chosen method (direct, iterative).

4 When the previous step has converged, the generalized eigenvalue problem $A\mathbf{x} = \sigma B\mathbf{x}$ is solved for the first m_e eigenvalues closest to the imaginary axis using a chosen eigenvalue solver (SIT, JDQZ).

5 Compute a desired number of test functions to monitor properties of the flow and to monitor whether bifurcations have occurred (real part of eigenvalues, test functions τ_{pq} as in (4.40), determinant of Jacobian matrix). Take action, if a bifurcation point is detected, for example proceed with branch switching.

The two-dimensional case of the example problem is considered for a liquid with $Pr = 1$ which is heated from below in a container of aspect ratio $A = 10$ (Fig. 4.17). For water, with $\kappa = 10^{-7}$ [m²s⁻¹] and $\nu = 10^{-6}$ [m²s⁻¹], the Prandtl number is about 10. Results for this problem have been presented exten-

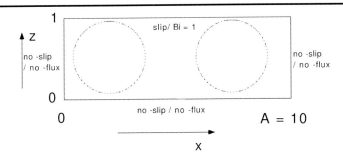

Figure 4.17.   Set-up of the two-dimensional configuration of the example problem.

sively in section 4 of Van Dorsselaer (1997) for the case $Bi \rightarrow \infty$ and no-slip conditions at all walls, using the original primitive-equation formulation with unknowns (u, w, p, T). In this paper, also results can be found on the performance of the iterative methods GMRES and BICGSTAB, either for the steady equations as well as within the JDQZ method. These results indicate that the methods used are indeed efficient for the example problem but since so many parameters are involved, they are not presented here; interested readers should consult Van Dorsselaer (1997).

For the two-dimensional case, a more efficient formulation of the example problem was used in Dijkstra *et al.* (1995). A streamfunction-vorticity formulation can be used, where the streamfunction ψ and the vertical component of the vorticity vector ζ are defined as

$$u = \frac{\partial \psi}{\partial z} \quad ; \quad w = -\frac{\partial \psi}{\partial x} \tag{4.114a}$$

$$\zeta = \frac{\partial w}{\partial x} - \frac{\partial u}{\partial z} \tag{4.114b}$$

$I \times J$ (Grid)	aspect ratio	Bi	lateral walls	Ra_c
128×16	$A = 10$	$Bi = 1$	no-slip	1589.76
256×16	$A = 10$	$Bi = 1$	no-slip	1566.30
512×16	$A = 10$	$Bi = 1$	no-slip	1563.78
16×16	$A = \pi/a_c$	$Bi = 1$	slip	1555.58
32×32	$A = \pi/a_c$	$Bi = 1$	slip	1544.50
64×64	$A = \pi/a_c$	$Bi = 1$	slip	1541.98
$\infty \times \infty$	$A = \pi/a_c$	$Bi = 1$	slip	1541.18
(Nield, 1964)	$A = \pi/a_c$	$Bi = 1$	slip	1541.14
256×16	$A = 10$	$Bi = 5$	no-slip	1620.10
256×16	$A = 10$	$Bi = 10$	no-slip	2019.02

Table 4.1. Grid test of the value of the first bifurcation point. The first three rows show the convergence of the value of Ra for the case considered here. In the next five rows, a comparison with analytically determined values can be made for a special aspect ratio $a_c = \pi/\sqrt{2}$ and slip conditions at the lateral boundaries (Nield, 1964). The last two rows show the sensitivity of the location of the first bifurcation point with Bi.

This reduces the number of unknowns per point from 4 (u, w, p, T) to 3 (ψ, ζ, T). The equations in this formulation are easily derived by taking the rotation of the momentum equations (4.4a) and the continuity equation become

$$Pr^{-1}\left[\frac{\partial \zeta}{\partial t} + \frac{\partial(u\zeta)}{\partial x} + \frac{\partial(w\zeta)}{\partial z}\right] = \nabla^2\zeta + Ra\frac{\partial T}{\partial x} \qquad (4.115a)$$

$$\zeta = -\nabla^2\psi \qquad (4.115b)$$

The boundary conditions are

$$x = 0, A \quad : \quad \frac{\partial T}{\partial x} = \psi = \gamma\frac{\partial\psi}{\partial x} + (1-\gamma)\zeta = 0 \qquad (4.116a)$$

$$z = 0 \quad : \quad T - 1 = \psi = \frac{\partial\psi}{\partial z} = 0 \qquad (4.116b)$$

$$z = 1 \quad : \quad \frac{\partial T}{\partial z} + Bi\, T = \psi = \zeta = 0. \qquad (4.116c)$$

where $\gamma = 0$ and $\gamma = 1$ give slip and no-slip conditions, respectively. Details of the discretization, on a non-staggered grid, can be found in Dijkstra (1992) and Dijkstra *et al.* (1995).

First aim of the computations is to find the critical temperature gradient (or critical value of Ra) for fixed Bi. Hence, we take $Bi = 1$, $\gamma = 1$, $\lambda = Ra$, start at the motionless solution (4.7) and prescribe the initial tangent as $(\dot{\mathbf{x}}_0, \dot{\lambda}_0) = (\mathbf{0}, 1)$. The latter can be used because the motionless solution is a solution for all values of $\lambda = Ra$. The version of the code applied has an iterative solver (BICGSTAB combined with ILU-preconditioning) and the SIT eigenvalue solver (Dijkstra *et al.*,

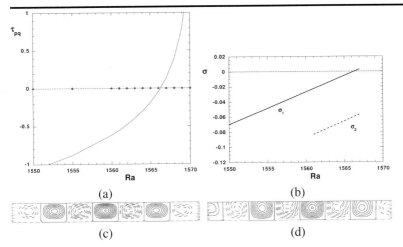

Figure 4.18.   (a) Computation of the test function τ_{pq} along the primary branch of motionless flow. A zero of this function may indicate a bifurcation point. The diamonds indicate the actual points computed along the motionless state. (b) First two eigenvalues as a function of Ra along the same branch. Indeed, a real eigenvalue passes the imaginary axis at the same location where the zero of τ_{pq} appears. (c) Pattern of the streamfunction of the eigenvector corresponding to σ_1, just at the point where the imaginary axis is crossed ($\sigma_1 = 0$). (d) Pattern of the temperature of the same eigenvector.

1995). In the latter paper, also the performance of the preconditioner (see Technical Box 4.4) can be found.

In Fig. 4.18, the computation along the primary branch (i.e., the motionless solution) is displayed using a 256×16 grid. Note that the dimension of the dynamical system is $3 \times 257 \times 17 = 13,107$. A particular test function τ_{pq} (4.40) for $p = q = 256 \times 16 + 1 = 4,097$, is shown in Fig. 4.18a and goes through zero near $Ra = 1565$. In this figure, the points actually computed are indicated by the diamonds. The first two eigenvalues, which are both real, are shown along this motionless solution in Fig. 4.18b, indicating that the motionless solution becomes unstable near $Ra = 1565$, since one eigenvalue crosses the imaginary axis. Patterns of the streamfunction and temperature perturbation which destabilize the motionless state (the eigenvector associated with σ_1) are plotted in Fig. 4.18c and Fig. 4.18d, respectively. The pattern consists of seven cells and the solution for the streamfunction is symmetric with respect to the mid-axis of the container. Note that the pattern with counter-rotating cells is also an eigenvector associated with σ_1.

For each application, it is recommended to check whether the chosen resolution is sufficient to obtain accurate results. If the discretization is consistent then, for an infinitely fine grid, the results of the continuous problem are approached. To check the convergence of the numerical discretization procedure and to be able to extrapolate the results to the continuous problem, the value of Ra at the first bifurcation is determined for a number of grid sizes; the result is shown in Table 4.1. One can see that there is convergence and that a 256×16 is a reasonable grid to perform the computations. In this case, a comparison with analytical solutions is also possible for a particular aspect ratio and value of Bi if the boundary conditions on the sidewalls (4.116a) are taken to be slip conditions ($\gamma = 0$). The sensitivity of the bifurcation point with Bi is illustrated in the last two rows of Table 4.1. Note that the value of Ra at these bifurcation points does not depend on Pr since the eigenvalues σ are all real.

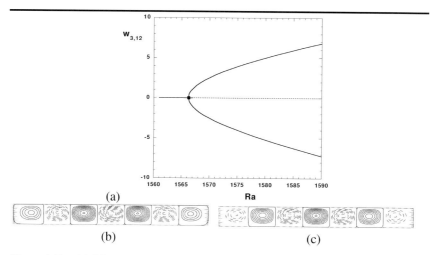

Figure 4.19.   (a) Bifurcation diagram and (b-c) streamfunction of the finite-amplitude cellular solutions (at $Ra = 1575$) arising at the first pitchfork bifurcation.

Because of the reflection symmetry through the mid-axis ($x = A/2$), a pitchfork bifurcation is expected to occur and it is found at $Ra = 1565$. The bifurcation structure for $Pr = 1.0$ is plotted in the weakly nonlinear regime in Fig. 4.19a. On the vertical axis, the vertical velocity at the gridpoint $(3, 12)$ - near the upper left corner - is plotted. The slightly supercritical streamfunction patterns on both branches near the primary bifurcation point are shown in the Figs. 4.19b-c. At the first primary bifurcation point ($Ra = 1565$) the motionless solution becomes unstable to the 7-cell pattern (Fig. 4.19b) which stabilizes for $Ra > 1565$. Also its symmetry related pattern stabilizes (Fig. 4.19c) and both patterns are stable up to $Ra = 1590$. For the three-dimensional case, similar results can be calculated and an overview of the complete solution to this problem is presented in Gelfgat (1999).

4.9. Exercises on Chapter 4

(E4.1) *Rayleigh-Bénard: linear stability boundary*

There are analytic results available for the stability of the motionless flow in the Rayleigh-Bénard problem under slightly different boundary conditions than in section 4.1. In this exercise, we determine one of these stability boundaries. Our starting point will be the dimensionless equations (4.9) and for simplicity, we will restrict to a two-dimensional situation, with velocities (u, w) in the (x, z) direction, respectively.

To investigate the stability of the solution (4.12), we perturb it according to

$$u = \bar{u} + \tilde{u} \; ; \; p = \bar{p} + \tilde{p} \; ; \; w = \tilde{w} \; ; \; T = \bar{T} + \tilde{T}$$

a. Linearize the equations (4.9) near the basic state given by (4.12) and determine the evolution equations for the perturbations.

Next, we consider slip conditions and isolating conditions ($Bi \rightarrow \infty$) at the top and bottom boundaries. The boundary conditions for the perturbations are then slightly modified from (4.10a-b) and become

$$z = 0, 1 : \frac{\partial \tilde{u}}{\partial z} = \frac{\partial \tilde{v}}{\partial z} = \tilde{w} = \tilde{T} = 0$$

and we consider the liquid layer to be unbounded in the horizontal direction.

b. Eliminate the pressure in the equations obtained in a., introduce a streamfunction ψ through $\tilde{u} = \partial \psi / \partial z$, $\tilde{w} = -\partial \psi / \partial x$ and derive the equations

$$Pr^{-1}\frac{\partial \nabla^2 \psi}{\partial t} = Ra\frac{\partial T}{\partial x} + \nabla^2 \nabla^2 \psi$$

$$\frac{\partial T}{\partial t} + \frac{\partial \psi}{\partial x} = \nabla^2 T$$

where the tilde on T has been omitted for convenience. Also formulate the boundary conditions for ψ and T.

As a next step, use normal mode expansions of the form

$$\psi(x, z, t) = \hat{\Psi} \sin n\pi z \sin kx \, e^{\sigma t}$$

$$T(x, z, t) = \hat{T} \sin n\pi z \sin kx \, e^{\sigma t}$$

Here, k is the wavenumber, σ the complex growth factor, $n = 1, 2, \cdots$, and $\hat{\Psi}$ and $\hat{T}$ are arbitrary amplitudes.

c. Determe the eigenvalues σ.

d. Determine the curve in the (k, Ra) plane on which $\sigma = 0$. This curve is called the neutral curve.

e. Show that the smallest value of Ra (say Ra_c) for which instability occurs if $Ra > Ra_c$, is given by

$$Ra_c = \frac{27}{4}\pi^4$$

Further reading: Chandrasekhar (1961), chapter 3.

(E4.2) *Staggerred grid*

In ocean models, several grids are used to discretize the primitive equations. These grids are indicated as Arakawa A, B, C, D and E grids. On the C-grid, the positioning of the variables is as depicted in Fig. 4.20. It shows the projections of the 3D cell on which the mass conservation law (continuity equation) is discretized. The four v-velocities surrounding a u-point are averaged to compute a v-velocity at a u-point.

The boundary at the east and west coast runs through the u-points.

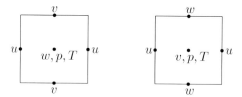

Figure 4.20.   Positioning of variables on the C-grid, topview (left) and vertical cross section (right)

a. Discretize the terms in the zonal momentum balance

$$-fv = -\frac{1}{\rho}\frac{\partial p}{\partial x} + A_H\frac{\partial^2 u}{\partial x^2}$$

at u-points on the C-grid using second-order central differences.

The positioning of the variables on the B-grid is shown in Fig. 4.21.

b. Discretize the zonal momentum balance above at u-points for the B-grid, also using second-order central differences.

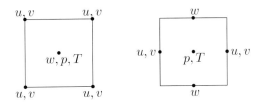

Figure 4.21.   Positioning of variables on the B-grid, topview (left) and vertical cross section (right)

c. Which discretization of the zonal momemtum balance is more susceptable to wiggles, a B- or a C-grid, and why?

Further reading: Adcroft et al. (1999).

(E4.3) *Pseudo-arclength continuation*

One of the most important advantages of the pseudo-arclength method is that one is able to trace a branch around a saddle-node bifurcation. To demonstrate this, consider system of nonlinear algebraic equations given by

$$\mathbf{F}(\mathbf{x}, \lambda) = 0$$

We first look at the so-called 'natural' continuation method by solving the equations

$$\mathbf{F}(\mathbf{x}(\lambda), \lambda) = 0$$

with the Newton Raphson method starting at the solution $(\mathbf{x}_0, \lambda_0)$, with $\lambda = \lambda_0 + \Delta\lambda$.

a. Why does this method cause problems near a saddle-node bifurcation?

Next, consider the pseudo-arclength method

$$
\begin{aligned}
\mathbf{F}(\mathbf{x}(s), \lambda(s)) &= 0 \\
\dot{\mathbf{x}}_0^T(\mathbf{x} - \mathbf{x}_0) + \dot{\lambda}_0(\lambda - \lambda_0) - \Delta s &= 0
\end{aligned}
$$

starting at the solution $(\mathbf{x}_0, \lambda_0)$, with $s = s_0 + \Delta s$.

b. Why is there no problem anymore at saddle-node bifurcations? (*Hint: Consider the extended Jacobian matrix of this system of equations at the saddle-node bifurcation*).

Further reading: Keller (1977) and Seydel (1994).

(E4.4) *Continuation of eigenvalue curves*

It is interesting that eigenvalue problems can be formulated as bifurcation problems. In this way, branches of eigensolutions and eigenvalues can be traced through parameter space using pseudo-arclength continuation. As shown in section 4.4, a typical linear stability problem is written as

$$Ax = \sigma Bx$$

with $x = x_R + ix_I$ the complex eigenvector and $\sigma = \sigma_R + i\sigma_I$ the complex eigenvalue. Assume that, as in many problems, the matrices A and B are real.

a. Write this eigenvalue problem as a nonlinear system of algebraic equations for the state vector $(x_R, x_I, \sigma_R, \sigma_I)$.

b. Formulate a pseudo-arclength scheme to follow a branch of eigensolutions in a particular parameter μ.

Further reading: Dijkstra and Steen (1991).

(E4.5) *Implicit time integration*

For the implicit time stepping scheme (4.75), also nonlinear systems of algebraic equations have to be solved per time step.

a. Formulate this nonlinear system of equations for the Crank-Nicholson method, with $\omega = 1/2$.

b. Formulate the Newton-Raphson scheme for these equations.

c. Why is the transient problem usually easier to solve than the steady state problem?

Further reading: Roache (1976).

(P4.1) *The JDQZ eigenvalue solver*

In this exercise, you learn to work with the JDQZ method using a relatively simple elliptic eigenvalue problem. This problem, with eigenvalue λ, is on a domain $x \in [0, 1], y \in [0, 1]$ and defined by the equations

$$\nabla^2 u = -\lambda u$$

$$x = 0 : u = 0 \quad ; \quad x = 1 : \frac{\partial u}{\partial x} = \frac{3}{4}u$$

$$y = 0 : \frac{\partial u}{\partial y} = 0 \quad ; \quad y = 1 : \frac{\partial u}{\partial y} = 0$$

a. Discretize these equations using central differences on a grid $[x_0 = 0, x_1, ..., x_{N-1}, x_N = 1] \times [y_0 = 0, y_1, ..., y_{M-1}, y_M = 1]$ and formulate the discrete eigenvalue problem

$$Au = -\lambda u$$

for the $(N + 1)(M + 1)$ vector of unknowns $u = (u_{0,0}, u_{1,0}, ..., u_{N,0}, u_{0,1}, ..., u_{N,M})^T$.

Download the JDQZ solver from http://www.math.uu.nl/~people/bomhof/jd.html, read the short manual and install the code on your machine.

b. For $N = M = 10$, solve for the first ten eigenvalues λ using JDQZ. Experiment with the choice of the target τ.

c. Study the convergence of the first five eigenvalues with increasing $N = M = 40, 80, 160$ and 320.

Further reading: Sleijpen and Van der Vorst (1996).

(P4.2) *Linear system solvers*

In this exercise, you will get experience with modern iterative linear systems solvers. To do this, you are going to solve the following problem defined by the equations

$$\nabla^2 T = 0$$
$$x = 0, L : T = T_0 \quad ; \quad y = 0 : T = T_0 , \ y = H : T = T_1$$

numerically for the temperature $T(x, y)$ on a domain $x \in [0, L], y \in [0, H]$, where $T_1 > T_0$.

One can show (by Fourier-series analysis) that the analytic solution to the problem is given by

$$T(x, y) = T_0 + \sum_{n=1}^{\infty} C_n \sin \frac{n\pi x}{L} \sinh \frac{n\pi y}{L}$$
$$C_n = \frac{2(T_1 - T_0)}{n\pi} \frac{1 - (-1)^n}{\sinh \frac{n\pi y}{H}}$$

a. Discretize the governing equations using central differences on a grid $[x_0 = 0, x_1, ..., x_{N-1}, x_N = L] \times [y_0 = 0, y_1, ..., y_{M-1}, y_M = H]$ and formulate the linear system of equations

$$Au = b$$

for the $(N + 1)(M + 1)$ vector of unknowns $\mathbf{u}$ = $(T_{0,0}, T_{1,0}, ..., T_{N,0}, T_{0,1}, ..., T_{N,M})^T$.

You can download different linear system solvers from the internet at http://www.netlib.org/. For this problem, download the LASPACK (written in C) package or the SPARSPAK (written in FORTRAN) package and install it on your machine. A short manual of each package is included in the distribution and the code contains most of the linear system solvers discussed in this chapter. If you do not like C or FORTRAN, codes of the methods in other languages are also available at http://www.netlib.org/.

Choose values: $T_0 = 0.0$, $T_1 = 1.0$, $L = 2.0$, $H = 1.0$.

b. Solve the linear system for $N = 20$, $M = 20$ using the following methods: Jacobi, GMRES and BiCGSTAB. Compare the convergence of the different methods to the exact solution.

c. Use also the classical Gauss Elimination (GE) method (section 4.6) to solve the linear system of equations.

d.   Compare the computational times for GE and GMRES when $N = M = 40, 80, 160$ and 320.

Further reading: Barrett et al. (1994).

Chapter 5

THE WIND-DRIVEN CIRCULATION

Sailing on the Gulf Stream, but not too rough !
English Suite, J.W. Duarte.

An important problem in physical oceanography is to understand the physics of the time-mean surface ocean circulation and the variability of this circulation on time scales from several months to several years. Focus of this chapter is on the Kuroshio in the North Pacific Ocean and the Gulf Stream in the North Atlantic Ocean which are the major northern hemispheric western boundary currents. The mean position of these currents is important for the global climate system and for both regions relatively many observations are available. The midlatitude surface ocean circulation has also been extensively studied theoretically using a wide range of ocean models (Kraus, 1996). Although the basin of the North Pacific has larger dimensions than that the North Atlantic, the time-mean wind-stress forcing is very similar and a close dynamical correspondence between both western boundary currents can be expected.

In section 5.1, a brief description is given of the flow phenomena in both the Gulf Stream and Kuroshio regions motivating the problems studied later on. These problems are (i) the separation of the Gulf Stream near the North American coast, (ii) the different time-mean paths of the Kuroshio near the Japanese coast, and (iii) the variability of both the Gulf Stream and Kuroshio on subannual-to-interannual time scales. At the end of this introductory section, the questions related to these problems are formulated from a dynamical systems perspective.

Section 5.2 introduces a hierarchy of ocean models of the wind-driven circulation (WDC) using a 'top-down' approach, i.e., starting with the most complex model and ending with a very elementary model. In the sections 5.3 to 5.7, bifurcation analysis is applied to this hierarchy of models, using a 'bottom-up' approach. In this way, the consequences of the relevant physical processes on the behavior of the circulation in the North Atlantic and North Pacific can be systematically determined. In the last sections 5.8 and 5.9, it will be evaluated whether the bifurcation analyses provide a framework to understand results from high-resolution ocean models and phenomena deduced from observations.

5.1. Phenomena

A sketch of the global surface ocean circulation was given in section 1.2.1 (Fig. 1.12). In the North Atlantic Ocean, the Gulf Stream is seen as an eastward jet forming part of two recirculating gyres, the subtropical and subpolar gyre. The Kuroshio takes the same role as the western boundary current in the North Pacific. In this section, more detail of the flows in the Gulf Stream and Kuroshio regions is provided. The description below is very limited and readers can consult Wunsch (1996) and WOCE (2001) for more details and references.

5.1.1. Gulf Stream

A sketch of the geography and bathymetry in the region of interest with the different currents is given in Fig. 5.1. The time-mean position of the Gulf Stream has fascinated oceanographers since its early description by Benjamin Franklin and Timothy Folger (Richardson, 1980). From the enormous amount of data obtained since then, from ships and satellites, the time-mean path of the Gulf Stream is now well-known (Auer, 1987; Lee and Cornillon, 1995). The southern part of the Stream (the Florida Current) flows almost parallel to the coastline (Fig. 5.1). At Cape Hatteras, the Gulf Stream leaves the North-American continent and moves further eastward along 40°N. It is accompanied by recirculation gyres to the north (Hogg *et al.*, 1986) and the south (Worthington, 1976). At 68°W, maximum zonal velocities of 2 ms^{-1} have been found near the surface, decreasing to about 0.2 ms^{-1} at 1000 m depth. A typical meridional width of the Gulf Stream at this location is about 150 km. Near Cape Hatteras, the volume transport is estimated to be about 50-65 Sv which increases to a total of about 145 Sv at 60°W (Johns *et al.*, 1995).

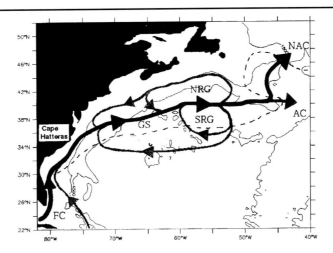

Figure 5.1. Sketch of the near-surface circulation in the Gulf Stream region (Dengg et al., 1996). Bold lines: Florida Current (FC) and Gulf Stream (GS), branching into the North Atlantic Current (NAC) and Azores Current (AC). The abbreviations NRG and SRG indicate Northern and Southern Recirculation Gyre, respectively.

As a typical example, a snapshot of the sea surface temperature (SST) field of the region is plotted in Fig. 5.2. The infrared data used to obtain this picture were obtained from observations from the Advanced Very High Resolution Radiometer (AVHRR, see http://fermi.jhuapl.edu/avhrr/index.html). Fig. 5.2 is a multipass image of the situation in May 1996 where the 'warmest' pixel is selected from each pass over (every three days).

The Gulf Stream transports warm water northward into the central Atlantic with typical surface temperatures of 25°C in the Florida Current and 15°C near the separation point at Cape Hatteras. Its northern boundary is characterized by a strong meridional temperature gradient, which has been referred to as the north wall or cold wall (Stommel, 1965). From the SST signature, it can also be seen that the current strongly meanders after it has separated from the North American coast. This meandering causes the position of the axis of the Gulf Stream to vary, leading to variations of the position of the cold wall (Auer, 1987). After crossing the Southeast Newfoundland Ridge (at about 50°W) the Gulf Stream splits into two branches: the northern branch becomes the North Atlantic Current, and the southern one the Azores Current (Kaese and Krauss, 1996).

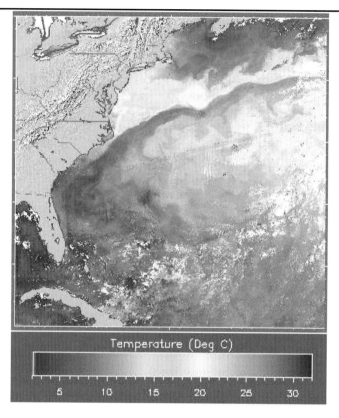

Figure 5.2. (in color on page 518). Multipass image of the SST field of the Gulf Stream region as determined by the Advanced Very High Resolution Radiometer (AVHRR) in May 1996 (obtained from http://fermi.jhuapl.edu/avhrr/gallery/sst/stream.html).

Although the mean separation location of the Gulf Stream has been fairly stable over the last decades, satellite images of SST of the North Atlantic have also revealed that the Gulf Stream near South Carolina can be in a weakly deflected or a strongly deflected state (Fig. 5.3). Bane and Dewar (1988) have presented observations which suggest that the seaward deflection of the Gulf Stream has a bimodal character and that the transitions between both states occur on inter-monthly time scales. Results from other studies of the Gulf Stream path using in situ measurements and very high resolution infrared satellite data also show indications of bimodal behavior. In Fig. 4 of Olson *et al.* (1983), a histogram of the cross-stream frontal position (over the period 1976-1980) indicates bimodal behavior in the area just before separation (at about 77°W). Similarly, in Fig. 8c of Auer (1987), using AVHRR data over the period 1980-1985, indications for bimodality are found in the area just after separation (at 71°W). The two peaks in the histogram are separated from each other by about 50 km in cross-stream direction and 0.5° in latitude.

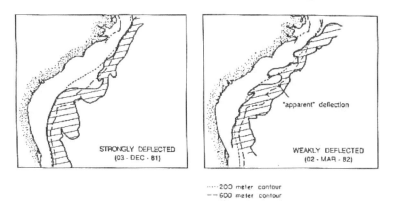

·····200 meter contour
– – 600 meter contour

Figure 5.3.   Locations of the Gulf Stream on December 3, 1981 and March 2, 1982, during the Gulf Stream Deflection and Meander Energetics Experiment (DAMEX). These images show the Stream in a typical strongly deflected state (left panel) and a weakly deflected state in the region with a topographic feature such as the Charleston Bump (figure from Bane and Dewar (1988)).

Concerning the variability of the Gulf Stream, there are now reasonably long datasets available (with enough spatial resolution) to study changes on time scales up to a few years in quite detail (Fu, 2001). An overview of meander characteristics and other type of variability is given in Kaese and Krauss (1996). East of Cape Hatteras, meanders with monthly variability exist with a typical spatial scale of 500 km. Large cold-core rings and warm-core rings are formed as a final stage of meander development, where the meander is detached from the main current. The signature of such a ring in the SST field can be seen in Fig. 5.2, in the middle of the picture just above the mean Gulf Stream. These rings have typical westward propagating speeds of 5 km day^{-1} and a diameter of about 200 km. Warm-core rings appear to be less frequent than cold-core rings, with typically five warm-core rings forming per year. Each of these rings has a life time of about 6 months.

Although the patterns on this smaller time scale (i.e., monthly) variability are reasonably clear, those associated with somewhat longer time scales are more difficult to identify. Satellite measurements of sea-level height (altimetry), in particular from the GEOSAT, TOPEX-POSEIDON and ERS1 missions, and the patterns of the SST field from AVHRR have contributed substantially to determine this type of variability. In many studies, variability on subannual to annual scales is found (Vazquez *et al.*, 1990; Wang and Koblinsky, 1995), but it has been difficult to clarify its physics.

For example, using AVHRR-derived infrared images for the period April 1982 through December 1989, Lee and Cornillon (1995) find two dynamically distinct modes of variability of the path of the Gulf Stream. The first mode of variability is associated with large-scale lateral shifts of the mean path having a near-annual period. These shifts are presumably caused by atmospheric forcing, through the changes in downward heat flux over the area (Wang and Koblinsky, 1996). The second type of variability, having a 9-month dominant periodicity, is associated with changes in meandering intensity. Lee and Cornillon (1995) suggest that this type of variability is related to internal oceanic dynamics. Using both ECMWF atmospheric forcing fields and GEOSAT data, Kelly *et al.* (1996) also find this type of variability in the Gulf Stream and attribute it to structural changes in the recirculation gyres. The variations in the Gulf Stream have largest amplitudes east of 62°W and are not directly related to heat flux variations over the area.

5.1.2. Kuroshio

The most intense current in the North Pacific Ocean is the Kuroshio and its time-mean path is now quite well-known. From observations, it is found that the Kuroshio path exhibits bimodal behavior to the south of Japan with transitions occurring between a small and a large meander state (Taft, 1972). Both states (Fig. 5.4a) can persist over a period ranging from a few years to a decade and transitions between them occur within a couple of months (Kawabe, 1986). At the southeast corner of Honshu, the Kuroshio separates from the Japanese coast and flows eastward, while meandering increases. However, the current keeps a mean latitudinal position of about 35°N up to 180°E.

The time-mean zonal geostrophic transport of the Kuroshio is estimated to be about 52 Sv at 137°E (Qiu and Joyce, 1992). Accurate measurements along the ASUKA observational line (at 137°E) combined with satellite date provide a mean Kuroshio transport of 42 Sv for 1992-1999 (Imawaki *et al.*, 2001). The variations in volume transport along the PN section in Fig. 5.4a, as deduced from observations in Kawabe (1995) are plotted (as thick dots) in Fig. 5.4b. During periods of the large-meander state (indicated by the horizontal lines), the transport is typically lower than average.

Similar to the 9-month variability in the Gulf Stream region (Lee and Cornillon, 1995), near-annual variability is also found in recent studies of the Kuroshio Extension. Wang *et al.* (1998) have used four years of data from the TOPEX/Poseidon (T/P) exact repeat mission (ERM) together with 2.3-yr data from the GEOSAT ERM. They have separated the low-frequency variability into subannual, annual and interannual variability through filtering processes. Hövmöller diagrams (their Figs. 9 and 14) show that the subannual sea level height fluctuations are primarily propagating westward and they weaken away from the Kuroshio axis (at about 35°N). Wang *et al.* (1998) speculate that instability and/or external forcing might be responsible for the generation of the subannual variability and that bottom topography plays a role as well.

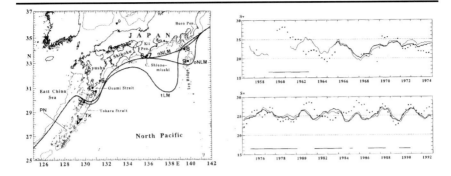

Figure 5.4. (a) Examples of paths of the Kuroshio in its small meander state (or non-large-meander (nNLM) path) and in its typical large-meander state (tLM). (b) Interannual variations of volume transport (dots) of the Kuroshio along the PN section in (a), as deduced from observations in Kawabe (1995). The horizontal lines show the large-meander state periods.

Kelly *et al.* (1996) also found variability with timescales of 5-9 months in the Kuroshio. There appear to be correlations between the height difference across the jet (surface transport) and the path itself. This suggests that there are structural changes in the recirculation gyres, associated with a path change. Although windstress (curl) is clearly correlated with surface transport, mean path and net surface heat flux in the Pacific ocean basin, there is no clear correlation between the net surface heat flux and the surface transport over the entire study region.

5.1.3. Central questions

Given the brief description of the surface flows in the Gulf Stream and Kuroshio regions, it is clear that both flows are very complex. Important features related to the time-mean path are the separation of the Gulf Stream from the North-American coast and the apparent bimodality of the Kuroshio path. What physics determines the spatial pattern of the time-mean path of these currents? We will approach this issue by first looking at the steady-state problem. For a hierarchy of models, the steady flow patterns will be determined versus parameters, such as the lateral friction or wind-stress strength. Special focus will be on the existence of different time-mean patterns under the same forcing conditions.

Having described the presence of meanders, rings and variations in the lateral extension of the western boundary currents, it can be anticipated that it is difficult to understand the physics of the transient flow. While several of the types of variability may be explained from the (highly variable) forcing from the atmosphere, some phenomena may be intrinsically caused by oceanic dynamical processes; an example is the 9-month variability in the Gulf Stream. The question is whether a clear candidate of such a mode of variability can be found. A second topic discussed below is the systematic determination of the internal modes of variability in models of the WDC. Identification of these modes may provide an understanding of how their nonlinear interactions leads to the complex behavior as seen in reality.

This sets the context for the approach in the following sections, which is motivated through the following questions.

(i) Are there different time-mean flow patterns of the midlatitude WDC under the same forcing conditions? Is different separation behavior of the Gulf Stream and Kuroshio dynamically possible?

(ii) Can temporal variability on subannual to interannual time scales in the Gulf Stream and Kuroshio be understood from instabilities of the time-mean flow?

5.2. Models of the Midlatitude Ocean Circulation

The full three-dimensional model as put forward in section 2.1.1 is the starting point of the hierarchy of models of the WDC. As a first reduction of this model, the vertical structure of the flow is idealized. The stratification of the ocean is thought as being built up of stacked layers of constant density of which the thickness varies. The simplest of these models is the homogeneous model, where the density of the ocean water is constant. This one-layer model will be considered in section 5.2.1 and subsequently be generalized to a multi-layer model in 5.2.2.

5.2.1. The homogeneous model

A typical domain of, for example, the North Atlantic is the region $[285°W - 350°W] \times [10°N - 60°N]$. The continental geometry can for example be determined from the ETOPO5 dataset (available through http://csep1.phy.ornl.gov/etopo5-doc/etopo5-doc.html), which provides the depth below the average sea level on a $0.5'$ grid. Within this domain, the flow of a constant density liquid is considered. This flow is driven by a wind stress $\tau_*(\phi, \theta) = \tau_0(\tau^\phi, \tau^\theta)$, where τ_0 is the amplitude and (τ^ϕ, τ^θ) provides the spatial pattern. Bottom topography is given by a function $z_* = r_* - r_0 = -D + h_{b*}(\phi, \theta)$ and the ocean-atmosphere interface is described by $z_* = \eta_*(\phi, \theta, t_*)$, where r_0 is the distance between the center of the earth and the average position of the ocean-atmosphere interface (see also Fig. 2.4).

The governing equations in spherical coordinates are obtained from (2.6a) - (2.6d) for constant density ρ_0. When the frictional terms (2.13), with constant mixing coefficients of momentum A_V and A_H, are written out in full, these equations become (e.g., Appendix 2 in Batchelor (2000))

$$\frac{Du_*}{dt_*} + \frac{u_* w_*}{r_*} - \frac{u_* v_*}{r_*} \tan\theta - 2\Omega \left(v_* \sin\theta - w_* \cos\theta \right) =$$

$$-\frac{1}{\rho_0 r_* \cos\theta} \frac{\partial p_*}{\partial \phi} + \frac{A_V}{r_*^2} \frac{\partial}{\partial r_*} \left(r_*^2 \frac{\partial u_*}{\partial r_*} \right) +$$

$$+ \frac{A_H}{r_*^2 \cos^2\theta} \left[L_H(u_*) - u_* + 2\cos\theta \frac{\partial w_*}{\partial \phi} - 2\sin\theta \frac{\partial v_*}{\partial \phi} \right] \quad (5.1a)$$

$$\frac{Dv_*}{dt_*} + \frac{w_* v_*}{r_*} + \frac{u_*^2}{r_*}\tan\theta + 2\Omega u_* \sin\theta =$$

$$-\frac{1}{\rho_0 r_*}\frac{\partial p_*}{\partial\theta} + \frac{A_V}{r_*^2}\frac{\partial}{\partial r_*}\left(r_*^2 \frac{\partial v_*}{\partial r_*}\right)$$

$$+\frac{A_H}{r_*^2 \cos^2\theta}\left[L_H(v_*) - v_* + 2\cos^2\theta\frac{\partial w_*}{\partial\theta} + 2\sin\theta\frac{\partial u_*}{\partial\phi}\right] \quad (5.1\mathrm{b})$$

$$\frac{Dw_*}{dt_*} - \frac{u_*^2 + v_*^2}{r} - 2\Omega u_* \cos\theta = -\frac{1}{\rho_0}\frac{\partial p_*}{\partial r_*} - g +$$

$$+\frac{A_V}{r_*^2}\frac{\partial}{\partial r_*}\left(r_*^2 \frac{\partial w_*}{\partial r_*}\right) + \frac{A_H}{r_*^2 \cos^2\theta} \times$$

$$\times\left[L_H(w_*) - 2w_* \cos^2\theta - 2\cos\theta\frac{\partial(v_* \cos\theta)}{\partial\theta} - 2\cos\theta\frac{\partial u_*}{\partial\phi}\right] \quad (5.1\mathrm{c})$$

$$\frac{\partial w_*}{\partial r_*} + \frac{2w_*}{r_*} + \frac{1}{r_* \cos\theta}\left[\frac{\partial(v_* \cos\theta)}{\partial\theta} + \frac{\partial u_*}{\partial\phi}\right] = 0 \quad (5.1\mathrm{d})$$

with the operators

$$\frac{D}{dt_*} = \frac{\partial}{\partial t_*} + \frac{u_*}{r_* \cos\theta}\frac{\partial}{\partial\phi} + \frac{v_*}{r_*}\frac{\partial}{\partial\theta} + w_*\frac{\partial}{\partial r_*} \quad (5.2\mathrm{a})$$

$$L_H = \cos\theta\frac{\partial}{\partial\theta}\left(\cos\theta\frac{\partial}{\partial\theta}\right) + \frac{\partial^2}{\partial\phi^2} \quad (5.2\mathrm{b})$$

The boundary conditions (2.18) at $z_* = \eta_*$ are given by

$$p_* = p_{a*} \quad (5.3\mathrm{a})$$

$$\rho_0 A_V r_* \frac{\partial}{\partial r_*}\left(\frac{u_*}{r_*}\right) + \frac{\rho_0 A_H}{r_* \cos\theta}\frac{\partial w_*}{\partial\phi} = \tau_0 \tau^\phi \quad (5.3\mathrm{b})$$

$$\rho_0 A_V r_* \frac{\partial}{\partial r_*}\left(\frac{v}{r}\right) - \frac{\rho_0 A_H}{r_*}\frac{\partial w_*}{\partial\theta} = \tau_0 \tau^\theta \quad (5.3\mathrm{c})$$

$$\frac{D}{dt_*}(r_* - r_0 - \eta_*) = 0 \quad (5.3\mathrm{d})$$

where p_{a*} is a constant atmospheric background pressure. At the bottom of the domain, $z_* = -D + h_{b*}$, the boundary conditions (2.16) become

$$u_* + \frac{w_*}{\cos\theta}\frac{\partial h_{b*}}{\partial\phi} = 0 \quad (5.4\mathrm{a})$$

$$v_* + w_*\frac{\partial h_{b*}}{\partial\theta} = 0 \quad (5.4\mathrm{b})$$

$$\frac{D}{dt_*}(z_* + D - h_{b*}) = 0 \quad (5.4\mathrm{c})$$

At the continental boundaries, no-slip boundary conditions are imposed unless otherwise specified.

5.2.2. Dominant balances

To determine whether each of the terms above are important in large-scale mid-latitude flows over a meridional sector of width $\Delta\theta$, their magnitude is estimated through scaling. Thereto scales U and W are introduced for the horizontal velocity and the vertical velocity, respectively. Similarly, $L = r_0\Delta\theta$ and D are typical horizontal and vertical length scales, respectively. A dimensionless coordinate z is introduced through

$$r_* = r_0(1 + \frac{z_*}{r_0}) = r_0(1 + z\frac{D}{r_0}) \tag{5.5}$$

From the continuity equation (5.1d), a relation between the vertical and horizontal velocity scales is obtained as

$$W = \frac{D}{L}U = \delta U \tag{5.6}$$

with $\delta = D/L$ is the ratio of vertical to horizontal length scale of the flow, which is very small. For $D = 10^3$ m and $L = 10^7$ m, the value of $\delta = \mathcal{O}(10^{-4})$.

For a central latitude $\theta_0 \neq 0$, let f_0 indicate the Coriolis parameter, with $f_0 = 2\Omega \sin \theta_0$. Because the vertical velocity is much smaller than the horizontal velocity, the magnitude of the horizontal components of the Coriolis acceleration (2.5) is $\rho_0 U f_0$. Since the magnitude of the inertial terms is $\rho_0 U^2/L$, the ratio of inertial and Coriolis terms is measured by the Rossby number $\epsilon = U/(f_0 L)$. The Rossby number is therefore very small in the large-scale ocean circulation and since the effects of frictional processes are uncertain, but very small, the only balance which seems possible is the geostrophic balance, in which horizontal pressure gradients balance the Coriolis acceleration. In this way, the dynamical pressure scale P can be anticipated from the dominant balance in (5.1b)

$$2\Omega u_* \sin\theta = -\frac{1}{\rho_0 r_*}\frac{\partial p_*}{\partial\theta} \tag{5.7}$$

which leads to $P = \rho_0 U f_0 L$.

The equations are now scaled by the introduction of the non-dimensional quantities u, v, w, t, η, h_b and p through

$$u_* = Uu \; ; \; v_* \;\; = \;\; Uv \; ; \; w_* = \delta Uw \; ; \; t_* = \frac{L}{U}t \tag{5.8a}$$

$$\eta_* \;\; = \;\; D\eta \; ; \; h_{b*} = Dh_b \tag{5.8b}$$

$$p_* \;\; = \;\; -gD\rho_0 z + \rho_0 U f_0 Lp \tag{5.8c}$$

where the advective timescale L/U is chosen as characteristic time scale. In (5.8), the hydrostatic part and dynamic part of the pressure are separated out. When (5.8) is substituted into the equations (5.1) and the shallow water limit ($\delta = D/L \to 0$ and $D/r_0 \to 0$) is taken, one obtains

$$\epsilon\left[\frac{Du}{dt} - \gamma uv \tan\theta\right] - v\frac{\sin\theta}{\sin\theta_0} = -\frac{1}{\cos\theta}\frac{\partial p}{\partial\phi} +$$

$$E_V \frac{\partial^2 u}{\partial z^2} + \frac{E_H}{\cos^2 \theta} \left[\cos \theta \frac{\partial}{\partial \theta} \left(\cos \theta \frac{\partial u}{\partial \theta} \right) + \frac{\partial^2 u}{\partial \phi^2} - u - 2 \sin \theta \frac{\partial v}{\partial \phi} \right] \quad (5.9a)$$

$$\epsilon \left[\frac{Dv}{dt} + \gamma u^2 \tan \theta \right] + u \frac{\sin \theta}{\sin \theta_0} = -\frac{\partial p}{\partial \theta} +$$

$$E_V \frac{\partial^2 v}{\partial z^2} + \frac{E_H}{\cos^2 \theta} \left[\cos \theta \frac{\partial}{\partial \theta} \left(\cos \theta \frac{\partial v}{\partial \theta} \right) + \frac{\partial^2 v}{\partial \phi^2} - v + 2 \sin \theta \frac{\partial u}{\partial \phi} \right] \quad (5.9b)$$

$$\frac{\partial p}{\partial z} = 0 \quad (5.9c)$$

$$\frac{\partial w}{\partial z} + \frac{1}{\cos \theta} \left[\frac{\partial (v \cos \theta)}{\partial \theta} + \frac{\partial u}{\partial \phi} \right] = 0 \quad (5.9d)$$

with

$$\frac{D}{dt} = \frac{\partial}{\partial t} + \frac{u}{\cos \theta} \frac{\partial}{\partial \phi} + v \frac{\partial}{\partial \theta} + w \frac{\partial}{\partial z}$$

The dimensionless boundary conditions at $z = \eta$ become

$$p = \epsilon Fr \, \eta \quad (5.10a)$$

$$E_V \frac{\partial u}{\partial z} = \alpha_{SW} \, \tau^\phi \quad (5.10b)$$

$$E_V \frac{\partial v}{\partial z} = \alpha_{SW} \, \tau^\theta \quad (5.10c)$$

$$\frac{D}{dt}(z - \eta) = 0 \quad (5.10d)$$

and those at $z = -1 + h_b$ become

$$u + \frac{w}{\cos \theta} \frac{\partial h_b}{\partial \phi} = 0 \quad (5.11a)$$

$$v + w \frac{\partial h_b}{\partial \theta} = 0 \quad (5.11b)$$

$$\frac{D}{dt}(z + 1 - h_b) = 0 \quad (5.11c)$$

In the equations above, several dimensionless parameters appear. They are the Rossby number ϵ, the inverse Froude number Fr, the wind-stress strength α_{SW}, a geometrical ratio γ and the horizontal and vertical Ekman numbers E_H and E_V. Expressions for these parameters are

$$\epsilon = \frac{U}{f_0 L} \; ; \; Fr = \frac{gD}{U^2} \; ; \; \alpha_{SW} = \frac{\tau_0}{f_0 \rho_0 DU}$$

$$E_H = \frac{A_H}{f_0 L^2} \; ; \; E_V = \frac{A_V}{f_0 D^2} \; ; \; \gamma = \frac{L}{r_0} \quad (5.12)$$

Note that only five of these parameters are independent, since the horizontal velocity can be chosen arbitrarily. For a characteristic horizontal velocity of $U = 0.1$ ms^{-1}, for later reference indicated by U_{SW}, and $\theta_0 = 45°$N, typical values of the

Parameter	Value		Parameter	Value	
$L = r_0$	6.4×10^6	m	τ_0	2.0×10^{-1}	Pa
D	1.0×10^3	m	ρ_0	10^3	kgm^{-3}
$f_0 = 2\Omega$	1.5×10^{-4}	s^{-1}	A_H	$10^{3\pm1}$	m^2s^{-1}
U_{SW}	10^{-1}	ms^{-1}	A_V	$10^{-3\pm1}$	m^2s^{-1}
Parameter	Value		Parameter	Value	
ϵ	1.0×10^{-4}		α_{SW}	1.4×10^{-2}	
γ	1.0		Fr	9.8×10^5	
E_H	$10^{-7\pm1}$		E_V	$7.0\ 10^{-6\pm1}$	

Table 5.1.   Typical values of dimensional and dimensionless parameters appearing in (5.12), with $g = 9.8\ ms^{-2}$ and $\theta_0 = 45°N$. For the representation of friction, ranges for the friction parameters A_V and A_H are indicated. The uncertainty in the value of these parameters is quite large.

dimensional and non-dimensional parameters for the North Atlantic basin (with $L = r_0$) are shown in Table 5.1.

By inspection of the equations (5.9) and the values of the dimensionless parameters, the coefficients before the highest order derivatives appear to be very small. One can therefore already anticipate that thin frictional boundary layers are likely to occur in these flows, both near horizontal and vertical boundaries. Since the value of the Rossby number is also small on the large scale, nonlinear effects are expected only to become important on smaller scales than the basin size or in flow regions where the horizontal velocity is substantially larger than U_{SW}.

5.2.3. The multi-layer model

The model of the previous section can be generalized to a two-layer model where two layers of constant density ρ_1 and ρ_2 (with $\rho_1 < \rho_2$) are separated by a material interface which is able to deform (Fig. 5.5). This interface is indicated by $z_* = -h_*$ and is sometimes referred to as a pycnocline (or a thermocline). In each of the layers, the equations (5.9) govern the flow and the top boundary conditions (5.10) hold for the upper layer, while the bottom boundary conditions (5.11) hold for the lower layer. What is additionally needed is a description of the evolution of the interface separating the layers. The conditions to be satisfied on this boundary are continuity of pressure and the (kinematic) requirement that the interface $z_* = -h_*$ is a material surface. When the top layer quantities are indicated by a subscript 1 and the bottom quantities by a subscript 2, then the dimensional conditions at $z_* = -h_*$ become

$$\frac{D}{dt_*}(z_* + h_*) = 0 \tag{5.13a}$$

$$p_{1*} = p_{2*} \tag{5.13b}$$

where the D/dt_* operator can be taken in either layer, since the vertical velocity is continuous over the thermocline (the horizontal velocities are discontinuous).

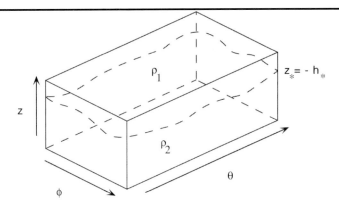

Figure 5.5. Sketch of the two-layer model set-up. Two layers having constant densities ρ_1 and ρ_2 are separated by an interface described by $z_* = -h_*$, referred to as a thermocline.

The equilibrium value of h_*, which occurs when there is no motion, is indicated by D_1. A dimensionless thermocline deviation h is defined as $h_* = D_1 + \mu Dh$, where μ is an *a priori* unknown dimensionless scale factor. Using the scaling for the pressure (5.8) in each layer, the condition (5.13b) becomes

$$g(\rho_1 - \rho_2)\mu h = f_0 LU(\rho_2 p_2 - \rho_1 p_1) \tag{5.14}$$

The relation (5.14) indicates that small dynamic pressure differences between the layers can lead to substantial deflections of the interface since the factor $\rho_2 - \rho_1$ is small compared to a typical reference density ρ_0. Hence, μ can be chosen as

$$\mu = \frac{\rho_0 f_0 UL}{gD(\rho_2 - \rho_1)} \tag{5.15}$$

and the dimensionless form of the equations (5.13) at $z = -D_1/D + \mu h$ is

$$w = -\mu \frac{Dh}{dt} \tag{5.16a}$$

$$h = p_1 - p_2 \tag{5.16b}$$

The quantity $g(\rho_2 - \rho_1)/\rho_0$ is called the reduced gravity and indicated by g'.

The two-layer model is formed by the homogeneous equations (5.9) in each layer together with boundary conditions (5.10) for the first layer, (5.11) for the second layer and the interface conditions (5.16). Understanding this set-up, it is easily generalized to a multi-layer model. More details on the derivation of layer models can be found in chapter 6 of Pedlosky (1987).

5.3. Shallow-water and Quasi-geostrophic Models

The multi-layer models defined in the previous section can be reduced in particular cases. The subsections that follow describe two of these simplifications, introducing the shallow-water (SW) models and the quasi-geostrophic (QG) models.

5.3.1. The spherical shallow-water model

A reduction of the homogeneous model in section 5.2.1 can be obtained by realizing that the horizontal and vertical Ekman numbers (E_H and E_V) and the Rossby number ϵ are very small numbers on the basin scale. Hence, the effects of inertia and friction are relatively small with respect to the Coriolis effect. Over most of the flow domain, the dominant balance in (5.9a-b) must be the geostrophic balance

$$v\frac{\sin\theta}{\sin\theta_0} = \frac{1}{\cos\theta}\frac{\partial p}{\partial\phi} \tag{5.17a}$$

$$u\frac{\sin\theta}{\sin\theta_0} = -\frac{\partial p}{\partial\theta} \tag{5.17b}$$

By differentiating these equations to z and using the hydrostatic relation (5.9c), it follows that the horizontal velocity field outside the vertical boundary layers is independent of depth. Of course, this solution cannot satisfy the boundary conditions at top and bottom and hence boundary layers occur, the so-called Ekman layers. The vertical structure of the solutions is sketched in Fig. 5.6; the thickness of the Ekman boundary layers is $\delta_E = D\sqrt{E_V}$. The latter can be easily deduced from the equations (5.9) since it is on this vertical scale that the vertical momentum exchange terms become $\mathcal{O}(1)$. The flow in the Ekman layer and its

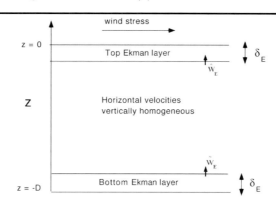

Figure 5.6.   Sketch of the vertical structure of the flow in the homogeneous model. In the region outside the Ekman layers, the horizontal velocities are depth independent. Within the Ekman layers, which have thickness $\delta_E = D\sqrt{E_V}$, vertical momentum exchange through friction become important in the momentum balance. In this way, the input of momentum by the wind stress is distributed over the bulk of the liquid.

coupling with the flow in the geostrophic interior are treated in many textbooks on geophysical fluid dynamics (Pedlosky, 1987; Cushman-Roisin, 1994) and will not be discussed here.

Shallow-water models are obtained by integrating over the total layer depth and simplifying (or neglecting) the momentum exchanges with the bottom Ekman layer. The input of momentum through the top Ekman layer is related to the wind stress, using (5.10), through

$$E_V \int_{-1+h_b}^{\eta} \frac{\partial^2 u}{\partial z^2}\, dz = E_V \frac{\partial u}{\partial z}\Big|_{-1+h_b}^{\eta} \approx \alpha_{SW}\, \tau^{\phi} \qquad (5.18)$$

with a similar approximation for the momentum transfer of the meridional wind stress. When this procedure is followed in the case of a spherical domain with $L = r_0$ ($\gamma = 1$), the shallow-water model becomes

$$\epsilon \left(\frac{\partial u}{\partial t} + \frac{u}{\cos\theta}\frac{\partial u}{\partial\phi} + v\frac{\partial u}{\partial\theta} - uv\,\tan\theta \right) - v\frac{\sin\theta}{\sin\theta_0} =$$
$$-\frac{\epsilon\, Fr}{\cos\theta}\frac{\partial\eta}{\partial\phi} + E_H \left(\nabla^2 u - \frac{u}{\cos^2\theta} - \frac{2\sin\theta}{\cos^2\theta}\frac{\partial v}{\partial\phi} \right) + \alpha_{SW}\frac{\tau^{\phi}}{h} \qquad (5.19a)$$

$$\epsilon \left(\frac{\partial v}{\partial t} + \frac{u}{\cos\theta}\frac{\partial v}{\partial\phi} + v\frac{\partial v}{\partial\theta} + u^2\,\tan\theta \right) + u\frac{\sin\theta}{\sin\theta_0} =$$
$$-\epsilon\, Fr\frac{\partial\eta}{\partial\theta} + E_H \left(\nabla^2 v - \frac{v}{\cos^2\theta} + \frac{2\sin\theta}{\cos^2\theta}\frac{\partial u}{\partial\phi} \right) + \alpha_{SW}\frac{\tau^{\theta}}{h} \qquad (5.19b)$$

$$\frac{\partial h}{\partial t} + \frac{1}{\cos\theta}\left(\frac{\partial(hu)}{\partial\phi} + \frac{\partial(hv\,\cos\theta)}{\partial\theta} \right) = 0 \qquad (5.19c)$$

where the total layer depth is written as $h = h_*/D = \eta + 1 - h_b$. The equilibrium value $h = 1$ appears for a motionless liquid in a flat-bottom basin. The model (5.19) together with boundary conditions is referred to as the spherical homogeneous shallow-water model.

5.3.2. The β-plane model

When the basin size extends only over a small angle $\Delta\theta$ in the meridional direction, a β-plane approximation of the equations can be made. Usually, one also assumes a small size of the basin in zonal direction and identifies $L = r_0\Delta\theta$ as the characteristic horizontal scale of the flow. When $\gamma = L/r_0 = \Delta\theta$ is small, local dimensionless Cartesian coordinates (x, y) are introduced through (Fig. 5.7)

$$x = (\phi - \phi_0)\cos\theta_0 \qquad (5.20a)$$
$$y = \frac{(\theta - \theta_0)}{\Delta\theta} \qquad (5.20b)$$

Locally near $\theta = \theta_0$ the Taylor series expansions

$$\sin\theta = \sin\theta_0 + (\theta - \theta_0)\cos\theta_0 + \mathcal{O}((\theta - \theta_0)^2) \qquad (5.21a)$$
$$\cos\theta = \cos\theta_0 - (\theta - \theta_0)\sin\theta_0 + \mathcal{O}((\theta - \theta_0)^2) \qquad (5.21b)$$

are used where the $\mathcal{O}$ symbol is short to indicate higher-order terms. In the β-plane approximation, the local variation of the Coriolis parameter is only taken

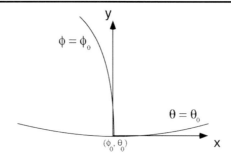

Figure 5.7. *Definition of local Cartesian coordinates (x, y) near a central point defined by longitude ϕ_0 and latitude θ_0.*

into account, whereas in the inertial and frictional terms, just the Cartesian form is taken. In this way, a new dimensionless parameter β arises, which is given by

$$\beta = \frac{\beta_0 L^2}{U} \text{ with } \beta_0 = \frac{2\Omega \cos \theta_0}{r_0} \tag{5.22}$$

A full derivation and discussion of the β-plane model can be found in chapter 6 of Pedlosky (1987) and is not given here. Substitution of the approximations (5.21) and the local coordinates (5.20) give directly the dimensionless equations of the β-plane model, which become

$$\epsilon \frac{Du}{dt} - v(1 + \beta\epsilon y) + \frac{\partial p}{\partial x} = E_H \left[\frac{\partial^2 u}{\partial x^2} + \frac{\partial^2 u}{\partial y^2} \right] + E_V \frac{\partial^2 u}{\partial z^2} \tag{5.23a}$$

$$\epsilon \frac{Dv}{dt} + u(1 + \beta\epsilon y) + \frac{\partial p}{\partial y} = E_H \left[\frac{\partial^2 v}{\partial x^2} + \frac{\partial^2 v}{\partial y^2} \right] + E_V \frac{\partial^2 v}{\partial z^2} \tag{5.23b}$$

$$\frac{\partial p}{\partial z} = 0 \tag{5.23c}$$

$$\frac{\partial w}{\partial z} + \frac{\partial v}{\partial y} + \frac{\partial u}{\partial x} = 0 \tag{5.23d}$$

$$\frac{D}{dt} = \frac{\partial}{\partial t} + u\frac{\partial}{\partial x} + v\frac{\partial}{\partial y} + w\frac{\partial}{\partial z} \tag{5.23e}$$

Note that for consistency, the term $\beta\epsilon = (\Delta\theta \cos \theta_0)/(\sin \theta_0)$ must be small compared to unity. The boundary conditions (5.10) at the ocean-atmosphere interface and the equations (5.11) at the bottom remain basically unchanged, except for their formulation in Cartesian coordinates. When the wind-stress field in the β-plane model is denoted by $\boldsymbol{\tau} = (\tau^x, \tau^y)$, the boundary conditions (5.10) at $z = \eta$ become

$$p = \epsilon \, Fr \, \eta \tag{5.24a}$$

$$E_V \frac{\partial u}{\partial z} = \alpha_{SW} \tau^x \qquad (5.24b)$$

$$E_V \frac{\partial v}{\partial z} = \alpha_{SW} \tau^y \qquad (5.24c)$$

$$\frac{D}{dt}(z - \eta) = 0 \qquad (5.24d)$$

and the equations (5.11) at $z = -1 + h_b(x, y)$ become

$$u + w\frac{\partial h_b}{\partial x} = 0 \qquad (5.25a)$$

$$v + w\frac{\partial h_b}{\partial y} = 0 \qquad (5.25b)$$

$$\frac{D}{dt}(z + 1 - h_b) = 0 \qquad (5.25c)$$

The β-plane model thus obtained is said to represent the dynamics in the tangent plane at the sphere at a certain latitude θ_0.

After vertical integration over the layer, the homogeneous shallow-water β-plane model results. The equations are

$$\epsilon(\frac{\partial u}{\partial t} + u\frac{\partial u}{\partial x} + v\frac{\partial u}{\partial y}) - (1 + \beta\epsilon y)v = \qquad (5.26a)$$
$$-\epsilon\, Fr\, \frac{\partial h}{\partial x} + E_H \nabla^2 u + \alpha_{SW}\frac{\tau^x}{h}$$

$$\epsilon(\frac{\partial v}{\partial t} + u\frac{\partial v}{\partial x} + v\frac{\partial v}{\partial y}) + (1 + \beta\epsilon y)u = \qquad (5.26b)$$
$$-\epsilon\, Fr\, \frac{\partial h}{\partial y} + E_H \nabla^2 v + \alpha_{SW}\frac{\tau^y}{h}$$

$$\frac{\partial h}{\partial t} + \frac{\partial(hu)}{\partial x} + \frac{\partial(hv)}{\partial y} = 0 \qquad (5.26c)$$

As for the shallow-water model in spherical coordinates, the problem of determining the flow has been reduced to a set of partial differential equations on a two-dimensional domain. Hence, only conditions at the continental boundaries, for example no-slip, have to be imposed to complete the model description.

5.3.3. Quasi-geostrophic β-plane models

When the Rossby number ϵ is small and the bottom topography variations are $\mathcal{O}(\epsilon)$, the shallow-water equations can be further simplified. In this regime, quasi-geostrophic (QG) theory is an adequate approximation to describe the flow. The QG theory is an asymptotic theory using ϵ as a small parameter with assumptions of the magnitude on the other parameters in terms of ϵ. For the homogeneous β-plane case, the starting point of the derivation are the equations (5.23).

One can immediately see that expansion of all quantities in ϵ leads to a zeroth order (indicated by the superscripts 0) balance

$$(u^0, v^0) = (-\frac{\partial p^0}{\partial y}, \frac{\partial p^0}{\partial x}) = (-\frac{\partial \psi}{\partial y}, \frac{\partial \psi}{\partial x}) \tag{5.27}$$

in the momentum equations. Here, $\psi = p^0$ is the geostrophic pressure, also called the geostrophic streamfunction. Consideration of the Ekman layers to satisfy boundary conditions at top and bottom of the domain and integration over the interior flow domain, where the flow is strictly geostrophic, finally leads to an evolution equation for the geostrophic streamfunction (Pedlosky, 1987).

In the homogeneous case, the quasi-geostrophic model leads to the so-called barotropic potential vorticity equation which is given by

$$\left[\frac{\partial}{\partial t} + u^0 \frac{\partial}{\partial x} + v^0 \frac{\partial}{\partial y}\right][\zeta - F\psi + \beta y] =$$

$$\frac{1}{Re}\nabla^2 \zeta + \alpha_{QG}(\frac{\partial \tau^y}{\partial x} - \frac{\partial \tau^x}{\partial y}) - r_{b1}\zeta \tag{5.28a}$$

$$\zeta = \nabla^2 \psi \tag{5.28b}$$

where $\zeta = \partial v^0/\partial x - \partial u^0/\partial y$ is the vertical component of the relative vorticity. The parameter β was already defined in (5.22) but additional parameters appear in (5.28): the bottom friction parameter r_b, the rotational Froude number F, the wind-stress coefficient α_{QG} and the Reynolds number Re. These new parameters have expressions

$$\alpha_{QG} = \frac{\tau_0 L}{\rho_0 DU^2} \quad ; \quad F = \frac{f_0^2 L^2}{gD}$$

$$r_{b1} = \frac{\sqrt{E_V}}{\epsilon} = \frac{L}{UD}\sqrt{A_V f_0} \quad ; \quad Re = \frac{E_H}{\epsilon} = \frac{UL}{A_H} \tag{5.29}$$

Values of both dimensional as well as dimensionless parameters for a basin of 1000×1000 km at $\theta_0 = 45°$N are shown in Table 5.2. In this case, the reference horizontal velocity U, for later reference indicated by U_{QG}, is chosen slightly smaller than the horizontal velocity scale in the shallow-water models.

A direct extension of the single-layer QG model is a two-layer QG model, where the densities are constant (ρ_1 and ρ_2) within the layers. Again, the interface between both layers is able to deform, the reduced gravity g' is given by $g' = g(\rho_2 - \rho_1)/\rho_0$ and the top and bottom layers have equilibrium depths D_1 and D_2 ($D = D_1 + D_2$), respectively. Within each layer, the homogeneous equations (5.28) hold and the motion of fluid in both layers is coupled through the continuity of pressure and vertical velocity. Both conditions are the same as in the two-layer shallow water model and were given in (5.13). When the QG-theory is used, the dimensionless two-layer quasi-geostrophic model is obtained as

$$\left[\frac{\partial}{\partial t} + u_1 \frac{\partial}{\partial x} + v_1 \frac{\partial}{\partial y}\right][\zeta_1 - F_1(\psi_1 - \psi_2) + \beta y] =$$

Parameter	Value		Parameter	Value	
L	1.0×10^6	m	τ_0	1.5×10^{-1}	Pa
D	6.0×10^2	m	β_0	$1.6 \ 10^{-11}$	$(ms)^{-1}$
f_0	1.0×10^{-4}	s^{-1}	A_H	$10^{3\pm1}$	$m^2 s^{-1}$
A_V	$10^{-3\pm1}$	$m^2 s^{-1}$	R_{ext}	1.0×10^6	m
ρ_0	10^3	kgm^{-3}	U_{QG}	1.6×10^{-2}	ms^{-1}
Parameter	Value		Parameter	Value	
α_{QG}	1.0×10^3		β	1.0×10^3	
F	1.0		Re	$1.6 - 160$	
r_{b1}	$1.0 - 100$				

Table 5.2.   Typical values of dimensional and dimensionless parameters for the single-layer quasi-geostrophic model. These parameters are defined in (5.12), (5.22) and (5.29). The quantity R_{ext} is the external Rossby radius of deformation, given by $R_{ext} = \sqrt{gD}/f_0$.

$$= \frac{1}{Re}\nabla^2 \zeta_1 + \alpha_{QG}\left(\frac{\partial \tau^y}{\partial x} - \frac{\partial \tau^x}{\partial y}\right) \quad (5.30a)$$

$$\zeta_1 = \nabla^2 \psi_1 \quad (5.30b)$$

$$\left[\frac{\partial}{\partial t} + u_2\frac{\partial}{\partial x} + v_2\frac{\partial}{\partial y}\right][\zeta_2 + \delta_F F_1(\psi_1 - \psi_2) + \beta y] =$$

$$= \frac{1}{Re}\nabla^2 \zeta_2 - r_2\zeta_2 \quad (5.30c)$$

$$\zeta_2 = \nabla^2 \psi_2 \quad (5.30d)$$

with additional parameters

$$F_1 = \frac{f_0^2 L^2}{g' D_1} , \ \delta_F = \frac{D_1}{D_2} ; \ r_{b2} = \frac{D}{D_2}\frac{r_{b1}}{2} \quad (5.31)$$

Again, typical values of the additional dimensional as well as dimensionless parameters for a basin of 1000×1000 km at $\theta_0 = 45°$N for the two-layer model (for typical depths D_1 and D_2) are shown in Table 5.3.

Parameter	Value		Parameter	Value	
D_1	6.0×10^2	m	D_2	1.4×10^3	m
g'	2.0×10^{-2}	ms^{-2}	R_{int}	3.5×10^5	m
Parameter	Value		Parameter	Value	
F_1	8.5×10^2		$\delta_F F_1$	4.5×10^2	
r_{b2}	$0.7 - 70$		δ_F	4.3×10^{-1}	

Table 5.3.   Typical values of additional parameters for the two-layer quasi-geostrophic model. The quantity R_{int} is the internal Rossby radius of deformation, given by $R_{int} = \sqrt{g'D_1}/f_0$.

When the bottom layer is motionless, i.e. $\psi_2 = 0$, (5.30a) reduces to a single equation for ψ_1. This equation is mathematically similar to (5.28a), but with a different value of the parameter F, i.e., F_1. This model is called an equivalent barotropic model or 1.5-layer QG model. In the same way, 1.5 layer SW models are obtained.

5.3.4. Overview of the SW and the QG models

The brief formulation of the SW models and QG models in the previous subsections completes the hierarchy of models which will be considered in the remainder of this chapter. The high-end member of the hierarchy is the multi-layer primitive equation model on the sphere and the low-end model member is the (single-layer) quasi-geostrophic barotropic vorticity equation on the β-plane.

With all the different models around for the WDC, it is desirable to have an easy reference system within the model hierarchy. There are many ways to do this and here only one of such possibilities is suggested. We can introduce dynamical classes, and use Q (quasi geostrophic), S (shallow water) and P (primitive equation) as subscripts to a W indicating the WDC. Then we add the number of layers and details in the bathymetry as subsequent superscripts. For the latter, we can use 0 for a rectangular basin, 1 for only continental geometry and 2 for full bathymetry. In this nomenclature, the model $W_Q^{1,0}$ will be the single-layer QG model with a flat bottom, while a 5-layer shallow-water model with real bathymetry has the notation $W_S^{5,2}$.

The advantage of this classification is that other dynamical classes, such as balanced models (McWilliams and Gent, 1980), can be added to the hierarchy. All ODE-type ad-hoc or reduced models can be put into one model class W_0 and superscripts can be added to indicate where the model was derived from. Finally, we can add an argument to the model to indicate the vector of parameters, say α, for example $W_Q^{1,0}[\alpha]$. Together with the coordinate system, the domain and the boundary conditions this then totally specifies a model configuration.

It is important to realize what type of assumptions underly each of the models and what physics is added/neglected when one goes up/down one level in the model hierarchy. The latter can be efficiently discussed in terms of mechanisms of production of vorticity, as presented in chapter 2. An overview of these assumptions and parameter restrictions of each of the homogeneous models is given in Table 5.4.

In the multi-layer spherical coordinates model, all vorticity production mechanisms, i.e., advection, diffusion, stretching, tilting and baroclinic vorticity production are present. The details of the baroclinic vorticity production which can be represented are limited by the choice of the vertical structure of the flow; this is basically set by the number of layers. The gradients in the density field can only be represented by gradients in the layer thicknesses. A priori, there is no restriction on the amplitude of the sea surface height and the layer thicknesses. In principle, the interface is allowed to intersect the ocean-atmosphere interface although problems can be anticipated when a layer thickness approaches zero. In the homogenous model, the baroclinic vorticity production is absent.

Overview of the type of single-layer models			
Model	Assumptions	Parameter Conditions	Momentum Equations
I. Spherical PE	hydrostatic limit	$\delta = \frac{D}{L} \ll 1$	(5.9)
II. β PE	as I & limited meridional extent	$\gamma = \frac{L}{r_0} \ll 1$	(5.23)
III. Spherical SW	as I & vertical homogeneity	$E_V \ll 1$	(5.19)
IV. β SW	as III & limited meridional extent	$\gamma = \frac{L}{r_0} \ll 1$	(5.26)
V. β QG	as III & small bottom topography	$\epsilon \ll 1$	(5.28)

Table 5.4.   Overview of the assumptions of the different homogeneous ocean (PE, SW and QG) models which form part of the hierarchy of models of the WDC described in this section.

In the SW models, there is the assumption of vertical homogeneity of the flow outside the frictional boundary layers. In the homogenous case, the mechanism of vortex tilting is eliminated, because there can be no vertical shear. However, in the multi-layer case, the vertical shear is represented by the differences in velocities between the layers reintroducing this mechanism. This is similar for either spherical or β-plane models. In the latter models, only the representation of the Coriolis acceleration is simplified and effects of Earth's curvature are neglected in the inertia and frictional terms.

In the QG models, the amplitudes of bottom topography and deformations of the equilibrium layer thicknesses are all assumed $\mathcal{O}(\epsilon)$, where ϵ is the Rossby number. The geostrophic balance is the dominant balance over the whole flow field. All ageostrophic terms are sources of potential vorticity and must balance in the QG models. In the homogeneous model, the stretching mechanism is only represented by changes in the sea surface height (the term $F\psi$). The length scale over which stretching is important is the external Rossby deformation radius R_{ext}, with $F = (L/R_{ext})^2$. In the multi-layer version, the changes in the depth of the interface separating the layers can contribute to the stretching. For example, in the two-layer model these are the terms involving $F_1(\psi_1 - \psi_2)$ and the relevant horizontal scale is the internal Rossby radius of deformation R_{int} with $F_1 = (L/R_{int})^2$.

The models can furthermore be categorized by the type of boundary conditions prescribed and the detail of the geometry taken into account. Below, rectangular

basins as well as realistic continental geometry will be considered. Having clarified the assumptions within this hierarchy models, next their solution structure is
described.

5.4. Classical Results

The quasi-geostrophic barotropic vorticity equation (5.28) is the cornerstone of
the classical explanation of the large-scale gyres, the intensification of the western
boundary currents and the theory of Rossby waves (Sverdrup, 1947; Stommel,
1948; Munk, 1950). Furthermore, the two-layer quasi-geostrophic model (5.30)
has been used extensively to understand instabilities in wind-driven flows. A brief
overview these results is given in the next subsections as a preparation for the
results which follow later on.

5.4.1. The Sverdrup-Munk-Stommel theory

The value of the horizontal velocity U_{QG} as in Table 5.2 is based on the choice

$$\alpha_{QG} = \beta \Rightarrow U = \frac{\tau_0}{\rho_0 D \beta_0 L} \tag{5.32}$$

In this case, the steady barotropic vorticity equation (5.28) can be written as

$$(\frac{\delta_{I*}}{L})^2 \left[\frac{\partial \psi}{\partial x} \frac{\partial}{\partial y} - \frac{\partial \psi}{\partial y} \frac{\partial}{\partial x} \right] \left[\nabla^2 \psi - F \psi \right] + \frac{\partial \psi}{\partial x} =$$
$$= \frac{\partial \tau^y}{\partial x} - \frac{\partial \tau^x}{\partial y} - \frac{\delta_{S*}}{L} \nabla^2 \psi + (\frac{\delta_{M*}}{L})^3 \nabla^4 \psi \tag{5.33}$$

where δ_{I*}, δ_{S*} and δ_{M*} are three internal length scales given by

$$\delta_{I*} = \sqrt{\frac{U}{\beta_0}}, \ \delta_{S*} = \frac{\sqrt{A_V f_0}}{2 D \beta_0} \text{ and } \delta_{M*} = (\frac{A_H}{\beta_0})^{\frac{1}{3}} \tag{5.34}$$

Taking values from Table 5.2, typical values of these internal length scales are
$\delta_{M*} = 18 - 85$ km, $\delta_{I*} = 32$ km and $\delta_{S*} = 5.2 - 52$ km.

Although both frictional length scales δ_{S*} and δ_{M*} can vary substantially in
magnitude due to the uncertainties in the values of A_H and A_V, all three internal
length scales are substantially smaller than the basin length L. Hence, on the basin
scale L, the dominant balance is the Sverdrup balance (Sverdrup, 1947), given by

$$\frac{\partial \psi}{\partial x} = \frac{\partial \tau^y}{\partial x} - \frac{\partial \tau^x}{\partial y} \tag{5.35}$$

Because (5.35) contains only first order derivatives to x, the Sverdrup solution can
only satisfy one boundary condition on the east-west boundaries. Hence, boundary layers are expected to appear at either eastern, western or both boundaries.
In example 5.1, the case considered in Stommel (1948) where the boundary layer
structure is provided by only bottom friction is described for a given wind-stress
forcing. In this case, bottom friction is assumed to be the dominant term in (5.33)

or $\delta_{S*} \gg \max\{\delta_{M*}, \delta_{I*}\}$. The Sverdrup-Stommel flow, as presented in Example 5.1, is the simplest flow containing a Sverdrup interior and a western boundary layer.

▶

Example 5.1: Sverdrup flow and Stommel boundary layer

Within a square basin $[0,1] \times [0,1]$, consider the wind-stress forcing

$$\tau^x(y) = -\frac{1}{2\pi} \cos 2\pi y \; ; \; \tau^y = 0$$

Under the conditions $\delta_{S*} \gg \max\{\delta_{M*}, \delta_{I*}\}$, the equation (5.33) becomes

$$\frac{\partial \psi}{\partial x} = -\frac{\partial \tau^x}{\partial y} - \frac{\delta_{S*}}{L}\nabla^2 \psi \tag{5.36}$$

As the bottom friction contains only second-order derivatives, we can only pre-scribe kinematic conditions on the lateral boundaries. In the simplest case $\psi = 0$ on all boundaries $x = 0, 1$ and $y = 0, 1$.

For the chosen wind-stress field, the solution to (5.35) satisfying the boundary conditions at $y = 0, 1$ is

$$\frac{\partial \psi}{\partial x} = -\sin 2\pi y \Rightarrow \psi(x,y) = (-x + C(y))\sin 2\pi y$$

The function $C(y)$ is determined by boundary conditions at $x = 0, 1$, but it is not clear a priori which one. Using the notation $\delta_S = \delta_{S*}/L$, a boundary layer coordinate $\mu_w = x/\delta_S$ is introduced near the western boundary and a coordinate $\mu_e = (1-x)/\delta_S$ near the eastern boundary. When transforming the equation (5.36) to obtain the boundary layer equations, note that the wind-stress curl is a slowly varying function on these scales. Let the boundary layer solutions at the western and eastern boundary be indicated by ψ_w and ψ_e, respectively.

The transformed equation (5.36) at the eastern boundary becomes

$$-\frac{\partial \psi_e}{\partial \mu_e} = -\delta_S \frac{\partial \tau^x}{\partial y} - \left[\frac{\partial^2 \psi_e}{\partial \mu_e^2} + \delta_S^2 \frac{\partial^2 \psi_e}{\partial y^2}\right]$$

The solution ψ_e of the $\mathcal{O}(1)$ balance, with respect to δ_S, in this equation is given by

$$\psi_e(\mu_e, y) = (C_1^e + C_2^e e^{\mu_e})\sin 2\pi y$$

with constants C_1^e and C_2^e. This solution has to match up at $\mu_e \to \infty$ with the Sverdrup solution ψ for $x \to 1$. However, it clearly cannot because it becomes unbounded for large μ_e unless $C_2^e = 0$. In the latter case, the condition $\psi_e(1,y) = 0$ gives also $C_1^e = 0$ and consequently $\psi_e \equiv 0$. Hence, the Sverdrup solution has

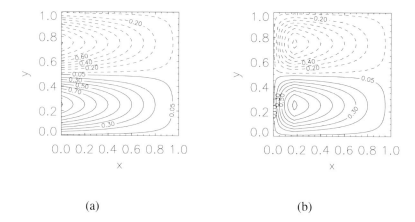

*Figure 5.8.   (a) Plot of the streamfunction of the Sverdrup solution $\psi(x,y) = (1-x)\sin 2\pi y$.
(b) Plot of the combined Sverdrup-Stommel solution, incorporating the western boundary layer flow
$\psi_w(\mu_w, y) = (1 - e^{-\mu_w})\sin 2\pi y$ with $\mu_w = x/\delta_S$ and $\delta_S = 0.1$.*

to satisfy $\psi(1,y) = 0$ and hence $C(y) = 1$. The Sverdrup flow then becomes

$$\psi(x,y) = (1-x)\sin 2\pi y$$

which is easily seen as consisting of a double-gyre flow (Fig. 5.8a).

At the western boundary, (5.33) transforms into

$$\frac{\partial \psi_w}{\partial \mu_w} = -\delta_S \frac{\partial \tau^x}{\partial y} - \left[\frac{\partial^2 \psi_w}{\partial \mu_w{}^2} + \delta_S^2 \frac{\partial^2 \psi_w}{\partial y^2}\right]$$

Again, the solution ψ_w of the $\mathcal{O}(1)$ balance in the equation above is given by

$$\psi_w(\mu_w, y) = (C_1^w + C_2^w e^{-\mu_w})\sin 2\pi y$$

The conditions $\psi_w(0, y) = 0$ and $\psi_w(\mu_w \to \infty, y) = \psi(x \to 0, y)$ determine C_1^w
and C_2^w to give

$$\psi_w(\mu_w, y) = (1 - e^{-\mu_w})\sin 2\pi y$$

The uniformly valid solution over the whole domain is plotted in (Fig. 5.8b) for
$\delta_S = 0.1$.

◄

A simple physical argument to explain why the boundary layer occurs in the
west instead of in the east was put forward by Stommel (1948). Consider only
the subtropical (southern) gyre of the flow in Fig. 5.8b, then the input of vorticity

by the wind stress is everywhere negative (clockwise). According to the Sverdrup flow, the interior meridional geostrophic velocity is everywhere negative on the domain $y \in [0, \frac{1}{2}]$. Hence, in either case of a boundary layer near the eastern or western wall, the flow must be northward to compensate the southward interior Sverdrup flow. Northward motion of fluid particles produces negative relative vorticity through the β-effect.

We now write the vorticity balance (5.33) with $\delta_{S*} \gg \max\{\delta_{M*}, \delta_{I*}\}$ in terms of velocities as

$$0 = -v - \frac{\partial \tau^x}{\partial y} - \delta_S \left(\frac{\partial v}{\partial x} + \frac{\partial u}{\partial y} \right) \tag{5.37}$$

The β induced negative vorticity due to a northward flow ($v > 0$) is the first term on the right hand side. The wind stress induced negative vorticity is represented by the second term and the last term is the vorticity induced by bottom friction. In the boundary layer, the first and the last term have to compensate. Also the zonal gradients in the meridional velocity are much larger than the meridional gradient of the zonal velocity and hence $| \partial v / \partial x | \gg | \partial u / \partial y |$.

As there are only kinematic conditions, the meridional velocity is maximal at the lateral boundaries. Hence, in the eastern boundary layer the zonal gradient of meridional velocity is positive ($\partial v / \partial x > 0$) and both the first and third term in (5.37) are negative and cannot compensate. In other words, the vorticity generated through bottom friction cannot balance that due to the β-effect. In the western boundary layer, the zonal gradient of the meridional velocity is negative ($\partial v / \partial x < 0$), the first and term in (5.37) have opposite sign, and hence the vorticity produced by both bottom friction and β-effect can compensate (Fig. 5.9). Hence, the boundary flow can only occur in the west, as follows directly mathematically from the boundary layer analysis (Example 5.1).

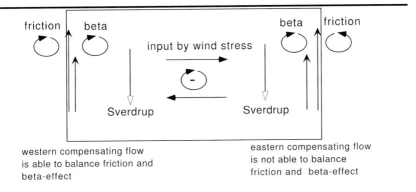

friction   beta   beta   friction

input by wind stress

Sverdrup   Sverdrup

western compensating flow is able to balance friction and beta-effect   eastern compensating flow is not able to balance friction and beta-effect

Figure 5.9. Sketch of the vorticity balances from which one can deduce that the compensating flow of the Sverdrup transport can only occur in the western boundary layer.

Nonlinearities due to advection of vorticity introduce north-south asymmetries and are responsible for strong recirculation regions (Veronis, 1966). Although

this regime will be explored in more detail in the following sections using bifurcation analysis, many classical theoretical results have been obtained. In the fully
inertially dominated regime, for example, special type of motion may occur, i.e.,
the Fofonoff inertial solution (Fofonoff, 1954) or the modon-type solution (Stern,
1975). An overview of these results is presented in chapter 5 of Pedlosky (1987).

5.4.2. Temporal variability

Apart from time-dependent forcing, which can be a major source of temporal
variability (for example the annual cycle), there are also internal sources of variability in large-scale ocean flows. There are at least two processes leading to this
variability: adjustment of the flow to changes in the forcing through the propagation of waves and the development of instabilities on the mean flow. Each of these
processes will be considered briefly in the following subsections.

5.4.2.1 Rossby waves

The homogeneous quasi-geostrophic model (5.28) has also been important to
understand wave phenomena in the ocean. Within this model, small amplitude
motions on a motionless background state in the unforced, frictionless model satisfy the dimensionless equation

$$\frac{\partial}{\partial t}(\nabla^2 \psi - F\psi) + \beta \frac{\partial \psi}{\partial x} = 0 \qquad (5.38)$$

In a horizontally unbounded domain, free wave solutions

$$\psi(x, y, t) = \Psi_0 \, e^{i(kx + ly - \sigma t)} \qquad (5.39)$$

exists for waves with wavenumber $\mathbf{k} = (k, l)$, phase speed $\mathbf{C} = (\sigma/k, \sigma/l)$ and
arbitrary amplitude Ψ_0. These waves are characterized by a dispersion relation

$$\sigma = -\frac{\beta k}{k^2 + l^2 + F^2} \qquad (5.40)$$

and represent westward propagating waves called Rossby waves. In a two-layer
quasi-geostrophic model, there are two types of Rossby modes: one barotropic
mode which has no vertical structure (both layers have the same response) and
a baroclinic mode (where the response in both layers is different). Dispersion
relations of these waves follow directly by considering small amplitude motions
in the unforced, frictionless model (5.30) and become

$$\sigma_{bt} = -\frac{\beta k}{k^2 + l^2} \quad : \quad \sigma_{bc} = -\frac{\beta k}{k^2 + l^2 + F_1(1 + \delta_F)} \qquad (5.41)$$

where the subscripts bc and bt refer to baroclinic and barotropic waves, respectively. In the latter equation, δ_F is again the depth ratio D_1/D_2.

▶

Example 5.2: Propagation of Rossby waves

The propagation mechanism of the barotropic Rossby wave can be understood, by writing (5.38), for $F = 0$, in the form

$$\frac{D}{dt}(\zeta + \beta y) = 0$$

which is a statement of conservation of the potential vorticity $\zeta + \beta y$ (see section 2.3). Consider now three initially motionless columns of liquid, I, II and III on a certain latitude. If column II moves slightly northward, the term βy increases and hence to conserve potential vorticity, the column starts to rotate clockwise since $\zeta_{II} < 0$. This motion induces downward velocities at column III and upward

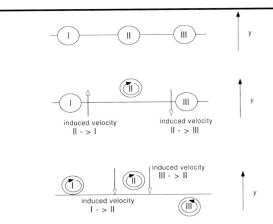

Figure 5.10.   Sketch to illustrate the mechanism of propagation of a barotropic Rossby wave.

velocities at column I (Fig. 5.10). Again by conservation of potential vorticity in each column the induced vorticities are $\zeta_I < 0$ and $\zeta_{III} > 0$, and both motions induce downward velocities on column II, which draws it back to its equilibrium position. In this way, a westward propagating wave motion results (Pedlosky, 1987).

◀

Consider Rossby waves in a zonal channel with a meridional extent of 1000 km and only those waves for which $l = \pi$. For the parameters as in Table 5.2, typical travel times of both barotropic and baroclinic Rossby waves over a zonal distance of 1000 km are provided in Table 5.5. Note that to convert to dimensional time, the factor is L/U_{QG}. In Table 5.5, the quantity λ_* is the dimensional zonal wavelength $\lambda_* = 2\pi L/k$, C_{x*} is the zonal phase speed ($C_{x*} = U_{QG}\sigma/k$) and

k	F	$\lambda_*(km)$	$C_{x*}(m/s)$	$T_{b*}(days)$
π	0	2000	0.811	14.2
2π	0	1000	0.324	35.6
π	100	2000	0.134	86.6
π	1200	2000	0.016	1031.3

Table 5.5. Typical values of basin crossing times T_{b*} for midlatitude Rossby waves (with $l = \pi$) traveling over a distance $L = 1000$ km. In this table, the second column is the quantity F (for the barotropic waves), as well as the quantity $F_1(1 + \delta_F)$ (for the baroclinic waves).

T_{b*} is the travel time $T_{b*} = L/C_{x*}$. This table illustrates that barotropic Rossby waves are much faster than baroclinic Rossby waves and that longer waves have shorter travel times. Typical time scales are in the order of days for the barotropic waves and years for the baroclinic waves.

5.4.2.2 Rossby basin modes

In a bounded basin, say the domain $[0, 1] \times [0, 1]$, the free solutions become so-called Rossby basin modes which satisfy kinematic boundary conditions, i.e., $\psi = 0$ at the lateral boundaries. For the barotropic case, solutions of (5.38) can be found of the form (Pedlosky, 1987)

$$\psi(x, y, t) = \Psi_0 \sin n\pi x \sin m\pi y \, e^{-i(\sigma_{nm}t + \frac{\beta x}{2\sigma_{nm}})} \qquad (5.42)$$

where Ψ_0 is again an arbitrary amplitude. The dispersion relation of these modes in a square basin is given by

$$\sigma_{nm} = -\frac{2\beta}{\sqrt{(n\pi)^2 + (m\pi)^2 + F^2}} \qquad (5.43)$$

where the indices (n, m) refer to the spatial structure of the basin modes as defined in (5.42). For the two-layer case, both barotropic and baroclinic Rossby basin modes exist.

Note that each σ_{nm} defines a period of oscillation T_{p*} given by

$$T_{p*} = 2\pi L/(\mid \sigma_{nm} \mid U_{QG})$$

The propagation direction can be seen through the pattern changes along one oscillation and is westward. For the parameters in the Tables 5.2 and 5.3, typical propagation periods of both barotropic and baroclinic basin modes are given in Table 5.6. For the gravest barotropic ocean basin mode $(n, m) = (1, 1)$, the propagation time scale is about 20 days. The pattern of this mode is shown in Fig. 5.11 at three instances during its propagation. The propagation mechanism of these modes is similar to that of Rossby waves, but their spatial structure is different. These modes can also be described by a sum of free Rossby waves where the coefficients are chosen such that the boundary conditions are satisfied.

(n, m)	F	$T_{p*}(days)$
$(1, 1)$	0	20.3
$(1, 1)$	100	454
$(2, 1)$	0	31.9
$(1, 1)$	1200	4500

Table 5.6. Typical values of travel times of Rossby basin modes over a distance of 1000 km in the North Atlantic. The value of F indicates both values for the barotropic modes, as well as the quantity $F_1(1 + \delta_F)$ for the baroclinic modes.

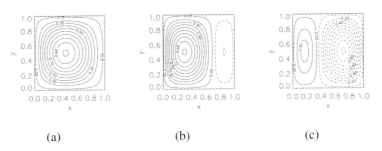

     (a)   (b)   (c)

Figure 5.11. Sketch of the streamfunction pattern of the $(1, 1)$ basin mode at three different instances during its propagation. In the contour plots, the contour values are with respect to the absolute maximum of the field ψ_m. (a) $\sigma_{11}t = 0$, $\psi_m = 0.867$, (b) $\sigma_{11}t = \frac{1}{2}\pi$, $\psi_m = 0.353$ and (c) $\sigma_{11}t = \frac{3}{4}\pi$, $\psi_m = 0.305$.

5.4.2.3 Basic instability mechanisms

Large-scale ocean flows are susceptible to instabilities. The growth of perturbations and their interaction with the background state (and with each other) leads to a rectification of the mean state and synoptic scale time-dependent features. Much of the understanding of the instability mechanisms has been obtained through the study of zonal flows in β-plane channels (Pedlosky, 1987). Two instability mechanisms are central in geophysical flows. The first mechanism is barotropic instability, where perturbations derive their kinetic energy from the horizontal shear of the basic state (Kuo, 1951). The second is baroclinic instability (Eady, 1949), in which the kinetic energy of the perturbations is drawn from the potential energy of the basic state associated with the existence of vertical shear.

Prototype situations to understand both instability mechanisms in detail exist. In Kuo (1951), the stability of a zonal jet is considered in a barotropic quasi-geostrophic model and stability bounds are derived numerically. Two prototype situations to understand baroclinic instability are the continuously stratified Eady model (Eady, 1949) and the two-layer Phillips model (Phillips, 1951); these case are extensively described in chapter 7 of Pedlosky (1987). Here, a typical case will be described where both instability mechanisms occur, usually referred to as a mixed barotropic/baroclinic instability.

An example, illustrating the essence of both mechanisms, is the stability of a zonal jet within a β-plane channel ($y \in [-1, 1]$) using the two-layer quasi-geostrophic model (Van der Vaart and Dijkstra, 1997). The dimensionless zonal velocity profile of the jet is prescribed analytically as

$$\bar{u}_1(y; \nu) = \frac{sech^2(\frac{y}{\nu}) - sech^2(\frac{1}{\nu})}{sech^2(\frac{1}{\nu})} \quad ; \quad \bar{u}_2(y) = \alpha\bar{u}_1(y; \nu) \tag{5.44}$$

where $sech(z) = 1/\cosh(z)$. This basic flow is such that $\bar{u}_1(-1; \nu) = \bar{u}_1(1; \nu) = 0$; the parameter ν measures the width of the jet. For a value $\nu = 0.3$, the zonal velocity profile $\bar{u}_1$ is plotted in Fig. 5.12. The basic state (5.44) is a solu-

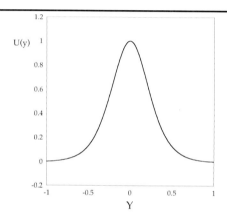

Figure 5.12.   Zonal velocity profile $\bar{u}_1 = U(y)$ of the basic state (5.44) for $\nu = 0.3$.

tion of the unforced, frictionless two-layer model (5.30) as can be easily verified. The difference in strength of the zonal velocity in both layers is monitored by α. A pressure difference $\psi_1 - \psi_2$ exists between the layers, which is immediately reflected in a slope of the interface between the layers according to (5.16b). This is a special case of the so-called thermal-wind balance in hydrostatic and geostrophic flows, which relates horizontal gradients in density to vertical gradients in the horizontal velocities.

In looking at the linear stability of the jet (5.44), the class of perturbations considered are those which satisfy only kinematic conditions at the channel walls, while being periodic in the zonal direction. Using the two-layer model (5.30), the equations for infinitesimal perturbations on the basic state (5.44) can be written in terms of the perturbation streamfunctions ϕ_j in both layers as

$$(\frac{\partial}{\partial t} + \bar{u}_1\frac{\partial}{\partial x})q_1 + \overline{\Pi}_1'\frac{\partial\phi_1}{\partial x} + r\nabla^2\phi_1 = 0 \tag{5.45a}$$

Parameter	Value		Parameter	Value	
L	1.0×10^5	m	D_2	1.4×10^3	m
D_1	5.0×10^2	m	β_0	$2.0 \ 10^{-11}$	$(\text{ms})^{-1}$
f_0	1.0×10^{-4}	s^{-1}	g'	2.0×10^{-2}	ms^{-2}
ρ_0	1.0×10^3	kgm^{-3}	U_{QG}	1.0×10^{-1}	ms^{-1}
Parameter	Value		Parameter	Value	
α	0.22		F_1	13.2	
δ_F	0.22		β	0.22	
r	0.2		ν	0.3	

Table 5.7.   Standard values of the parameters for the Gulf Stream regime of a zonal channel flow of horizontal dimension $L = 100 \ km$. The meaning of the quantities is the same as in Table 5.3.

$$(\frac{\partial}{\partial t} + \bar{u}_2 \frac{\partial}{\partial x})q_2 + \overline{\Pi}_2' \frac{\partial \phi_2}{\partial x} + r\nabla^2 \phi_2 \quad = \quad 0. \qquad (5.45b)$$

where the potential vorticity of the perturbations q_j and the potential vorticity gradient of the basic state $\overline{\Pi}'$ are given by

$$q_1 \quad = \quad \nabla^2 \phi_1 - F_1(\phi_1 - \phi_2) \qquad (5.46a)$$
$$q_2 \quad = \quad \nabla^2 \phi_2 + \delta_F F_1(\phi_1 - \phi_2) \qquad (5.46b)$$
$$\overline{\Pi}_1'(y) \quad = \quad \beta + F_1(1 - \alpha)\bar{u}_1 - \frac{d^2\bar{u}_1}{dy^2} \qquad (5.46c)$$
$$\overline{\Pi}_2'(y) \quad = \quad \beta - \delta_F F_1(1 - \alpha)\bar{u}_1 - \frac{d^2\bar{u}_2}{dy^2} \qquad (5.46d)$$

and only linear friction has been taken into account in both layers through the coefficient r. These equations are complemented with boundary conditions at the channel walls in both layers

$$y = \pm 1 : \frac{\partial \phi_j}{\partial x} = 0 \qquad (5.47)$$

corresponding to the kinematic condition of zero normal flow at the channel walls.

The system of equations (5.45) allows for traveling wave solutions in the x-direction with wavenumber k, complex growth factor σ and unknown meridional structure, i.e.

$$\Phi(x, y, t) = \hat{\Phi}(y)e^{ikx + \sigma t} \qquad (5.48)$$

with the complex function $\hat{\Phi} = (\phi_1, \phi_2)$. The eigenvalue σ is written as $\sigma = \lambda + i\omega$ and considered as a function of the wavenumber k and a control parameter, in this case chosen as β^{-1}, of the system. If $\lambda > 0$ for a particular wavenumber k, the basic state is unstable. The neutral curve, $\lambda(k, \beta^{-1}) = 0$ in the (k, β^{-1}) plane separates linearly stable basic states from unstable ones.   The parameters in Table 5.7 are used in the two-layer model to represent the Gulf Stream in a zonal channel; choices of parameters are discussed in Flierl (1978). The neutral curve in the (k, β^{-1}) plane is shown in Fig. 5.13 and has a minimum at (k_c, β_c^{-1}). These critical values occur at $\beta_c^{-1} = 2.2$ and $k_c = 2.35$, the latter corresponding to a wavelength of 260 km. Since the standard value $\beta^{-1} = 4.5$ for this regime (Table 5.7) is larger than β_c^{-1}, the zonal jet described by (5.44) is unstable.

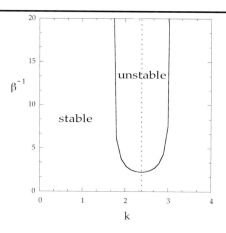

Figure 5.13.   Neutral curve for the parameter values of the Gulf Stream regime as given in Table 5.7. The parameter β^{-1} is used as control parameter and k is the dimensionless wavenumber (Van der Vaart and Dijkstra, 1997).

The perturbation streamfunction of the most unstable mode at one phase of the oscillation (the solution (5.48) at t = 0) is for both layers shown in Fig. 5.14 with dark (light) shading indicating positive (negative) values. The spatial patterns have a symmetric structure in both layers with maximum amplitude at the center line of the jet ($y = 0$) in the upper layer and off the axis for the lower later. There exists a phase shift between ϕ_1 and ϕ_2 characteristic of baroclinic effects in the instability.

One way to understand the nature of the instability is to investigate changes in the mechanical energy balance. For the two-layer model this balance can be derived by multiplying equation (5.45a) by ψ_1, equation (5.45b) by ψ_2 and integrate the sum of both results over the domain to give (Pedlosky, 1987)

$$\frac{\partial E}{\partial t} = I_1 + I_2 + C_p - D \tag{5.49}$$

where the total energy E is given by the sum of kinetic and potential energy over both layers

$$E = \frac{1}{2} \int_{-1}^{1} \left(\delta_F(|u_1|^2 + |v_1|^2) + |u_2|^2 + |v_2|^2 + \delta_F F_1 |\phi_1 - \phi_2|^2 \right) dy, \tag{5.50}$$

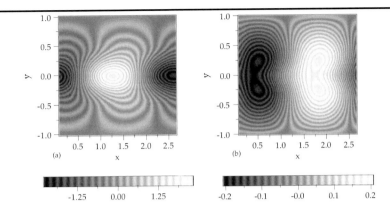

Figure 5.14. *Perturbation streamfunction* $\hat{\mathbf{\Phi}}$ *of the neutral mode at the minimum of the curve in Fig. 5.13 in layer 1 (ϕ_1) and layer 2 (ϕ_2). Dark (light) shading indicates positive (negative) values (Van der Vaart and Dijkstra, 1997).*

with perturbation velocities in both layers represented by $(u_j, v_j), j = 1, 2$. The dissipation D is given by

$$D = \frac{r}{2} \int_{-1}^{1} (\, \delta_F(|u_1|^2 + |v_1|^2) + |u_2|^2 + |v_2|^2 \,) \, dy, \qquad (5.51)$$

The energy production terms I_1 and I_2 due to Reynolds' stresses in both layers are given by

$$I_1 = \int_{-1}^{1} (\Re(u_1 v_1^*)) \frac{d}{dy} \overline{u}_1(y) \, dy \qquad (5.52a)$$

$$I_2 = \int_{-1}^{1} (\Re(u_2 v_2^*)) \frac{d}{dy} \overline{u}_2(y) \, dy \qquad (5.52b)$$

where a $*$ superscript indicates complex conjugate and $\Re$ indicates the real part of the expression which follows. Finally, the potential energy conversion C_p is expressed by

$$C_p = \delta_F F_1 (1 - \alpha) \int_{-1}^{1} \overline{u}_1(y) \Re(-v_1 \phi_2^*) \, dy \qquad (5.53)$$

These terms nicely illustrate the instability mechanisms at work. Clearly, the dissipation is always positive and is stabilizing. First, suppose that the mean flow has no vertical shear ($\alpha = 0$), then the potential energy conversion term is zero. Energy production sufficient to overcome the dissipation can then only be generated by the Reynolds stress terms through horizontal shear. In this barotropic instability mechanism, the horizontal shear of the basic state is able to transfer energy through the Reynolds stress to the perturbations.

When $\alpha \neq 0$, but the horizontal shear of the mean state is absent, then $I_1 = I_2 = 0$ and no barotropic instability can occur. However, the zonal jet can still become unstable because the potential energy conversion term C_p may overcome the dissipation. Because the vertical shear of the zonal jet is associated with a slope in the interface between the layers, it is possible that potential energy is released through the associated stratification and transferred to the kinetic energy of the perturbations. This is the baroclinic instability mechanism. One can indeed describe this instability in more detailed mechanistic terms; this is done in several textbooks (Holton, 1992; Pedlosky, 1987).

The net perturbation energy term $\partial E / \partial t = \Re(\sigma)E = \lambda E$ is zero at neutral conditions and the other quantities in (5.50) can be computed as β is decreased through its critical value β_c along the dotted curve in (Fig. 5.15). Note that instability occurs for values of $\beta/\beta_c < 1$ and that both I_1 and C_p are positive. The perturbation patterns used to compute these terms are those in Fig. 5.14. When β decreases, the term C_p is fairly constant, the term I_1 slightly decreases, but the dissipation term decreases more rapidly. For values of β slightly smaller than β_c, both barotropic and baroclinic mechanisms provide enough energy to the perturbations to cause exponential growth.

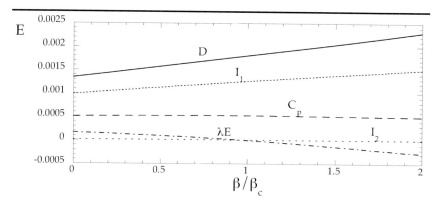

Figure 5.15.   Terms of the mechanical energy balance (5.50) computed with the perturbations in Fig. 5.14, when β is changed through its critical value (Van der Vaart and Dijkstra, 1997).

5.5. Regimes of Double-Gyre QG Flows

With the knowledge of the linear theory of the steady ocean circulation and the basic mechanisms of temporal variability, the bifurcation behavior of the full nonlinear circulation is presented systematically in the following sections. As a 'bottom-up' approach is followed, we start in section 5.5.1 with the equivalent barotropic model and look at the transition to complex flows when the friction is decreased in this model. In section 5.5.2, flows modeled by two- and three-layer quasi-geostrophic models are considered.

5.5.1. Equivalent barotropic flows

The simplest model of this type is the quasi-geostrophic barotropic vorticity equation (5.28) in a rectangular basin ([0, 1] × [0, A]), where A is the aspect ratio of the basin, and other dimensionless parameters as in Table 5.2. For convenience, the equations of this model are again written out in full and become

$$
(\frac{\partial}{\partial t} + u\frac{\partial}{\partial x} + v\frac{\partial}{\partial y})(\zeta - F\psi + \beta y) = \frac{1}{Re}\nabla^2\zeta - r_b\nabla^2\psi
$$
$$
+ \alpha_{QG}(\frac{\partial \tau^y}{\partial x} - \frac{\partial \tau^x}{\partial y}) \quad (5.54a)
$$
$$
\zeta = \nabla^2\psi \quad (5.54b)
$$

No-slip boundary conditions are prescribed at the east-west boundaries and slip conditions at the north-south boundaries, i.e.

$$
x = 0, 1 \quad : \quad \psi = 0, \quad \frac{\partial \psi}{\partial x} = 0 \quad (5.55a)
$$

$$
y = 0, A \quad : \quad \psi = 0, \quad \frac{\partial^2 \psi}{\partial y^2} = 0 \quad (5.55b)
$$

The unforced system of equations (5.54) has a reflection symmetry with respect to the mid-axis ($y = A/2$) of the basin. This reflection $\mathcal{R}_{QG}$ has a representation

$$
\mathcal{R}_{QG}(\psi(x, A - y)) = -\psi(x, y);
$$
$$
\quad (5.56)
$$
$$
\mathcal{R}_{QG}(u(x, A - y)) = u(x, y) \quad ; \quad \mathcal{R}_{QG}(v(x, A - y)) = -v(x, y)
$$

Hence, if the applied wind forcing is symmetric with respect to $y = A/2$, the equations (5.54) are invariant with respect to $\mathcal{R}_{QG}$. This symmetry will put constraints on the bifurcations which can occur. The double-gyre wind-stress forcing is chosen as

$$
\tau^x(y) = -\frac{A}{2\pi}\cos(2\pi\frac{y}{A}) \; ; \; \tau^y = 0 \quad (5.57)
$$

and since it is symmetric with respect to $y = A/2$, symmetry-breaking bifurcations of pitchfork type are expected.

5.5.1.1 Basic bifurcation diagrams

Cessi and Ierley (1995) determined stationary solutions of the equations (5.54) over a large range of conditions for the case $A = 1, F = r_b = 0$ using a classical Newton-Raphson scheme. In the parameter plane of the Munk western boundary layer thickness δ_{M*} and the inertial western boundary layer thickness δ_{I*} (as in (5.34)), regions of multiple equilibria are found. Even seven steady solutions exist in a small area of parameter space. At some value of δ_{I*}, with fixed δ_{M*}, the symmetric double-gyre circulation becomes unstable to anti-symmetric perturbations leading to asymmetric double-gyre flows. There are at least two such pitchfork bifurcations and the second one occurs in the inertially dominated regime (at relatively large δ_{I*}). Although Cessi and Ierley (1995) investigate the stability to stationary perturbations, they exclude oscillatory perturbations, and hence no complete picture of the stability of the steady states is given.

Dijkstra and Katsman (1997) computed bifurcation diagrams for the barotropic QG-model, also for the case $A = 1, F = r_b = 0$, using a pseudo-arclength continuation technique. The structure of the steady solutions is shown through the bifurcation diagram in Fig. 5.16, where the value of the streamfunction at a point in the southwest of the domain (ψ_{SW}) is plotted versus $Re = U_{QG}L/A_H$. Each point on the curves represents a steady state and its stability is indicated by the linestyle, with solid (dashed) curves indicating stable (unstable) solutions. At large values of A_H (small Re), the symmetric double-gyre flow (Fig. 5.17a) is a unique state. Note that although the streamfunction is anti-symmetric with respect to the mid-axis of the basin, we call the pattern in Fig. 5.17a the symmetric double-gyre flow because it satisfies (5.56).

When lateral friction is decreased, this flow becomes unstable at the pitchfork bifurcation P_1 and two stable asymmetric states appear for smaller values of A_H (larger Re). The solutions on these branches have the jet displaced either southward (Fig. 5.17b) or northward (Fig. 5.17c) and are exactly symmetrically related for the same value of Re. For even smaller friction, the symmetric flow becomes inertially dominated (Fig. 5.17d) and ψ_{SW} increases rapidly. Note that this is not visible when a value of ψ on the mid-axis of the basin is chosen as indicator of the solution, since this value is zero for the symmetric double-gyre solution for all values of Re. The asymmetric solution branches arising from the bifurcation P_2 are not shown; all these solutions are unstable.

The existence of the bifurcation P_1 (Fig. 5.16) captures the heart of the physics of symmetry breaking in these flows. The physical mechanism of the instability can be analyzed with help of the approach outlined in section 3.7.3.2. Thereto we use the patterns of the steady state and the eigenvector of the linear stability analysis which has a zero growth rate just at P_1. The streamfunction and vorticity field of the steady state are presented in Fig. 5.18. The spatial patterns of the streamfunction and vorticity perturbation (determined from the eigenvector) are shown in Fig. 5.19.

The streamfunction perturbation has a tripole-like structure with a negative vorticity center along the jet-axis and two positive vorticity centers at either side (Fig. 5.19). The special property of these patterns is that the center negative vorticity lobe is exactly localized within the vorticity extrema of the symmetric basic state (Fig. 5.18). If we consider the region just above the symmetry line of the eastward jet ($y = 0.5$), the perturbation zonal flow is eastward, and therefore in the same direction as that of the basic state. More northward (above $y = 0.7$), the perturbation flow is westward and therefore also in the same direction as the steady flow. If we consider the flow just below the symmetry line of the steady jet, it is observed that the flow perturbations are in the opposite direction to that of the basic state. Hence, the flow perturbation weakens the subtropical gyre and strengthens the subpolar gyre. The asymmetric change in the strength of the basic flow due to the perturbations leads to increased horizontal shear in the eastward

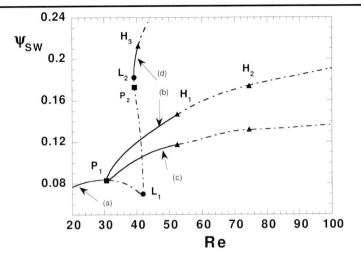

Figure 5.16. Bifurcation diagram (from Dijkstra and Katsman (1997)) for the barotropic QG-model for a square basin ($L = 1000$ km) with $Re = U_{QG}L/A_H$ as the control parameter. Other parameters are as in Table 5.2.

jet, which leads to an additional negative vorticity. This extra vorticity just amplifies the original perturbation flow in this region leading to instability (Fig. 5.19). A similar mechanism can, for example, be put forward for the region of positive vorticity with its maximum near $x = 0.3$, $y = 0.7$ in Fig. 5.19. In each region, the destabilization of the flow is induced by horizontal shear and hence is barotropic.

The symmetry-breaking mechanism leads to two stable asymmetric steady states for values of Re slightly larger than that at P_1. However, these states also become unstable at larger values of Re due to the occurrence of Hopf bifurcations. The first two on each asymmetric branch are labelled H_1 and H_2. The pattern of the oscillatory mode which destabilizes the asymmetric double-gyre flow at each Hopf bifurcation can be determined from the solution of the linear stability problem. At the Hopf bifurcation, a complex conjugate pair of eigenvalues $\sigma = \sigma_r \pm i\sigma_i$ crosses the imaginary axis. The corresponding complex eigenfunction $\hat{\mathbf{x}} = \hat{\mathbf{x}}_R + i\hat{\mathbf{x}}_I$ provides the disturbance structure $\Phi(t)$ with angular frequency σ_i and growth rate σ_r to which the steady state becomes unstable, i.e.,

$$\Phi(t) = e^{\sigma_r t}\left[\hat{\mathbf{x}}_R \cos(\sigma_i t) - \hat{\mathbf{x}}_I \sin(\sigma_i t)\right] \tag{5.58}$$

Propagation features of a neutral eigenmode ($\sigma_r = 0.0$) can be determined by first looking at $\Phi(-\pi/(2\sigma_i)) = \hat{\mathbf{x}}_I$ and then at $\Phi(0) = \hat{\mathbf{x}}_R$. The period $\mathcal{P}$ of the oscillation is given by $\mathcal{P} = 2\pi/\sigma_i$ with dimensional value $\mathcal{P}_* = \mathcal{P}L/U_{QG}$.

The patterns of this perturbation at H_1 are shown for four phases within half a period of the oscillation (which is at $\sigma_i t = \pi$) in Fig. 5.20a-d, whereas the steady

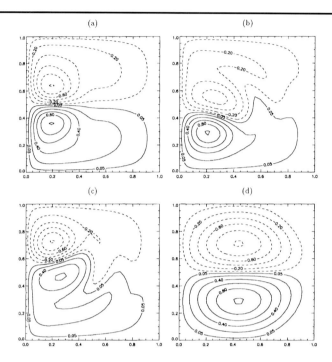

Figure 5.17.   Steady solutions for the streamfunction ψ at labelled points (a)-(d) in Fig. 5.16. All contour plots are scaled with the absolute maximum and the contour levels are with respect to this maximum. Along the horizontal and vertical axes, the dimensionless quantities $x = x_*/L$ and $y = y_*/L$ are shown (from Dijkstra and Katsman (1997)).

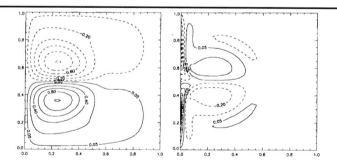

Figure 5.18.   Contour plots of the steady state at the pitchfork bifurcation P_1 in Fig. 5.16 with the streamfunction ($\bar{\psi}$) in the left panel and the vorticity ($\bar{\zeta}$) in the right panel (from Dijkstra and Katsman (1997)).

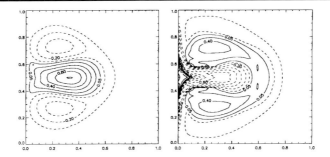

Figure 5.19. Contour plots of the perturbation destabilizing the steady state of Fig. 5.18 with the streamfunction ($\tilde{\psi}$) in the left panel and the vorticity ($\tilde{\zeta}$) in the right panel (from Dijkstra and Katsman (1997)).

state which is destabilized is close to that in Fig. 5.17c. The oscillation consists of a basin-scale perturbation that propagates through the basin. Its dimensionless frequency $\sigma_i = 75.0$ which corresponds to an intermonthly period. When this oscillatory mode is followed to zero forcing, it can be shown to 'deform' into a stable Rossby-basin mode as considered in section 5.4.2.2. The propagation mechanism of the oscillatory pattern therefore can be deduced from that of the (inviscid) Rossby-basin modes, which is in turn similar to that of free Rossby waves. The growth rate of the perturbation is determined by the horizontal shear within the asymmetric double-gyre flow.

The patterns of the oscillatory mode which destabilizes the solutions on the asymmetric branch in Fig. 5.16 containing the flow labelled (b) at H_2, are shown in Fig. 5.21a-d; the steady state at H_2 is similar to that in Fig. 5.17b. This mode has an interannual period ($\sigma_i = 5.43$) and the perturbations strengthen and weaken the jet during both phases of the oscillation (Figs. 5.21). The interannual modes do not have their origin in the spectrum of the linear operator (5.38) related to free Rossby-wave propagation. The patterns of these modes already indicate a close orientation within the gyres and the interannual time scale of these modes is related to the circulation time scale within the gyres which is in the order of 3 years. Although these modes can obtain positive growth factors by horizontal shear at small friction they disappear when the amplitude of the wind forcing becomes very weak; these therefore have been called gyre modes (Dijkstra and Katsman, 1997).

Simonnet and Dijkstra (2002) clarified the spectral origin of the gyre mode and presented a physical mechanism of the propagation of the pattern of this mode. For a barotropic QG model with slightly different parameters as in Table 5.2, the first pitchfork bifurcation P_1 occurs at $Re = 29.4$ and the gyre mode destabilizes the asymmetric solutions at the Hopf bifurcation H_2 at $Re = 83.2$. The gyre mode, therefore has a negative growth factor for $Re < 83.2$. In the right panels

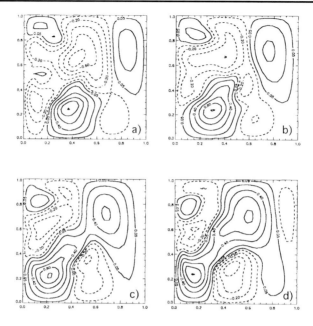

Figure 5.20.   Contour plot of the streamfunction of the of the oscillatory mode at the Hopf bifurcation H_1 in Fig. 5.16 at several phases of the oscillation (from Dijkstra and Katsman (1997)). (a) $\sigma_i t = 0.0$; (b) $\sigma_i t = \pi/4$; (c) $\sigma_i t = \pi/2$; (d) $\sigma_i t = 3\pi/4$.

of Fig. 5.22, the patterns of the real and imaginary parts of this eigenmode are shown near $Re = 40$. The path of the gyre mode (the dash-dotted curve in Fig. 5.22) ends at the point M, where it splits into two stationary eigenmodes. These stationary modes exist up to the point P_1 where the asymmetric solutions cease to exist.

Also shown in Fig. 5.22 are the leading eigenmodes on the symmetric solution branch. The non-oscillatory mode responsible for the first pitchfork bifurcation (P_1) has a symmetric tripolar structure (Fig. 5.22, upper-left panel), similar to the streamfunction pattern in Fig. 5.19. At P_1, the growth rate σ_r of this mode becomes positive which means that the symmetric steady flow becomes unstable to this perturbation pattern. Because this mode is responsible for multiple equilibria and asymmetric states under symmetric forcing conditions, Simonnet and Dijkstra (2002) called it the P-mode. The P-mode streamfunction keeps its symmetric structure unchanged over the whole Re-range.

The non-oscillatory mode responsible for the saddle-node bifurcation at L_1 in Fig. 5.16 has a dipolar anti-symmetric structure (Fig. 5.22, lower-left panel). It thus acts on both gyres simultaneously so that they either increase or decrease in intensity. Simonnet and Dijkstra (2002) called this non-oscillatory mode (involved

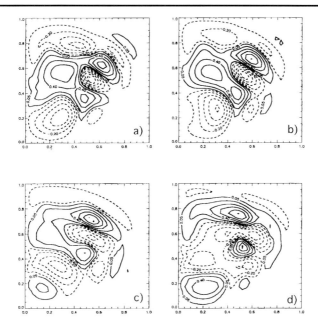

Figure 5.21.   Contour plot of streamfunction of the of the oscillatory mode at the Hopf bifurcation H_2 in Fig. 5.16 at several phases of the oscillation (from Dijkstra and Katsman (1997)). (a) $\sigma_i t =$ 0.0; (b) $\sigma_i t = \pi/4$; (d) $\sigma_i t = \pi/2$; (e) $\sigma_i t = 3\pi/4$.

in the saddle-node bifurcation, or limit point) the L-mode. At P_1, the L-mode is damped (with a dimensionless growth rate $\sigma_r \approx -5$) and σ_r becomes positive at the saddle-node bifurcation. The L-mode perturbation (as shown) increases the energy of both gyres so that the symmetric steady flow will reach a more energetic state. On the contrary, if the perturbation has opposite sign, both gyres will decrease simultaneously so that the basic flow reaches a less energetic state.

Relevant for the spectral origin of the gyre mode is the path of both the P-mode and L-mode on the asymmetric branches for $Re > 29.4$. For Re slightly above P_1, both modes are still non-oscillatory and have negative growth factor since the asymmetric branch is stable. The paths of the eigenvalues of both modes are indicated by the thick lines in Fig. 5.22 (starting at $Re = 29.4$). The growth factor of the P-mode decreases with Re, whereas that of the L-mode increases. Both modes meet at the point M (Fig. 5.22), which Simonnet and Dijkstra (2002) called the *merging point*, and give birth to the gyre mode.

Simonnet and Dijkstra (2002) showed that this mode merging also occurred in a much simpler truncated QG-model, having only two degrees of freedom, as formulated in Jiang *et al.* (1995). From this, the physical mechanism of the exis-tence of the gyre mode could be described. The growth and decay of the energy

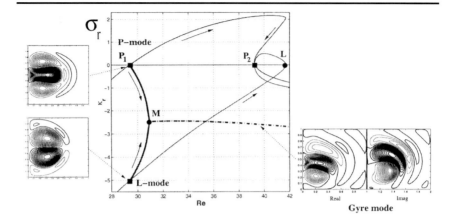

Figure 5.22.   Real part σ_r of the eigenvalues closest to the imaginary axis of the linear stability of the symmetric double-gyre flow (thin lines) and along one of the asymmetric steady states (thick lines). The P-mode (streamfunction pattern in the upper left panel) destabilizes the symmetric state at the pitchfork P (from Simonnet and Dijkstra (2002)). Along the asymmetric states, however, it deforms and merges with the L-mode (streamfunction pattern in the lower left panel) at the point M. This gives rise to the gyre mode (streamfunction patterns in the right panel).

of the gyre mode in the different stages of the oscillation is determined by a pure shear mechanism and free of Rossby-wave dynamics. The combined effect of this shear, the asymmetry of the steady flow and the symmetry-breaking mechanism – which induces temporal constraints between the symmetric and anti-symmetric components of the perturbations – causes the low-frequency oscillation. In this way, the oscillatory behavior of the gyre mode is linked to the same mechanism leading to the appearance of multiple equilibria at the (first) pitchfork bifurcation. The gyre mode tries to break the 'symmetry' of the steady solution on the asymmetric branch (growth of the P-mode) but the resulting flow does not reach a steady equilibrium state because the background flow is asymmetric (growth of the L-mode).

So far, the horizontal scale of the basin ($L = 1000$ km) was kept smaller than that of typical ocean basins. Primeau (2002) computed the bifurcation diagrams for a large, 2800×3600 km, ocean basin using an equivalent barotropic QG model (with $g' = 0.02$ ms^{-2}). Instead of Laplacian friction, represented by the term with $\nabla^2 \zeta$ in (5.54a) with the lateral friction coefficient A_H, he used a biharmonic representation of lateral friction. The friction term is this case becomes $\nabla^4 \zeta$ with a biharmonic friction coefficient A_B. A linear friction term is also included in his model. The instabilities of the symmetric double-gyre flow are followed with decreasing A_B and a sequence of symmetry-breaking pitchfork bifurcations is found. The steady flow patterns at the first four pitchfork bifurcations are plotted in Fig. 5.23a and the patterns of the eigenmodes which destabilize these steady states are shown in Fig. 5.23b.

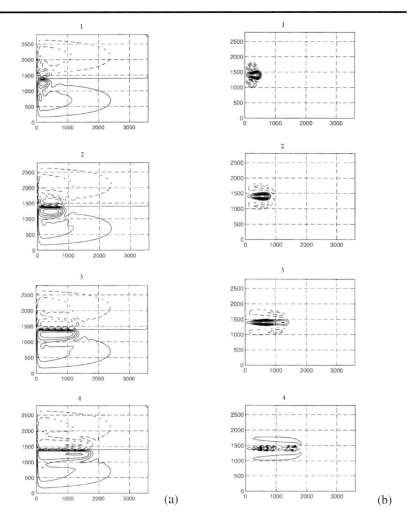

Figure 5.23. Results from the bifurcation study of the double-gyre flow in a large ocean basin of Primeau (2002). (a) Sequence of the double-gyre flows at successive symmetry breaking pitchfork bifurcations $P_1 - P_4$ when the biharmonic friction coefficient A_B is decreased. (b) Patterns of the eigenmodes destabilizing the double-gyre flows in (a) at each pitchfork bifurcation.

The pattern of the symmetric double-gyre flow at the first pitchfork bifurcation P_1 and that of the eigenmode (upper panels in Fig. 5.23) can be identified with the ones in Fig. 5.18a and Fig. 5.19a for the smaller basin. When the basin size increases, the zonal extension of the gyres increases and additional symmetry-breaking instabilities occur. These subsequent instabilities can be classified according to the zonal extent of the P-mode, but the mechanism of instability is the same as that at P_1. Hence, there is a quantization of the instabilities and the structure of equilibria and instabilities becomes therefore much more complicated in larger basins. However, it can in principle be understood from the bifurcation diagrams for the small-basin case which repeats itself at lower friction with spatial patterns having a richer structure in the zonal direction.

5.5.1.2 Transient flows

From the bifurcation analyses of the (equivalent) barotropic QG models, it can be expected that for small lateral friction the flows will be transient and highly complicated, in particular in the large-basin case. To understand the route to complexity it seems best to start with relatively small basins and then proceed to the larger ones. Trajectories computed with the barotropic QG-model for the 1000×1000 km basin show that indeed intermonthly variability first occurs with increasing Re in the form of periodic oscillations. Subsequently, when Re is increased, a quasi-periodic orbit is obtained with both interannual and intermonthly frequencies. This is nicely in agreement with the bifurcation diagram in Fig. 5.16. Soon after $Re = 80$, the flow becomes irregular.

In Meacham (2000), transient flows in a basin of 1024×2048 km with no-slip boundary conditions on the lateral walls are considered. For parameters and forcing similar as in Dijkstra and Katsman (1997), transient solutions are computed using the frictional boundary layer thickness $\delta_M = \delta_{M*}/L$ as control parameter. An overview of the different solutions (with the kinetic energy as norm) is shown in Fig. 5.24. As can be seen, steady (diamonds), periodic (plusses) and aperiodic solutions (squares) are found. The structure of steady states and periodic orbits can be understood with help of the bifurcation diagram Fig. 5.16, where the periodic orbits are coming from the Hopf bifurcations. In some aperiodic solutions, large excursions are made and ultra low-frequency variability arises. This does not seem to occur as a movement of trajectories between unstable steady states and/or limit cycles; Meacham (2000) suggests that it arises through a homoclinic orbit.

In Nadiga and Luce (2001), the location of the homoclinic orbit in the double-gyre flows is precisely located for flows in a basin of size 1000×2000 km. Many transient computations are performed for different parameters and spectra are plotted versus the dimensionless inertial layer thickness δ_I. In this way, they find evidence for the occurrence of a homoclinic orbit of Shilnikov (1965) type (see section 3.5). This behavior is characterized by specific periodic and aperiodic orbits that can be observed in the spectrum of the time series. Nadiga and Luce (2001) also demonstrated the importance of this dynamical phenomenon in explaining low-frequency variability in these flows.

For a 2560×2560 km basin, Chang *et al.* (2001) show that the flow transitions, when the ratio δ_I/δ_M is increased, are qualitatively similar to those in the smaller basins. The symmetric flow destabilizes through a pitchfork bifurcation and the asymmetric double-gyre flows subsequently destabilize through Hopf bifurcations. The first periodic orbit that appears has a subannual time scale and interannual variability occurs at slightly larger values of δ_I/δ_M. They monitor the

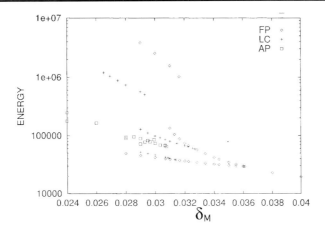

transition to aperiodicity in detail by plotting the transport difference $\Delta\Phi$ between the subtropical and subpolar gyre versus the basin kinetic energy E of the flow for different ratios δ_I/δ_M (Fig. 5.25). The quantity $\Delta\Phi$ is defined as

$$\Delta\Phi = \frac{-\psi_{po} - \psi_{tr}}{\max|\psi|} \tag{5.59}$$

where $\psi_{po} < 0$ is the maximum transport of the subpolar gyre and $\psi_{tr} > 0$ the maximum transport of the subtropical gyre. Note that $\Delta\Phi = 0$ for a symmetric flow, with $\psi_{po} = -\psi_{tr}$.

In Fig. 5.25a, the projection of a periodic orbit around an asymmetric steady state can be seen and it has a period of about 148 days. As δ_I/δ_M increases, the periodic orbit at some instant of time reaches the symmetric double-gyre solution, for which $\Delta\Phi = 0$ (Fig. 5.25b-d). For slightly larger values the flow becomes aperiodic while the trajectory now attains both positive and negative values of $\Delta\Phi$ (Fig. 5.25e-f). It appears as though the periodic orbit makes a connection with the branch of steady symmetric solutions and then connects to the periodic orbit which is present around the symmetry-related asymmetric state: this is characteristic of the presence of a homoclinic bifurcation.

The connection between the pitchfork bifurcation, the gyre modes and the occurrence of the homoclinic bifurcation was clarified in Simonnet *et al.* (2005) and an overview of the bifurcation behavior leading to the homoclinic orbit is plotted in Fig. 5.26. The symmetry-breaking pitchfork bifurcation P is responsible for the asymmetric states; the P-mode is involved in this instability. The merging of the P-mode and the L-mode on the branches of the asymmetric states (at the points M) is responsible for the Hopf bifurcations H associated with the gyre

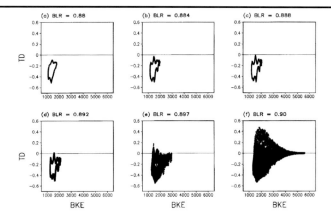

Figure 5.25.   *Phase projections of trajectories computed with a barotropic QG model by Chang* et al. *(2001). On the horizontal axis, the basin averaged kinetic energy (BKE) of the flow and on the vertical axis, the asymmetry of the flow measured through $TD = \Delta\Phi$ is plotted. The different panels are for several values of the ratio of δ_I/δ_M. (a) 0.880, (b) 0.884, (c) 0.888, (d) 0.892, (e) 0.897 and (f) 0.900.*

modes. Finally, the periodic orbits arising from these Hopf bifurcation points on both asymmetric branches connect with the unstable symmetric steady state at the point A; this gives rise to the homoclinic orbit. As explained in section 3.6, the type of homoclinic orbit depends on the eigenvalues associated with the linear stability of the symmetric state at the connection point A. In case there are only real eigenvalues, there is a homoclinic connection of Lorenz-type and when the second and third eigenvalue form a complex-conjugate pair, there is a homoclinic bifurcation of Shilnikov type. Simonnet *et al.* (2005) show that both types can occur and that Shilnikov is more likely to occur in the small lateral friction limit, in accordance with the results in Nadiga and Luce (2001).

   McCalpin and Haidvogel (1996) investigated the time-dependent solutions of an equivalent-barotropic QG model for a basin of realistic size (3600×2800 km), as well as the sensitivity of solutions to the magnitude of the wind stress and its meridional profile. They classified solutions according to their basin-averaged kinetic energy, and found three persistent states in their simulations (Fig. 5.27a). High-energy states are characterized by near-symmetry with respect to the mid-axis, weak meandering, and large jet penetration into the basin interior (Fig. 5.27b, left panel). Low-energy states have a strongly meandering jet that extends but a short way into the basin (Fig. 5.27b, right panel), while intermediate-energy states resemble the time-averaged flow and have a spatial pattern somewhere between high- and low-energy states (not shown). The persistence of the solutions near either state is irregular but can last for more than a decade of simulated time (Fig. 5.27a).

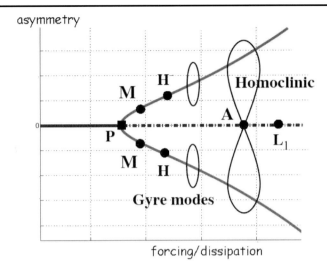

Figure 5.26. Schematic bifurcation diagram of an equivalent-barotropic QG model, plotted in terms of a measure of the asymmetry of the solution (for example, $\Delta\Phi$) versus either wind-stress intensity, the ratio δ_I/δ_M or simply the Reynolds number Re (from Simonnet et al. (2005)).

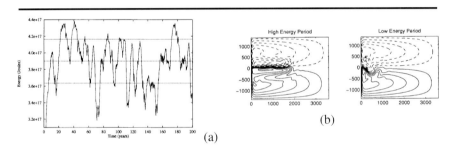

Figure 5.27. (a) Typical variation of the basin averaged kinetic energy of the double-gyre flow in a large basin for the high-forcing or low-dissipation regime (from McCalpin and Haidvogel (1996)). (b) Typical patterns of the streamfunction for the high-energy state (left panel) and the low-energy state (right panel).

Primeau (1998) reproduces the time-dependent behavior of the multiple regimes as found by McCalpin and Haidvogel (1996). To test whether unstable steady states act to steer the model trajectories, the steady states of the system are determined using continuation methods. Multiple equilibria, both symmetric and asymmetric, are found for the parameter values used for the transient computations. By projecting the instantaneous flow fields onto four of the steady solutions, he found that a significant amount of the low-frequency variability of the trajectories are associated with transitions between these steady solutions. Furthermore, he explained that the reduction of the low-frequency variability associated with an increased asymmetry of the wind forcing is a result of the fact that some of the steady states cease to exist. However, knowing that there are many branches of steady solutions, many gyre modes and possibly several homoclinic connections, more work is needed to figure out the precise dynamics causing the different energy states found in McCalpin and Haidvogel (1996).

5.5.2. Baroclinic flows

In this section, we follow the same approach as in the previous section, but now for two- and three-layer quasi-geostrophic models.

5.5.2.1 Basic bifurcation diagrams

As was shown in section 5.3, an eastward jet is susceptible to mixed baroclinic/barotropic instabilities, which may lead to a strong modification of the mean flow through the formation of eddies. It is therefore expected that additional instabilities will appear in multi-layer quasi-geostrophic models, where baroclinic effects can be represented. As in the previous section, we again consider the small basin and high friction case first and then proceed to flows in larger basins at small friction.

Dijkstra and Katsman (1997) first studied the bifurcation behavior of the double-gyre flows in a two-layer quasi-geostrophic model. Equations of the two-layer model were given in (5.30) and the standard parameter values used were provided in Table 5.3; these are for a small basin of dimension 1000×1000 km. Boundary conditions (5.55) and wind forcing (5.57) are similar as for the single-layer model. Since there is no interfacial friction between the layers, the second layer is unforced and the steady solutions for the baroclinic case have a motionless second layer.

The bifurcation structure (Fig. 5.28) shows the presence of three (new) Hopf bifurcations H_1, H_2 and H_3, which are already on the symmetric double-gyre solution. The patterns of the modes which destabilize the steady state (Fig. 5.29) represent baroclinic modes having time scales of variability of 4.3, 8.2 and 6.0 months, respectively. For each mode, there is a phase difference between the response in both layers, illustrating the baroclinic nature of the instability. These modes are the equivalent of the baroclinic modes destabilizing a zonal jet (section 5.4.2.3.), but which now have to satisfy the lateral boundary conditions.

In Fig. 5.28, the window of multiple equilibria is rather small in this case as the equilibria coming from P_2 are connected to those of P_1. The connection between the bifurcation diagram in the two-layer and single-layer model (Fig. 5.16) was investigated by subsequently reducing the layer thickness ratio δ_F and the rotational Froude number F_1 to zero (Dijkstra and Katsman, 1997). For decreasing δ_F (or increasing the second layer depth D_2), the Hopf bifurcation points H_1 and H_2 move towards higher values of Re, and closer to each other. They can be detected for values of δ_F down to $\delta_F = 0.08$, but not for smaller values of

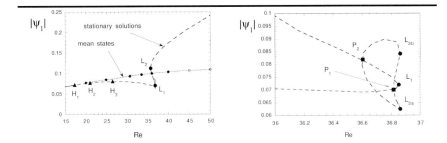

Figure 5.28. (a) *Bifurcation diagram (Dijkstra and Katsman, 1997) for the baroclinic two-layer QG-model of a square basin with* $\delta_F = 0.4$ *and control parameter* Re. *On the vertical axis, the value of the upper layer streamfunction* ψ_1 *at the same gridpoint as in Fig. 5.16 is plotted. Solid (dashed) lines represent stable (unstable) solutions. Marked are three Hopf bifurcations (* H_1 *to* H_3 *) and two limit points (* L_1 *and* L_2 *). The dotted line connects the mean states of time integrations (marked by circles) performed at specific values of* Re. *(b) Detail of the bifurcation diagram for* $Re \in [36.5, 37]$. *Marked are two pitchfork bifurcations (* P_1 *and* P_2 *) and three limit points (* L_1, L_{3a} *and* L_{3b} *).*

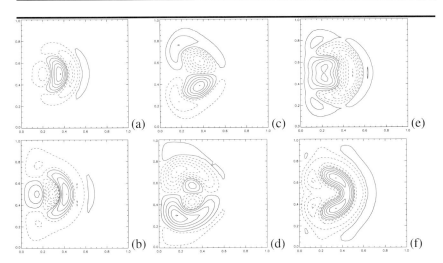

Figure 5.29.   Transition patterns at the Hopf bifurcations H_1 *(a/b),* H_2 *(c/d) and* H_3 *(e/f), showing both the upper and lower layer response at phase* $t = 0$ *of the oscillation (from Dijkstra and Katsman (1997)).*

δ_F. For $\delta_F = 0.08$, the spatial structure of the most unstable modes is still the same, but the time scales of both oscillatory modes approach 6 months, pointing to the existence of a point in parameter space where the instabilities cease to exist. Decreasing F_1 mainly affects the region where multiple asymmetric equilibria exist. For standard parameter values for the two layer model ($F_1 = 850$), the asymmetric solutions branching off at pitchfork bifurcations P_1 and P_2 are connected through the limit points L_{3a} and L_{3b} (Fig. 5.28b). With decreasing F_1, the sadde-node bifurcations L_{3a} and L_{3b} move towards larger values of Re, and disappear for $F_1 \approx 450$. For smaller values of F_1, the asymmetric branches are disconnected, as was the case for the equivalent barotropic model (Fig. 5.16) and a large interval of multiple steady states exists.

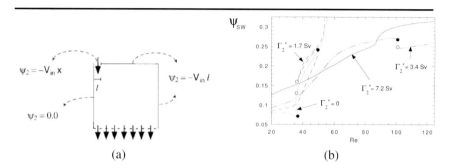

(a)   (b)

Figure 5.30.   (a) Plan view of the boundary conditions in the lower layer where V_{in} is a control parameter defining the dimensionless inflow velocity of the DWBC over a dimensionless width l; the transport $\Gamma == \Gamma_2^* = V_{in} U l L D_2$. (b) Bifurcation diagrams (from Katsman et al. (2001)) showing only the branches of steady states for different values of $\Gamma = \Gamma_2^*$; note that for $\Gamma = 0$ the diagram is the same as in Fig. 5.28.

Katsman *et al.* (2001) investigated the impact of the presence of a deep western boundary current (DWBC) in the lower layer within the same two-layer model. The DWBC is modeled through the inflow in the lower layer with a certain strength Γ (Fig. 5.30a). As it introduces an asymmetry into the problem, the pitchfork bifurcations in Fig. 5.28 no longer exist when $\Gamma \neq 0$. The bifurcation diagrams due to this imperfection are shown for several values of Γ in Fig. 5.30b. For a value of $\Gamma \approx 7$ Sv, there is only a single branch of solutions.

Steady flows for different values of Γ and for $Re = 31$ (Fig. 5.31) show that the zonal jet is shifted southwards and that the upper-layer flow becomes strongly asymmetric. Katsman *et al.* (2001) also follow the Hopf bifurcations on the branches versus Γ. They show that the gyre modes are destabilized by the DWBC and that they are responsible for the low-frequency variability in the flow at large values of Re. According to the mechanism of the gyre mode, merging of the P-mode and the L-mode can only occur when the asymmetry of the mean state is large enough. It is therefore expected that the appearance of gyre modes is favored when the DWBC strengthens. In this way, the gyre modes may be at the origin of the decadal variability found in the model study by Spall (1996a,b) on the interaction of the Gulf Stream and the DWBC.

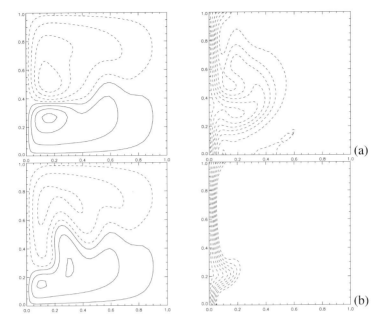

Figure 5.31.   Streamfunction of steady upper (left panel) and lower (right panel) layer flows for $Re = 31$ and two values of the strength Γ of the DWBC. (a) $\Gamma = 5.0$ Sv; (b) $\Gamma = 10.0$ Sv.

5.5.2.2 Transient flows

For the two-layer small-basin case, transient flow computations were performed for different values of Re in Katsman *et al.* (1998). The norm of the time-mean states of these flows were shown as circles in Fig. 5.28. The periodic orbit coming from the first Hopf bifurcation H_1 is stable up to $Re \approx 40$. By comparing the mean state of this periodic orbit, which has a non-zero response in the lower layer, with the unstable steady circulation with a motionless lower layer, the forcing of the second layer through the baroclinic instabilities was analyzed. Self-interactions of the most unstable mode turn out to be the main forcing for the mean flow in the lower layer.

Nauw *et al.* (2004b) investigated the different flow regimes in a 2000×2000 km basin which appear when the lateral friction coefficient A_H is decreased from $A_H = 2400$ m^{2}s^{-1} to $A_H = 300$ m^{2}s^{-1}, using a three-layer quasi-geostrophic model (for specific parameter values, see Nauw *et al.* (2004b)). Four flow regimes are identified by the analysis of a combination of the maximum northward transport of the time-mean flow Ψ_{max} and the normalized transport difference between the subtropical and subpolar gyre ($\Delta\Phi$). In this case, $\Delta\Phi$ is defined as in (5.59) in which the streamfunction ψ is that of the depth averaged flow. With decreasing A_H, the regimes found are the viscous symmetric regime (for $A_H \geq 2100$ m^{2}s^{-1}), the asymmetric regime (for $1400 \leq A_H \leq 2100$ m^{2}s^{-1}), the quasi-homoclinic regime (for $700 \leq A_H \leq 1400 \text{ m}^2\text{s}^{-1}$) and the inertial symmetric regime (for $A_H \leq 700 \text{ m}^2\text{s}^{-1}$).

For four different values of A_H, the value of $\Delta\Phi$ is plotted versus time in Fig. 5.32 and time-mean plots of the barotropic transport streamfunction are shown in Fig. 5.33. The time series of $\Delta\Phi$ in the viscous symmetric regime (Fig. 5.32a) displays a low-frequency modulation of a high-frequency signal, while the time-mean state (Fig. 5.33a) is symmetric. A transition to an asymmetric regime occurs at smaller A_H and a typical time series of $\Delta\Phi$ in that regime is shown in Fig. 5.32b. The value of $\Delta\Phi$ remains positive after a spin-up of slightly more than 25 years and the amplitude of the high-frequency oscillation changes on a decadal time-scale. The time-mean barotropic transport streamfunction is asymmetric and displays a jet-down solution (Fig. 5.33b) in correspondence with the positive value of $\Delta\Phi$.

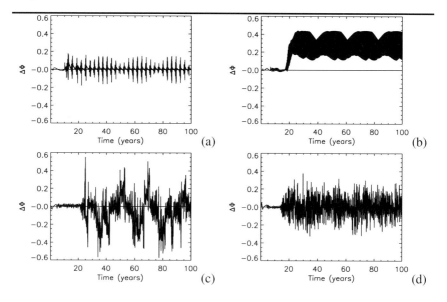

Figure 5.32.   *Time series (including spin-up) of the transport difference, $\Delta\Phi$, for different values of A_H in the large basin case. (a) $A_H = 2400 \ m^2s^{-1}$, viscous symmetric regime; (b) $A_H = 1600$ m^2s^{-1}, asymmetric regime; (c) $A_H = 900 \ m^2s^{-1}$, quasi-homoclinic regime; (d) $A_H = 600$ m^2s^{-1}, inertial symmetric regime.*

For the flow in the quasi-homoclinic regime, several intervals can be distinguished in which there is a preference for either positive or negative values (Fig. 5.32c). The time-mean flow in this regime is slightly asymmetric (Fig. 5.33c). The time series of the case in the inertial symmetric regime (Fig. 5.32d) consists of a mainly high-frequency signal. The time-mean flow in this regime (Fig. 5.33d) is also symmetric, but the midlatitude jet is much stronger than in the symmetric viscous regime. Moreover, the large-scale gyres are accompanied by small-scale subgyres near the northern and southern boundary (Fig. 5.33d).

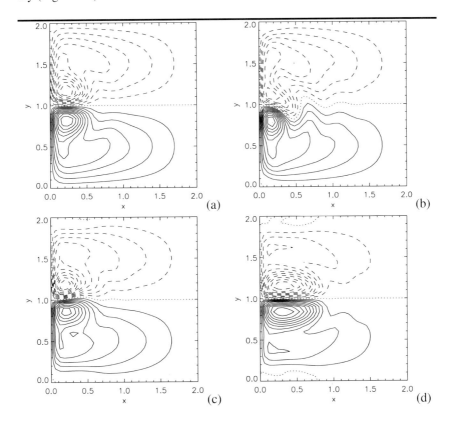

Figure 5.33. The patterns of the barotropic transport function, Ψ, averaged over the final 75 years of integration for selected values of A_H. (a) $A_H = 2400\ m^2 s^{-1}$, viscous symmetric regime; (b) $A_H = 1600\ m^2 s^{-1}$, asymmetric regime; (c) $A_H = 900\ m^2 s^{-1}$, quasi-homoclinic regime; (d) $A_H = 600\ m^2 s^{-1}$, inertial symmetric regime.

The four regimes are also characterized by different types of variability. In Nauw *et al.* (2004b), the spatio-temporal variability of the flows (of which time series were shown for different values of A_H in Fig. 5.32) was analyzed with the M-SSA technique (Plaut *et al.*, 1995). In Fig. 5.34, a histogram is shown of the variance explained by each of the statistical modes, classified into groups that can be related to an internal mode. Case (a) is for a symmetric wind-stress forcing, while case (b) is for a slightly asymmetric wind stress. Most of the variance in the viscous symmetric regime can be explained by two classical baroclinic modes (CB1 and CB2), both with a period of about 3 months. In the inertial symmetric regime, the variability is controlled by basin-wide westward travelling Rossby basin modes (RB) with intermonthly periods. Part of the variance in the asymmetric and quasi-homoclinic regimes can be explained by a gyre mode (G). It causes low-frequency variability with a period of about 3 years. The case with asymmet-

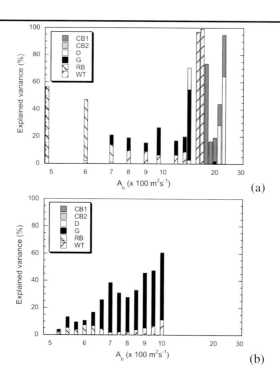

Figure 5.34.   Histograms of explained variances (%) for each of the different statistical oscillations, (a) symmetric wind-stress forcing and (b) asymmetric wind-stress forcing. CB1 = classical baroclinic mode, causing meandering of the midlatitude jet; CB2 = classical baroclinic mode causing strengthening and weakening of the midlatitude jet; D = dipole oscillation; G = gyre mode; RB = Rossby basin mode; WT = wall-trapped mode.

ric wind-stress forcing demonstrates that the presence of the gyre mode is linked to the asymmetry of the time-mean state (Fig. 5.34b).

Two classes of patterns of variability are found as statistical modes in the M-SSA analysis but have not been identified (yet) with specific Hopf bifurcations in the double-gyre problem. One class are the high-frequency wall-trapped modes (WT) (Cessi and Ierley, 1993; Sheremet *et al.*, 1997) with a period of about 3 months. Upper-layer streamfunction anomalies of such a wall-trapped mode originate along the northern and southern boundaries and are subsequently advected by the western boundary currents (Fig. 5.35). The other class are dipole modes (D), which have a spatial pattern much like that of the P-mode.

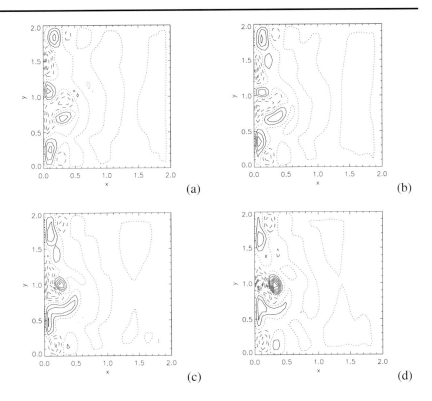

Figure 5.35.   Reconstructed component of the wall-trapped statistical mode in the upper layer at four phases during half of a period of an oscillation (panels are 1/8th period apart). It has a dominant period of 3 months and explains 99.3% of the variance of the solution at $A_H = 1600 \, m^2 s^{-1}$.

Nauw *et al.* (2004b) study the transitions between the different regimes. The transition from the viscous symmetric regime to the asymmetric regime is associated with the symmetry-breaking pitchfork bifurcation. A homoclinic bifurcation, caused by the merging of two mirror-symmetric low-frequency relaxation oscillations and the unstable symmetric steady state, marks the transition from the asymmetric regime to the quasi-homoclinic regime. Indications of the nearby presence of such a homoclinic bifurcation are that the circulation alternates between a jet-up and a jet-down solution on interdecadal time-scales, similar to the results in Chang *et al.* (2001). The transition from the quasi-homoclinic regime to the inertial symmetric regime occurs through symmetrization of the zonal velocity field of the time-mean state. The interaction of high-frequency modes (such as CB1 and CB2) introduces forcing terms that oppose the wind-stress forcing, thereby moving the system towards a regime where both multiple equilibria and the gyre mode cease to exist. The results in Nauw *et al.* (2004b) indicate that the study of the steady states, the bifurcations and the internal modes of variability provide an interpretation framework for complex time-dependent flows. But, as the bifurcation diagrams become more complicated for 'realistic' size basins, much work is needed to obtain a more detailed dynamical interpretation of time-dependent flows in these basins (Berloff and McWilliams, 1999a,b).

Berloff and McWilliams (1999a) computed numerical solutions of the WDC in a two-layer QG model for five values of the lateral friction coefficient A_H in a basin of realistic size (3200 × 2800 km). For $A_H = 1200$ m^{2}s^{-1}, an asymmetric steady state was found. When the lateral friction was decreased, the flow first displays variability and two modes of variability could be identified on basis of statistical (EOF-) analysis. The so-called 'primary' mode corresponds to inter-monthly variability and is characterized by the presence of Rossby waves in the interior. The 'secondary' mode introduces variability on an interannual time-scale and is associated with a fluctuating envelope surrounding a standing Rossby wave. At $A_H = 1000$ m^{2}s^{-1}, quasi-periodic variability was found containing two dominant frequencies in the sub- and interannual range, which could still be related to the two modes discussed earlier. At even smaller friction, a broadband spectrum appears, with the spectral power of the total energy increasing towards lower frequencies. At $A_H = 800$ m^{2}s^{-1}, the behavior of the solutions is called 'chaotic', while at $A_H = 600$ m^{2}s^{-1} the flow patterns hover near three states with distinct total energy. These states are characterized by a different penetration length of the eastward jet and the presence or absence of dipole-pattern oscillations in the recirculation region.

Characteristics of the asymmetric and quasi-homoclinic regimes are found in Berloff and McWilliams (1999a). The spatial pattern of the 'secondary' mode in the symmetrically forced equivalent barotropic case at $A_H = 1000$ m^{2}s^{-1} (their Fig. 15) is similar to that of the gyre mode. The meridional position of the separation point of their time-dependent solution at $A_H = 600$ m^{2}s^{-1} alternates between locations to the north and to the south of the mid-axis of the basin on a decadal time-scale (their Fig. 18). This indicates an alternation between a jet-up and a jet-down solution and provides support for the nearby presence of a homoclinic orbit. Hence, this solution is likely to reside in a quasi-homoclinic regime. Berloff and McWilliams (1999b) investigate the double-gyre flows at even smaller values of A_H in a three-layer QG model and find a destabilization of the western boundary current at small values of A_H. The patterns of variability associated with this instability have a strong similarity with the WT modes (cf. Fig. 5.35) which suggests (but this has not been shown yet) that these can also be associated with a Hopf bifurcation on a branch of asymmetric steady states.

In Siegel *et al.* (2001), flows at very high resolution and small values of A_H are computed in a six-layer quasi-geostrophic model. In Fig. 5.36, upper layer streamfunction plots are shown from several numerical experiments differing in the values of A_H. The displayed sequence goes from relatively low Reynolds numbers Re (see definition in caption of Fig. 5.36) in panel A to very high (by modern GCM standards) Reynolds numbers in panel D. In panel D, numerous small-scale coherent vortices are displayed. Comparable features occasionally

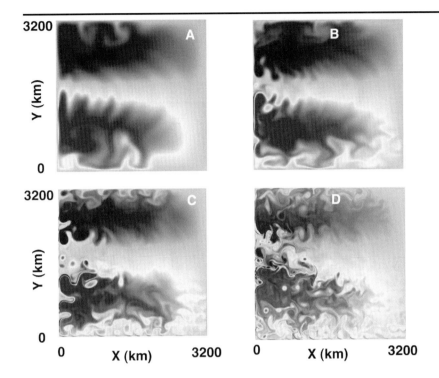

Figure 5.36. (in color on page 519). Upper layer streamfunction snapshots of the ocean in a 3200 km square basin for varying Reynolds numbers, Re, with $Re = 0.375, 1.5, 6.0, 24$ for the panels (A)-(D), respectively. Here, $Re = UL/A_H$, with $U = 10^{-3}\ ms^{-1}$ and $L = 3200$ km. The time-mean flow consists of an anticyclonic midlatitude subtropical gyre and a cyclonic subpolar gyre. The resolution in the computations increases from 25 km in (A) to 1.56 km in (D). Note the appearance of coherent vortices throughout the circulation in the highest value of Re results (from Siegel et al. (2001)).

appear in the flow in panel C, but are essentially absent in the panels A and B. Importantly, for the experiment in C, the vortices are far sparser and do not survive for 'long' times (relative to vortex turnover time scales). The highest Reynolds number computations possess eddy kinetic energies approaching values like those observed in the open ocean. The dynamical origin of this so-called coherent vortex regime is still unknown.

5.6. Regimes of Double-Gyre SW Flows

For shallow-water models, we follow the same approach as for the quasi-geostrophic models, starting with the (equivalent) barotropic case (section 5.6.1) followed by the baroclinic case (section 5.6.2).

5.6.1. The equivalent barotropic case

A first systematic analysis of possible equilibria of the WDC using a 1.5-layer shallow-water β-plane model was carried out by Jiang *et al.* (1995). Complementary to the dimensionless formulation in section 5.3.2, and because of the slightly different set-up, the dimensional formulation of the model is given below. If the velocities in eastward and northward directions are indicated by $\mathbf{u}_* = (u_*, v_*)$, respectively and h_* is the thickness of the upper layer (with equilibrium value D), the upper-layer mass-flux vector $\mathbf{U}_*$ is given by $\mathbf{U}_* = (U_*, V_*) = (h_* u_*, h_* v_*)$ and the equations describing the flow are (Jiang *et al.*, 1995)

$$\frac{\partial U_*}{\partial t_*} + \nabla \cdot (\mathbf{u}_* U_*) - (f_0 + \beta_0 y_*)V_* =$$
$$-g'h_*\frac{\partial h_*}{\partial x_*} + A_H \nabla^2 U_* - r_b U_* + a_\tau \frac{\tau_0 \tau^x}{\rho_1} \tag{5.60a}$$

$$\frac{\partial V_*}{\partial t_*} + \nabla \cdot (\mathbf{u}_* V_*) + (f_0 + \beta_0 y_*)U_* =$$
$$-g'h_*\frac{\partial h_*}{\partial y_*} + A_H \nabla^2 V_* - r_b V_* + a_\tau \frac{\tau_0 \tau^y}{\rho_1} \tag{5.60b}$$

$$\frac{\partial h_*}{\partial t_*} + \frac{\partial U_*}{\partial x_*} + \frac{\partial V_*}{\partial y_*} = 0 \tag{5.60c}$$

Most of the results in Jiang *et al.* (1995) were presented for no-slip conditions on each lateral boundary and a similar wind-stress forcing was taken as in (5.57), i.e.

$$\tau^x(y_*) = -\cos\frac{2\pi y_*}{2L} \; ; \; \tau^y = 0 \tag{5.61}$$

where L is the length of the basin and $2L$ is its width. Standard values of the dimensional parameters used are listed in Table 5.8.

Parameter	Value		Parameter	Value	
L	1.0×10^6	m	τ_0	1.0×10^{-1}	Pa
D	5.0×10^2	m	β_0	$2.0 \ 10^{-11}$	$(ms)^{-1}$
f_0	5.0×10^{-5}	s^{-1}	A_H	3.0×10^2	$m^2 s^{-1}$
g'	3.0×10^{-2}	ms^{-2}	R_{int}	75.0×10^3	m
ρ_1	1.0×10^3	kgm^{-3}	r_b	5.0×10^{-8}	s^{-1}

Table 5.8.   Typical values of dimensional parameters used in Jiang et al. *(1995); for this case, the internal Rossby deformation radius* $R_{int} = \sqrt{(g'D)}/f_0$ *is about 75 km.*

Jiang *et al.* (1995) integrate the equations (5.60) in time for different values of the wind-stress forcing strength a_τ, and monitor the position of the confluence point, i.e., the merging point of the northward and southward moving currents near the western boundary. The horizontal resolution used is 20 km. For small values of a_τ, a unique nearly symmetric (with respect to mid-axis of the basin at $y_* = L$)

flow is found for which the confluence point is displaced slightly northward of $y_* = L$. For larger values of a_τ, multiple equilibria are found and also a solution having a southward displaced confluence point exists. Patterns of the thickness anomaly h_* for both solutions are shown in Fig. 5.37 for $a_\tau = 0.9$.

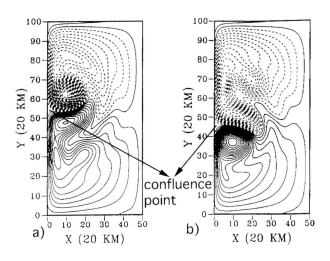

Figure 5.37. *Patterns of the layer thickness anomaly of two steady-state solutions of the 1.5-layer shallow water model at $a_\tau = 0.9$ (slightly modified from Jiang* et al. *(1995)). The position of the confluence point is indicated and $x_* = 20X$ km, $y_* = 20Y$ km.*

Based on the transient results and the steady states found, Jiang *et al.* (1995) proposed the bifurcation diagram as in Fig. 5.38. Here, drawn lines are actually computed values of the confluence point, but dotted lines represent unstable states, which were guessed based on generic situations in bifurcation theory, but which could not be computed using the forward time integration method. The steady solution structure corresponds to that of an imperfect pitchfork bifurcation, because the different branches are unconnected but have the appearance of a pitchfork (see section 3.2.3). Both steady solutions become unstable at larger wind-stress strength, and no stable steady states are found beyond $a_\tau = 1.1$.

A direct follow-up of the Jiang *et al.* (1995) study was the work by Speich *et al.* (1995), in which the bifurcation structure of the 1.5-layer shallow-water model was determined using continuation techniques. The exact position of the saddle-node bifurcation in Fig. 5.38 and the Hopf bifurcations destabilizing the steady states were determined. The bifurcation diagram, using a horizontal resolution of 20 km, is shown in Fig. 5.39, where the minimum of the upper layer thickness $h_{min} = min\ h_*/D$ is plotted versus the dimensionless wind-stress strength $\bar{\sigma} = a_\tau \tau_0/(\rho f_0 D U_{SW})$, with $U_{SW} = 1.0$ ms^{-1} as a characteristic horizontal

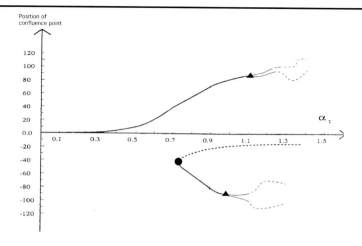

*Figure 5.38. Bifurcation diagram (slightly modified from Jiang et al. (1995)) to explain the so-
lutions found in the shallow-water model by forward time integration. On the vertical axis, the
position of the confluence point (see Fig. 5.37) is plotted in km, whereas on the horizontal axis,
a measure of the wind-stress strength is shown. Again, markers indicate bifurcation points with
dots being saddle-node bifurcations and triangles Hopf bifurcations. The periodic orbits arising
from each Hopf bifurcation are monitored by plotting the minimum and maximum excursion of the
confluence point.*

velocity. The locations of the values of $\bar{\sigma}$ marking the transition from steady to
time-dependent behavior were computed accurately in Speich *et al.* (1995). In
Fig. 5.39, these can be seen as the two Hopf bifurcations, marked as H_u and H_l.
At H_u, an oscillatory mode having a period of 28 months becomes unstable and
at H_l, a mode with an intermonthly time scale destabilizes.

Jiang *et al.* (1995) find periodic behavior in the shallow-water model for values
of a_τ just above Hopf bifurcation. For such a periodic orbit, having a 2.8 year
period, the anomalies in the h-field, with respect to the time-mean field over one
period of the oscillation, resemble an elongated wave pattern confined to the recirculation regions of the mean flow. In the plots of the trajectories (Figs. 5-6 in their
paper), one can also observe smaller time scale oscillations. Hence, both interannual and intermonthly modes contribute to the temporal variability of the solution
at supercritical conditions, with the interannual mode being most dominant. For
even larger a_τ, Jiang *et al.* (1995) find that the interannual mode is still dominant,
but the time scale of variability has slightly increased to just over 3 years. With
decreasing friction, two distinct interannual periods were found (of about 3 and 6
years) with signatures of strongly asymmetric (relaxation) oscillations.

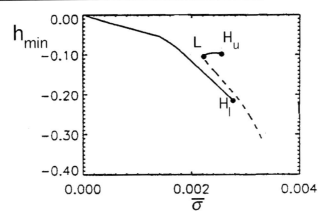

Figure 5.39. Bifurcation diagram for the 1.5-layer shallow-water model (slightly modified from Speich et al. *(1995)) for the same parameters as in Table 5.8 using the parameter* $\bar{\sigma} = a_\tau \tau_0/(\rho f_0 D U_{SW})$ *as a control parameter, with* $U_{SW} = 1 \ ms^{-1}$. *On the vertical axis, the minimum (dimensionless) anomaly of the upper layer depth,* h_{min}, *is plotted.*

5.6.2. Connection: SW- and QG-models

The apparent correspondence of the steady-state structure of the SW-model and the QG-model motivates to look at this connection in more detail. Dijkstra and Molemaker (1999) investigated this connection for a basin of 1000×2000 km with no-slip boundaries with parameters are as in Table 5.8. The control parameter is the dimensionless lateral friction coefficient, the Ekman number $E = A_H/(f_0 L^2)$. This parameter can be related to the Reynolds number Re, used in the QG-model, through $E = \epsilon U_{QG}/(U_{SW} Re)$ and $\epsilon = U_{SW}/(f_0 L)$, where $U_{SW} = 1$ ms^{-1} is again the characteristic velocity used in the SW-model. As an indicator of the solutions, the northward dimensional volume transport ϕ (in Sv), i.e.,

$$\phi = \max_{y_*, x_{e*}} \left[\int_0^{x_{e*}} v_* h_* \, dx_* \right] \tag{5.62}$$

was used.

In Fig. 5.40, ϕ is plotted versus E for steady state solutions of both models. Solid (dotted) branches indicate steady states for the QG (SW) model, whereas bifurcation points are indicated by markers. The bifurcation for the SW-model is qualitatively similar (imperfect pitchfork with Hopf bifurcations) to that in Fig. 5.39, whereas the bifurcation for the QG-model is qualitatively similar to that in Fig. 5.16 (pitchfork bifurcation). The Hopf bifurcations on the asymmetric branches occur at smaller values of E for the QG model and are not shown. In the limit of large E, the models have the same (Sverdrup) transport since they both

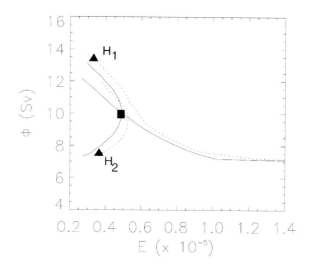

Figure 5.40.   Bifurcation diagram (from Dijkstra and Molemaker (1999)) using the Ekman number E as control parameter for the QG-model (drawn) and SW-model (dashed) in the rectangular domain of 1000 × 2000 *km with no-slip boundaries. On the vertical axis, the volume transport* ϕ, *as in (5.62), is plotted.*

become approximately linear. It can easily be shown that the SW-equations (even with the idealized wind stress (5.57)) are not invariant to the reflection symmetry with respect to the mid-axis of the basin (Dijkstra and Molemaker, 1999). Hence, no perfect pitchfork bifurcation can occur. Small deviations from the equilibrium thermocline depth and advection induce the imperfection, illustrated by the break-up of the pitchfork bifurcation in the SW-model. The branch of the stable symmetric double-gyre solution (present in the QG-model) is connected continuously to one of the asymmetric solutions. The other asymmetric solution has connected to the branch originating from the unstable part of the symmetric solution in the QG-model, resulting in a saddle-node bifurcation. The occurrence of multiple equilibria in the SW-model hence has its origin in the same physical processes which cause the symmetry breaking within the QG-model (discussed in section 5.5.1.1).

As explained in section 3.7.3.3, Hopf bifurcations mark the transition to transient behavior. As the lateral friction is decreased in the QG-model, the asymmetric steady states become unstable to an oscillatory instability at the Hopf bifurcations H_1 and H_2 (Fig. 5.16). Also in the SW-model, Hopf bifurcations destabilize the steady solutions for decreasing friction (Fig. 5.39 and Fig. 5.40). For the shallow water model, three modes become unstable nearly at the same con-

ditions at the Hopf bifurcation H_1 in Fig. 5.40. The real part and imaginary part
of the eigenvector of the first two modes are shown in Fig. 5.41. The first mode
(Fig. 5.41a) is neutral and its period is approximately 5 months. The pattern shows
a maximum response located to the north of the mid-axis and the mode propagates
westward. The second mode (Fig. 5.41b) has a period of about 1.5 years, it is
slightly damped and its pattern shows a propagation of the perturbations south-
westward. The third mode (not shown) is also slightly damped and has a similar
period ($\mathcal{P}_* = 3.7$ months) as the first mode. The modes at the Hopf bifurcation
H_2 in Fig. 5.40 have very similar time scales and patterns as those at H_1. The

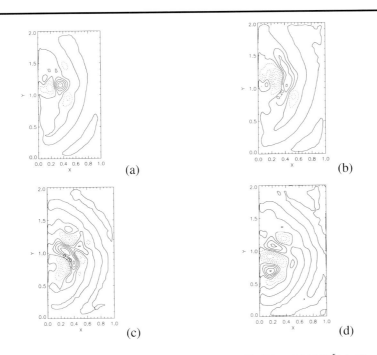

*Figure 5.41.   First two eigenmodes at the Hopf bifurcation H_1 ($E = 0.34\ 10^{-5}$) in Fig. 5.40
(Dijkstra and Molemaker, 1999). On the axes, the dimensionless quantities $x = x_*/L$ and $y =
y_*/L$ are again shown. Layer thickness perturbations h for the real (a) and imaginary part (b) of
the most unstable mode with $\sigma_{r*} = 0.0\ yr^{-1}$ and $\sigma_{i*} = 16.4\ yr^{-1}$. Here $\sigma_* = \sigma_{r*} + i\sigma_{i*}$ is
the complex growth factor of the eigenmode. (c) and (d) as above but for the 2nd mode ($\sigma_{r*} = -
1.1\ yr^{-1}, \sigma_{i*} = 4.1\ yr^{-1}$).*

modes of variability of the SW model and QG model are therefore closely linked
(Dijkstra and Molemaker, 1999). The mode which destabilizes at H_1 in Fig. 5.16
has an intermonthly time scale whereas at H_2, an interannual mode destabilizes.
The Rossby-basin mode (Fig. 5.20) has a basin wide pattern and when the internal
Rossby deformation radius and basin size are adjusted, it can be shown For the

gyre mode (Fig. 5.21), the perturbation has the characteristic asymmetry of the steady-state gyre structure, similar to that in Fig. 5.41b.

5.6.3. The baroclinic case

Nauw and Dijkstra (2001) investigated the steady solutions and their successive bifurcations in a two-layer SW model for a 1000×2000 km basin, located on a β-plane centered at $45°$N. Bottom topography may be present giving a total depth of $H_0 - H_b(x, y)$, where H_0 represents the constant depth in the flat bottom case. The top and bottom layer have a constant density ρ_1 and ρ_2, respectively, with $\rho_2 > \rho_1$. Laplacian lateral friction is present with (eddy) lateral friction coefficient A_H. The flow is driven by a stationary wind-stress pattern (τ_1^x, τ_1^y). For each layer $i = 1, 2$, the dimensional (for convenience, the $*$-subscript has now been omitted) governing equations are (Holland and Lin, 1975)

$$\frac{\partial U_i}{\partial t} + \boldsymbol{\nabla} \cdot (\boldsymbol{u}_i U_i) - f V_i = -\frac{h_i}{\rho_0}\frac{\partial p_i}{\partial x} + A_H h_i \nabla^2 u_i + \frac{\tau_i^x}{\rho_0} \quad (5.63a)$$

$$\frac{\partial V_i}{\partial t} + \boldsymbol{\nabla} \cdot (\boldsymbol{u}_i V_i) + f U_i = -\frac{h_i}{\rho_0}\frac{\partial p_i}{\partial y} + A_H h_i \nabla^2 v_i + \frac{\tau_i^y}{\rho_0} \quad (5.63b)$$

$$\frac{\partial h_i}{\partial t} + \boldsymbol{\nabla} \cdot \boldsymbol{U_i} = 0 \quad (5.63c)$$

where $U_i = u_i h_i$ and $V_i = v_i h_i$ are the volume fluxes per unit length, (u_i, v_i) the zonal and meridional velocities and h_i is the thickness of layer i. In the β-plane approximation, $f = f_0 + \beta_0 y$, where f_0 is the Coriolis parameter and β_0 is the local meridional derivative of f. The rigid-lid approximation and the condition of continuity of normal stress at the interface of the layers become

$$h_1 + h_2 = H_0 - H_b(x, y) \quad (5.64a)$$

$$\boldsymbol{\nabla} p_2 = \boldsymbol{\nabla} p_1 - \rho_0 g' \boldsymbol{\nabla} h_1 \quad (5.64b)$$

The reduced gravity is defined by $g' = g(\rho_2 - \rho_1)/\rho_0$, where ρ_0 is a reference density. No-slip boundary conditions are applied at all lateral walls. The standard values of the parameters can be found in Table 5.9 and are similar to those used in Speich *et al.* (1995). The equilibrium depth of each layer in the flat-bottom case is indicated by H_1 and H_2.

The stress term in the right hand side of equations (5.63a,b) represents the wind-stress forcing in the top layer and a bottom friction in the second layer. For the top layer, a double-gyre wind stress is chosen

$$\tau_1^x = -\tau_0 \cos(2\pi y/B) \quad (5.65a)$$

$$\tau_1^y = 0 \quad (5.65b)$$

where B is the width of the basin. For the second layer, a linear bottom friction is used, with

$$\tau_2^x = -\rho_0 R U_2 \quad (5.66a)$$

$$\tau_2^y = -\rho_0 R V_2 \quad (5.66b)$$

There is no interfacial friction and hence in steady state, the lower layer is motionless. Non-zero flow in the lower layer can only occur through transient effects, such as the presence of and interaction between baroclinic instabilities (Pedlosky, 1996).

Parameter	Value		Parameter	Value	
L	1.0×10^6	m	B	2.0×10^6	m
g	9.8	ms^{-2}	ρ_0	1.0×10^3	kgm^{-3}
f_0	1.0×10^{-4}	s^{-1}	A_H	3.0×10^2	$m^2 s^{-1}$
H_1	7.0×10^2	m	H_2	3.3×10^3	m
ρ_1	1000	kgm^{-3}	ρ_2	1002	kgm^{-3}
β_0	$1.8 \ 10^{-11}$	$(ms)^{-1}$	R	5.0×10^{-8}	s^{-1}

Table 5.9. Standard values of the dimensional parameters in the two-layer shallow-water model as used in Nauw and Dijkstra (2001). The wind-stress amplitude τ_0 is used as the control parameter.

With a flat bottom, the equilibrium layer thicknesses are $H_1 = 700 \ [m]$ and $H_2 = 3300 \ [m]$ as shown in Table 5.9, giving a total thickness $H_0 = 4000 \ [m]$. Steady states are computed, using τ_0 as control parameter, and as a norm of the solution the minimum upper layer thickness, $h_{1,min}$, is plotted. In Fig. 5.42, the drawn (dashed) lines represent stable (unstable) steady states. An imperfect pitchfork bifurcation is found with a unique steady solution for values below $\tau_0 = 6.7 \times 10^{-2} \ [Pa]$ and multiple steady states beyond this value.

Along the upper branch in Fig. 5.42, the results of the linear stability analysis indicate that the steady state becomes unstable through a sequence of Hopf bifurcations (indicated by triangular markers). Two of these are classical baroclinic modes (H_1 and H_3), two others are Rossby-basin modes (H_2 and H_4) and a baroclinic gyre mode destabilizes the steady state at H_5. The oscillation periods of these modes are 4.8 months, 2.8 months, 5.6 months, 2.6 months and 12.7 years, respectively.

The steady-state upper-layer streamfunction at a wind-stress strength of $\tau_0 = 8.7 \times 10^{-2} \ [Pa]$ and the imaginary and real parts of the streamfunction of the (12.7 year oscillating) gyre mode are shown in Fig. 5.43. The spatial pattern of the mode displays a strong alignment with the direction of the steady-state jet (Fig. 5.43a); its extrema are located along a line through the extrema in the steady-state streamfunction of the subtropical and subpolar gyre. The mode has a baroclinic character and its upper-layer pattern (Figs. 5.43b-c) strongly resembles that of the gyre mode in the 1.5-layer shallow-water models (Fig. 5.41c-d) and in quasi-geostrophic equivalent barotropic models (Fig. 5.21).

Time-dependent flows were calculated for several distinct values of the wind-stress parameter τ_0 in Nauw and Dijkstra (2001). In Fig. 5.44, the bifurcation diagram together with the minimum upper-layer thickness of the transient flows are plotted; the marker shows the time-mean average of $h_{1,min}$ and the 'error' bar indicates its variability. For the values up to $\tau_0 = 6.0 \times 10^{-2} \ [Pa]$, a

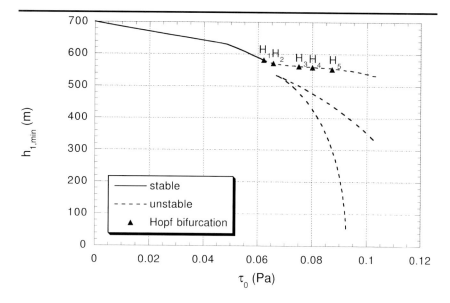

Figure 5.42. Bifurcation diagram of the two-layer shallow-water model. The minimum upper layer thickness $h_{1,min}$ is plotted versus the wind-stress strength τ_0. Drawn (dashed) lines indicate stable (unstable) steady states and the markers labelled $H_1, \cdots, H_5$ indicate Hopf bifurcations (from Nauw and Dijkstra (2001)).

steady double-gyre upper-layer flow is found with a motionless lower layer. For $\tau_0 = 6.3 \times 10^{-2}$ $[Pa]$, oscillatory behavior is found which contains a high-frequency periodic orbit on top of an oscillation with a lower frequency, the latter is damping out in time. The period of this orbit is indeed similar to the period at Hopf bifurcation H_1 ($\mathcal{P}_1 = 4.79$ $[mo]$). The variability of the flow for $\tau_0 = 7.0 \times 10^{-2}$ $[Pa]$ consists of two oscillations, which originate through the instabilities at the Hopf bifurcations H_1 and H_2. The variability of the flow at $\tau_0 = 8.5 \times 10^{-2}$ $[Pa]$ displays a low-frequency oscillation with a period of 18 $[yr]$ together with high-frequency oscillations. From M-SSA analysis, again a low-frequency statistical mode is found with a pattern resembling the baroclinic gyre mode.

For the same size basin, 1000×2000 km, Simonnet *et al.* (2003a) also find the imperfect pitchfork bifurcation diagram in a 2.5-layer SW model (Fig. 5.45). The first Hopf bifurcations on each of these branches are computed and also inter-annual and intermonthly modes of variability are found. Simonnet *et al.* (2003b) study in detail the time-dependent behavior of these flows for values of the wind-stress strength above the Hopf bifurcations. They again find strong evidence for the existence of a homoclinic orbit, where a gyre mode type limit cycle connects to an unstable steady state (Fig. 5.45). In this figure, $\Delta\Phi$ is again the measure of the asymmetry of the flow as in (5.59) and both homoclinic connections from the

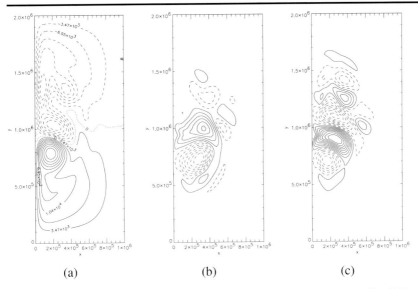

Figure 5.43.   Panel (a) shows the upper layer streamfunction of the steady state at Hopf bifurcation H_5 at a wind stress of $\tau_0 = 8.7 \times 10^{-2}$ $[Pa]$ (Nauw and Dijkstra, 2001). Panel (b) and (c) show the imaginary and real parts of the perturbation of the upper layer streamfunction; the mode has a period of 12.7 years.

limit cycle of the subtropical gyre branch as well as from the subpolar gyre branch are suggested. With the help of this bifurcation diagram, Simonnet *et al.* (2003b) are able to interpret quite complex behavior of trajectories.

5.7. Continental Geometry

Certainly, these bifurcation studies would be quite academic if all features would be lost in more realistic settings, for example when continental geometry, bottom topography and better representations of friction were included. In this section, this matter of robustness is addressed only with respect to continental geometry within the (equivalent) barotropic shallow-water models.

5.7.1. Continents within a β-plane SW model

In Dijkstra and Molemaker (1999), a finite element discretization of the the β-plane SW-model equations was used and realistic geometry was incorporated. To represent a more realistic wind-stress forcing, the annual mean wind stress of the Hellerman and Rosenstein (1983) data set was used. These data consist of values on a $2° \times 2°$ grid that was interpolated onto the computational grid.

For a maximum value of the wind-stress amplitude of 0.05 Pa, which is slightly less than that in reality, the bifurcation diagram is shown in Fig. 5.46a using again $E = A_H/(f_0 L^2)$ as control parameter. Other parameters are similar to those

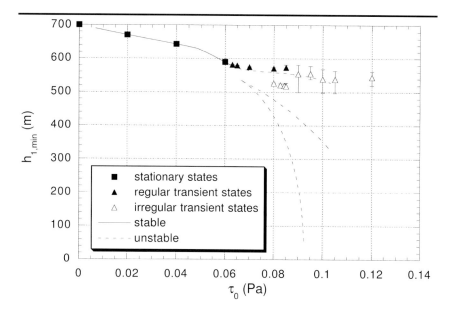

Figure 5.44. Bifurcation diagram showing the minimum upper layer thickness $h_{1,min}$ versus the strength of the wind stress τ_0 for the steady states (as in Fig. 5.42). The markers indicate the average minimum upper layer thickness $h_{1,min}$ of the computed time-dependent solutions and the bars indicate their amplitude range.

in Table 5.8. Main result is the existence of multiple equilibria, just as in the rectangular geometry considered in the previous section (cf. Fig. 5.40). Two solution branches are found, on which solutions are unstable for $E < 7.5 \times 10^{-6}$. One of the branches is connected to the large E regime and a solution on this branch is shown for $E = 3.0 \times 10^{-6}$ in Fig. 5.46b. The model Gulf Stream flows northward along Cape Hatteras and turns into the open ocean at slightly northerly latitudes giving a transport ϕ (as in (5.62)) of about 21.5 Sv. There is a very weak northern recirculation region at this value of E. The other branch exists only for values of E smaller than $3.2 \ 10^{-6}$, which is the position of the saddle-node bifurcation on this branch. The solution at $E = 3.0 \times 10^{-6}$ (Fig. 5.46c) displays a model Gulf Stream which turns into the open ocean near Cape Hatteras. There is now a northern recirculation region, although too much concentrated near the coast compared to reality. At this value of E, the maximum transport ϕ of this 'separated' model Gulf Stream (Fig. 5.46c) is about 26.2 Sv, which is larger than that of the 'deflected' model Gulf Stream (Fig. 5.46b).

The first Hopf bifurcation is found at $E = 7.5 \times 10^{-6}$ on the branch of the deflected Gulf Stream solution (H in Fig. 5.46a). At $E = 3.0 \times 10^{-6}$, the deflected solution (Fig. 5.46b) is unstable to only one oscillatory mode. The real and imaginary part of the eigenvector of this mode are shown in Fig. 5.47a-b; the period of

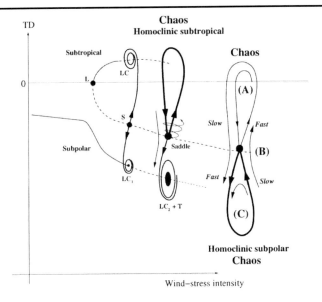

Figure 5.45. Sketch of the local and global bifurcations in the 2.5-layer shallow-water model studied in Simonnet *et al.* (2003b). The abbreviations LC stand for limit cycles, and specific trajectories going along states (A), (B) and (C) are indicated; again TD = $\Delta\Phi$ is a measure of the asymmetry of the solution.

oscillation is about 5 months. The center of action for the oscillation is located in the Gulf Stream region and the scale of the perturbations is about $500\ km$. The orientation of the perturbations does not appear directly related to the orientation of the jet itself. The disturbances propagate south-westward against the flow direction of the steady state flow. The separated solution (Fig. 5.46c) is also unstable to only one oscillatory mode (Fig. 5.47c-d) having a period of about 4 months. Thickness perturbations with a scale of about 400 km are again localized in the jet and the response outside the jet is weak. Because of its time scale and pattern, this mode of variability is a Rossby-basin mode, whose pattern is deformed by the continental geometry.

5.7.2. Continents on the sphere

The shallow-water model on the sphere enables one to look at the impact of more realistic continental geometry. In the results below, both the North Atlantic and the North Pacific basin are considered.

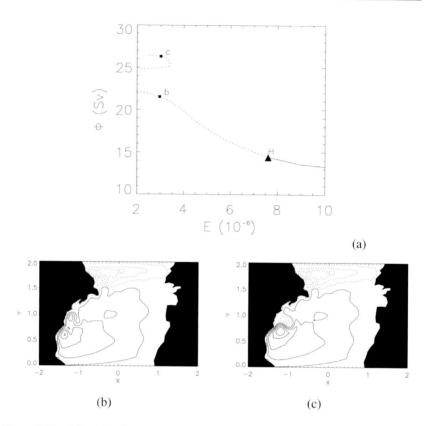

Figure 5.46. Bifurcation diagram for the 1.5-layer shallow water model (Dijkstra and Mole-maker, 1999), with a representation of the continents and with realistic wind forcing. Again, the Ekman number $E = A_H/(f_0 L^2)$ is the control parameter; drawn (dotted) branches indicate stable (unstable) steady states. (b) Contour plot of the layer thickness anomaly for the 'deflected' Gulf Stream at $E = 3.0 \times 10^{-6}$. The field is scaled with its absolute maximum and contours are with respect to this maximum. The lateral coordinates x and y are scaled by the basin length scale L. (c) Contour plot of the layer thickness anomaly for the 'separated' Gulf Stream at $E = 3.0 \times 10^{-6}$.

5.7.2.1 North Atlantic basin

The full extent of the North Atlantic basin over the domain [85°W-5°W, 10°N-65°N] with constant depth $D = 1000$ m is considered in Schmeits and Dijkstra (2000). The barotropic shallow-water model on the sphere (5.19) is used and the flow is driven by a realistic wind-stress field (Trenberth *et al.*, 1989). For a horizontal resolution of 0.5°, the bifurcation diagram is shown in Fig. 5.48a using the Ekman number E as control parameter. Here, the Ekman number is defined as $E = A_H/(2\Omega r_0^2)$, where Ω and r_0 are the angular frequency and radius of the Earth, respectively. Again an imperfect pitchfork bifurcation is found, with

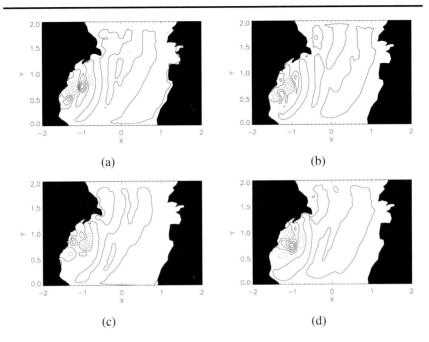

Figure 5.47. Eigenfunctions corresponding to the unstable mode at the marked points in Fig. 5.46a for $E = 3.0 \times 10^{-6}$; shown are the layer thickness perturbations (from Dijkstra and Mole-maker (1999)). (a-b): Real and imaginary part of the unstable oscillatory mode on the steady state shown in Fig. 5.46b, with $\sigma_{r*} = 2.2\ yr^{-1}$ and $\sigma_{i*} = 16.1\ yr^{-1}$. (c-d): As in (a-b) but for the steady state shown in Fig. 5.46c, with $\sigma_{r*} = 0.47\ yr^{-1}$ and $\sigma_{i*} = 19.2\ yr^{-1}$.

two steady solution branches, on which solutions are unstable for $E < 2.5 \times 10^{-7}$. A solution on the lower branch is shown as a contour plot of layer thickness anomalies for $E = 1.6 \times 10^{-7}$ in Fig. 5.48b, with a detail in Fig. 5.48d. It displays the double-gyre circulation, typical for the North-Atlantic Ocean, with a 'deflected' model Gulf Stream which separates too far north, compared to reality. Moreover, there is a weak southern recirculation region and at this value of E the transport ϕ is about 46 Sv.

The second branch exists only for $E < 2.2 \times 10^{-7}$, which is the position of the saddle-node bifurcation on this branch. The solution at $E = 1.6 \times 10^{-7}$ (Figs. 5.48c and 5.48e) shows a 'separated' model Gulf Stream that actually seems to separate twice. First, it separates too far south compared to reality, and later on it separates too far north (at about the same latitude as in Figure 5.48d). There is now a strong southern and a weak northern recirculation region. At this value of E, the transport ϕ is about 70 Sv, which is 1.5 times larger than that in Figure 5.48b, and somewhat larger than estimates of Gulf Stream transport near Cape Hatteras of about 50-65 Sv (Johns *et al.*, 1995). By comparing Figure 5.48b and

5.48c, it can be concluded that the circulation patterns outside the region of the western boundary current are very similar, and hence the multiple equilibria are related to the different separation patterns of the model Gulf Stream.

On the lower branch in Figure 5.48a, a Hopf bifurcation H_1 occurs at $E = 2.5 \times 10^{-7}$. The steady state flow pattern at this value of E is close to Fig. 5.48b and therefore not shown. The perturbation is shown at two phases within half a period of the oscillation in Fig. 5.49a-b. The maximum amplitude of the mode is located around the axis of the western boundary current and propagates southwestward, i.e. upstream. It has a period of 6 months and a wavelength of about 550 km. From Figs. 5.48d and 5.49 we can deduce that the perturbation adds cross-stream components to the flow in the western boundary current, which causes the Gulf Stream to meander. The mode has strong similarities with the first unstable mode in the β-plane model (Fig. 5.47a-b). A second Hopf bifurcation (H_2) occurs on the branch of the 'separated' Gulf Stream solutions (Fig. 5.48a). The period of this oscillation is 2 months and the maximum response is found in the high-shear region to the southeast of Greenland.

5.7.2.2 North Pacific domain

Part of the North Pacific basin [120°E,150°W] × [10°N,55°N] is considered using a horizontal resolution of $5/12° \times 5/12°$. The 100 m depth contour was taken as the continental boundary because otherwise the model Kuroshio would enter the East China Sea. In reality, it is steered by bottom topography, so that it follows a more or less straight path from Taiwan to Japan. The bifurcation diagram for the North Pacific domain (Fig. 5.50a) consists also of an imperfect pitchfork bifurcation and clearly shows that multiple equilibria exist when the lateral friction is small enough. Note that there is quite a range of Ekman numbers where two equilibria are (barotropically) stable. Down to $E = 1.8 \times 10^{-7}$, there is a unique steady solution for each value of the Ekman number (Fig. 5.50a). The stationary solution at location **b** in Fig. 5.50a displays a model Kuroshio path south of Japan, quite similar to the observed small meander state (cf. Fig. 5.4a). The modeled Kuroshio eventually separates too far north, compared to reality. At this value of E, the transport ϕ is about 66 Sv.

The upper branch of solutions in Fig. 5.50a continues to exist for values smaller than $E = 1.8 \times 10^{-7}$, but the solutions loose stability at $E = 1.2 \times 10^{-7}$. These steady states become unstable to one oscillatory mode at a Hopf bifurcation H_1 and to another oscillatory mode at Hopf bifurcation H_2 (Fig. 5.50a). The former oscillatory mode has a timescale of 2 months and has its maximum amplitude in the high-shear region to the south of Kamchatka; the latter oscillatory mode has a timescale of 3 months and is located in the separation region of the Kuroshio. A stable stationary solution on this branch (location **c**) is shown in Fig. 5.50b for $E = 1.5 \times 10^{-7}$, which is in the multiple equilibria regime. It displays a Kuroshio path south of Japan which is different from the observed meander states (Fig. 5.4a), with a transport ϕ of about 86 Sv. Compared to the solution at location **b**, to which it is continuously connected, the anti-cyclonic recirculation gyre to

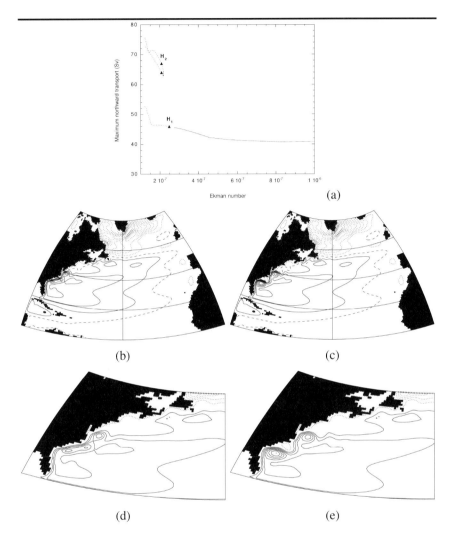

Figure 5.48. (a) *Bifurcation diagram for the barotropic shallow-water model on the sphere (Schmeits and Dijkstra, 2000) with the Ekman number as control parameter. Drawn (dotted) branches indicate stable (unstable) steady states, whereas the Hopf bifurcation points are indicated by triangles. The intersection of the upper branch does not indicate a bifurcation; it is due to the choice of norm.* (b) *Contour plot of the layer thickness anomaly for the 'deflected' model Gulf Stream at $E = 1.6 \times 10^{-7}$. The contour levels are scaled with respect to the maximum value of the field.* (c) *Contour plot of the layer thickness anomaly for the 'separated' model Gulf Stream at $E = 1.6 \times 10^{-7}$.* (d) *Same as (b), but for part of the domain $[85, 45]°W \times [24, 51]°N$.* (e) *Same as (c), but for part of the domain $[85, 45]°W \times [24, 51]°N$.*

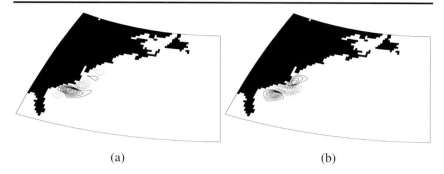

(a)   (b)

Figure 5.49. Patterns of the neutral mode at H_1 in Fig. 5.48a, for part of the domain [85°W,45°W] × [24°N,51°N]; shown are the perturbation layer thickness anomalies at (a) $\sigma_i t = 0$ and at (b) $\sigma_i t = \pi/2$ (from Schmeits and Dijkstra (2000)).

the south of Japan has intensified and has caused the Kuroshio to deviate from the coast.

The second branch of solutions exists only for $E < 1.8 \times 10^{-7}$, which is the position of the saddle-node bifurcation on this branch. Steady states on the upper part of this branch are stable down to $E = 1.2 \times 10^{-7}$ and become unstable at Hopf bifurcation H_1 to the same oscillatory mode as the steady states on the upper branch. The steady solution at location **d** at $E = 1.5 \times 10^{-7}$ (Fig. 5.50c) displays a model Kuroshio path south of Japan vaguely resembling the observed large-meander state (Fig. 5.4). The flow eventually separates at the southeast corner of Honshu, which roughly corresponds to the observed separation point. At this value of E, the transport ϕ is about 84 Sv. The third steady state at $E = 1.5 \times 10^{-7}$ (location **e**) is unstable and it has different separation behavior than the other states. The circulation patterns outside the region of the western boundary current are very similar for each solution. Hence, the multiple equilibria are related to the different meandering structures of the Kuroshio. The transports for the steady solutions in the multiple equilibria regime are slightly larger than estimates of the Kuroshio transport to the south of Japan of about 50-55 Sv (Qiu and Joyce, 1992).

5.7.3. Summary

Before turning to the results of explicit high-resolution ocean models, it is time briefly summarize the results so far. Within a hierarchy of (equivalent) barotropic and baroclinic QG and SW models, it is found that multiple equilibria exist for the North Atlantic and North Pacific WDC. In the case with realistic continental geometry, these multiple steady states are associated with different separation paths of the western boundary current. While the different steady states are well known for the small basin case (typically with a zonal scale of $1000 - 2000$ km), it is expected (Primeau, 2002) that more isolated branches exist in SW models in the realistic case. These branches still await computation and will indicate whether

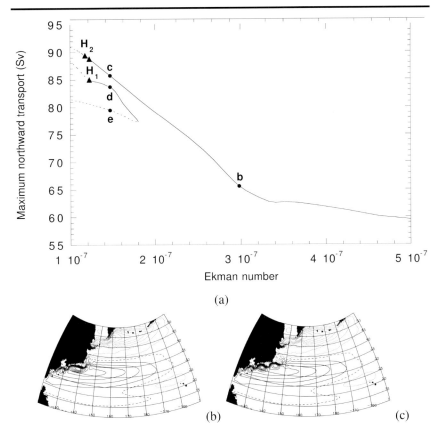

Figure 5.50.   (a) Bifurcation diagram for the barotropic shallow-water model on the North Pacific domain with the Ekman number as control parameter. Drawn (dotted) branches indicate stable (unstable) steady states, whereas the Hopf bifurcation points are indicated by triangles. (b) Contour plot of sea-surface height deviations for the steady state solution at location (c) $E = 1.5 \times 10^{-7}$ on the upper branch in (a). (c) Contour plot of sea-surface height deviations for the steady state solution at location (d) $E = 1.5 \times 10^{-7}$ on the upper branch in (a). The contour levels are scaled with respect to the maximum value of the field (from Schmeits and Dijkstra (2001).

multiple separation patterns with more elongated (Gulf Stream or Kuroshio) jets are possible.

The origin of the multiple equilibria can be traced to a symmetry-breaking pitchfork bifurcation in the QG model for flow in a rectangular basin. Due to intrinsic asymmetries, the pitchfork bifurcation becomes imperfect in the SW models. The 'jet up' and 'jet down' solution, i.e., the multiple equilibria in the QG-case, deform into solutions with different separation behavior of the Gulf Stream and Kuroshio under realistic continental geometry and wind-stress forcing. No new equilibria appear to be introduced by deformation of the geometry from rect-

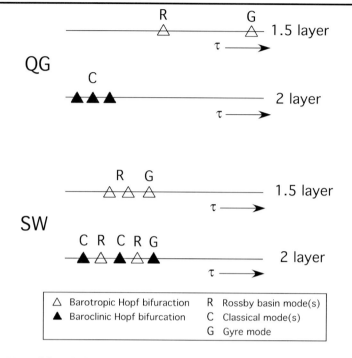

Figure 5.51.   Schematic figure of the different oscillatory internal modes identified in bifurcation diagrams of a hierarchy of QG and SW models.

angular to realistic. The simultaneous existence of the deflected and separated model Gulf Stream and the different meander states of the model Kuroshio are hence not due to continental geometry, but to internal ocean dynamics. In a way, the symmetry breaking found in the QG-model in rectangular geometry is still present, but it is localized in the high-shear Gulf Stream and Kuroshio regions in case of the SW model with realistic continental geometry and wind-stress forcing.

The modes of variability arising as instabilities on the mean states are closely linked within the hierarchy of the models. An overview of the ordering and types of modes found in QG and SW models of the WDC is presented (schematically) in Fig. 5.51, using the wind-stress strength (τ) as parameter. There are only a few internal modes that contribute to the transient dynamics: classical baroclinic modes, Rossby-basin modes and gyre modes. In statistical analysis of complex time-dependent flows, the wall-trapped modes are found as an additional pattern, but these modes have not yet been associated with a particular Hopf bifurcation of a steady double-gyre flow.

In the two-layer shallow-water model, all the different modes in the spectrum destabilize close to each other in parameter space, contrary to the two-layer quasi-geostrophic model, which shows a strong separation of the location where these modes destabilize the steady states. Therefore, in the two-layer shallow-water model complex, time-dependent behavior is already found at relatively low wind-stress forcing and can be analyzed near criticality. The actual ordering of the modes is sensitive to details in the models. The oscillatory time scale and even the pattern of each mode does not depend much on the continental geometry and wind-stress shape. The reason is that already in rectangular geometry, these modes are strongly localized within the high-shear region of jet, which basically does not change through the hierachy of the models.

By going through the hierarchy of models one has gotten an idea of the robustness of these phenomena and has an understanding of their physics. A legitimate question is, however, how relevant these results are for the actual ocean circulation. In the next section, we first turn to the 'virtual' ocean circulation as simulated with high-resolution ocean models.

5.8. High-Resolution Ocean Models

Ocean General Circulation Models (OGCMs) are widely used in physical oceanography in order to increase our understanding of the oceans by providing simulations of the large-scale circulation. The relatively high-resolution models (with a horizontal resolution of $1°$ or smaller) are roughly divided (McWilliams, 1996) into three classes depending on their capabilities of resolving the development of baroclinic instabilities of the mean flow (the so-called mesoscale eddies). Non-eddy resolving models typically have a $1°$ horizontal resolution and cannot capture the mesoscale eddies. Eddy-permitting models, having a resolution down to $1/6°$, can capture the baroclinic instabilities but represent only part of the interactions of these eddies. The models at the highest resolution (presently about $0.1°$) are called eddy-resolving OGCMs, as they capture the full spectrum of interactions of the eddies. We will refer below to the last two classes of models as eddy-representing OGCMs (ER-OGCMs).

Over the last decade, ER-OGCMs have been developed which are able to produce many observed features in the ocean, e.g., the meandering of the Gulf Stream and formation and propagation of rings. A fairly complete overview of the models currently in use, with links to the websites where these models are available, is given in http://stommel.tamu.edu/~baum/ocean_models.html. Although it is impossible to give an overview of the capabilities of all the different ER-OGCMs, below are a few words about the models which are around.

First, the M(odular) O(cean) M(odel), developed at the Geophysical Fluid Dynamics Laboratorium (Princeton, USA) has been developed since the early 1960's (Bryan *et al.*, 1974; Cox, 1987). Versions of the MOM model have developed into the P(arallel) O(cean) C(limate) M(odel), with home base at the Naval Postgraduate School (Monterey, USA) (Stammer *et al.*, 1994). Third, the isopycnal model M(iami) I(sopycnal) C(limate) O(cean) M(odel) is a layer model, with mixing along isopycnals (Bleck and Chassignet, 1994). A fourth model is the P(arallel) O(cean) M(odel), which has been developed in Los Alamos (Dukowicz and Smith, 1994). Many of the models are discussed in detail in the books by Haidvogel and Beckmann (1999) and Griffies (2004). Some of the typical simulations are sum-

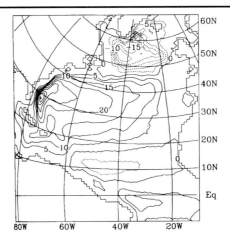

Figure 5.52. Streamfunction of the vertically averaged circulation as in Gerdes and Köberle (1995) using the MOM model at 1° horizontal resolution.

marized in Table 5.10, where also the horizontal resolution and simulation period are given.

The results of these numerical simulations show large internal variability on a wide range of space and time scales and the influence of the eddies on the mean flow (Schmitz and Holland, 1982; Holland and Schmitz, 1985; Miller *et al.*, 1987; McWilliams, 1996). The results are, however, not always easier to interpret than the observations, with which they share many physical processes and scales of motion. For example, model resolution strongly affects the mean flow path of the Gulf Stream, in particular its separation near Cape Hatteras; this is an interesting problem to which we turn below. Focus will be on the results of a global simulation with the POCM model (Stammer *et al.*, 1994) and a North Atlantic basin simulation with the MICOM model (Chassignet and Garaffo, 2001).

5.8.1. Typical results

In models with a 1° horizontal resolution (Holland and Bryan, 1994; Gerdes and Köberle, 1995), separation of the Gulf Stream is very diffuse between Cape Hatteras and Newfoundland and no recirculation regions are present. An example of the streamfunction for the vertically integrated transport of the reference run in Gerdes and Köberle (1995) is shown in Fig. 5.52. The Gulf Stream north of Cape Hatteras appears as a rather broad current following the coastline. Further east, there is a strong component of transport between 40°N and 55°N across the basin. In a high-resolution MOM simulation with 1/3° horizontal resolution (Beckmann *et al.*, 1994; Bryan *et al.*, 1995), the time-mean state shows a large anti-cyclonic gyre north of Cape Hatteras giving an actual separation north of the

Typical high-resolution OGCM studies			
OGCM	Reference	Resolution	Time
MOM	Holland and Bryan (1994)	$1° \times 1.2°$	100 years
	Gerdes and Köberle (1995)	$1° \times 1°$	10 years
MICOM	New *et al.* (1995)	$1° \times 1°$	30 years
	Piava *et al.* (2000)	$1° \times 1°$	64 years
POCM	Semtner and Chervin (1992)	$1/2° \times 1/2°$	32.5 years
	Stammer *et al.* (1994)	$1/4° \times 1/4°$	19 years
MOM	Bryan *et al.* (1995)	$1/3° \times 2/5°$	25 years
	Beckmann *et al.* (1994)	$1/6° \times 1/5°$	5 years
POP	Chao *et al.* (1996)	$1/6° \times 1/6°$	22 years
	Maltrud *et al.* (1998)	$0.28° \times 0.28°$	10 years
MICOM	Smith *et al.* (2000a)	$1/3° \times 1/3°$	25 years
MICOM	Chassignet and Garaffo (2001)	$1/12° \times 1/12°$	20 years
POP	Smith *et al.* (2000b)	$1/10° \times 1/10°$	11 years
NLOM	Hurlburt and Hogan (2000)	$1/64° \times 1/64°$	13 years

Table 5.10.   Some typical studies of the three-dimensional ocean circulation using high-resolution General Circulation Models.

observed position (Fig. 5.53). With a horizontal resolution of $1/6°$ and 37 vertical levels, the POP model shows a more reasonable separation near Cape Hatteras (Chao *et al.*, 1996). The clockwise circulation region as in Fig. 5.53 is weaker and the angle of separation of the Gulf Stream near the North American coast has improved, but the current still separates at the wrong location. Only at a horizontal resolution of $0.1°$ or higher, at which eddies are well resolved, the Gulf Stream does tend to separate at the correct position (Smith *et al.*, 2000b). The root-mean square sea-surface height variability for this simulation is in good agreement with the one reconstructed by blending altimeter data from the TOPEX-POSEIDON and the ERS1–2 satellites (Le Traon *et al.*, 1998).

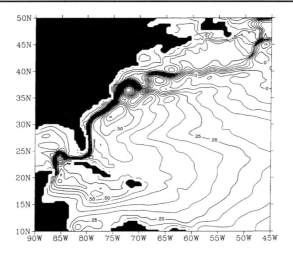

Figure 5.53.   Mean surface pressure for a typical simulation of the North Atlantic simulation using a version of MOM in Bryan et al. *(1995) on* $\frac{1}{3}^\circ$ *horizontal resolution.*

The OGCM simulations clearly show that it is necessary to have a mesh size that is well below the internal Rossby deformation radius in order to simulate a Gulf Stream separation at the correct location. Recent very-high resolution simulations show a surface ocean flow field much resembling the one reconstructed from observations (WOCE, 2001). An example of the two-year time mean North Atlantic sea-surface height field, as computed by MICOM using a $1/12^\circ$ horizontal resolution (and 20 isopycnal layers in the vertical) is plotted in Fig. 5.54. The meridional boundaries of the domain used are 28°S and 70°N and the domain includes the Mediterranean Sea. In this model simulation, the Gulf Stream nicely separates at Cape Hatteras and meanders eastward into the North Atlantic (Chassignet and Garaffo, 2001).

Still, it is far from clear which physical processes control the separation process. Haidvogel *et al.* (1992) and Dengg *et al.* (1996) have reviewed the problem of Gulf Stream separation. Both external factors — such as bottom topography or the wind-stress field — and internal ones — such as adverse pressure gradients (Tansley and Marshall, 2001), vorticity crisis (Kiss, 2002) or the outcropping of isopycnals (Gangopadhyay *et al.*, 1992; Chassignet and Bleck, 1993) — play an important role. Several studies using idealized models have tried to isolate only one or two factors as decisive but the relation between these results and those obtained by using more realistic models and observations is hard to establish. The tools of dynamical systems theory are thus essential to explore systematically the results across the full hierarchy of ocean models.

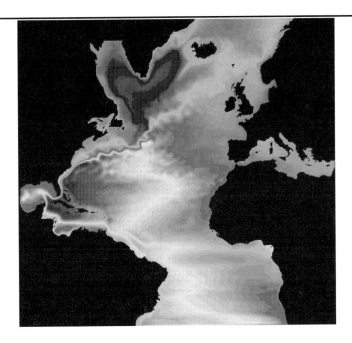

Figure 5.54. (in color on page 520). Plot of the two-year mean of the sea-surface height of a high-resolution (1/12°) simulation with MICOM (Chassignet and Garaffo, 2001).

As we know from the bifurcation diagrams of the shallow-water models, multiple mean states appear dynamically possible for both the Gulf Stream and Kuroshio. Are signatures of these multiple mean states, with different separation characteristics, found in high-resolution ocean models?

5.8.2. Analysis of POCM results

In Schmeits and Dijkstra (2000) and Schmeits and Dijkstra (2001), monthly mean sea surface height (SSH) data over the years 1979 to 1997 of the Parallel Ocean Climate Model (POCM) (Semtner and Chervin, 1992; Stammer *et al.*, 1994) are analyzed. SSH is a prognostic variable in POCM because of the incorporation of a free surface formulation. The POCM output used was from run 4C having an average horizontal resolution of $\frac{1}{4}°$ and 20 non-equidistant levels in the vertical direction. The global simulation was performed for the period 1979-1997 and the ocean model was forced by either ECMWF reanalysis (1979-1993) or ECMWF operational (1994-1997) fields of heat fluxes, freshwater fluxes and wind stress. For more details about this simulation the reader is referred to http://vislab-www.nps.navy.mil/~rtt.

5.8.2.1 Gulf Stream

The Gulf Stream region analyzed was the domain 90°W-30°W, 23°N-48°N. In Fig. 5.55a, the latitudinal position of the maximum zonal geostrophic velocity at 75°W is plotted over this period. During the first year, the maximum zonal geostrophic velocity is at a latitude of about 34.5°N and a contour plot of (SSH) anomalies is shown for January 1979 in Fig. 5.55b. The Gulf Stream separates too far north compared to reality, and has a weak southern recirculation region and a strong anti-cyclonic cell near Cape Hatteras. A similar mean state is also found in other high resolution models (Beckmann *et al.*, 1994; Bryan *et al.*, 1995). Within the QG and SW model context, as discussed in the previous sections, one would call this a 'deflected' Gulf Stream.

In early 1980, a significant shift southward occurs in the latitudinal position of the maximum zonal geostrophic velocity (Fig. 5.55a). The SSH anomalies for January 1981 (Fig. 5.55c) display a Gulf Stream which actually seems to separate twice. First, it separates too far south compared to reality, and later on it separates too far north (at about the same latitude as in Fig. 5.55b). It is characterized by a strong southern and a weak northern recirculation region, centered at about 74°W, 32°N and 73°W, 34°N, respectively, and by a strong anti-cyclonic cell near Cape Hatteras. Within the QG and SW model context, one would call this a 'separated' Gulf Stream.

The southerly position is maintained for about two years and then the earlier position is retained (Fig. 5.55a).   Shaded contour plots of SST indicate that at 75°W, the cold wall (region of large temperature gradient) of the 'deflected' Gulf Stream is situated at a more northerly latitude (Fig. 5.55b) than that of the 'separated' Gulf Stream (Fig. 5.55c). The difference in annual mean SST between the 'deflected' and 'separated' Gulf Stream paths in POCM reveals that the northern recirculation region is up to 2°C warmer in the case of the 'deflected' than in the case of the 'separated' path. Analysis of the temperature fields at depth in POCM indicates that the signatures of the different separation paths are also found in the deeper ocean.

Stammer *et al.* (1994) discuss the temporal variability of POCM fields and compare these to available observations. In Schmeits and Dijkstra (2000), non-seasonal SSH anomalies were analyzed using the (M-SSA) statistical technique (Vautard *et al.*, 1992). A 9-month statistically significant (with respect to a red noise null-hypothesis) mode is found. The anomalies are concentrated around the mean axis of the Gulf Stream, have a maximum amplitude of 2 cm, propagate upstream, and have a wavelength of about 500 km. The pattern of this oscillatory mode extends from 75°W up to 60°W.

5.8.2.2 Kuroshio

The Kuroshio region analyzed was the domain 120°E-160°E, 25°N-50°N. In Fig. 5.56a, the latitudinal position of the maximum zonal geostrophic velocity at 136° E is plotted over the period 1979-1998. This plot indicates that several transitions between two states occur on interannual timescales. In Figs. 5.56b

(a)

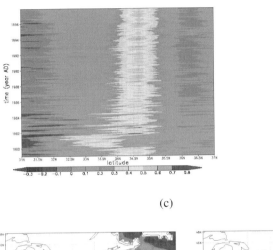

(b)   (c)

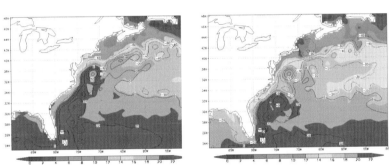

Figure 5.55.   (in color on page 521).   (a) Contour plot of zonal geostrophic velocity (ms^{-1}), calculated from SSH gradients in POCM, at 75° W as a function of latitude for the period 1979-1997. Contour plot of monthly mean SSH deviations (cm), superimposed on a shaded contour plot of monthly mean SST for (b) January 1979, representing the 'deflected' Gulf Stream in POCM, characterized by a northerly position of its cold wall at 75°W and (c) January 1981, representing the 'separated' Gulf Stream in POCM, characterized by a southerly position of its cold wall.

and c, SSH anomalies are superimposed on SST for the two states. One state (Fig. 5.56b) resembles the observed small meander state (Fig. 5.4a). It is characterized by a northerly position of the maximum zonal geostrophic velocity at a latitude of about 33°N (Fig. 5.56a), accompanied by an anti-cyclonic recirculation cell to the south. The other state (Fig. 5.56c), resembles the observed large meander state (Fig. 5.4a). It is characterized by a southerly position of the maximum zonal geostrophic velocity at a latitude of about 31.5°N (Fig. 5.56a), accompanied by a cyclonic recirculation cell to the north. In both cases, the main separation latitude of the Kuroshio is too far north (at about 40°N), as is also a problem in other high-resolution models (Maltrud *et al.*, 1998). Shaded contour plots of SST (Figs. 5.56b and c) indicate that the Kuroshio advects more heat northward in the case of small meander state (Fig. 5.56b) than in the case of large meander state (Fig. 5.56c). The simulated Kuroshio has a larger eastward geostrophic transport when it is in its small meander state than when it is in its large meander state, at least at 136°E (Fig. 5.56a). Observations point at geostrophic transports of equal magnitude for both states at 137°E (Qiu and Joyce, 1992).

(a)

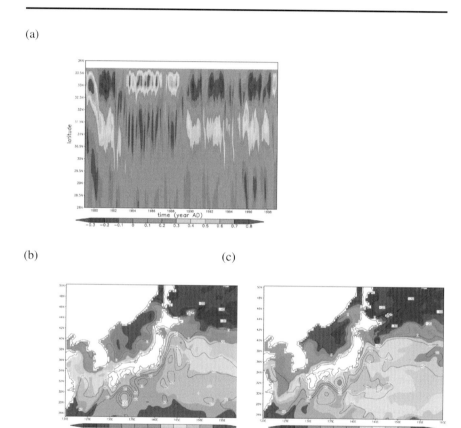

(b)   (c)

Figure 5.56.   (in color on page 522). (a) Contour plot of zonal geostrophic velocity (m.s⁻¹), *calculated from sea surface height (SSH) gradients in POCM, at 136° E as a function of latitude for the period 1979-1998. (b-c) Contour plot of monthly mean SSH deviations (cm), superimposed on a shaded contour plot of monthly mean SST for (b) January 1988, representing the Kuroshio small meander state in POCM and (c) January 1996, representing the Kuroshio large meander state in POCM.*

Schmeits and Dijkstra (2002) also analyzed the variability of simulated temperature fields by POCM in the Kuroshio region at three depth levels — 160 m, 310 m and 610 m — using the M-SSA technique. In all datasets, there exists an oscillatory mode of variability with a timescale of 7 months. The anomalies are rotating in the Kuroshio separation region of POCM and have a maximum amplitude of 2.3 K at 160 m, 2.2 K at 310 m, and 1.0 K at 610 m.

5.9. Synthesis

The results from the analysis of POCM for both Gulf Stream and Kuroshio are encouraging since features can be found which look like those seen in the QG and SW models. However, there is still a huge gap between the understanding of the behavior of the solutions of the QG and SW models and those of the ER-OGCMs. In this last section of the chapter, the relation between solutions of the different models in the hierarchy and the connections to observations are discussed.

5.9.1. Summary

With the dynamical systems approach as presented in this chapter, the idea is that an understanding of the physics of the observed complex ocean flows can be obtained by approaching the 'real' situation from particular limiting flows. One path proceeds from simple to complex situations through a hierarchy of models. Here, the lowest member of the model hierarchy describing the WDC is the barotropic QG model in a square basin with a flat bottom. In this chapter, the hierarchy of models proceeded upwards by inclusion of stratification (multi-layer models) and continental coastlines. A second path was taken within one particular member of the model hierarchy where we proceeded from steady, highly-dissipative or weakly-forced flows to irregular, weakly dissipative or strongly forced flows by varying parameters.

By proceeding along both paths, two important issues have become apparent. The first issue is the existence of multiple steady flow patterns in wind-driven midlatitude ocean flows. These multiple states have been robust in the model hierarchy and their origin is a symmetry breaking shear instability, most apparent in the barotropic QG model. In realistic continental geometry, these multiple states are associated with different separation patterns of the western boundary currents. The second issue is that a classification of internal modes of variability is appearing. These modes can be considered as 'dynamics modes', since they are related to the operator associated with the momentum equations. From a mathematical point of view, there are two types of modes. One type of modes comes from the basic linear operator arising from the linear stability analysis of the no-flow state. These are the Rossby-basin modes (RB) in QG models which also exist in SW models. The other types of modes have no origin in this basic linear operator. The oscillatory classical baroclinic modes (CB) arise when vertical shear is present in the background state. The low-frequency gyre modes (G) arise through a merger process of stationary modes, and the wall-trapped (WT) modes likely arise through an instability of the viscous boundary layer. We do not know yet whether these four classes form all the 'dynamics' modes, but no other types have been found at this point.

A posteriori, the double-gyre wind-forcing has not been a bad choice as a starting point for studying the midlatitude WDC. An alternative would have been to use the much studied single-gyre case of which the bifurcation structure and the analysis of the linear stability of barotropic steady states has been presented in Ierley and Sheremet (1995), Kamenkovich *et al.* (1995) and Sheremet *et al.* (1997). One finds a back-to-back saddle-node bifurcation structure very similar to the behavior of the symmetric solution branch in Fig. 5.19. The solution becomes inertially dominated at large values of Re, a phenomenon called 'inertial runaway' in Ierley and Sheremet (1995).

In fact, the symmetric double-gyre solution in a 1000 × 2000 km basin can be seen as a superposition of two single-gyre systems, each occupying a 1000 × 1000 km basin. The stability of the symmetric double-gyre state is similar to that of the single-gyre flow if only symmetric perturbations are allowed. The bifurcation behavior of the double-gyre system is different because of the symmetry-breaking pitchfork bifurcations (as in Fig. 5.19) which give rise to additional branches.

It is good to realize that when the path and analysis of the model hierarchy would have been started at the single-gyre flows, one would eventually end up with the same bifurcation diagram for the SW model in the North Atlantic (Fig. 5.48) and North Pacific basin (Fig. 5.50). However, the imperfect pitchfork bifurcation in the double-gyre flows is much easier to compute and understand since it derives from a perfect pitchfork bifurcation. In a single-gyre approach, the additional isolated branch also appears when approaching the most realistic configuration, but it is more difficult to understand where it comes from.

In this view, it is not surprising that there is a connection between modes of variability of the double-gyre flow (as discussed here) and those of the single-gyre flows (Sheremet *et al.*, 1997). For the barotropic single-gyre case, also Rossby-basin modes, wall-trapped modes and so-called recirculation gyre modes are found. The precise connection between the internal modes in the single- and double-gyre flows, however, has not been investigated yet. Also here, in the realistic configuration of the North Atlantic and North Pacific in a SW-model, the modes coming from the analysis of both configurations should eventually become similar.

It is less clear at the moment, why the transition behavior to complexity differs in both single- and double-gyre flow. The time-dependent behavior of the single-gyre flows has been studied extensively (Berloff and Meacham, 1997, 1998a; Meacham and Berloff, 1997, 1998). In Berloff and Meacham (1998b), it is suggested that the route to chaos in the baroclinic case is the classical three-frequency route (Holmes *et al.*, 1996; Ruelle and Takens, 1970). which appears different from the Lorenz (Simonnet *et al.*, 2005) and Shilnikov (Nadiga and Luce, 2001) route (through a homoclinic connection) as found in double-gyre flows. The study of the route to complex flows over the model hiearchy, however, is in its infancy and many exciting new results can be expected in the near future.

5.9.2. Interpretation framework

As most ER-OGCMs perform simulations in the complex flow regime, and observations come also from ocean flows in such a regime, there is at the moment quite a limited interpretation framework that the dynamical systems analyses of SW and QG models have to offer. However, on the positive side, there is a start of a systematic qualitative theory which provides new concepts (e.g., the gyre mode) and also new diagnostic tools for looking at both ER-OGCM output and observations (e.g., the subannual variability in western boundary currents and the different separation patterns).

5.9.2.1 Multiple mean paths?

With respect to multiple mean paths, observations indicate that the Kuroshio clearly shows bimodality in the form of the large and small meander states, but there seem to be less clear indications for bimodality in the Gulf Stream. The question arising from these observations is: what are the physical processes responsible for this bimodal behavior and why is this behavior more pronounced for the Kuroshio than for the Gulf Stream?

The existence of multiple equilibria was a central element in the classical explanation of the bimodality of the Kuroshio (Charney and Flierl, 1981; Masuda, 1982). However, in most of the previous modeling studies in which multiple paths of the Kuroshio were found, regional models were used with in/outflow boundary conditions (Chao, 1984). As Qiu and Miao (2000) have pointed out, regional models with in/outflow boundary conditions may not be able to realistically capture the Kuroshio's recirculation gyre, which is an inseparable part of the Kuroshio current system. Based on calculations with a high-resolution 2-layer primitive-equation model, Qiu and Miao (2000) propose that the observed alternations of the Kuroshio's two states are due to a self-sustained internal oscillation involving the evolution of the southern recirculation gyre and the stability of the Kuroshio current system.

The barotropic shallow-water model for a Pacific basin, as used in the bifurcation studies above, contains only a very small part of the physics governing the actual flow. It contains no bottom topography, no baroclinic effects and a very simple parameterization of mixing of momentum, i.e., downgradient Laplacian diffusion. The latter is represented by a single Ekman number E which is used as control parameter. It is standard practise in high-resolution ocean modelling to decrease the coefficients measuring the magnitude of the dissipative processes (mixing of momentum and heat/salt) simultaneously with an increase in resolution. From an oceanographic point of view, this is easily justified, since the major part of the mixing is produced by the eddies itself and not by a background downgradient diffusion, which is thought to be very small. However, from a dynamical systems point of view one changes two things at the same time. The degree of approximation to the continuous equations is changed with changing resolution, which introduces shifts in location of bifurcation points. Simultaneously, also parameters in the model are changed, which also induces movement in the locations of these bifurcation points. Hence, the correspondence with changes in E in the SW model and the increase in resolution (and corresponding decrease in mixing coefficients) in ocean models is admittedly vague. It seems, however, reasonable that a decrease in E mimics this change in the ocean model configuration.

Bifurcation analysis of this model for the North Pacific gives the following results

(i) There is unique state when E is too large, with a small meander mean state. This would indicate that ocean models with too coarse resolution will find a similar state.

(ii) When E is smaller than a certain value, a different flow regime is entered where more than one stable steady equilibrium exists. The existence of such a flow regime would imply that switches between mean states are possible, for example induced by stochastic 'noise' in the wind-stress field.

The results certainly indicate that the multiple equilibrium regime is dynamically possible in the Pacific and that the equilibria differ only in the local separation behavior of the Kuroshio. A qualitatively similar result was obtained for the North Atlantic version of this model, where the unique state (at large E) has a separation of the western boundary current which is too far northward. The other equilibrium found at smaller E has an (early) separated Gulf Stream. This result implies that an ocean model has to be in this flow regime to get correct Gulf Stream separation, which requires sufficiently low values of E and hence a sufficiently high resolution.

The details on the transition to complex flows are still unknown for the more realistic configurations, but one can try to anticipate the behavior from the studies in idealized configurations (McCalpin and Haidvogel, 1996; Primeau, 2002; Simonnet *et al.*, 2003b). One would expect that low-frequency behavior can occur at even smaller lateral friction due to the appearance of homoclinic connections. Eventually, a trajectory of the system may can come close to each of the steady states which may explain the suggested 'internal oscillation' leading to bimodality by Qiu and Miao (2000). Of course, there is still a long way to go since the effect of the baroclinic eddies and bottom topography has hardly been studied. The latter may be a main factor why the bimodality of the Kuroshio is observed while the Gulf Stream bimodality is limited. Bottom topography, combined with a rectification effect on the mean flow due to the baroclinic eddies can certainly have a large impact on the transitions between the multiple mean flow patterns.

5.9.2.2 Modes of variability?

With respect to variability, a central problem in comparing theory with observations is the lack of accurate long time series of data which have sufficient spatial resolution. The satellite data from TOPEX and the AVHRR maybe satisfy the requirements for accuracy, but the time period is only just over a decade. This limits the time scale of phenomena associated with internal variability of the ocean circulation, which can be investigated with some confidence, certainly to one year.

The 9-month variability in the Gulf Stream region is within this range and it seems to be the first period of variability which can be analyzed with reasonable confidence beyond the baroclinic eddy variability (with a time scale of a 3-4 months). Let's take this as an example for using the QG and SW model results to interpret ER-OGCM results and observations.

As mentioned in section 5.8, dominant patterns of variability in SSH observations were determined in Schmeits and Dijkstra (2000), using the M-SSA technique. The data set used in this study were anomalous sea-surface height observations from the NASA TOPEX/POSEIDON Altimeter Pathfinder (T/P) dataset (http://neptune.gsfc.nasa.gov/~krachlin/opf/algorithms.html). A statistical significant 9-month mode, which shows anomalous power against a red-noise null-hypothesis was found. When nothing else would be available, the result would be meager since it might be that techniques like M-SSA overestimate the power at this particular frequency. After all, the total period of the data analyzed is about 5 years and the frequency is also close to the annual cycle. Luckily, there is more observational work using other types of satellite and in-situ data, which strongly indicates that there indeed seems to be a preferred period of variability and that it is difficult to attribute its origin to external (in this case) atmospheric forcing (Lee and Cornillon, 1995; Kelly *et al.*, 1996). The latter can never be excluded, however, because stochastic noise in an ocean where advection occurs can also introduce variability on preferred time scales (Saravanan and McWilliams, 1998). This motivates to look at internal ocean dynamics as a likely cause of this preferred time scale of variability. But what is the physics of this phenomenon ?

Although output from only one high-resolution model was analyzed in Schmeits and Dijkstra (2000) (the POCM model), it is encouraging that in this model, also statistical significant modes (against a red noise null-hypothesis) are found in the SSH field with a near 9-month time scale. The spatial patterns do not well correspond to those in the TOPEX data, because the mean flow in POCM has a Gulf Stream which also separates at too northerly latitudes. However, the general characteristics with respect to location and propagation are not that different. It seems reasonable to state that the POCM configuration contains all the physical processes needed to explain this 9-month variability in observations up to sufficient detail. Through further analysis of the POCM output, it turns out that one can trace this 9-month signal down to about 1 km, with hardly any distortion in spatial structure. But again, what are the physical processes responsible for this type of variability ?

Bifurcation analysis of the SW model for the North Atlantic shows that when the Ekman number E is decreased, there is a transition where certain preferred patterns are amplified through energy transfer from the mean flow. These patterns correspond to the eigenvectors associated with the stability problem of the steady states just at Hopf bifurcation. The perturbations are localized in the region of large-horizontal shear of the mean flow. The latter is easy to understand, since only the Reynolds' stresses can be responsible for the energy transfer. Clearly, this is a barotropic instability giving patterns which have a subannual time scale for reasonable choices of the average basin depth. Hence, a clear 'candidate' for explaining the origin of the subannual variability is available. Again by going through the model hierarchy, this mode can be traced to a barotropic instability associated with one of the Rossby basin modes.

This leads to the next question: how about the relevance of the gyre modes, having interannual-to-decadal periods? Recall from section 1.3.2 that in the Central England Temperature series, there is a peak in the spectrum at 7.7 years. This 7-8-yr peak has been reported in North-Atlantic SST and SLP data by Moron *et al.* (1998) and Da Costa and Colin de Verdiére (2004). The SST pattern associated with this variability shows downstream (with respect to the Gulf Stream) propagation in a northward direction, similar to the variability presented in Sutton and Allen (1997). In SW models of the double-gyre circulation in idealized basins, also this period of variability is found in the so-called quasi-homoclinic regime, which is located after the appearance of the homoclinic orbit (Speich *et al.*, 1995; Nauw *et al.*, 2004b; Simonnet *et al.*, 2005). In this regime, the oscillation period saturates with decreasing friction to about twice the period at the Hopf bifurcation. The variability introduced by the gyre mode dynamics is therefore a nice prototype for this 7-8-yr variability, but the connection is still not very strong.

For the understanding of the multiple mean flows of the Kuroshio, the subannual variability of the Gulf Stream, and the interannual-to-decadal variability in North-Atlantic SST, the path through the model hierarchy with analysis techniques from bifurcation theory directly connects to, builds on and extends the work of the early dynamical oceanographers. There is still a long journey to go, in terms of model complexity, effects of stochastic noise in the forcing (e.g., Sura *et al.* (2000)) and in the analysis of complex behavior of the particular model solutions, but the path is clear and it will certainly be a great adventure.

5.10. Exercises on Chapter 5

(E5.1) *Weakly nonlinear single-gyre flow*

In this exercise, we consider the weakly nonlinear extension of the Stommel boundary solution. The stationary dimensionless barotropic vorticity equation on the domain $(x, y) \in [0, 1] \times [0, 1]$ can be written (with $\delta_M = 0$) as

$$\lambda \left[\frac{\partial \psi}{\partial x} \frac{\partial}{\partial y} - \frac{\partial \psi}{\partial y} \frac{\partial}{\partial x} \right] \nabla^2 \psi + \frac{\partial \psi}{\partial x} = -\frac{\partial \tau^x}{\partial y} - \nabla^2 \psi$$

where $\lambda = (\delta_I / \delta_S)^2$. Let the wind forcing be given by

$$\tau^x = -\frac{1}{\pi} \cos \pi y$$

We will consider λ as a small parameter of the system.

a. Show that the Stommel-Sverdrup solution is obtained for $\lambda = 0$. Indicate this solution as ψ^0.

Now expand
$$\psi = \psi^0 + \lambda \psi^1 + \mathcal{O}(\lambda^2)$$

b. Determine the solution ψ^1.

c. Sketch the resulting flow for small λ.

Further reading: Pedlosky (1987), chapter 5 and Salmon (1998).

(E5.2) *Effect of friction representation in SW models*

In a 1.5-layer SW model (section 5.6.1), different forms of lateral friction are considered in the literature. The one employed in the model in (5.60) is of the form
$$\mathcal{F}_F = \nabla^2 (h\mathbf{u})$$

where $\mathbf{u}$ is the horizontal velocity field. Other forms of lateral friction used are $\mathcal{F}_E$ and $\mathcal{F}_M$ given by

$$\begin{aligned} \mathcal{F}_E &= \nabla.(h\nabla \mathbf{u}) \\ \mathcal{F}_M &= h\nabla^2 \mathbf{u} \end{aligned}$$

In this exercise, we are going to investigate the impact of these different friction formulations.

Consider a simple parallel zonal flow in a channel with a width of $2L_c$. The velocity field is $u = u(y), v = 0$, the layer thickness field is $h = h(y)$ and the equilibrium layer thickness is H_0.

a. Show that under these approximations, the 1.5-layer shallow-water model equations (5.60), with $a_\tau = 1$ and $r_b = 0$, reduce to

$$0 = A_H \mathcal{F}^x(u, h) + \frac{\tau_0 \tau^x(y)}{\rho_1 h}$$

$$(f_0 + \beta_0 y)u = -g' \frac{dh}{dy}$$

and show that the frictional terms become

$$\mathcal{F}_M^x(u, h) = \frac{d^2 u}{dy^2}$$

$$\mathcal{F}_E^x(u, h) = \frac{1}{h} \frac{d}{dy} \left[h \frac{du}{dy} \right]$$

$$\mathcal{F}_F^x(u, h) = \frac{1}{h} \frac{d^2(uh)}{dy^2}$$

With no-slip boundary conditions at the channel walls, i.e., $y = \pm L_c : u = 0$, we need a condition to ensure overall mass conservation.

b. Show that this condition is provided by

$$\frac{1}{2L_c} \int_{-L_c}^{L_c} h(y)\, dy = H_0$$

Suppose we prescribe a symmetric velocity profile and a purely anti-symmetric thermocline profile and ask: what wind-stress shape is needed to force this as a steady flow in the channel? In the simplest case, neglecting the β-effect, we choose

$$h(y) = H_0 \left(1 - \frac{1}{4} \sin\left(\frac{\pi y}{2L_c} \right) \right) \; ; \; u(y) = U_0 \cos\left(\frac{\pi y}{2L_c} \right)$$

with $U_0 = \pi g' H_0/(8 f_0 L_c)$.

c. Show that for this flow field, the general solution for the wind stress profile is

$$\tau^x(y) = \left(\cos\left(\frac{\pi y}{2L_c} \right) - A_{as} \sin\left(\frac{\pi y}{L_c} \right) \right)$$

and determine A_{as} for each of the frictional representations.

d. Under a prescribed symmetric wind forcing, for example,

$$\tau^x(y) = \cos\frac{\pi y}{2L_c}$$

which of the frictional parameterizations will lead to the strongest asymmetric (with respect to $y = 0$) flow?

Further reading: Nauw et al. (2004a) and references therein.

(E5.3) *Fofonoff inertial circulation*

In this exercise, we will study a pure inertial quasi-geostrophic flow in a homogeneous (constant density) ocean. Consider the barotropic vorticity equation (5.33) for $\mathbf{T} = \mathbf{0}$ and $\delta_S^* = \delta_M^* = F = 0$.

a. Show that the resulting equation can be written as

$$\mathbf{u} \cdot \nabla\left(\left(\frac{\delta_I}{L}\right)^2 \nabla^2\psi + y\right) = 0$$

where $\mathbf{u} = (u, v)$ is the horizontal velocity vector.

b. Show that a pure zonal (horizontally unbounded) flow is a solution of the equation above.

Next consider the situation where the ocean is bounded by a continent at $x = 0$, while for $x \rightarrow \infty$, the flow is still pure zonal with dimensionless velocity field $u = -1$ and $v = 0$.

c. Show that, in this case, the absolute vorticity is a linear function of the streamfunction.

d. Consider in the problem above explicitly the western boundary layer (with $\delta_I \ll L$) and show that the solution is given by

$$\psi_W(x, y) = (y - y_0)(1 - e^{-xL/\delta_I})$$

where y_0 is an arbitrary meridional coordinate at which $\psi = 0$ for all x.

From now on take $y_0 = 0$. As a next step, we consider also a continent at the eastern boundary $x = 1$; between the zonal boundaries, the flow is still zonal.

e. Show that the total solution (eastern plus western boundary layer) is given by

$$\psi_{WE}(x, y) = y(1 - e^{-xL/\delta_I} - e^{-(1-x)L/\delta_I})$$

Finally, we look at a closed basin bounded meridionally by $y = -1$ and $y = 1$, where both $\psi = 0$.

f. Show that the total solution is given by

$$\psi_B(x, y) = \psi_{WE}(x, y) + c_1 e^{-(1-y)L/\delta_I} + c_2 e^{-(y+1)L/\delta_I})$$

and determine the constants c_1 and c_2.

g. Provide a sketch (plot) of the so-called Fofonoff inertial circulation in the closed basin.

Further reading: Fofonoff (1954) and Pedlosky (1987), chapter 5.

(E5.4) *Boundary conditions*

Consider the dimensional problem for the equivalent barotropic vorticity equation for flow in a square $L \times L$ basin described by

$$\frac{\partial}{\partial t}(\nabla^2 \psi - \frac{1}{R^2}\psi) + \beta_0 \frac{\partial \psi}{\partial x} = A_H \nabla^2 (\nabla^2 \psi - \frac{1}{R^2}\psi) + \frac{\tau_0}{\rho L} G(x, y, t)$$

The lateral friction in this model is represented by $A_H \nabla^2 q$, where $q = \nabla^2 \psi - \psi/R^2 + \beta_0 y$ is the potential vorticity and $R = \sqrt{g'D}/f_0$ is the internal Rossby deformation radius. The curl of the wind-stress forcing pattern, G, is a prescribed function.

a. Use scales L, $L/(\beta_0 R^2)$ and $\tau_0/(\beta_0 R)$ for length, time and streamfunction and show that the dimensionless equation becomes

$$\frac{\partial}{\partial t}(\epsilon \nabla^2 \psi - \psi) + \frac{\partial \psi}{\partial x} = \delta \nabla^2 (\epsilon \nabla^2 \psi - \psi) + G(x, y, t)$$

and determine δ and ϵ.

In QG models, there is subtlety in the boundary condition of mass conservation. The correct conditions on the boundary S of the domain are (McWilliams, 1977)

$$\psi = \psi_0(t) \; ; \; \int_S \psi \, dx \, dy = 0$$

which are more general than the condition $\psi = 0$.

b. Use the dimensional model above to show that for steady states of the barotropic vorticity equation with $R \to \infty$, both boundary conditions give the same solutions.

It is illustrative for the difference in the boundary conditions to investigate the energy balance of the model above.

c. Multiply the barotropic vorticity equation above by $\nabla \psi$, integrate over the domain $(x, y) \in [0, L] \times [0, L]$ and derive the energy balance

$$\frac{\partial \mathcal{E}}{\partial t} = -\nabla.\mathbf{F} - \psi G - \mathcal{D}$$

where $\mathcal{E} = (\epsilon \nabla \psi.\nabla \psi + \psi^2)/2$. Derive the expressions for $\mathbf{F}$ and $\mathcal{D}$.

d. Integrate the energy equation over the flow domain $(x, y) \in [0, 1] \times [0, 1]$, use the correct boundary conditions above (and assume no-slip) and derive the expression

$$\frac{\partial < \mathcal{E} >}{\partial t} = \psi_0 \oint \mathbf{H.n} \, ds - < \psi G + \mathcal{D} >$$

and determine $\mathbf{H}$. Here, $<>$ indicates integration over the flow domain, the contour integral is over the boundary over the domain and $\mathbf{n}$ is the outer normal to the boundary.

e. Interpret the difference in the energy balance for the correct boundary condition and for the boundary condition $\psi = 0$.

Further reading: McWilliams (1977) and Cessi and Primeau (2001).

(E5.5) *QG - SW transition*

It is useful to think how branches of steady solutions can be continuously traced between QG and SW models. Suggest at least two methods by which this can be accomplished.

(P5.1) *The nonlinear Munk model*

In this exercise, we will consider solutions of the barotropic vorticity equation for the case when bottom friction can be neglected, i.e. $\delta_S = 0$. In the western boundary layer, the derivatives to the zonal coordinate x are much larger than those to y.

a. Show that the boundary layer equation can be written as

$$\lambda \left[\frac{\partial \psi}{\partial x} \frac{\partial}{\partial y} - \frac{\partial \psi}{\partial y} \frac{\partial}{\partial x} \right] \frac{\partial^2 \psi}{\partial x^2} + \frac{\partial \psi}{\partial x} = \frac{\partial^4 \psi}{\partial x^4}$$

with $\lambda = (\delta_I/\delta_M)^2$.

Assume now that the flow far away from the boundary layer is paralell with velocity field

$$\psi = y$$

We now try to find solutions $\psi = \phi(x)y$.

b. Show that the equation for $\phi(x)$ becomes

$$\phi''' = \lambda((\phi')^2 - \phi\phi'') + \phi - 1$$

where the prime denotes differentiation to x.

c. Formulate the three boundary conditions for ϕ.

Write the equations as a first order system of equations and implement them into the AUTO continuation code (see the beginning of chapter 4 on how to obtain the code). The condition at infinity can be handled in the simplest way by defining $x = \alpha\,\tilde{x}$, with $\tilde{x} \in [0, 1]$ and by choosing α large.

d. Determine the solution for $\lambda = 0$. (*Hint: Choose a smart starting point by introducing an additional 'homotopy' parameter*)

e. Calculate the bifurcation diagram in λ and show that there is a saddle-node bifurcation at $\lambda = -0.79$.

Further reading: Pedlosky (1996), chapter 2.

(P5.2) *A 4-mode model of the wind-driven circulation*

In this exercise, we will investigate the steady equilibria of the double-gyre flow in a square basin and study their linear stability within a low-order model using the AUTO software (see beginning of chapter 4 how to obtain this software).

The low-order model derivation starts from the barotropic vorticity equation by taking zero lateral mixing ($Re \to \infty$) and keeping non-zero bottom-friction. On the boundaries only no normal flow conditions, i.e., $\psi = 0$ are applied. Next step is to project the equations using suitable expansion functions. In order to account for the existence of the western boundary layer, a decaying exponential in the x direction is introduced while a sine expansion is retained in the y direction, i.e.

$$
\begin{aligned}
\psi &= A_1(t)G(x)\sin y + A_2(t)G(x)\sin 2y + \\
&+ A_3(t)G(x)\sin 3y + A_4(t)G(x)\sin 4y \\
G(x) &= e^{-sx}\sin x
\end{aligned}
$$

where $s = 1.3$ is chosen such that $G(x)$ fits with the zonal asymmetric structure of a typical flow. The truncated equations are obtained by projecting the barotropic voriticity equation onto the orthogonal basis $(G(x) \sin ky$, k=1,4) using the inner product $< f,g > = \int_0^\pi \int_0^\pi fg \, dxdy$ such that the energy of the truncated system is conserved.

This leads to the following set of ODEs

$$\frac{dA_1}{dt} = c_1 A_1 A_2 + c_2 A_2 A_3 + c_3 A_3 A_4 - a_1 A_1$$

$$\frac{dA_2}{dt} = c_4 A_2 A_4 + c_5 A_1 A_3 - c_1 A_1^2 - a_2 A_2 + c_7 \sigma$$

$$\frac{dA_3}{dt} = c_6 A_1 A_4 - (c_2 + c_5) A_1 A_2 - a_3 A_3$$

$$\frac{dA_4}{dt} = -c_4 A_2^2 - (c_3 + c_6) A_1 A_3 - a_4 A_4$$

with $c_1 = 0.020736$, $c_2 = 0.018337$, $c_3 = 0.015617$, $c_4 = 0.031977$, $c_5 = 0.036673$, $c_6 = 0.046850$ and $c_7 = 0.314802$. Furthermore, $a_1 = 0.0128616$, $a_2 = 0.0211107$, $a_3 = 0.0318615$ and $a_4 = 0.0427787$. We consider only σ, which represents the strength of the wind-stress forcing, as a control parameter.

a. Start at the zero solution for $\sigma = 0$ and determine a branch of fixed points in σ. Determine the σ value of the first pitchfork bifurcation.

b. What is the internal symmetry of the system giving rise to the pitchfork bifurcation?

c. Determine the branches of asymmetric solutions. Plot the bifurcation diagram as value of A_1 versus σ.

d. Determine the value of σ at the Hopf bifurcation on one of the branches of asymmetric solutions. How do you determine whether the Hopf bifurcation is subcritical or supercritical?

e. Calculate a branch of periodic solutions from the Hopf bifurcation. Are these periodic orbits stable?

f. Follow the periodic orbits up to very large period and find the approximate value of σ where the homoclinic connection occurs. How would you determine whether this is a Shilnikov or Lorenz type of homoclinic connection? Compute a few trajectories just beyond the homoclinic connection.

g. Make a sketch of the different solutions of the model versus σ up to the homoclinic connection.

Further reading: Simonnet et al. *(2005).*

Chapter 6

THE THERMOHALINE CIRCULATION

Excitement from down the bottom right to the top.
Asturias, Isaac Albeniz.

In chapter 1, a caricature was provided of the three-dimensional circulation as a 'conveyor belt' driven by both the wind-stress forcing and buoyancy fluxes at the ocean-atmosphere interface. In this chapter, focus will be on the thermohaline component of this circulation, i.e., that associated with the transport of heat and salt. The motivation to study the thermohaline circulation (THC) in isolation is that the ocean flow transitions which may have been responsible for rapid climate changes in the past, strongly involved this component of the circulation.

In section 1.2, the mean state of the present global ocean circulation and its associated heat and freshwater fluxes were presented. This chapter starts with a slightly more detailed description of long term variability of this circulation, with focus on the North Atlantic (section 6.1). As in previous chapters, the description of observations is far from complete and other sources should be consulted to obtain an adequate feeling for the complexity of this circulation (Schmitz, 1995; Wunsch, 1996; Ganachaud and Wunsch, 2000; WOCE, 2001). In section 6.2, potential mechanisms of changes in the THC, both in the time-mean state as well as variability around this mean state are considered. The next sections, 6.3 to 6.9, follow a path through the model hierarchy of the THC, touching on two-dimensional models in the sections 6.3 to 6.5, zonally averaged models in section 6.6 and ending with low-resolution general circulation models in the sections 6.8 to 6.9.

6.1. North Atlantic Climate Variability

Over the years and with use of data from international measurement pro-grammes, such as the World Ocean Circulation Experiment (WOCE, see http://oceanic.cms.udel.edu/woce/), a more and more detailed picture of the global ocean circulation is emerging. In WOCE (2001), many of the results from this programme are nicely summarized and put into context with historic observations and results from ocean models.

6.1.1. Observations

Properties of the zonally integrated time-mean flow at several sections in the Atlantic were briefly summarized in section 1.2. The strength of the Atlantic meridional overturning circulation (MOC) at 25°N is estimated to be 16 ± 2 Sv. The heat transport associated with the MOC is positive at every latitude in the Atlantic with a maximum of 1.3 PW at about 25°N. The freshwater transport is southwards in the Atlantic with a typical amplitude of 0.9×10^9 kgs^{-1} at 25°N.

The seasonal variability of the MOC and the physical mechanisms causing this variability are discussed in Jayne and Marotzke (2001). From the very few obser-vations available, estimates of the peak-to-peak seasonal meridional heat transport variations in the midlatitude North Atlantic are about 0.6 PW $\pm$ 0.1 PW (Baringer and Molinari, 1999). At 36°N, the wind (Ekman) induced heat transport is the dominant contribution (Sato and Rossby, 2000), but at 24°N baroclinic processes are likely to be important since this latitude is near the node of the seasonal cycle of wind stress.

There are no indications for the occurrence of different time-mean states in the Atlantic in the instrumental record. Proxy data (briefly mentioned in section 1.1) indicate, however, that rapid changes in deep water temperatures have occurred during the last glacial. The Dansgaard-Oeschger oscillations with a time scale of 1000-2000 year are typical examples of this variability. There are now many indications (Clark *et al.*, 2002) that these variations involved changes in the time-mean state of the THC.

Concerning the variability of the present THC on decadal and longer time scales, not much is known through direct (sufficiently accurate) data. Most observations are available in the North Atlantic Ocean, with measurements of the overflow from the Nordic Seas (Dickson and Brown, 1994), convective activity (Schlösser *et al.*, 1991), repeated ship measurement over the same section (Bryden *et al.*, 1996) and ocean weather stations (Sy *et al.*, 1997; Joyce and Robbins, 1995). Deep water formation was interrupted over the period 1967-1972 due to the presence of a Great Salinity Anomaly (GSA) (Dickson *et al.*, 1988). A low-salinity patch of water travelled along a path around the south of Greenland (late 1960's), through the Labrador Sea (early 1970's) in the subpolar gyre of the North Atlantic and was found in the Norwegian Sea in the late 1970's (Fig. 6.1). The salinity anomaly influenced the velocity field along its way, but it is not known how this has influenced the Atlantic MOC. Mysak *et al.* (1990) suggest that the GSA is part of an approximately 20-year climate cycle, involving changes in the atmosphere, ocean and sea-ice.

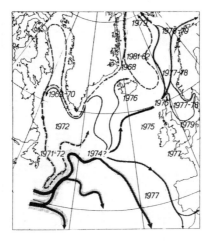

Figure 6.1.   Transport scheme for the 0-1000 m layer of the northern North Atlantic with dates of the salinity minimum superimposed (Dickson et al., *1988).*

From the Bermuda station 'S' data, Joyce and Robbins (1995) find changes in temperature and salinity at different depths on decadal time scales. Temperature and salinity anomalies appear highly correlated in the thermocline layer (500-1000 m) and may be associated with vertical displacements of up to 50 m. There is a warming trend in the deepest layer, but it is difficult to relate this to either actual changes in water-mass characteristics or vertical displacement of density surfaces (isopycnals). Decadal changes of water-mass characteristics have also been described at 24°N in Bryden *et al.* (1996) through data from three cruises over this latitude band in the North Atlantic. The differences in potential temperature and salinity over the period 1957-1992 display negative temperature and salinity anomalies at 500-750 depth over the whole section; at deeper levels, the signs of these anomalies reverse. The changes in both temperature and salinity are very small and, apart from that, it is again difficult to relate them to changes in water mass characteristics. If more intermediate water is formed (for example through increased convection more northward), this could show up as an increased thickness of a particular water mass. There are measurements supporting a relatively fast spreading of these newly formed intermediate water masses from the Labrador Sea into the North Atlantic (Sy *et al.*, 1997; Curry *et al.*, 1998). Isopycnals may also move up and down due to changes in the wind forcing, as the gyre circulation changes intensity.

To study interannual-to-multidecadal variability in the Atlantic climate system there is quite a long sea-surface temperature (SST) data set available: the C(omprehensive) O(cean) A(tmosphere) D(ata) S(et) (http://ingrid.ldgo.columbia.edu/SOURCES/). A nice overview of the results of this data analysis the methods and terminology used can be found in Moron *et al.* (1998). In Deser and Blackmon (1993), the first empirical orthogonal function (EOF) of wintertime mean SST anomalies in the North Atlantic over the period 1900-1989 displays a basin scale SST pattern with strongest positive anomalies in the Gulf Stream region. The time series of this EOF indicates that this region was colder than average over the period 1900-1940 and warmer over the remaining period. The second EOF is a dipole-like pattern with positive (negative) anomalies in the northern (southern) part of the basin with variability in the time series on decadal scales. Using more than 100 years of SST, SLP and wind data from the COADS dataset, Kushnir (1994) showed that the SST exhibits multidecadal variability, with a basin scale SST pattern having maxima in the Labrador Sea and northeast of Bermuda (see Fig. 1.20).

More recently, Tourre *et al.* (1999) and Delworth and Greatbatch (2000) have identified North Atlantic SST and sea-level pressure (SLP) variability with a dominant time scale of about 50 year. In Fig. 6.2, six phases of the pattern of variability of SST and SLP are presented, each about 4.3 years apart such that half of the oscillation is shown. At phase 0° in Fig 6.2a, positive SST anomalies are seen in the North Atlantic. These fill up the basin (Fig 6.2b-c) up to phase 90°, where there is a positive maximum south of Greenland (Fig 6.2d). Cooling of the central part is seen at phase 120° (Fig 6.2e) as the pattern develops to the original pattern but with opposite sign (Fig 6.2f). The SST anomalies do not seem to be caused directly by the overlying atmosphere, since an enhanced meridional gradient of

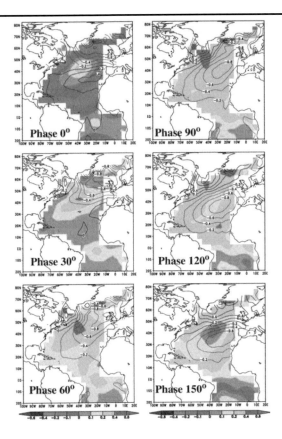

Figure 6.2. (in color on page 523). Reconstruction of the approximately 52-year signal (panels are about 4.3 years apart) in SST and SLP from Delworth and Greatbatch (2000). Units of SST are in °C (from -0.6°C (blue) to 0.6°C (red)) and that of SLP in hPa.

SLP (and hence stronger westerlies) are accompanied by positive SST anomalies (e.g., Fig 6.2a).

The changes in heat content over the upper 300 m, as compiled in Levitus *et al.* (2000) for the North, South, and whole Atlantic are plotted in Fig. 6.3a. This plot clearly shows the warming trend of the upper ocean over the last century. The difference in the North Atlantic's heat content between the pentads 1988–1992 and 1970–74 is shown for two reference depths (300 m and 3000 m) in Figs. 6.3b and 6.3c. Both patterns are strongly aligned with the Gulf Stream, with the heat content decreasing north of it and increasing south of it. The heat content of the subtropical gyre has thus increased substantially from 1970 to 1990, while that of the subpolar gyre has decreased.

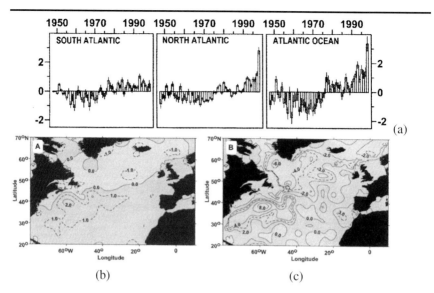

Figure 6.3. *(in color on page 524). (a) Time series of ocean heat content (10^{22} J) in the upper 300 m of the Atlantic for the half-century 1948–1998. For comparison, the climatological range of upper ocean heat content for the North Atlantic is about 5.6×10^{22} J. (b-c) Heat storage difference (Wm^{-2}) for the North Atlantic between 1988–1992 and 1970–1974 within (b) the upper 300 m and (c) the upper 3000 m; warming is indicated by pink and cooling in light blue (from Levitus et al. (2000)).*

In summary, there are strong indications of long term (decadal-to-multidecadal) variability in the North Atlantic climate system. The processes that control these changes, however, are poorly understood so far.

6.1.2. Central questions and approach

A theory of the THC should contain an explanation of the physical processes involved in the time-mean flow and the variability of the flow on decadal and longer time scales. In the context of the description of the THC above, this leads to the following more specific questions

(i)   Are different (global) patterns of the three-dimensional circulation dynamically possible under the same (surface) forcing conditions? Is the present
time-mean THC in such a multiple equilibria regime?

(ii) Does variability on decadal and longer time scales originate from instabilities
of the mean flow associated with the THC?

(iii) How stable is the present THC to perturbations? When rapid transitions are
indeed possible, this question is rather important and one would like to know,
how close the present THC is to the relevant stability boundaries.

In the following sections an attempt is made to present theoretical results, from the perspective of dynamical systems theory, aimed to tackle the questions above. The approach is the same as in the previous chapter, where the wind-driven ocean circulation (WDC) was considered, and bifurcation diagrams are presented for a hierarchy of models of the THC. The potential mechanisms of changes in the thermohaline flows are discussed in the next section using the lowest order models in the hierarchy, i.e., box models. Next, the model hierarchy is traversed, proceeding from the box models, via pure two-dimensional models, zonally averaged models to low-resolution three-dimensional ocean models.

6.2. Potential Mechanisms

In early work of Stommel (1961), the THC was viewed as being driven by the density gradient between high and low latitudes. Within this view, the present circulation is thermally driven since the largest densities are found in the cold, fresh high-latitude regions instead of in the salty, warm equatorial areas. However, in the multiple equilibria regime of the Stommel (1961) model also a salinity-driven state is possible under the same forcing conditions. In this section, potential mechanisms for changes in mean state and for oscillatory behavior are discussed.

6.2.1. What drives the THC?

What is central in the Stommel (1961) model for the surface buoyancy fluxes to drive the THC is that one assumes a strong vertical mixing; in fact, the 'mixers' are explicitly shown in Fig. 3.2. In absence of this mixing, the surface buoyancy fluxes alone cannot drive a steady-state THC. This is an old result in Sandstrom (1908), who performed laboratory experiments to investigate the flows arising from heat and cold sources put at different depths in a water column. For the ocean to have a steady closed circulation, it must do work against friction and this can only be accomplished if the heating occurs at a larger pressure (hence at larger depth) than the cooling (Huang, 1999). In reality, the heating in the tropics occurs at a lower pressure than the cooling at high-latitudes and a very weak and shallow THC would be expected. A modern and more rigorous view of Sandstrom's theorem is the 'anti-turbulence theorem' of Paparella and Young (2002).

To drive the circulation, there must be input of mechanical energy which is provided by the winds and the tides (Munk and Wunsch, 1998). This mechanical energy can directly contribute to the large-scale motion by generating upwelling and it causes turbulent mixing such that heat is transferred downward over isopycnals; the latter is the so-called diapycnal mixing. The surface buoyancy fluxes are still important, however, since they control where deep water is formed (Nilsson and Walin, 2001). Munk and Wunsch (1998) suggest that the planetary-scale ocean circulation is a passive consequence of a circulation which is really driven by energy input by the winds and tides. The strength of the MOC is set by the diapycnal mixing in the upwelling branch. More recent work has provided estimates of the different energy sources and sinks and conversion routes (Wunsch, 1998; Wunsch and Ferrari, 2004).

The role of both diapycnal mixing and wind-induced upwelling in the strength of the THC has recently received much attention. In one extreme view (Toggweiler and Samuels, 1995, 1998), the MOC is completely driven by the wind-induced upwelling and Ekman transport in the Southern Ocean. The northward Ekman transport of water basically drives the deep water formation in the North Atlantic which acts as a return flow of the surface flow. In another extreme view, diapycnal mixing is dominant but quite a large diapycnal diffusivity is needed to obtain a reasonable strength of the MOC. Mixing near bottom topography may play an important role here, as the interior diapycnal mixing is very small. There are many studies which have investigated the effect of both mechanisms on the strength of the MOC (and its scaling behavior with mixing parameters and surface forcing). Apart from giving several references (Marotzke and Scott, 1997; Gnanadesikan, 1999; Marotzke and Scott, 1999a,b; Zhang *et al.*, 1999; Marotzke, 2000; Vallis, 2000; Klinger *et al.*, 2003), these studies are not further considered here. In this chapter, we will assume that a diapycnal mixing coefficient is given (without considering its origin) and this mixing (and the wind) will drive a THC.

6.2.2. Advective feedback

Using a two-box model, as introduced in section 3.1, the possibility of multiple equilibria under similar surface forcing conditions was discovered (Stommel, 1961) about 40 years ago. Responsible for this non-uniqueness is a nonlinear feedback between the flow and the density structure, called the (salt) advection feedback. Consider in Fig. 6.4 a zonally averaged (overturning) circulation from the equator towards northern latitudes. The surface forcing saltens/warms the low latitude region and freshens/cools the high-latitude region and the circulation is driven by the meridional density gradient. Since there is northern sinking, the circulation is thermally driven. If the circulation strengthens, then more salt is transported northward. This enhanced salt transport will increase the density in the north and consequently amplify the original perturbation in the circulation. The strengthening of the circulation also transports more heat northward, which will weaken the flow by lowering the density. Heat transport therefore provides a negative feedback on the circulation.

In addition to the advection feedback, a central ingredient to the existence of multiple steady states are the different damping times of salinity and temperature anomalies. The atmosphere exerts quite a strong control on the sea surface temperature anomalies, but salinity in the ocean does not affect the freshwater flux at all. In the two-box model in section 3.1, these different response time scales of salinity and temperature, with $\tau_S = 1/R_S$ and $\tau_T = 1/R_T$, were taken into account by the coefficient $\eta_3 = R_S/R_T = \tau_T/\tau_S$, which was smaller than unity. In general, the different surface boundary conditions for temperature and salinity are referred to as mixed boundary conditions (Haney, 1971; Welander, 1986; Tziperman *et al.*, 1994b). The extreme case is a prescribed surface temperature ($\tau_T \ll 1$) and prescribed surface freshwater flux ($\tau_S \gg 1$) for which surface

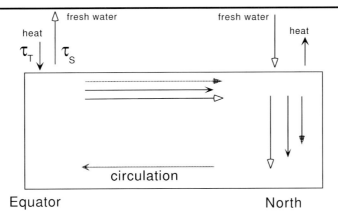

Figure 6.4.   Sketch of the physics of the salt advection feedback. The mean circulation is indicated by the closed arrows. The upper ocean temperature and salinity fields can be inferred from the surface forcing of heat and freshwater. A perturbation which strengthens the circulation leads to a northward salt transport, which leads to amplification (open arrows) of the circulation (positive feedback). The perturbation in the circulation also leads to increased heat transport which opposes (negative feedback) the original perturbation (closed arrows).

temperature perturbations are essentially zero. As seen in chapter 3, multiple equilibria arise if indeed the ratio $\eta_3 < 1$ and only under certain forcing conditions (Fig. 3.4), i.e. a particular area in the (η_1, η_2) plane.

Together, the advective feedback and the different response time scales provide a potential mechanism of change of the THC. Consider again the thermally driven circulation as in Fig. 6.4 and imagine that a surface freshwater anomaly is suddenly present in the north part of the domain. Because the density is lowered in the north, the meridional buoyancy gradient decreases and hence the strength of the circulation decreases. The effect is that both the northward salt and heat transport decrease. Now, the negative heat anomaly is rapidly damped at the sea surface, but the freshwater anomaly is not damped at all and hence amplifies the original freshwater perturbation. This positive feedback is able to rapidly weaken the thermally driven overturning circulation.

6.2.3. Convective feedback

A convective feedback may also be responsible for multiple equilibria (Welander, 1982; Lenderink and Haarsma, 1994). Consider in Fig. 6.5 a box model with time-varying temperature T_* and salinity S_* due to a surface heat fluxes $F_T = \alpha(T_a - T_*)$ and surface salinity flux F_S in the surface box, coupled to a box with constant temperature T_i and S_i and constant prescribed flow rate q. Convective exchange with time constant τ^{-1} occurs if the surface water becomes denser than the deep water, which has constant temperature T_b and salinity S_b. For $q = 0$,

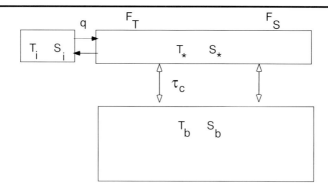

Figure 6.5. Sketch of the box model set-up to illustrate the convective feedback. An active box of temperature T_*, S_* is coupled to boxes of constant temperature T_i, S_i and T_b, S_b. Advective exchange takes place with flow rate q and vertical (convective) exchange occurs, on a time scale τ_c, if the surface water is denser than the bottom water.

the model reduces to the model used by Welander (1982), and when there is no vertical exchange, the model can be considered as a limit of the Stommel (1961) model, for which the surface forcing in the equatorial box is adjusted such that temperature and salinity T_i and S_i remain constant.

The equations for the evolution of the temperature T_* and S_* are

$$\frac{dT_*}{dt_*} = \alpha(T_a - T_*) + q(T_i - T_*) + \tau_c \mathcal{H}(\rho_* - \rho_b)(T_b - T_*) \quad (6.1\text{a})$$

$$\frac{dS_*}{dt_*} = F_S + q(S_i - S_*) + \tau_c \mathcal{H}(\rho_* - \rho_b)(S_b - S_*) \quad (6.1\text{b})$$

with $\mathcal{H}$ being the Heaviside function. With the equation of state

$$\rho_*(T_*, S_*) = \rho_0 - \alpha_T T_* + \alpha_S S_* \quad (6.2)$$

the steady states can be easily solved and become

$$T_* = \frac{qT_i + \alpha T_a + \tau_c \mathcal{H}(\rho_* - \rho_b)T_b}{q + \alpha + \tau_c \mathcal{H}(\rho_* - \rho_b)} \quad (6.3\text{a})$$

$$S_* = \frac{qS_i + F_S + \tau_c \mathcal{H}(\rho_* - \rho_b)S_b}{q + \tau \mathcal{H}(\rho_* - \rho_b)} \quad (6.3\text{b})$$

Two types of equilibria can be distinguished. Those for which the argument of the Heaviside function is positive are called convective equilibria, and those for which it is negative are called non-convective equilibria. With the new parameters

$$\Phi_T = -\alpha_T(\alpha(T_a - T_b) + q(T_i - T_b)) \quad (6.4\text{a})$$

$$\Phi_S = \alpha_S(F_S + q(S_i - S_b)) \quad (6.4\text{b})$$

$$\kappa(\tau) = \frac{q + \tau}{q + \tau + \alpha} \quad (6.4\text{c})$$

three different solution regimes exist (Fig. 6.6).

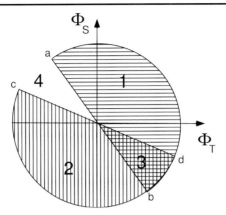

Figure 6.6.   Sketch of the regimes of convective and non-convective equilibria in the box model in the (Φ_T, Φ_S) *parameter plane. In regime* **1**, *there are convective states, in regime* **2** *there are non-convective states, whereas in regime* **3** *both states are present. In regime* **4** *no steady states exist.*

The condition that a convective equilibrium exists can be written as $\Phi_S > -\kappa(\tau)\Phi_T$ (indicated as the line $a - b$ in Fig. 6.6) which defines regime **1** in Fig. 6.6. Similarly, the condition for a non-convective equilibrium to exists can be written as $\Phi_S < -\kappa(0)\Phi_T$ (indicated as the line $c - d$ in Fig. 6.6) which defines regime **2**. In regime **3**, both convective and non-convective equilibria exist and transitions between these solutions can occur under the same forcing conditions. Consider a non-convective state with cold/freshwater above warm/salty water which is only marginally stable and an atmospheric forcing which is cooling and freshening the upper box. A finite-amplitude positive density perturbation is able to induce convection and if this occurs, warmer and saltier water is mixed to the surface. The heat in the surface layer is quickly lost to the atmosphere but the surface salinity is increased and hence convection is maintained, leading to a convective state.

For the particular case $\Phi_T = 1.0$, $q/\alpha = 0.5$ and $\tau/\alpha = 2.0$, the bifurcation diagram of the model (6.1) is plotted in Fig. 6.7. In this diagram, both the dimensionless temperature $T = \alpha_T(T_* - T_b)$ and salinity $S = \alpha_S(S_* - S_b)$ are plotted versus the control parameter Φ_S. Two saddle node bifurcations (L_1 and L_2) occur at $\Phi_S = -5/7$ and $\Phi_S = -1/3$. These are exactly the values $-\kappa(\tau)$ and $-\kappa(0)$ bounding the regions of convective and non-convective regimes, respectively. Hence, the high temperature and salinity states are convective and exist for $\Phi_S > -5/7$ (regime **1**) whereas the low salinity and temperature states are non-convective and exist for $\Phi_S < -1/3$ (regime **2**). Regime **3** is exactly located in the interval $-5/7 < \Phi_S < -1/3$ and in this regime, multiple equilibria exist. Note that regime **4** is not reached here, because $\Phi_T > 0$.

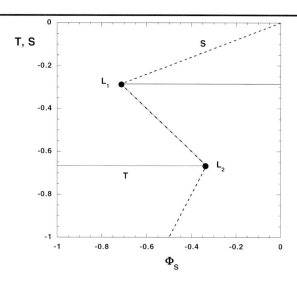

Figure 6.7.   Bifurcation diagram for the box model (6.1) with $\Phi_T = 1.0$, $q/\alpha = 0.5$ and $\tau/\alpha = 2.0$ and Φ_S as control parameter. The branch connecting L_1 and L_2 arises through the numerical approximation of the Heaviside function $\mathcal{H}$ as in (3.10) and is absent when $\epsilon \to 0$.

6.2.4. The flip-flop oscillation

Within simple box models, two types of oscillatory phenomena can be found. One is associated with propagation of perturbations along the mean thermohaline flow which will be subject of the next subsection and another is associated with transitions between convective and non-convective states. This type of oscillation was found by Welander (1982) in a box model which is a special case of the model in section 6.2.3. Only two boxes which exchange heat and salt vertically are considered and moreover the ocean-atmosphere salinity flux is chosen as $F_S = \beta(S_a - S_*)$, with different restoring times for freshwater and heat. With $q = 0; T_b = T_0; S_b = S_0; \rho_b = \rho_0$ in (6.1) and with a more general form of the convective exchange function $k(\rho_*, \rho_0)$ the model equations become

$$\frac{dT_*}{dt_*} = \alpha(T_a - T_*) + k(\rho_*, \rho_0)(T_0 - T_*) \qquad (6.5a)$$

$$\frac{dS_*}{dt_*} = \beta(S_a - S_*) + k(\rho_*, \rho_0)(S_0 - S_*) \qquad (6.5b)$$

with the equation of state (6.2).

One of the cases considered (Welander, 1982) is $T_0 = S_0 = 0$ and

$$k(\rho_*) = 0 \, , \rho_* \le \varepsilon$$
$$k(\rho_*) = k \, , \rho_* \ge \varepsilon$$

which means that if the surface density ρ_* becomes slightly larger than the density in the bottom box, an exchange flux $(-kT_*)$ is generated for both temperature and salinity with constant exchange coefficient k.

For the particular choice of parameters

$$\frac{\alpha_T T_a}{\alpha_S S_a} = 0.2 \; ; \; \frac{\alpha}{\beta} = 10 \; ; \; \frac{k}{\alpha} = 5 \; ; \; \frac{\varepsilon}{\alpha_S S_a} = -0.01 \tag{6.6}$$

a trajectory is plotted in Fig. 6.8a. All fields oscillate and the oscillation seems to be sustained. In Welander (1982), it is shown that it damps for the case $\varepsilon = 0.0$, hence a nonzero ε is essential to the existence of the oscillation.

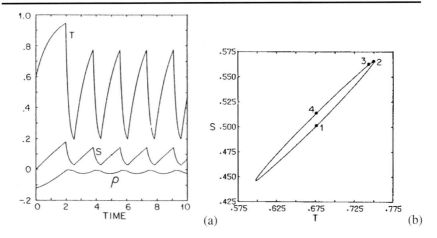

Figure 6.8. *(a) Trajectory of the box model (6.5) for $\varepsilon \ne 0$ in the convective exchange function (6.5) showing the oscillation in temperature $T = T_*/T_a$, salinity $S = S_*/S_a$ and density $\rho = \rho_*/\rho_a$. (b) Phase plane picture of the oscillation in (a), where T_*/T_a is plotted versus S_*/S_a (Welander, 1982).*

The advantage of these type of models is that the oscillation can be understood in quite detail. In the particular case above, the model has two steady states. The non-convective state is given by

$$\bar{\rho}_* \le \varepsilon : \bar{T}_* = T_a \; ; \; \bar{S}_* = S_a \tag{6.7}$$

and the convective state by

$$\bar{\rho}_* > \varepsilon : \bar{T}_* = \frac{\alpha T_a}{\alpha + k} \tag{6.8a}$$

$$\bar{S}_* = \frac{\beta S_a}{\beta + k} \tag{6.8b}$$

with
$$\bar{\rho}_* = -\alpha_T \bar{T}_* + \alpha_S \bar{S}_*$$

There exists a parameter regime (similar to regime **4** in Fig. 6.6), where both steady states cannot be reached. The trajectory then oscillates between both steady states without actually reaching them. A phase-plane picture of the oscillation in Fig. 6.8a is plotted in Fig. 6.8b. At point 1, T and S are such that $\rho_* < \varepsilon$ so that the trajectory is attracted towards the non-convective steady state and both T and S increase (towards point 2). This steady state is never reached, because at point 3, the boundary $\rho_* = \varepsilon$ is crossed and convection occurs. Then the trajectory is attracted towards the convective steady state, but it also does not reach this state because at some point, convection will stop. Hence, the oscillation can be described as a 'flip-flop' between convective and non-convective states, where during the oscillation neither of these states is actually reached.

6.2.5. The loop oscillation

The most elementary box model which includes a loop oscillation is the four-box model originally used by Huang *et al.* (1992) and analysed in more detail in Tziperman *et al.* (1994b). It differs from the two-box model by including two deep boxes and vertical exchange of heat and salt (Fig. 6.9a). The surface and deep boxes may have different volumes, but their ratio is fixed and Tziperman *et al.* (1994b) mostly consider the case of equal volumes. The surface boundary

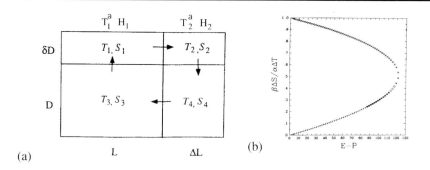

Figure 6.9.   (a) Sketch of the box model to illustrate the loop oscillation (Tziperman et al., 1994b). (b) Bifurcation diagram for the 4-box model in (a) as used in Tziperman et al. (1994b). The markers correspond to the stability of the flow: '+' indicates stability, '*' indicates stability but the least stable mode is oscillatory, 'o' corresponds to an oscillatory instability and 'x' to instability.

conditions consist of fresh-water fluxes H_1 and H_2, atmospheric temperatures T_1^a and T_2^a and the transport q_* is related to the average north-south density difference. The governing equations of this box model can be found in Tziperman *et al.* (1994b), where also values of the parameters are provided.

A typical bifurcation diagram of this model is plotted in Fig. 6.9b with the (dimensionless) freshwater flux $E-P = H_1+H_2$ as control parameter. On the vertical axis, the buoyancy ratio is plotted with $\Delta T = T_{1*} - T_{2*}$ and $\Delta S = S_{1*} - S_{2*}$. The basic thermally driven state is stable for small $E - P$, but it looses stability due to a Hopf bifurcation (the point marking the transition between '*' and 'o' in Fig. 6.9b at $E - P \approx 110$). The pair of complex conjugated eigenvalues which have crossed the imaginary axis become real for slightly larger values of $E - P$ (at the transition of 'o' and 'x') and then one of them moves into the left complex plane at the saddle node bifurcation (at $E - P \approx 125$) and a branch of unstable solutions exists at smaller values of $E - P$. The mechanism of the oscillatory instability was investigated in Tziperman *et al.* (1994b) and shown to be related to the propagation of a salinity anomaly along the mean flow. This mechanism is very similar to that of the Howard-Malkus loop oscillation discussed in Welander (1986). It will be considered in more detail in the context of a somewhat more complex model below (section 6.4.3).

6.2.6. Models of the THC

Certainly box models are useful to illustrate basic physical phenomena, but more complex models are needed to capture the full spatial-temporal behavior of the THC. Contrary to the wind-driven circulation in the previous chapter, now also the modelling of the heat and salt transport is essential. Starting point of all the models are the full (Boussinesq) primitive equations, with velocity vector $\mathbf{v}_*$ and pressure p_*, that were presented in chapter 2. They are repeated here for convenience,

$$\rho_0 \left[\frac{D\mathbf{v}_*}{dt_*} + 2\mathbf{\Omega} \wedge \mathbf{v}_* \right] = -\nabla p_* - g\rho_* \mathbf{e}_3 + \rho_0 \mathcal{F}_{I*} \qquad (6.9a)$$

$$\nabla . \mathbf{v}_* = 0 \qquad (6.9b)$$

$$\rho_0 C_p \frac{DT_*}{dt_*} = \mathcal{F}_{T*} \qquad (6.9c)$$

$$\rho_0 \frac{DS_*}{dt_*} = \mathcal{F}_{S*} \qquad (6.9d)$$

$$\rho_* = \rho_0 (1 - \alpha_T (T_* - T_0) + \alpha_S (S_* - S_0)) \qquad (6.9e)$$

where a simple linear equation of state is assumed with reference temperature T_0, salinity S_0 and density ρ_0. Vertical and horizontal mixing of momentum and of heat and salt is represented by eddy diffusivities with coefficients A_H and A_V for momentum and K_H and K_V for heat and salt, respectively. As the eddy-diffusivities of heat and salt are determined by turbulent processes, there is a good reason to take them equal. The ocean circulation is driven by a wind stress τ and by heat and fresh-water fluxes at the surface.

Within our limited view of the mixing processes, the hierarchy of models to study transition phenomena of the THC is as follows:

(i) Strictly two-dimensional ocean models, which completely ignore the effects of wind-stress forcing and rotation. In these models, focus is on the different flows which can be realized under a given surface buoyancy forcing.

(ii) Zonally averaged models, in which the effects of rotation and wind-stress forcing are somehow parameterized, but for which the equations actually solved
have two spatial dimensions.

(iii) Low-resolution three-dimensional ocean general circulation models, in which
full physics of rotation and wind-stress forcing is present, but because of the
low resolution there is a cut-off in spatial and temporal scales, which limits the
representation of physical processes.

Just as for the WDC (section 5.2), it is desirable to have an easy reference system within the model hierarchy of the THC. The dynamical classes here can be
indicated as B (2D Boussinesq), Z (zonally averaged) and P (3D primitive equation). These can be used as subscripts to a T, indicating the THC. As the surface
boundary conditions for salt and heat play an important role in the behavior of the
THC, it is useful to introduce superscripts R, M and C for restoring, mixed and
coupled conditions, respectively. Finally, we can add a superscript to indicate the
details in the bathymetry using 0 for a rectangular basin, 1 for only continental
geometry and 2 for full bathymetry. In this nomenclature, the model $T_B^{R,0}$ would
be a 2D Boussinesq model with restoring boundary conditions and idealized continents, while the model $T_P^{M,2}$ would be a 3D primitive equation model with real
bathymetry under mixed boundary conditions. Other dynamical classes could be
added, such as the planetary geostrophic models (Salmon, 1986). All ODE-type
ad-hoc or reduced models can be put into one model class T_0 and we can add
an argument to the model to indicate the vector of parameters α, for example,
$T_B^{R,0}[\alpha]$. Together with the coordinate system, the forcing, the domain and the
boundary conditions this then totally specifies a model configuration.

In the next sections, bifurcation analyses will be performed on some of these
models and focus will be on the structure of equilibria in parameter space and
their stability.

6.3. Two-dimensional Boussinesq Models

In two-dimensional Boussinesq models (Thual and McWilliams, 1992; Quon
and Ghil, 1992; Cessi and Young, 1992) rotation and wind-stress forcing are neglected. The use of these models is motivated by the understanding of the basic
fluid mechanics of thermohaline flows. In principle, laboratory experiments could
be devised to check the results of these models.

6.3.1. Formulation

The simplest of these models employ a Cartesian coordinate system, with
meridional coordinate y_* and vertical coordinate z_* (Fig 6.10). For two-
dimensional flow fields which do not depend on the zonal coordinate, the governing equations for the meridional velocity v_*, vertical velocity w_*, pressure p_*,
density ρ_*, temperature T_* and salinity S_* are given by

$$\rho_0 \frac{Dv_*}{dt_*} = -\frac{\partial p_*}{\partial y_*} + \rho_0 A_H \frac{\partial^2 v_*}{\partial y_*^2} + \rho_0 A_V \frac{\partial^2 v_*}{\partial z_*^2} \qquad (6.10a)$$

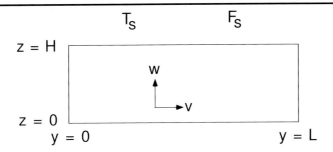

Figure 6.10.   Sketch of the two-dimensional model set-up, where T_S and F_S refer the surface forcing patterns.

$$\rho_0 \frac{Dw_*}{dt_*} = -\frac{\partial p_*}{\partial z_*} + \rho_0 A_H \frac{\partial^2 w_*}{\partial y_*^2} + \rho_0 A_V \frac{\partial^2 w_*}{\partial z_*^2} - \rho_* g \quad (6.10b)$$

$$\frac{\partial v_*}{\partial y_*} + \frac{\partial w_*}{\partial z_*} = 0 \qquad \qquad \quad (6.10c)$$

$$\frac{DT_*}{dt_*} = K_H \frac{\partial^2 T_*}{\partial y_*^2} + K_V \frac{\partial^2 T_*}{\partial z_*^2} \qquad (6.10d)$$

$$\frac{DS_*}{dt_*} = K_H \frac{\partial^2 S_*}{\partial y_*^2} + K_V \frac{\partial^2 S_*}{\partial z_*^2} \qquad (6.10e)$$

$$\rho_* = \rho_0(1 - \alpha_T(T_* - T_0) + \alpha_S(S_* - S_0)) \qquad (6.10f)$$

with

$$\frac{D}{dt_*} = \frac{\partial}{\partial t_*} + v_* \frac{\partial}{\partial y_*} + w_* \frac{\partial}{\partial z_*}$$

The downward heat flux Q_{oa} on the ocean surface is assumed to be proportional to the temperature difference between the ocean surface temperature and a steady prescribed atmospheric temperature T_S, i.e. $Q_{oa} = B_T(T_S - T_*)$, with B_T being a surface exchange coefficient of heat. It is positive if heat is transferred from the atmosphere to the ocean. The freshwater flux is converted into a equivalent salt flux $F_0 F_S$ and is simply a prescribed function. The boundary conditions for temperature and salinity at the ocean surface ($z_* = H$) then become

$$\rho_0 C_p K_V \frac{\partial T_*}{\partial z_*} = B_T(T_S - T_*) \qquad (6.11a)$$

$$K_V \frac{\partial S_*}{\partial z_*} = F_0 F_S(y_*) \qquad (6.11b)$$

Salt and heat fluxes are assumed to be zero at the bottom and lateral boundaries and slip conditions are applied at all boundaries. In this way, the other boundary conditions become

$$y_* = 0, L \quad : \quad \frac{\partial T_*}{\partial y_*} = \frac{\partial S_*}{\partial y_*} = 0, v_* = 0, \frac{\partial w_*}{\partial y_*} = 0 \qquad (6.12a)$$

$$z_* = 0 \quad : \quad \frac{\partial T_*}{\partial z_*} = \frac{\partial S_*}{\partial z_*} = 0, w_* = 0, \frac{\partial v_*}{\partial z_*} = 0 \qquad (6.12b)$$

$$z_* = H \quad : \quad w_* = 0, \frac{\partial v_*}{\partial z_*} = 0 \qquad (6.12c)$$

6.3.2. Nondimensional equations

When these equations are non-dimensionalized, a number of non-dimensional parameters appear dependent on the type of scaling. In Dijkstra and Molemaker (1997), scales H, K_H/H, $\rho_0 K_H A_H/L^2$, ΔT and ΔS for length, velocity, pressure, temperature and salinity are used, with ΔT and ΔS being characteristic meridional temperature and salinity differences. In this case, the Prandtl number Pr, the thermal Rayleigh number Ra, the buoyancy ratio λ, the aspect ratio A, the salt flux strength $\tilde{\sigma}$, the ratio of vertical and horizontal diffusivities for momentum R_{HV}^M and for heat and salt R_{HV}^T, and the Biot number Bi appear. These parameters are defined as

$$Pr = \frac{A_H}{K_H} \; ; \; Ra = \frac{g \, \alpha_T \, \Delta T H^3}{A_H K_H} \; ; \; A = \frac{L}{H}$$

$$R_{HV}^M = \frac{A_V}{A_H} \; ; \; R_{HV}^T = \frac{K_V}{K_H}$$

$$Bi = \frac{H B_T}{\rho_0 C_p K_V} \; ; \; \lambda = \frac{\alpha_S \Delta S}{\alpha_T \Delta T} \; ; \; \tilde{\sigma} = \frac{F_0 H}{K_V} \qquad (6.13)$$

With this scaling, the non-dimensional equations become

$$\frac{1}{Pr} \frac{Dv}{dt} = -\frac{\partial p}{\partial y} + \frac{\partial^2 v}{\partial y^2} + R_{HV}^M \frac{\partial^2 v}{\partial z^2} \qquad (6.14a)$$

$$\frac{1}{Pr} \frac{Dw}{dt} = -\frac{\partial p}{\partial z} + \frac{\partial^2 w}{\partial y^2} + R_{HV}^M \frac{\partial^2 w}{\partial z^2} + Ra(T - \lambda S) \qquad (6.14b)$$

$$\frac{\partial v}{\partial y} + \frac{\partial w}{\partial z} = 0 \qquad (6.14c)$$

$$\frac{DT}{dt} = \frac{\partial^2 T}{\partial y^2} + R_{HV}^T \frac{\partial^2 T}{\partial z^2} \qquad (6.14d)$$

$$\frac{DS}{dt} = \frac{\partial^2 S}{\partial y^2} + R_{HV}^T \frac{\partial^2 S}{\partial z^2} \qquad (6.14e)$$

and the surface boundary conditions for temperature and salinity at $z = 1$ become

$$\frac{\partial T}{\partial z} = Bi \, (T_S - T) \qquad (6.15a)$$

$$\frac{\partial S}{\partial z} = \tilde{\sigma} \, F_S(y) \qquad (6.15b)$$

Apart from parameters in the forcing functions, the system of equations contains eight parameters (those in (6.13)). However, only seven of these are independent. If the salinity is rescaled with λ, then the buoyancy forcing becomes $Ra\,(T-S)$, (6.14e) remains the same and in (6.15b), the coefficient measuring the magnitude of the freshwater flux becomes $\sigma = \tilde{\sigma}\lambda$. This reduces the number of free parameters by one; we will use σ in subsequent bifurcation diagrams as control parameter.

6.4. Diffusive Thermohaline Flows

Solutions of the THC within this two-dimensional model were first obtained for the most simple case of prescribed equatorially symmetric temperature T_S ($Bi \to \infty$) and salt flux F_S, isotropic eddy diffusivities $R_{HV}^M = R_{HV}^T = 1$ and relatively large depth to width ratio (A up to 10). Since the vertical length scale of this flow is smaller than the horizontal scale, isotropic diffusivities imply that vertical diffusion is a dominant transport mechanism and hence this regime is labelled the 'diffusive regime'.

Although this does not resemble the actual regime of the ocean circulation, these studies were aimed to investigate whether the elementary phenomena as found in box models would persist in a model in which the transport of heat and salt and its effect on the flow are modelled more realistically. In this case, a dynamical system with four parameters (Pr, A, σ and Ra) remains of which the bifurcation behavior is considered next.

6.4.1. Basic bifurcation diagram

In Dijkstra and Molemaker (1997), full bifurcation diagrams were calculated using the continuation techniques as in chapter 4. It is convenient to present this material first and then put main results from earlier studies using these models (Thual and McWilliams, 1992; Quon and Ghil, 1992; Cessi and Young, 1992) into context. The surface forcing functions used in Dijkstra and Molemaker (1997) are

$$T_S(y) = \frac{1}{2}(\cos\,2\pi\left(\frac{y}{A}-\frac{1}{2}\right)+1) \tag{6.16a}$$

$$F_S(y) = 3\,\cos\,p\pi\left(\frac{y}{A}-\frac{1}{2}\right)-\frac{6}{p\pi}\sin\frac{p\pi}{2} \tag{6.16b}$$

where p is an additional parameter controlling the shape of the salt flux. For two values of p, this shape is plotted in Fig. 6.11. The temperature profile simply mimics the warm equator versus colder poles and the salt flux mimics the evaporation at low latitudes and precipitation at higher latitudes. For the parameter values $Ra = 4 \times 10^4$, $Pr = 2.25$, $p = 2.6$ and $A = 5$, the bifurcation diagram, with σ as control parameter, is shown in Fig. 6.12. The value at the vertical axis (ψ_{RM}) is that of the streamfunction at a particular gridpoint ($y = 0.851$, $z = 0.5$), such that the different branches can be well distinguished in a plot. Bifurcation points along the branches are indicated by markers: squares indicate pitchfork bifurcations and triangles Hopf bifurcations. Saddle node bifurcations are not marked

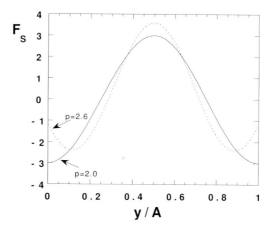

Figure 6.11.   Shape of the salt flux (6.16b) for two values of p.

but are obvious from the shape of the branches. Solid branches indicate (linearly) stable solutions whereas the solutions along dashed branches are unstable. Corresponding flow patterns (through contour plots of the streamfunction ψ) and salinity fields at several marked points along the branches in Fig. 6.12 are shown in Fig. 6.13. The same terminology TH, SA, NPP and SP, introduced in section 3.1, is used to indicate the different solutions.

The symmetric thermally dominated 2-cell state (Fig. 6.13a) – along the TH-branch – becomes unstable at the supercritical pitchfork P_1 near $\sigma = 0.13$ at which two symmetry related asymmetric states stabilize. Both asymmetric states (the southward sinking SPP solution is shown in Fig. 6.13b) remain stable up to the Hopf bifurcations H_1 at $\sigma = 1.06$ where they become unstable. At slightly larger σ, these states stabilize again at the Hopf bifurcations H_2 ($\sigma = 1.44$), but they eventually cease to exist at the saddle node L_3 ($\sigma = 1.47$). Along an unstable branch, the 1-cell patterns then deform towards 2-cell solutions with equatorial downwelling (Fig. 6.13c) and connect at a second pitchfork bifurcation P_2 ($\sigma = 0.24$) to the stable branch of a salinity dominated 2-cell (Fig. 6.13d) solution, the SA-branch. The stability properties of this branch are difficult to detect very near P_2. The unstable 2-cell TH-solution also connects up to the SA-branch at P_2, after it has gone through two limit points L_1 and L_2, the latter being very close to P_2. Hence, there are three σ-intervals where a unique stable steady state appears, and three intervals where there are multiple stable steady states. The latter intervals are given by the σ-values between both pitchfork bifurcations P_1 and P_2, between the pitchfork P_2 and the Hopf bifurcation H_1 and between the Hopf bifurcation H_2 and the limit point L_3.

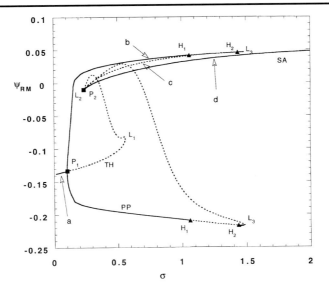

Figure 6.12. Bifurcation diagram obtained by Dijkstra and Molemaker (1997) at small aspect ratio $A = 5$ for $Ra = 4 \times 10^4$, $Pr = 2.25$ and $p = 2.6$.

Integral properties of the solutions were considered in Dijkstra and Molemaker (1997), in particular the mechanical energy balance. This balance can be obtained by multiplying the dimensionless momentum balances by $\mathbf{v}$ and integration over the flow domain (section 2.1.4). This gives with the buoyancy $B = T - S$,

$$\frac{dE}{dt} = < wB > - \mathcal{D} \qquad (6.17)$$

where E is the dimensionless volume averaged kinetic energy, $< . >$ indicates volume integration, $< wB >$ the volume averaged buoyancy production and $\mathcal{D}$ is the dissipation. Whether a steady flow is saline or thermally driven can be decided from (6.17), since $< wB > = < wT - wS >$ has to balance the dissipation. The steady solutions can hence be distinguished according to the sign of $< wT >$ and $< wS >$. A state is thermally driven and inhibited by freshwater forcing, if $< wT > \geq 0$ and $< wS > \geq 0$. It is saline driven and inhibited by thermal forcing if $< wT > \leq 0$ and $< wS > \leq 0$ and it is driven by both mechanisms if $< wT > \geq 0$ and $< wS > \leq 0$. If $< wT > \leq 0$ and $< wS > \geq 0$ then there can be no steady flow. Along the TH branch, both $< wT >$ and $< wS >$ are positive, indicating that the solutions are thermally driven and inhibited by freshwater forcing. On the NPP- and SPP-branches originating from the symmetry breaking bifurcation, $< wS >$ is negative and $< wT >$ is positive and the pole to pole solutions are therefore driven by both thermal and saline forcing.

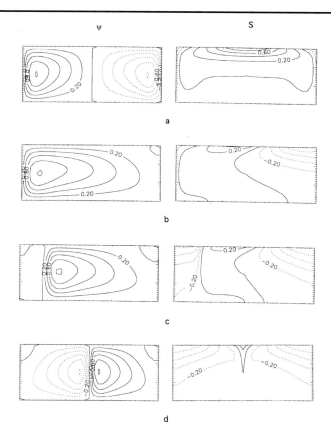

Figure 6.13.   Solutions of the streamfunction and salinity at labelled points in Fig. 6.12. In these and following contour plots, the fields are scaled with their maximum values and contour levels with respect to this maximum are indicated.

6.4.2. Physical mechanisms

From the steady states and eigenvectors corresponding to eigenvalues which cross the imaginary axis in a bifurcation point, one can attempt to describe the instability mechanism of this transition as sketched in section 3.7.

6.4.2.1 Symmetry breaking

Just as in the wind-driven ocean circulation, pitchfork bifurcations associated with the symmetry breaking are at the basis of multiple equilibria in this model. Both the TH-branch and the SA-branch undergo these bifurcations and both 2-cell symmetric solutions become unstable to a perturbation having a particular spatial pattern. For example, for values of σ slightly larger than the value at P_1 in

Fig. 6.12, the symmetric TH solution will be unstable to a particular pattern even if this disturbance has a very small amplitude initially. The instability mechanism can be analysed within this model using the framework as sketched in section 3.7.3, i.e. by looking at the steady state and the destabilizing perturbation just at bifurcation. The steady state streamfunction, temperature and salinity are plotted

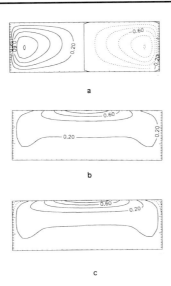

Figure 6.14. Basic state (a) streamfunction, (b) temperature and (c) salinity at the pitchfork bifurcation P_1 in Fig. 6.12

in the Figs. 6.14a-c, respectively. The patterns of the destabilizing perturbation (at neutrality) are plotted in Fig. 6.15a-d for streamfunction, temperature, salinity and density, respectively.

The instability of the 2-cell symmetric TH-solution at the bifurcation point P_1 (in Fig. 6.12) can be described as follows. Consider the salinity perturbation in Fig. 6.15c as the initial disturbance. This salt perturbation is positive over most of the northern part of the basin and negative over the southern part, with substantial gradients near the equator. From the density perturbation (Fig. 6.15d), it follows that salinity mainly determines the sign of the surface density perturbation. The perturbation salt gradient therefore drives the flow which is seen in Fig. 6.15a. The temperature perturbation is compatible with this flow and the surface temperature perturbation is zero. North of the equator, water in the upper layers is replaced by slightly warmer water from the south and hence the temperature perturbation is positive. South of the equator, upper ocean water is replaced by water from the south which has a slightly smaller temperature; a negative temperature perturbation results. This describes the initiation stage as in section 3.7.3.2.

To describe the growth stage, it is first realized that the northern cell of the steady state (Fig. 6.14a) is strengthened by the flow disturbance (Fig. 6.15a) and the southern cell is weakened. The horizontal perturbation velocities at the surface induce a horizontal salt transport (Fig. 6.14c), because of the meridional salinity gradient of the steady state. Note that there is no surface perturbation meridional heat transport, because the temperature perturbation is zero at the surface. In the salinity equation for the perturbations, which can be derived by linearizing (6.10e) around the steady state, the tendency is proportional to

$$\frac{\partial \tilde{S}}{\partial t} \approx -\tilde{v}\frac{\partial \bar{S}}{\partial y} \qquad (6.18)$$

where the bar refers to the steady state and the tilde to the perturbation.

In the northern part of the basin, the term in the right hand side of (6.18) is positive since the basic state salt gradient is negative and $\tilde{v} > 0$. This leads to an amplification of the original positive salinity disturbance during the growth stage. Similar reasoning holds for the southern part of the basin, where the sign of the right hand side of (6.18) is negative, amplifying the original negative salinity perturbation. Note that the temperature perturbation plays a rather passive role in this mechanism, except that it weakens the perturbation flow, because of its influence on the perturbation density field. This explains why there is a positive critical value of σ, since the flow due to the salinity anomaly has to overcome the damping effect of temperature. However, the thermal field itself is crucial since the temperature field maintains the circulation of the equilibrium state. When the flow is too weak, the basic state salinity gradient is too weak to cause any amplification. It is clear that this mechanism involves the salt advection feedback as described in section 6.2, and several nearly equivalent, but less accurate, descriptions have appeared in the literature (Walin, 1985; Welander, 1986; Marotzke *et al.*, 1988; Thual and McWilliams, 1992; Quon and Ghil, 1992; Cessi and Young, 1992; Vellinga, 1996).

6.4.2.2 Transition to time-dependence

A Hopf bifurcation, for example H_1 along the asymmetric branch in Fig. 6.12, marks the location in parameter space where time periodic disturbances are about to be amplified through their interaction with the steady state. In this section, we concentrate on the Hopf bifurcation H_1 along the southward sinking branch in Fig. 6.12; the equilibrium state at this point is plotted in Fig. 6.16. The Hopf bifurcation is supercritical, since a periodic orbit is found for parameter values σ slightly larger than the value at bifurcation. This periodic orbit disappears at H_2 (Fig. 6.12) in a subcritical Hopf-bifurcation. Near the Hopf bifurcation H_1 the corresponding complex eigenfunction, say $(\mathbf{x}_R + i\mathbf{x}_I)$, shows exactly the time-periodic disturbance structures to which the steady state becomes unstable. At neutrality this disturbance structure $\phi(x, y, t)$ is given by (section 3.7.3.3)

$$\phi(x, y, t) = \mathbf{x}_R(x, y) \; \cos \; \omega t - \mathbf{x}_I(x, y) \; \sin \; \omega t \qquad (6.19)$$

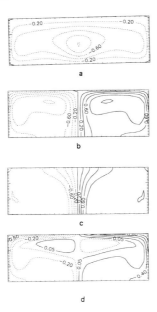

Figure 6.15.   Contour plots of the (a) streamfunction, (b) temperature, (c) salinity and (d) density of the perturbation destabilizing the basic state at the pitchfork bifurcation P_1 in Fig. 6.12

where ω is the angular frequency, corresponding to the imaginary part of the eigenvalue at the Hopf bifurcation. For $\omega t = 0, \pi/4, \pi/2$ and $3\pi/4$ the stream-function, density, temperature and salinity corresponding to ϕ are plotted in Fig. 6.18.

Before discussing the oscillation in terms of the changes of the spatial patterns of the perturbation, it is illustrative to monitor the changes of relevant integral quantities along the oscillation. In Fig. 6.17, the terms $< wS >'=< \bar{w}S'+w'\bar{S} >$, $< wT >'=< \bar{w}T' + w'\bar{T} >$ and $< wB >'=< \bar{w}B' + w'\bar{B} >$ are plotted along one period of the oscillation. Here the prime refers to the perturbation quantities (as in Fig. 6.18) and the bar to the steady state (as in Fig. 6.16). Within the linearized volume averaged energy balance for the perturbations, the term $< \bar{w}B' + \bar{B}w' >$ appears as the buoyancy forcing. The buoyancy forcing is mainly determined by the salinity forcing (Fig. 6.17), indicating that the salinity perturbation field mainly drives the oscillatory flow. The thermal contribution to the forcing, becoming positive over half a cycle of the oscillation, introduces the phase difference between salinity and buoyancy forcing. In this way, the oscillation resembles a 'thermohaline loop' oscillation as presented earlier in the four-box model of Tziperman *et al.* (1994b)

A more detailed mechanism, compatible with the results in Fig. 6.17, can be described with help of the patterns of the steady state (Fig. 6.16) and perturbation

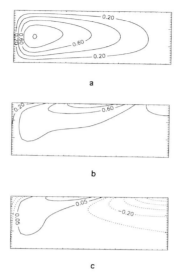

Figure 6.16.   Contour plots of the streamfunction, temperature and salinity of the basic state at the Hopf bifurcation H_1 in Fig. 6.12

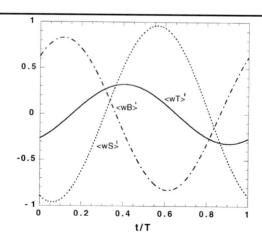

Figure 6.17.   Plot of the volume integrated perturbation quantities determining the buoyancy forcing of the perturbation flow at the Hopf bifurcation H_1, versus one period of the oscillation with period $2\pi/\omega$.

structures (Fig. 6.18), but it is not easy. Here is a try ! Within the initiation stage one needs to describe the causal chain of why the perturbations return, after half a cycle to the initial pattern but with negative sign, i.e. this stage is concerned with the essence of the oscillation. The growth phase is concerned with the amplification of the perturbation after one cycle of the oscillation. Suppose that the salt perturbation in Fig. 6.18d, applied at t = 0, is the initial disturbance. Since it controls the density anomaly (Fig. 6.18b, 0) it causes liquid to sink in the south and to rise over the rest of the basin, thereby giving the perturbation flow structure (Fig. 6.18a, 0). The temperature perturbation is consistent with this flow giving relatively warmer water in the south through advection and colder water over the central part in the basin (Fig. 6.18c, 0). Because the flow perturbation strengthens the equilibrium state circulation (Fig. 6.16a), it transports salt to the central part of the basin (Fig. 6.18d, 0.125). The presence of this heavier water substantially weakens the flow in the northern part (Fig. 6.18c, 0.125). In this region of weak flow, the heat transport is dominated by diffusion and colder water appears in the northern region (Fig. 6.18c, 0.25). This induces a reverse flow in the northern part of the basin (Fig. 6.18a, 0.25). The reverse flow affects the salt perturbation (Fig. 6.18d, 0.25) and the perturbation flow (Fig. 6.18a, 0.25) transports salt water northwards in the central part of the basin. Hereby, the positive salt perturbation is extended over the whole basin from south to north (Fig. 6.18d, 0.375). As a consequence, the salinity is reduced near the southern boundary and this strengthens the reverse flow perturbation in the basin at half a period of the oscillation (Fig. 6.18a, 0.375) and the reverse cycle starts. In summary, the oscillation is based on a combined advective salt/diffusive heat transport, where the salt perturbation drives the oscillation (compatible with Fig. 6.17). The temperature perturbation becomes important only in regions where the perturbation flow is weak, inducing the phase difference (Fig. 6.17) between salinity and buoyancy forcing.

6.4.3. Model-model comparison

Other studies, which appeared much earlier than the results discussed above, can now be put easily into perspective. In Thual and McWilliams (1992), the parameters a, b and k are used, instead of Ra, σ and A, because of a slightly different non-dimensionalization and the case of infinite Prandtl number $Pr \to \infty$ is studied. The correspondence between these parameters and the ones introduced in (6.13) is

$$a = \frac{Ra}{A^2} \; ; \; b = \sigma a \; ; \; k = \frac{2\pi}{A} \qquad (6.20)$$

The idealized equatorially symmetric surface forcing imposed is

$$T_S(y) = a \, \cos \, y; F_S(y) = b \, \cos \, y \qquad (6.21)$$

Bifurcation diagrams are constructed in the (a, b) parameter plane and a typical result is shown in Fig. 6.19. The positions of limit points, marked by e_1, e_2, f_1 and f_2 are shown in the (a, b)-plane for $k = 0.4$. The limit point e_1 and e_2 correspond to the endpoints of the branches of the pole to pole solutions and, when comparing

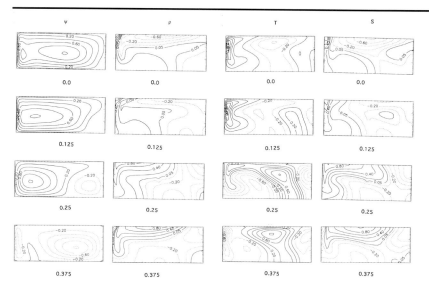

Figure 6.18. Contour plots of (a) streamfunction ψ, (b) density ρ, (c) temperature T and (d) salinity S for the periodic disturbance ϕ destabilizing the steady solution in Fig. 6.16 at H_1. The patterns are shown for four different times $\frac{\omega t}{2\pi}$ along the orbit. The dimensionless angular frequency $\omega = 0.326$ which implies a period in the order of the overturning time scale of the steady SPP-solution.

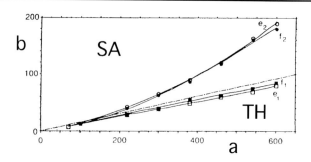

Figure 6.19. Regime diagram as found in Thual and McWilliams (1992) in the (a, b) parameter plane for $k = 0.4$. The dotted line is the curve of the no-flow solution. The other curves are explained in the text and represent paths of limit points (e_2, f_1 and f_2) and a path of a pitchfork bifurcation (e_1).

with Fig. 6.12, we can make an identification $e_1 = P_1$ and $e_2 = L_3$. The other limit points are those on the symmetric branch and hence $f_1 = L_2$ and $f_2 = L_1$. With this identification, the steady state structure is qualitatively similar to that in

Fig. 6.12. Consider for example $a = 400$ (note that the corresponding value of a with parameters in Fig. 6.12 is 1600), and increase b, the latter being equivalent to an increase in σ. The sequence of bifurcation points (excluding Hopf bifurcations) encountered in Fig. 6.12 is P_1, $P_2 = L_2$, L_1 and L_3 which indeed corresponds to the sequence e_1, f_1, f_2 and e_2 in Fig. 6.19.

Three other parameters, i.e. Ra_T, γ and d, are used in Quon and Ghil (1992) because of a slightly different non-dimensionalization. The correspondence between these parameters and the ones introduced above is

$$Ra_T = A^3 \, Ra \; ; \; d = \frac{1}{A} \; ; \; \gamma = \frac{\sigma}{\lambda} \qquad (6.22)$$

In Quon and Ghil (1992), the freshwater flux is not prescribed but determined from a steady state situation in which an equatorially symmetric salinity is prescribed. This freshwater flux is then used as boundary condition for salinity and contains a free parameter γ, which measures its strength. Quon and Ghil (1992) compute the stability boundaries for the TH-flow pattern by using time integration and identify the symmetry breaking bifurcation through a change in steady state from symmetric (open circles in Fig.6.20a) to asymmetric (closed circles in Fig. 6.20a). Part of this bifurcation diagram was also recalculated in Dijkstra and Molemaker

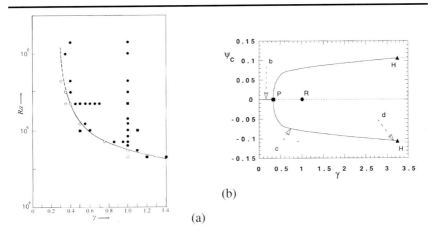

(b)

(a)

Figure 6.20. (a) 'Guessed' regime diagram obtained by Quon and Ghil (1992). All the markers represent asymmetric solutions and the curve monitors the position of the pitchfork bifurcation. (b) Computation of a bifurcation diagram at a fixed value of $Ra_T = 5 \times 10^6$ by Dijkstra and Molemaker (1997)

(1997) for slightly different lateral boundary conditions and is shown for $Ra_T = 5 \times 10^6$ in Fig. 6.20b. The value of the pitchfork bifurcation P is indeed close to that determined by Quon and Ghil (1992).

Technical box 6.1:
Amplitude equation approach

In Cessi and Young (1992) and Fleury and Thual (1997), reduced simpler models are derived from the full governing equations, which describe the two-dimensional thermohaline flows in limiting cases. In Cessi and Young (1992), the governing two-dimensional equations are written in streamfunction ψ and vorticity ζ formulation, with

$$v = \frac{\partial \psi}{\partial z} \; ; \; w = -\frac{\partial \psi}{\partial x} \; ; \; \zeta = \frac{\partial v}{\partial z} - \frac{\partial w}{\partial x}$$

where $x = -\pi + 2y\epsilon$ is a new meridional coordinate on the domain $[-\pi, \pi]$ and the aspect ratio $\epsilon = \pi D/L$. In this way, the dimensionless equations

$$\frac{1}{Pr}(\zeta_t + J(\psi, \zeta)) = T_x - S_x + \zeta_{zz} + \epsilon^2 \zeta_{xx}$$

$$\zeta = \psi_{zz} + \epsilon^2 \psi_{xx}$$

$$T_t + J(\psi, T) = T_{zz} + \epsilon^2 T_{xx}$$

$$S_t + J(\psi, S) = S_{zz} + \epsilon^2 S_{xx}$$

are obtained for $z \in [0, 1]$. Here $J(f, g) = f_x g_z - f_z g_x$ is the Jacobian operator. The boundary conditions at the surface are

$$z = 1 : T = aT_S(x) \; ; \; S_z = bF_S(x) \; ; \; \psi = \zeta = 0$$

and at all other boundaries satisfy slip and no-flux conditions are prescribed. The parameters a and b are related to those in (6.13) as

$$a = 4Ra \; \epsilon^2 \; ; \; b = 4\sigma Ra \; \epsilon^2$$

In the limit of small ϵ, expansions

$$(\psi, T, S) = \epsilon(\psi_1, T_1, S_1) + \epsilon^2(\psi_2, T_2, S_2) + \cdots$$

$$a = \epsilon a_1 + \cdots \; ; \; b = \epsilon^3 b_3 + \cdots \; ; \; \tau = \epsilon^2 t$$

are considered. At $\mathcal{O}(\epsilon)$, the governing equations become

$$T_{1,x} - S_{1,x} - \psi_{1,zzzz} = 0$$

$$T_{1,zz} = 0$$

$$S_{1,zz} = 0$$

with boundary conditions at $z = 1$:

$$T_1 = a_1 \; ; \; S_{1,z} = 0 \; ; \; \psi = 0 \; ; \; \psi_{1,zz} = 0$$

The solution to this problem is straightforward and given by

$$
\begin{aligned}
T_1(x) &= a_1 T_S(x) \\
\psi_1(x, z, \tau) &= W(z)(S_1(x, \tau) - a_1 T_S(x))_x \\
W(z) &= \frac{z^4 + 2z^3 - z}{24}
\end{aligned}
$$

where $S_1(x, \tau)$ is still undetermined. At $\mathcal{O}(\epsilon^2)$, the equations become

$$
\begin{aligned}
\frac{1}{Pr} J(\psi_1, \psi_{1,zz}) &= T_{2,x} - S_{2,x} + \psi_{2,zzzz} \\
-\psi_{1,z} T_{1,x} = J(\psi_1, T_1) &= T_{2,zz} \\
-\psi_{1,z} S_{1,x} = J(\psi_1, S_1) &= S_{2,zz}
\end{aligned}
$$

with $S_{2,z} = 0$ and $T_2 = 0$ at $z = 1$. One integration in z on the temperature and salinity equations and application of $S_{2,z} = T_{2,z} = 0$ at $z = 0$ gives the relations

$$
-\psi_1 T_{1,x} = T_{2,z} ; \quad -\psi_1 S_{1,x} = S_{2,z}
$$

In principle, ψ_2, T_2 and S_2 could be determined explicitly, but this is not necessary to obtain the reduced model. At $\mathcal{O}(\epsilon^3)$, the salinity equation becomes

$$
S_{1,\tau} + J(\psi_2, S_1) + J(\psi_1, S_2) = S_{3,zz} + S_{1,xx}
$$

with boundary conditions

$$
z = 0 : S_{3,z} = 0 ; \quad z = 1 : S_{3,z} = b_3 F_S
$$

Integration of the salinity over the vertical and using the boundary conditions and the fact that $S_1 = S_1(x, \tau)$ gives

$$
S_{1,\tau} - \int_0^1 \psi_{2,z} S_{1,x} dz + \int_0^1 S_{2,z} \psi_{1,x} dz - \int_0^1 \psi_{1,z} S_{2,x} dz = b_3 F_S + S_{1,xx}
$$

The first integral is zero because $\psi_2 = 0$ at top and bottom boundaries. Through one partial integration of the third integral it can be combined with the second integral to give

$$
S_{1,\tau} - \int_0^1 (\psi_1^2 S_{1,x})_x dz = b_3 F_S + S_{1,xx}
$$

Using the expression for ψ_1 and defining

$$
\hat{S} = \frac{S_1}{a_1} ; \quad \mu^2 = a_1^2 \int_0^1 W^2(z) dz ; \quad r = \frac{b_3}{a_1}
$$

equation (6.29), for $\delta = 0$, is obtained.

Cessi and Young (1992) explore asymptotically the large aspect ratio regime $A \rightarrow \infty$, as described in more detail in Technical box 6.1. The parameters a, b and ϵ are used in Cessi and Young (1992), which correspond to

$$a = 4\,Ra\,\epsilon^2;\ b = 4\,\sigma\,Ra\,\epsilon^2\ ;\ \epsilon = \pi/A \tag{6.28}$$

A regular expansion using the small parameter ϵ is pursued which leads to a one-dimensional evolution equation for the first order vertically averaged salinity $\hat{S}$. With $x = -\pi + 2\,y\,\epsilon$, this equation becomes

$$\hat{S}'' + \mu^2[\hat{S}'(\hat{S}' - T_S')^2]' + r\,F_S = \delta^2\hat{S}'''' \tag{6.29}$$

with parameters defined in Technical box 6.1 and and the primes indicate differentiation to x. The parameter δ serves to allow for boundary layers in regions of the flow with steep gradients. Cessi and Young (1992) analyse the limit $\delta \rightarrow 0$, where analytical progress can be made, but in this limit the solutions are not globally defined on the whole x–interval. For nonzero δ, and the choice $F_S = T_S = \cos x$, the equation (6.29) can be integrated once in x to give the boundary value problem

$$\delta^2\chi'' = r\sin x + \mu^2\chi(\chi + \sin x)^2 + \chi \tag{6.30a}$$

$$\chi(-\pi) = \chi(\pi) = 0 \tag{6.30b}$$

with $\chi = \hat{S}'$. Solutions of this boundary value problem were computed numerically in Dijkstra and Molemaker (1997), by the AUTO software (section 4.1) package (Doedel, 1980). For $\delta = 0.1$ and $\mu^2 = 7$ the bifurcation diagram with respect to r is shown in Fig. 6.21a. This diagram is characterized by two pitchfork bifurcations P_1 and P_2, which connect TH, PP and SA branches. The symmetric branch has two limit points and the path of these limit points in the (r, μ) plane for several values of δ is shown in Fig. 6.21b. Only two parameters are necessary to obtain this cusp (a codimension-two singularity). For smaller values of δ this cusp shifts to smaller values of r and μ and converges in the limit $\delta \rightarrow 0$ to analytic results in Cessi and Young (1992), with the cusp located at $r = \frac{8}{9}$ and $\mu^2 = 3$.

The results in the previous sections provide the elementary bifurcation behavior and the corresponding physical mechanisms at work. They show that symmetry breaking can occur in case of equatorially symmetric forcing leading to asymmetric pole to pole solutions. Overturning oscillations may arise through propagation of salinity anomalies with the mean flow, with phase differences caused by the phase lagged effects of heat and salt on the buoyancy production. These results will be subsequently extended to more and more realistic models in the following sections.

6.5. Convective Thermohaline Flows

The results of the previous section were rather qualitative and illustrated the complex interplay between heat and salt transport and the THC. There is no a priori reason, however, why these results should be of any relevance to the ocean circulation which is operating in a different parameter regime. Hence, no effort

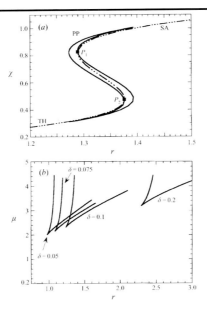

Figure 6.21. a). Bifurcation diagram for the model given by (6.30) with δ = 0.1 as calculated in Dijkstra and Molemaker (1997). Here the parameter r represent the strength of the salinity forcing and ξ is the vertically integrated meridional salinity gradient. The symmetric branch (connecting TH and SA) is dash-dotted, whereas the asymmetric NPP and SPP branches are drawn. b). Regime diagram in the (r, μ) plane, showing the path of the limit points in this plane. For δ = 0.1 and μ = √7 ≈ 2.65, the two limit points on the PP branches can be found at r = 1.28 and r = 1.39.

was made to use realistic surface fluxes and to relate results to actual oceanic quantities. In this section, a first step is taken into this direction.

In reality, the aspect ratio A is very large and the ratio of vertical and horizontal diffusivities is also considered to be very small. The usual argument is that the mixing coefficients, as turbulent eddy viscosities/diffusivities, scale with a characteristic length scale, which is very much larger horizontally than vertically. As argued by Quon and Ghil (1995), small aspect ratio geophysical fluid systems can only be convective when the ratio of vertical and horizontal diffusivity (viscosity) is very small. From the governing equations (6.14), a simple rescaling of the meridional coordinate shows that vertical and horizontal diffusive transports are of the same order of magnitude when the ratio of the diffusivities are taken to be of order A^2. Much used values in ocean models for example K_H and K_V are of the order 10^3 and 10^{-4} m^2s^{-1}, respectively, yielding a ratio of the same order of magnitude as A^2 ($\approx 10^7$).

6.5.1. Basic bifurcation diagrams

In Quon and Ghil (1995), the small aspect ratio regime was explored in the same way as in Quon and Ghil (1992). For a choice $R_{HV}^M = R_{HV}^T = 0.01$ and $A = 100$, the behavior of the model (with increasing Ra) is investigated for two different sets of surface boundary conditions. Again they find both symmetric and asymmetric steady states in this case, similar to the diffusive case. For large surface salinity forcing they also find a robust transition to oscillatory behavior through a Hopf bifurcation.

Bifurcation diagrams in this regime were computed in Weijer *et al.* (1999) for quite realistic shapes of the surface buoyancy forcing. The zonally averaged profile of Atlantic ocean surface temperatures (Levitus, 1982) can be approximated reasonably well by a cosine function of the form (for $y \in [0, A]$)

$$T_S(\frac{y}{A}) = \cos\ (\pi(-1 + 2\ \frac{y}{A})) \tag{6.31}$$

again with y the dimensionless meridional coordinate and A the dimensionless aspect ratio. The profile of T_S is plotted as the solid curve in Fig. 6.22a with on the horizontal axis the coordinate $\theta = 60(-1 + 2y/A)$.

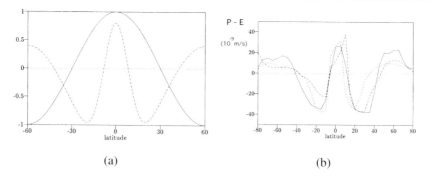

(a)   (b)

Figure 6.22.   (a) Plot of the dimensionless forcing functions T_S (solid) and $-F_S$ (dashed) as used in Weijer et al. *(1999). (b) Three different approximations of the zonally averaged freshwater flux over the Atlantic basin. The solid curve is from Baumgartner and Reichel (1975), the dotted from ECMWF data (Zaucker* et al., *1994) and the dash-dotted curve is from the Oort (1983) climatology. The function F_S in (a) is the symmetric part of the average of the three curves in (b).*

Figure 6.22b shows three estimates of surface freshwater flux over the Atlantic ocean, which equals the difference between precipitation and evaporation $(P-E)$. The solid line is obtained from the Baumgartner and Reichel (1975) dataset, by assuming that the volumes of freshwater, given in km^3 per $5°$ latitude bands, apply to a $60°$ wide ocean basin, in order to obtain the units of ms^{-1}. This same assumption is made for the other profiles, which are derived by Zaucker et al. (1994) from ECMWF data (long-dashed line), and from the Oort (1983) climatology (short-dashed line). Although the three profiles differ in detail, they agree on the particular form of the $P - E$ profiles in the Atlantic basin: they all show excess precipitation at high latitudes, excess evaporation at midlatitudes, and a high precipitative maximum over equatorial regions, at latitudes of the Inter Tropical Convergence Zone (ITCZ). This precipitative maximum may have considerable impact on the stability of the overturning circulation.

When the three profiles are averaged and symmetrized with respect to the equator, a profile for the salt flux over the domain $[60°S, 60°N]$ can be obtained as

$$F_S(\frac{y}{A}) = \cos\left(\frac{\pi}{60}\theta\right) - 2.4\exp\left[-\left(\frac{\theta}{12}\right)^2\right] + 0.6 \qquad (6.32)$$

(with again $\theta = 60(-1 + 2y/A)$) and an amplitude of the strength (6.11b) of $F_0 = 3.3 \times 10^{-7}$ ms^{-1}. This flux is plotted as the dashed line in Fig. 6.22a. Using this value of F_0, a realistic value of the salt-flux strength is obtained as $\sigma_s = 9.24$, the subscript s referring to the surface forcing.

Parameter	Value		Parameter	Value	
L	1.0×10^7	m	α_T	1.9×10^{-4}	K^{-1}
H	5.0×10^3	m	α_S	7.6×10^{-4}	
K_V	1.0×10^{-4}	m^{2}s^{-1}	K_H	1.0×10^3	m^{2}s^{-1}
A_V	2.2×10^{-4}	m^{2}s^{-1}	A_H	2.2×10^3	m^{2}s^{-1}
C_p	4.2×10^3	Jkg^{-1}K^{-1}	ρ_0	1.0×10^3	kgm^{-3}

Table 6.1.   Standard values of dimensional parameters for the two-dimensional ocean model used in Weijer et al. (1999).

For the surface forcing functions T_S and F_S and the parameters as in Table 6.1, the bifurcation diagram is shown in Fig. 6.23. The dimensionless values of parameters are $Ra = 10^4$, $Pr = 2.25$ and $A = 2 \times 10^3$. In Fig. 6.23, the norm plotted is the maximum value of the dimensionless meridional overturning streamfunction ψ_{max}, being a measure of the overturning strength. Due to this choice, the asymmetrical pole-to-pole solutions are projected onto the same curve and are not distinctly visible. Solid lines denote stable solutions, while dashed lines denote linearly unstable solutions. Contour plots of ψ of four solutions at several (labelled) locations in Fig. 6.23 are plotted in Fig. 6.24.

The solution for $\sigma_s = 0.0$ (Fig. 6.24a) is a thermally driven two-cell TH-solution. For $\sigma_s = 25.0$ two stable asymmetric solutions are depicted: the southern-sinking SPP solution (Fig. 6.24b) is characterized by major downwelling in the southern part of the basin and upwelling in the rest of the basin, while the northern-sinking NPP solution (Fig. 6.24c) is characterized by strong downwelling in the northern part of the basin. These solutions are connected to the branch of symmetrical solutions through the symmetry-breaking pitchfork bifurcations, indicated by the squares in (Fig. 6.23). Note that these are now subcritical pitchforks (contrary to the supercritical pitchfork in the diffusive thermohaline flows). For $\sigma_s = 37.6$ the SA solution is shown in Fig. 6.24d, which turns out to

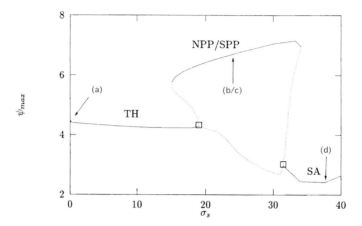

Figure 6.23.   Bifurcation diagram under symmetric forcing conditions in the convective regime as obtained by Weijer et al. (1999). Through the norm chosen, the NPP and SPP branches are indistinguishable.

be a four cell circulation pattern. This is due to the additional equatorial maximum and mid-latitude minima in the surface freshwater flux profile.

6.5.2. Imperfections

In the previous sections, we have presented the bifurcation diagrams for the Atlantic THC in a very idealized two-dimensional equatorially-symmetric config-uration. Apart from the three-dimensionality of the flow in reality, it is also not equatorially symmetric. The dominant asymmetries in the present Atlantic Ocean are summarized as follows.

(i) The present surface freshwater flux appears to be slightly asymmetric. When the zonally averaged profile for the Atlantic is considered (Oberhuber, 1988; Zaucker *et al.*, 1994), there is slightly more net evaporation in the northern part than in the southern part of the basin.

(ii) The continental geometry of the Atlantic is quite asymmetric, with the north-ern basin closing towards the north and the southern basin opening towards the south. The area of ocean-atmosphere interaction therefore considerably changes in meridional direction and this has consequences for the strength of the atmospheric feedback on SST anomalies (Marotzke and Stone, 1995).

(iii) The salt and heat input into the Atlantic basin due to interbasin exchanges is strongly asymmetric. In the Northern Hemisphere, an important component of this inflow comes from the Mediterranean and the Arctic ocean. Compensation

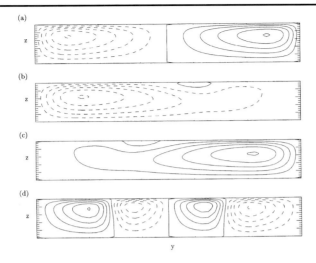

Figure 6.24.   Solutions for the streamfunction at several points in the diagram of Fig. 6.23. (a)
$\sigma_s = 0$, *TH-branch. (b)* $\sigma_s = 25.0$, *SPP-branch. (c)* $\sigma_s = 25.0$, *NPP-branch. (d)* $\sigma_s = 37.6$,
SA-branch.

(Gordon, 1986; Schmitz, 1995) of the outflow of North Atlantic Deep Water is
accomplished through water coming from the Indian Ocean (the 'warm' water
path) and that coming through Drake Passage (the 'cold' water path). Both
compensation routes provide a complicated structure of heat and salt input
into the South Atlantic (Weijer *et al.*, 2001).

(iv) The opening in the Southern Ocean impedes meridional flow since there is no
topography to support meridional flow above the sill depth of Drake Passage.
Moreover, a strong Antarctic Circumpolar Current (ACC) is present at the
southern part of the basin and absent in the north. The presence of such a
current influences the structure of the density field which may influence the
equilibria in the Atlantic basin (Toggweiler and Samuels, 1995).

(v) The winds over the Atlantic are fairly asymmetric about the equator. From
the zonally-averaged wind-stress shape, one observes that the winds over the
Southern Ocean are much stronger than those over the rest of the basin (Rahmstorf and England, 1997).

The presence of each of the asymmetries will lead to the break-up of the pitchfork bifurcations, according to the imperfection theory described in chapter 3. The
question addressed here is: does any preference for the type of pole-to-pole solutions exist, when the equatorially symmetric system is perturbed with asymmetric
effects?

Of the asymmetries summarized above, only the effect of surface fluxes, the lateral fluxes and the air-sea interaction can be considered in a two-dimensional configuration. The impact of the lateral heat and salt fluxes on the Atlantic overturning was systematically studied in Weijer *et al.* (1999) and Weijer (2000). The results indicate that these asymmetries lead to a strong preference for the NPP solution. In Dijkstra and Neelin (2000), the impact of asymmetric surface fluxes and air-sea interaction was considered. The effects of these asymmetries will be considered next.

6.5.2.1 Coupled model

To investigate the effect of the asymmetric distribution of the continents, the two-dimensional equatorially symmetric Boussinesq ocean model is coupled to a one-dimensional energy balance atmosphere model (North *et al.*, 1981) along lines developed by Stocker *et al.* (1992) and Chen and Ghil (1996). The equation for the surface temperature ϑ_* of the atmosphere is

$$R_a \frac{\partial \vartheta_*}{\partial t_*} = Q_s - (A_l + B_l \vartheta_*) + \frac{\partial}{\partial y_*} \left[D_a \frac{\partial \vartheta_*}{\partial y_*} \right] - \gamma Q_{oa} \quad (6.33\text{a})$$

$$y_* = 0, L \quad : \quad \frac{\partial \vartheta_*}{\partial y_*} = 0 \quad (6.33\text{b})$$

In the equation above, R_a is the (very small) thermal inertia of the atmosphere, A_l and B_l are two constants parameterizing the effect of long wave radiative cooling, D_a parameterizes the effect of baroclinic eddies on the meridional heat transport, γ is the fraction of the earth covered by the ocean basin and Q_{oa} is the (downward) ocean-atmosphere heat flux. The latter is positive when heat is transferred from the atmosphere to the ocean. The short wave radiation at the top of the atmosphere is prescribed as

$$Q_s(y_*) = Q_a^* S_a\left(\frac{y_*}{L}\right) \quad (6.34\text{a})$$

$$S_a\left(\frac{y_*}{L}\right) = 1 - 0.239(3(2\frac{y_*}{L} - 1)^2 - 1) \quad (6.34\text{b})$$

with $Q_a^* = \frac{1}{4}\Sigma_0(1 - \alpha)$, where Σ_0 is the solar constant and α the planetary albedo. The function S_a parameterizes the latitudinal dependence of the short wave radiation (North *et al.*, 1981).

The formulation of the downward heat flux Q_{oa} requires some care in interpretation since ϑ is the atmospheric surface temperature. As considered by Haney (1971), the net downward heat flux into the ocean Q_{oa} can be approximated by

$$Q_{oa} = Q_1 + Q_2(\vartheta_* - T_*) \quad (6.35)$$

if it is assumed that the air-sea temperature difference is small. The quantity Q_1 models the net downward heat flux of solar radiation across the ocean surface, minus the upward flux of longwave radiation and latent heat from an ocean surface at a temperature ϑ_*. The term Q_2 represents the net upward flux of long wave radiation and sensible and latent heat per degree excess of ocean surface temperature T_* over the atmospheric surface temperature ϑ_*. The downward heat flux Q_{oa} into the ocean, obtained from the surface heat parameterization as in Haney (1971), leads to

$$Q_{oa} = Q_o^* S_a\left(\frac{y_*}{L}\right) + \mu_{oa}(\vartheta_* - T_*) \tag{6.36}$$

where $Q_o^* < Q_a^*$ is the amplitude of the short wave flux absorbed by the ocean and μ_{oa} is the air-sea heat exchange coefficient. There are some important physical and quantitative differences between this formulation of the air-sea exchange used here, and some implementations of coupled energy balance models. In (6.36) the heat flux reaching the ocean surface has a large solar component, with the remainder related to air-sea interaction. If ϑ_* is interpreted as surface air temperature, then this solar contribution to the heat flux has to be taken into account. This is consistent with Stocker *et al.* (1992) but contrasts with equation (2) in Chen and Ghil (1996), where effectively it is assumed that the solar heat flux is absorbed in the atmosphere. Boundary conditions for the ocean at the surface, $z_* = H$, become again (6.11), with Q_{oa} now given by (6.36). The dimensionless model obtained in this way has several additional parameters which can be found in Dijkstra and Neelin (2000). The dimensional values used are shown in Table 6.2.

Parameter	Value		Parameter	Value	
L	1.5×10^7	m	α_T	1.6×10^{-4}	K^{-1}
H	4.0×10^3	m	α_S	7.6×10^{-4}	
R_a	10^7	$Jm^{-2}K^{-1}$	A_l	216	Wm^{-2}
D_a	10^{13}	WK^{-1}	B_l	1.5	Wm^{-2}
Q_a^*	240	Wm^{-2}	K_H	10^3	m^2s^{-1}
K_V	7.3×10^{-5}	m^2s^{-1}	A_H	2.5×10^5	m^2s^{-1}
C_p	4.2×10^3	$Jkg^{-1}K^{-1}$	A_V	1.8×10^{-2}	m^2s^{-1}
Q_o^*	180	Wm^{-2}	μ_{oa}	10.5	Wm^{-2}
ρ_0	10^3	kgm^{-3}	F_0	3.3×10^{-7}	ms^{-1}

Table 6.2.   Standard values of dimensional parameters for the coupled ocean-atmosphere model in Dijkstra and Neelin (2000).

6.5.2.2 Asymmetric air-sea interaction

With the function γ in the model, the effect of the continental asymmetry can be modelled and for constant $\gamma = \gamma_0$ and S_a and F_S symmetric around the equator, the model is equatorially symmetric. In this case, for the values of the parameters as in Table 6.2 and the freshwater flux (6.32), the bifurcation diagram of the coupled model is very similar to that in Fig. 6.23 (see Fig. 8 in Dijkstra and Neelin (2000)). Over the whole domain, the ocean is a few degrees warmer than the atmosphere, in agreement with observations (Peixoto and Oort, 1992).

The effect of continental asymmetry is idealized to a latitudinally asymmetric meridional distribution of the relative area of ocean and land by the γ function in Fig. 6.25a. The larger region of air-sea interaction in the Southern ocean and the reduced area in the northern ocean are taken into account by a piecewise linear shape of γ. At the southern boundary the value of γ is taken as 0.4, which effectively assumes that heat transports in the Atlantic overturning circulation influence the atmosphere over 40% of the latitude circle, while the rest of the latitude circle is passive and behaves like a land surface. Asymmetry due to the Southern Ocean is likely underestimated by this, but even without this the effects are already substantial.

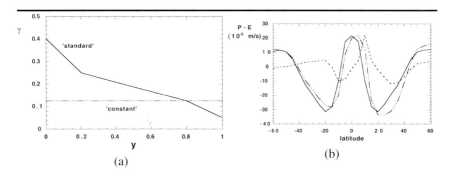

(a)   (b)

Figure 6.25.   (a) Shape of γ used in the standard case asymmetry case, the former including an increase in the area of air-sea interaction due to the presence of the Southern Ocean.   For completeness, the value of $\gamma_0 = 0.125$ used in the symmetric case is also plotted. (b) The $P - E$ curve using the data of zonally averaged profiles (dash-dotted). The symmetric component of this flux is shown as the solid curve and the difference between the two is plotted as the dashed curve.

The zonally averaged freshwater flux over the Atlantic basin (Baumgartner and Reichel, 1975; Zaucker *et al.*, 1994) as in (6.32) is replotted in Fig. 6.25b as the dash-dotted curve. Its symmetric component is shown as the solid curve and the dashed curve is the difference between the two. The latter curve suggests that asymmetry is introduced mainly in the tropics and subtropics. The degree of asymmetry in the freshwater flux is described by a homotopy parameter p, with $p = 0$ indicating the symmetric profile and $p = 0.4$ a near realistic $E - P$ pattern, according to Fig. 6.25b.

The bifurcation diagrams with varying p and the function γ of the 'standard' case in Fig. 6.25a are shown in Fig. 6.26. The overturning streamfunction value in the centerpoint of the domain (ψ) is used as an indicator of the flow. This value is plotted against the dimensionless salt flux strength $\sigma_r = \frac{\sigma}{\sigma_c}$, where σ_c is the value of σ at the first pitchfork bifurcation in the equatorially symmetric case. Along the branches, stability to stationary perturbations is indicated by markers along the branches, $-$ (+) indicating an stable (unstable) branch of solutions.

The case $p = 0$ are the solutions for which the freshwater flux is still symmetric, but the effect of continental asymmetry is taken into account; this case is

referred to as the 'weakly' asymmetric case. The bifurcation diagram is shown
as the dashed curve in Fig. 6.26. Due to the asymmetry of the continents, the
pitchfork bifurcation has disappeared and the NPP and TH have reconnected into
a branch, which is labelled NPP/TH. Similarly, the SPP branch and TH branch
have reconnected into the SPP/TH branch.

With an asymmetric freshwater flux, the southern part of the basin is freshened
with respect to the northern part, which tends to favor northern sinking. The saddle
node on the SPP/TH therefore moves to the right with increasing p. For $p = 0.2$,
the dot marks the position of this point and for $p = 0.4$, it is located at $\sigma_r = 0.87$.
Both asymmetries (continents and $E - P$) cooperate in limiting the interval in σ_r
for which the southern-sinking branch exists. For larger p eventually the southern-
sinking branch moves quite far from the region of "realistic" σ_r values, which is
around $\sigma_r \approx 0.6$.

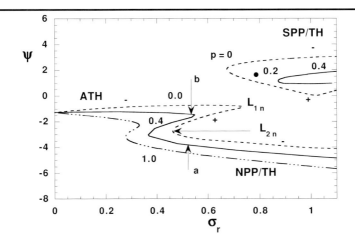

*Figure 6.26. Bifurcation diagram as a function of the parameter controlling the strength of fresh-
water flux σ_r, for standard values of the parameters given in Table 6.2. The asymmetric 'standard'
case continental configuration ($\gamma(y)$ as in Fig. 6.25a) is considered for several values of p, measuring the degree of asymmetry of the freshwater flux. On the vertical axis, the dimensionless overturning streamfunction value ψ at the center of the grid ($y_* = L/2, z_* = H/2$) is shown. Points P
and L denote pitchfork bifurcation and saddle node bifurcation, respectively. NPP and SPP denote
northern and southern pole-to-pole circulation branches, and TH the thermally driven branch. The
dot marks the position of the saddle node bifurcation on the SPP/TH branch for p = 0.2. Letters
a–b show points for which solutions are shown in Fig. 6.27.*

Because of the salinification of the northern part of the basin with increasing
p, the limit points (L_{1n} and L_{2n}) on the northern-sinking branch move to smaller
values of σ_r. This opens a window in σ_r where a unique northern-sinking branch
appears. Hence, both asymmetric air-sea interaction and asymmetric freshwater
flux induce a preference for the northern sinking branch. However, for $p = 0.4$
and in the range of realistic σ_r, multiple equilibria still occur due to the two limit

points. Two solutions at marked locations in Fig. 6.26 are shown in Fig. 6.27 and correspond to a strong (Fig. 6.27a) and a weak (Fig. 6.27b) northern-sinking solution. The latter solution is a slightly asymmetric version of the TH solution and therefore labelled ATH. It is significant that the ATH solution branch has almost a pole to pole flow, but with weaker overturning circulation than the NPP/TH solution.

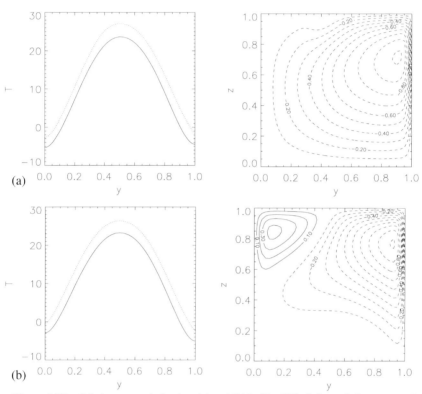

Figure 6.27.   Solutions at marked points (a) and (b) in Fig. 6.26. Left panel shows sea surface temperature (dotted curve) and atmospheric surface temperature (solid curve). Right panel shows latitude-depth plots of the streamfunction, scaled by its absolute maximum; contour levels are with respect to this maximum. (a) NPP/TH branch, $p = 0.4$, $\sigma_r = 0.51$. (b) NPP/TH branch, $p = 0.4$, $\sigma_r = 0.54$.

Transitions between these states due to finite-amplitude perturbations are possible for a small interval of σ_r. These transitions would thus be between states that both have northern sinking, but simply weaker and stronger values. The zonally averaged SST differs by less than $1\,^\circ$C between the two solutions. When $p = 1$, i.e., strong asymmetry in the freshwater flux, both limit points have shifted to small σ (dash-dotted curve in Fig. 6.26) and the multiple equilibria are less likely to be relevant. In the latter case, a large interval of σ_r appears where the NPP/TH solution is the only steady state.

6.5.2.3 Regime diagram

The results can be summarized into a schematic regime diagram (Fig. 6.28) with on the horizontal axis the strength of the freshwater flux forcing σ. Three different regimes can be distinguished and are listed below in the order that they would be encountered while increasing the magnitude of the freshwater flux, contrasting the symmetric case, the weakly asymmetric case, and the realistically asymmetric case. Panel (a) is the 'symmetric' case, corresponding to the bifurcation diagram in Fig. 6.23. There is only a unique TH solution at small σ, and multiple equilibria for larger σ where both SPP and NPP exist. The weakly 'asymmetric' case in panel (b) corresponds to symmetric $E - P$ but asymmetric air-sea exchange and hence with the bifurcation diagram for $p = 0$ in Fig. 6.26. Here, the limit point on the SPP/TH branch is situated between L_{1n} and L_{2n} which gives three different regimes of multiple equilibria. Panel (c) corresponds to the bifurcation diagram with increasing p in Fig. 6.26 modelling an asymmetric $E - P$ field. Here, a regime with a unique NPP solution appears. The qualitative properties of

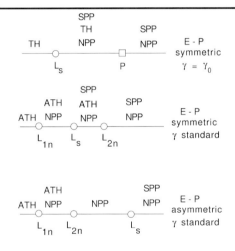

Figure 6.28.   Regime diagram of the coupled ocean atmosphere model according to the asymmetry introduced through the continents (modelled by γ) and through a slightly asymmetric freshwater flux. Three regimes can be distinguished which are named (a) 'symmetric', with (1) TH only; (2) NPP+TH+SPP; (3) NPP+SPP, (b) 'weakly asymmetric', with (1) ATH only; (2) NPP+ATH; (3) NPP+ATH+SPP; (4) NPP+SPP and (c) 'asymmetric' with (1) ATH only; (2) NPP+ATH; (3) NPP only; (4) NPP+SPP.

these regimes will tend to be robust, whereas it is possible that other parameters could affect which regime falls into the realistic range. Using this regime diagram, one can summarize:

(i) Asymmetry about the equator due to continental configuration in the Atlantic and due to the freshwater flux both tend to produce a preference for northern-sinking solutions.

(ii) The separation of the SPP branch from the NPP/TH branch by these asymmetries can create a significant region of parameter space where there is no southern-sinking branch, but where there is a northern-sinking branch. The salt advection feedback that was responsible for the bifurcation in the symmetric case acts to enhance the overturning in this NPP solution.

(iii) There is a region in the realistic part of parameter space with coexistence of the NPP branch and an asymmetric version of the thermally driven branch. However, when full asymmetry is included, there can also be a significant range of parameters where the NPP solution is unique and no multiple equilibria occur.

(iv) The role of the fractional region of air-sea interaction at each latitude on the heat flux feedback of the THC has been neglected in many studies, but is substantial in this model. This effect of the continental configuration is important, especially in the North Atlantic.

6.6. Zonally Averaged Models

In the two-dimensional models above important physics was neglected, such as the effect of wind-stress forcing and rotation. In addition, in the non-hydrostatic formulation used, solutions were obtained which are unstably stratified in some regions convection is not resolved. At the next level of models, which is still far from fully three-dimensional ocean models, the zonally averaged models appear. In this section, an overview is given of the different models and the results these models provide on the stability of the THC.

6.6.1. Scaling of the equations

In deriving the equations for the zonally averaged models and to prepare for results of low resolution three-dimensional models in later sections, starting point are the full equations in spherical coordinates as presented in section 2.1.2. In the boundary conditions for the temperature and salinity (2.19), the downward heat flux is prescribed as $Q_{oa} = B_T(T_S - T_*)$ and the freshwater flux $E - P$ is converted to a prescribed salinity flux $F_0 F_S$. For large scale motions with typical horizontal scale the radius of the earth r_0 and vertical scale D, with $D \ll r_0$, these equations can be reduced substantially. The equations (2.6) are non-dimensionalized using scales

$$p_* = -gD\rho_0 z + 2\Omega\rho_0 U r_0 p \tag{6.37a}$$

$$\rho_* = \rho_0(1 + \frac{2\Omega U r_0}{gD}\rho) \tag{6.37b}$$

for pressure and density. Horizontal and vertical velocity are scaled with U and DU/r_0, respectively while temperature and salinity are scaled with typical values ΔT and ΔS. In the shallow water limit $D/r_0 \to 0$, the non-dimensional equations become

$$\epsilon \left[\frac{Du}{dt} - uv \tan \theta \right] - v \sin \theta =$$

$$-\frac{1}{\cos \theta} \frac{\partial p}{\partial \phi} + \frac{1}{2} E_V \frac{\partial^2 u}{\partial z^2} + E_H L_u(u,v) \qquad (6.38\text{a})$$

$$\epsilon \left[\frac{Dv}{dt} + u^2 \tan \theta \right] + u \sin \theta =$$

$$-\frac{\partial p}{\partial \theta} + \frac{1}{2} E_V \frac{\partial^2 v}{\partial z^2} + E_H L_v(u,v) \qquad (6.38\text{b})$$

$$\frac{\partial p}{\partial z} = Ra(T - \lambda S) \qquad (6.38\text{c})$$

$$\frac{\partial u}{\partial \phi} + \frac{\partial (v \cos \theta)}{\partial \theta} + \cos \theta \frac{\partial w}{\partial z} = 0 \qquad (6.38\text{d})$$

$$\frac{DT}{dt} = L_T(T; P_H) + \frac{\partial}{\partial z}\left(P_V \frac{\partial T}{\partial z} \right) \qquad (6.38\text{e})$$

$$\frac{DS}{dt} = L_T(S; P_H) + \frac{\partial}{\partial z}\left(P_V \frac{\partial S}{\partial z} \right) \qquad (6.38\text{f})$$

where a linear equation of state has been assumed and

$$\frac{D}{dt} = \frac{\partial}{\partial t} + \frac{u}{\cos \theta} \frac{\partial}{\partial \phi} + v \frac{\partial}{\partial \theta} + w \frac{\partial}{\partial z}$$

$$L_T(\Phi; P_H) = \frac{1}{\cos \theta} \left[\frac{\partial}{\partial \phi}\left(\frac{P_H}{\cos \theta} \frac{\partial \Phi}{\partial \phi} \right) + \frac{\partial}{\partial \theta}\left(P_H \cos \theta \frac{\partial \Phi}{\partial \theta} \right) \right]$$

$$L_u(u,v) = \nabla_H^2 u - \frac{u}{\cos^2 \theta} - \frac{2 \sin \theta}{\cos^2 \theta} \frac{\partial v}{\partial \phi}$$

$$L_v(u,v) = \nabla_H^2 v - \frac{v}{\cos^2 \theta} + \frac{2 \sin \theta}{\cos^2 \theta} \frac{\partial u}{\partial \phi}$$

representing the horizontal diffusion (mixing) of a scalar Φ and frictional terms in the horizontal momentum equations.

With this scaling, the boundary conditions at bottom of the domain ($z = -1$) become

$$u = v = w = 0 \,, \quad \frac{\partial T}{\partial z} = \frac{\partial S}{\partial z} = 0 \qquad (6.40)$$

The surface boundary conditions (2.18) at $z = \epsilon F \eta$ can be written as

$$\alpha_\tau \tau^\phi = E_V \frac{\partial u}{\partial z} \qquad (6.41\text{a})$$

$$\alpha_\tau \tau^\theta = E_V \frac{\partial v}{\partial z} \qquad (6.41\text{b})$$

$$w = \epsilon F \frac{D\eta}{dt} \tag{6.41c}$$

$$\frac{\partial T}{\partial z} = -Bi(T - T_S) \tag{6.41d}$$

$$\frac{\partial S}{\partial z} = \frac{\sigma}{\lambda} F_S \tag{6.41e}$$

where the scaling of the sea surface height is derived from continuity of pressure. The parameters in these equations are the Rayleigh number Ra, the vertical and horizontal Ekman number E_V and E_H, the wind-stress coefficient α_τ, the vertical and horizontal inverse Péclet numbers P_V and P_H, the Biot number Bi, the rotational Froude number F and the freshwater flux strength σ. Expressions for these parameters are

$$
\begin{aligned}
Ra &= \frac{\alpha_T \Delta T g D}{2\Omega U r_0} \; ; \; E_V = \frac{A_V}{2\Omega D^2} \; ; \; E_H = \frac{A_H}{2\Omega r_0^2} \; ; \; \alpha_\tau = \frac{\tau_0}{2\Omega \rho_0 D U} \\
\sigma &= \frac{\alpha_S D F_0}{\alpha_T \Delta T K_V} \; ; \; \lambda = \frac{\alpha_S \Delta S}{\alpha_T \Delta T} \; ; \; P_H = \frac{K_H}{U r_0} \; ; \; P_V = \frac{K_V r_0}{U D^2} \\
Bi &= \frac{B_T r_0}{U D} \; ; \; \epsilon = \frac{U}{2\Omega r_0} \; ; \; F = \frac{4\Omega^2 r_0^2}{g D}
\end{aligned}
\tag{6.42}
$$

With values of $U = 10^{-2}$ ms^{-1}, $L = 1000$ km and $D = 5$ km, typical values of $\epsilon = 10^{-5}$ and $F = 10^2$ are obtained. Hence, the value of ϵ is very small which justifies neglecting the effects of inertia on these scales. Moreover, the product ϵF is also a small parameter. Hence, at these scale the deformation of the ocean atmosphere boundary does not play a role which justifies the application of the surface boundary conditions at $z = 0$ and also the 'rigid-lid approximation' $w = 0$ (Huang, 1993).

6.6.2. Zonal averaging

With the approximations as mentioned above, the *dimensional* equations (2.6) for a flat bottom ocean basin, which form the starting point of the zonally averaged equations become

$$-fv_* = -\frac{1}{\rho_0 r_0 \cos\theta} \frac{\partial p_*}{\partial \phi} + A_H L_u(u_*, v_*) + A_V \frac{\partial^2 u_*}{\partial z_*^2} \tag{6.43a}$$

$$fu_* = -\frac{1}{\rho_0 r_0} \frac{\partial p_*}{\partial \theta} + A_H L_v(u_*, v_*) + A_V \frac{\partial^2 v_*}{\partial z_*^2} \tag{6.43b}$$

$$\frac{\partial p_*}{\partial z_*} = -\rho_0 g(1 + \rho_r) \tag{6.43c}$$

$$\frac{\partial w_*}{\partial z_*} + \frac{1}{r_0 \cos\theta}\left(\frac{\partial u_*}{\partial \phi} + \frac{\partial(v_* \cos\theta)}{\partial \theta}\right) = 0 \tag{6.43d}$$

$$\frac{\partial T_*}{\partial t_*} + \frac{u_*}{r_0 \cos\theta}\frac{\partial T_*}{\partial \phi} + \frac{v_*}{r_0}\frac{\partial T_*}{\partial \theta} + w_*\frac{\partial T_*}{\partial z_*} =$$

$$\frac{1}{r_0^2} L_T(T_*; K_H) \;+\; \frac{\partial}{\partial z_*}\left(K_V \frac{\partial T_*}{\partial z_*}\right) \tag{6.43e}$$

$$\frac{\partial S_*}{\partial t_*} + \frac{u_*}{r_0 \cos\theta}\frac{\partial S_*}{\partial \phi} \;+\; \frac{v_*}{r_0}\frac{\partial S_*}{\partial \theta} + w_*\frac{\partial S_*}{\partial z_*} =$$

$$\frac{1}{r_0^2} L_T(S_*; K_H) \;+\; \frac{\partial}{\partial z_*}\left(K_V \frac{\partial S_*}{\partial z_*}\right) \tag{6.43f}$$

$$\rho_* = \rho_0(1 \;-\; \alpha_T(T_* - T_0) + \alpha_S(S_* - S_0)) \tag{6.43g}$$

with $f = 2\Omega\sin\theta$ and the reduced density $\rho_r = (\rho_* - \rho_0)/\rho_0$. At the surface $z_* = 0$, the dimensional boundary conditions are

$$\rho_0 A_V \frac{\partial u_*}{\partial z_*} \;=\; \tau_0 \tau^\phi \tag{6.44a}$$

$$\rho_0 A_V \frac{\partial v_*}{\partial z_*} \;=\; \tau_0 \tau^\theta \tag{6.44b}$$

$$w_* \;=\; 0 \tag{6.44c}$$

$$\rho_0 C_p K_V \frac{\partial T_*}{\partial z_*} \;=\; B_T(T_S - T_*) \tag{6.44d}$$

$$K_V \frac{\partial S_*}{\partial z_*} \;=\; F_0 F_S \tag{6.44e}$$

At the bottom boundary ($z_* = -D$) and continental boundaries, no-slip boundary conditions are prescribed and there is no-flux of heat and salt.

6.6.2.1 Procedure

Within this section, dimensional quantities are used throughout but the star subscript is suppressed because of clarity. For zonal averaging over a basin with zonal extent $\Delta\phi = \phi_e - \phi_w$, it is convenient to introduce local variables x, y such that $x = \phi r_0 \cos\theta$ and $y = r_0\,\theta$. Then a zonal averaging operator is introduced through

$$\bar{u} = \frac{1}{L}\int_{x_w}^{x_w + L} u\, dx \tag{6.45}$$

where $x_w = \phi_w r_0 \cos\theta$ and $L = r_0 \cos\theta(\phi_e - \phi_w)$. In terms of the local coordinates, the zonally averaged momentum and continuity equations become

$$-f\bar{v} \;=\; -\frac{\Delta p}{\rho_0 L} + \overline{(A_H u_x)_x} + \overline{(A_H u_y)_y} + \overline{(A_V u_z)_z} \tag{6.46a}$$

$$f\bar{u} \;=\; -\frac{1}{\rho_0}\frac{\partial \bar{p}}{\partial y} + \overline{(A_H v_x)_x} + \overline{(A_H v_y)_y} + \overline{(A_V v_z)_z} \tag{6.46b}$$

$$\frac{\partial \bar{p}}{\partial z} \;=\; -\rho_0 g(1 + \bar{\rho}_r) \tag{6.46c}$$

$$0 \;=\; \frac{\partial \bar{w}}{\partial z} + \frac{1}{\cos\theta}\frac{\partial(\bar{v}\cos\theta)}{\partial y} \tag{6.46d}$$

with subscripts (x, y and z) indicating differentiation. The quantity Δp is the pressure difference between the eastern and western boundary. The system of equations (6.46) is not closed and the quantities $\overline{(A_H u_x)_x}$, $\overline{(A_H v_x)_x}$ and Δp have to be parameterized. In addition, an expression for the zonal velocity $\overline{u}$ in (6.46b) has to be obtained. If each of these quantities is parameterized in terms of the zonally averaged meridional and vertical velocity and density field, then effectively a model with two space dimensions is obtained which incorporates the effects of rotation and wind-stress forcing.

Technical box 6.2: Closure problem

The different types of closure are explained in detail in Wright *et al.* (1998). The first term to parameterize in (6.46b) is the zonally averaged zonal mixing of momentum represented by $\overline{(A_H v_x)_x}$. Immediately, this term can be written as

$$\overline{(A_H v_x)_x} = \frac{A_H}{L} \frac{\partial v}{\partial x} \Big|_{x_w}^{x_w+L}$$

Consider in Fig. 6.29 a sketch of the flow field, with an approximately inviscid interior coupled to a frictional western boundary layer, the latter having a thickness δ_M. Clearly, meridional velocity gradients are largest near the western boundary and by defining a boundary layer averaged velocity $\overline{v}^\delta$ through

$$\overline{v}^\delta = \frac{1}{\delta_M} \int_{x_w}^{x_w+\delta_M} v \, dx$$

the zonal mixing of meridional momentum can be approximated by

$$\overline{(A_H v_x)_x} \approx -\frac{A_H}{L} \Gamma_1 \frac{\overline{v}^\delta - 0}{\delta_M}$$

where Γ_1 is a $\mathcal{O}(1)$ constant. It remains to relate $\overline{v}^\delta$ to $\overline{v}$, but this can be easily done through the overall mass balance

$$\delta_M \overline{v}^\delta \approx \overline{v} L$$

The second term to parameterize in (6.46b) is the zonally averaged meridional mixing of meridional momentum represented by the term $\overline{(A_H v_y)_y}$. Assuming that effects of variations on the horizontal basin scale are small in evaluating the dominant contribution of this term in the western boundary layer, then

$$\overline{(A_H v_y)_y} \approx A_H \overline{v}_{yy}$$

The last term in (6.46b) to parameterize is the term $f\overline{u}$ since (6.46b) is dropped.

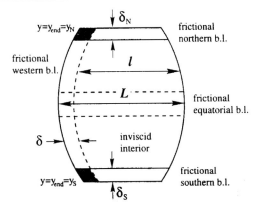

Figure 6.29. Sketch of the flow domain with the interior flow and the western boundary layer having a thickness $\delta = \delta_M$ (Wright et al., 1998).

This is done by writing first

$$f\bar{u} + \frac{1}{\rho_0}\frac{\partial \bar{p}}{\partial y} = -f(\bar{u}_g - \bar{u})$$

where $\bar{u}_g$ is the geostrophic velocity $f\bar{u}_g = -\bar{p}_y/\rho_0$. Next, the term in the right hand side is written as

$$\bar{u}_g - \bar{u} \approx \frac{(L - \delta_M)(\bar{u}_g^i - \bar{u}^i) + \delta_M(\bar{u}_g^b - \bar{u}^b)}{L}$$

where the superscripts i and b indicate interior flow and boundary layer flow, respectively. In the interior, the velocity is in geostrophic balance and hence the first term in the nominator is zero. Defining $\Gamma_2 = (\bar{u}_g^b - \bar{u}^b)/\bar{u}_g$, the term $f(\bar{u} - \bar{u}_g)$ can be rewritten as

$$f(\bar{u} - \bar{u}_g) = f\frac{\delta_M \Gamma_2}{L}\bar{u}_g = -\frac{\delta_M \Gamma_2}{\rho_0 L}\frac{\partial \bar{p}}{\partial y}$$

Using (6.46c) differentiated to y, i.e.

$$\frac{\partial^2 \bar{p}}{\partial y \partial z} = -\rho_0 g\frac{\partial \bar{\rho}_r}{\partial y}$$

differentiation of (6.46b) to the vertical coordinate z leads to (6.47) with $A_H^* = A_H L/(\Gamma_2 \delta_M)$ and $A_V^* = A_V L/(\Gamma_2 \delta_M)$.

The approach followed by Wright and Stocker (1992) is to drop (6.46a) and to rewrite (6.46b), under reasonable approximations of the western boundary layer structure (see Technical Box 6.3), as

$$A_H^* \overline{v}_{zyy} + A_V^* \overline{v}_{zzz} - \frac{\Gamma_1 A_H^*}{\delta_M^2} \overline{v}_z = -g \frac{\partial \overline{\rho_r}}{\partial y} \qquad (6.47)$$

where $\Gamma_1 = \mathcal{O}(1)$ is constant, δ_M is a measure of the thickness of the western boundary layer and A_V^* and A_H^* are modified mixing coefficients.

In the approach taken by Marotzke *et al.* (1988), effectively two terms in (6.47) are neglected to give

$$A_V^* \overline{v}_{zzz} = -g \frac{\partial \overline{\rho_r}}{\partial y} \qquad (6.48)$$

which is referred to as the 'frictional' closure. This equation can be directly obtained by putting $\overline{u} = 0$ in (6.46b), neglecting $\overline{(A_H v_x)_x}$ and eliminate the pressure from (6.46b) and (6.46c). It is therefore equivalent to neglecting the effect of rotation a priori and using a modified vertical friction coefficient.

In the approach taken by Wright and Stocker (1992), the two frictional terms are neglected in (6.47) giving

$$\frac{\Gamma_1 A_H^*}{\delta_M^2} \frac{\partial \overline{v}}{\partial z} = g \frac{\partial \overline{\rho_r}}{\partial y} \qquad (6.49)$$

which is written in Wright and Stocker (1992) as

$$\frac{\partial \overline{v}}{\partial z} = \epsilon_0 \frac{g}{r_0 \Omega} \frac{\partial \overline{\rho_r}}{\partial y} \; ; \; \epsilon_0 = \frac{r_0 \Omega \delta_M^2}{\Gamma_1 A_H^*} \qquad (6.50)$$

This closure is referred below as the 'geostrophic' closure, since elements of the geostrophic balance and western boundary current structure are taken into account. The usefulness of this parameterization has been shown in Wright and Stocker (1992) and from a comparison with three-dimensional models, a value of $\epsilon_0 \approx 0.1$ is suggested. An implication of this closure is that there exists a linear relation between the zonal density difference $\Delta \rho_r$ over the basin and the meridional density gradient. In a more rigorous analysis based on vorticity dynamics, Wright *et al.* (1995) have demonstrated that a relation between the east-west density difference and the zonally averaged meridional density gradient exists more generally, but it is a more complicated one than (6.49).

In principle, 2-dimensional models can also be considered as special cases of zonally averaged models with $\overline{u} = 0$ and all zonal derivatives being equal to zero. Models used by Sakai and Peltier (1995) and Vellinga (1996) are then obtained which are hydrostatic versions of the two-dimensional Boussinesq models.

6.6.2.2 Convective adjustment

In hydrostatic ocean models, adaptations are needed when the stratification becomes locally (statically) unstable. The effects of convection have to be parameterized and several ad-hoc procedures, generally referred to as convective adjustment, are used. In all these procedures, the temperature and salinity fields are locally adjusted in such a way that a stable stratification is achieved. In the first procedure, which we will indicate by classical adjustment (CA), the temperature and salinity are explicitly mixed in adjacent vertical levels of the water column if the density stratification is unstable resulting in equal temperature and salinity for those levels. This procedure has to be repeated a number of times at each time step in the evolution of the flow, as an iteration towards complete removal of static instabilities (Cox, 1984). A variation of this technique is suggested by Rahmstorf (1995a): groups of levels in the water column are treated as one convective region. The latter procedure guarantees that the liquid is stably stratified after the procedure is terminated. This occurs within one time step of the model, and hence it is assumed that the time scale of convective mixing is at most equal to the time step of the numerical model. Furthermore, it is assumed that the convection only mixes quantities vertically; no horizontal mixing is involved.

The other procedure of convective adjustment is indicated by implicit mixing (IM) and assumes that the effect of convection on a sub-grid scale can be modelled by a large vertical diffusion coefficient for heat and salt (Cox, 1984). For example, the vertical mixing of heat in (6.9c) is parameterized as

$$\mathcal{F}_{T*} = \frac{\partial}{\partial z_*}((K_{V0} + K_{Vc})\frac{\partial T_*}{\partial z_*}) \tag{6.51}$$

where K_{V0} is the background value of the vertical mixing coefficient and K_{Vc} is an extra vertical diffusion representing convection, which becomes large in areas with an unstable stratification and zero otherwise.

6.6.3. Bifurcation diagrams

Results for a zonally averaged model using the 'geostrophic' closure were first presented in Wright and Stocker (1991), using a domain [-80°S,80°N] and a fixed ocean basin depth of 5000 m. Under equatorially symmetric restoring conditions for temperature and salinity, a steady state is obtained very similar to the TH state. Next, the freshwater flux is diagnosed from this solution and used as forcing under mixed boundary conditions. The TH state is unstable and eventually a pole-to-pole solution is found.

A bifurcation analysis of the model as used in Wright and Stocker (1991) was performed by Vellinga (1996). The dimensional surface boundary condition for salinity was written as (cf. (6.11b))

$$K_V \frac{\partial S_*}{\partial z_*} = F_0 F_S(y_*) \tag{6.52}$$

To obtain the shape of the freshwater flux, steady states of the model are first obtained with a restoring condition for salinity using standard values of parameters. Next, the salt flux was diagnosed and used as the forcing function F_S. The dimensionless parameter $\sigma = F_0 H/(K_V \Delta S)$ was used as control parameter, where ΔS is a characteristic salinity difference. The value $\sigma = 1$ corresponds to the solution for which the freshwater flux was obtained under restoring conditions.

The TH state thus obtained becomes unstable through a pitchfork bifurcation at $\sigma = 1.07$. The fact that the bifurcation point need not always be situated below $\sigma = 1$ means that a switch from restoring to mixed boundary conditions is not necessarily accompanied by loss of stability of the TH state. The general features of the symmetric and asymmetric solutions are much like those of the 2D-Boussinesq model. When the pole-to-pole cell has grown to a size that it occupies nearly all of the basin, a further increase of the salt flux hardly alters the flow near its rising branch. This qualitatively similar behaviour suggests that the essentials of symmetry breaking of the thermally-dominated circulation are not in the dynamics, but can be deduced from the transport equations for heat and salt. This is compatible with the physical mechanism of symmetry breaking as described in section 6.4.2. Only a sufficiently strong meridional velocity response to a meridional density gradient is required to break the symmetry.

No further bifurcation analysis has been performed on the these type of models because of the dynamic similarities to the 2D-Boussinesq type models. However, much modelling has been done using the zonally averaged models with a 'geostrophic' closure. In Stocker and Wright (1992), such a model was extended to a two-basin situation of the Atlantic and the Pacific. Under 'realistic' (restoring) salinity forcing, a global thermohaline circulation is found with strong interbasin exchange. The 'conveyor-like' state is stable under mixed boundary conditions and is maintained by the net evaporation in the Atlantic and net precipitation in the Pacific. By perturbing this state under mixed boundary conditions it is shown that if perturbations are big enough, different steady states can be reached.

Further extensions of these models have been described in Wright and Stocker (1992), where the sensitivities of a three-basin version are investigated and in Stocker *et al.* (1992) where a climate model, containing a zonally averaged model as ocean component, is developed. Such a global ocean model has also been coupled to a simple atmosphere model and thermodynamic sea-ice model (Ganopolsky *et al.*, 1998). Much work is currently being performed with these type of models to study climate changes on long time scales.

Bifurcation analysis has also been performed on the 2D hydrostatic Navier-Stokes model (Vellinga, 1996) using the IM-type convective adjustment (6.51). The dimensional equations of this model are

$$v_* \frac{\partial v_*}{\partial y} + w_* \frac{\partial v_*}{\partial z_*} = -\frac{1}{\rho_0} \frac{\partial p_*}{\partial y_*} \; +$$
$$+ A_H \frac{\partial^2 v_*}{\partial y_*^2} \; + \; A_V \frac{\partial^2 v_*}{\partial z_*^2} \tag{6.53a}$$
$$0 = -\frac{1}{\rho_0} \frac{\partial p_*}{\partial z_*} - \left(\frac{\rho_* - \rho_0}{\rho_0} \right) g \tag{6.53b}$$
$$\frac{\partial w_*}{\partial z_*} + \frac{\partial v_*}{\partial y_*} = 0 \tag{6.53c}$$
$$\frac{\partial T_*}{\partial t_*} + v_* \frac{\partial T_*}{\partial y_*} + w_* \frac{\partial T_*}{\partial z_*} =$$

$$\frac{\partial}{\partial y_*}\left(K_H\frac{\partial T_*}{\partial y_*}\right) \; + \; \frac{\partial}{\partial z_*}\left(K_V\frac{\partial T_*}{\partial z_*}\right) \tag{6.53d}$$

$$\frac{\partial S_*}{\partial t_*} + v_*\frac{\partial S_*}{\partial y_*} + w_*\frac{\partial S_*}{\partial z_*} \; =$$

$$\frac{\partial}{\partial y_*}\left(K_H\frac{\partial S_*}{\partial y_*}\right) \; + \; \frac{\partial}{\partial z_*}\left(K_V\frac{\partial S_*}{\partial z_*}\right) \tag{6.53e}$$

These equations are solved on a rectangular geometry with lateral walls at $y_* = \pm L$, a flat bottom at $z_* = -D$ and the ocean-atmosphere interface at $z_* = 0$. Slip conditions are assumed on the lateral and bottom boundaries and except at the surface, no flux conditions are applied. The dimensional surface boundary conditions for salinity are similar as in (6.52).

The pitchfork bifurcation point P is located at $\sigma = 0.96$ (remember that $\sigma = 1$ corresponds to the restoring solution) and only one of the asymmetric branches is shown (Fig. 6.30a). A more detailed inset (dashed box in (a)) is plotted in

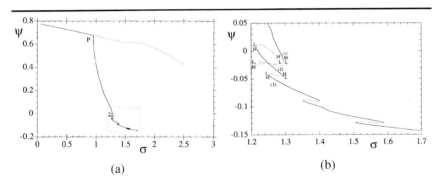

Figure 6.30. (a) Bifurcation diagram as presented in Vellinga (1998) for the hydrostatic 2D Navier-Stokes equations. (b) Inset of the region inside the dashed box of (a). Stable states are indicated by solid curves, unstable states by dotted curves. The pitchfork bifurcation point is indicated by 'P', Hopf bifurcations by 'H' and saddle-node bifurcations by 'L'.

Fig. 6.30b. It appears that for stronger flows (larger σ), convective adjustment causes problems; these were investigated in more detail in Vellinga (1998). Tracing the branch of asymmetric solutions in σ, the steady solution becomes unstable to oscillatory modes in several Hopf bifurcation points 'H', the first one occurring at $\sigma = 1.3$ in Fig. 6.30b. Shortly after, a saddle-node bifurcation is encountered (marked 'L' in Fig. 6.30b). This point is the beginning of a parameter regime in which the structure of the bifurcation diagram becomes rather complicated: the system undergoes a big loop through parameter space in which closely knit pairs of Hopf and saddle-node bifurcations occur regularly. Eventually, the solution reaches a more saline dominated regime along the indicated sections (2) and (3). This problematic and incorrect bifurcation behavior is not seen in the 2D-Boussinesq models.

The origin of these artificial multiple steady states was further investigated in Vellinga (1998). The regular occurrence of saddle-node bifurcations is not reflecting a real physical process but is caused by the way in which convection is represented in the model. It demonstrates the extreme sensitivity of the flow to finite-amplitude perturbations under the convective adjustment procedure (which only mixes heat and salt in columns downward). This sensitivity can already be deduced from the simple Welander (1982) model (section 6.2), where a finite-amplitude perturbation can induce a transition between a non-convective and convective state under only vertical transport. These problems do not occur in the 2D-Boussinesq model. The fact that the solutions are not completely stably stratified in the latter models, does not seem to have any effect on the overall bifurcation diagram. The results give a warning that convective adjustment may be responsible for spurious equilibria (Vellinga, 1998). Similar problematic model behavior due to convective adjustment was discussed in Cessi (1996).

6.7. Three-Dimensional Models

Much is now known about the bifurcation diagrams of three-dimensional models of the THC. For the presentation of these results and to connect them to results from traditional modeling approaches, it is useful to define a hierarchy of geometrical situations (Fig. 6.31). The single-hemispheric (SH) configuration, with and without a representation of continental geometry (Fig. 6.31a-b), has been used extensively to study decadal-to-multidecadal variability in the North Atlantic and the dynamics of the flows will be presented in the sections 6.7.1 and 6.7.2. Both this configuration and the next configuration in the hierarchy, the double-hemispheric (DH) configuration, with an without a representation of the Antarctic Circumpolar Current (Fig. 6.31c-d), have been studied to investigate the equilibria of the Atlantic circulation (section 6.7.3). In the multibasin (MB) and global geometry (Fig. 6.31e-f), results have focussed on the different multibasin flow patterns and transitions between these flows; the results are presented in section 6.7.4. Note that in all the three-dimensional models considered below, the full equations (6.38) with boundary conditions (6.40) and (6.41) are solved. In nearly all cases, the surface forcing is distributed as a body forcing in the top layer of the model.

6.7.1. The SH configuration: thermal flows

The simplest flows are those that are only forced by a steady zonally-independent meridional temperature gradient (no wind-stress forcing) and in which the salinity is constant. Because there can be no salt-advection feedback (section 6.2) in these flows, one does not expect multiple equilibria to occur. However, there are some interesting Hopf bifurcations through which these flows go unstable and spontaneous oscillatory behavior occurs. This transitional behavior was first noticed in Greatbatch and Zhang (1995) and Chen and Ghil (1995) and later extensively investigated by Huck and co-workers in a series of papers (Huck *et al.*, 1999; Colin de Verdière and Huck, 1999; Huck and Vallis, 2001). The linear stability analyses showing that the Hopf bifurcations actually existed were

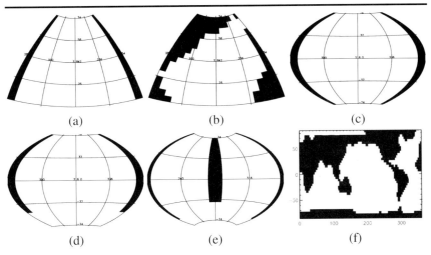

Figure 6.31. Sketch of the hierachy of geometrical configurations considered in 3D models of the THC. (a) single-hemispheric basin (SH) without continents, (b) single-hemispheric basin with continents, (c) double-hemispheric (DH) basin without a southern channel, (d) double-hemispheric basin with an southern channel, (e) multi-basin (MB) configuration, and (f) global (low-resolution) configuration.

presented in Huck and Vallis (2001) and Te Raa and Dijkstra (2002). We will discuss many of these results after presenting the bifurcation analysis proposed in Te Raa and Dijkstra (2002).

6.7.1.1 The multidecadal mode

The geometrical configuration used in Te Raa and Dijkstra (2002) is an idealization of the Atlantic sector basin with $(\phi, \theta) \in [\phi_W = 286°, \phi_E = 350°] \times [\theta_S = 10°N, \theta_N = 70°N]$ having a constant depth $D = 4$ km. The values of dimensional and dimensionless parameters that are used are provided in Table 6.3. These parameters are fairly standard in low-resolution three-dimensional models, except for the value of A_H which is a factor 100 larger because of numerical reasons. The IM variant of convective adjustment was applied with K_V^c as in Table 6.3 (it is also relatively small for numerical reasons). The horizontal resolution of the model is 4° horizontally and the model has 16 layers in the vertical. The prescribed surface temperature T_S is idealized as

$$T_S(\theta) = 10 \, \cos(\pi \, \frac{\theta - \theta_S}{\theta_N - \theta_S}) \qquad (6.54)$$

such that there is a meridional temperature difference of 20°C over the basin.

Along the branches of steady states computed, the meridional overturning streamfunction Ψ_M and zonal overturning streamfunction Ψ_Z are monitored.

2Ω	=	$1.4 \cdot 10^{-4}$	$[s^{-1}]$	r_0	=	$6.4 \cdot 10^6$	$[m]$
D	=	$4.0 \cdot 10^3$	$[m]$	U	=	$1.0 \cdot 10^{-1}$	$[ms^{-1}]$
ρ_0	=	$1.0 \cdot 10^3$	$[kgm^{-3}]$	g	=	9.8	$[ms^{-2}]$
α_T	=	$1.0 \cdot 10^{-4}$	$[K^{-1}]$	B_T	=	$6.3 \cdot 10^{-4}$	$[ms^{-1}]$
A_H	=	$1.6 \cdot 10^7$	$[m^2 s^{-1}]$	A_V	=	$1.0 \cdot 10^{-3}$	$[m^2 s^{-1}]$
K_H	=	$1.5 \cdot 10^3$	$[m^2 s^{-1}]$	K_V	=	$2.3 \cdot 10^{-4}$	$[m^2 s^{-1}]$
T_0	=	15.0	$[K]$	K_V^c	=	$3.3 \cdot 10^{-3}$	$[m^2 s^{-1}]$
Ra	=	$4.2 \cdot 10^{-2}$		P_H	=	$2.3 \cdot 10^{-3}$	
E_H	=	$2.7 \cdot 10^{-3}$		P_V	=	$9.2 \cdot 10^{-4}$	
E_V	=	$4.3 \cdot 10^{-7}$		Bi	=	$1.0 \cdot 10^1$	

Table 6.3. Standard values of parameters used in the numerical calculations of Te Raa and Dijk-stra (2002). The parameter B_T is given by $B_T = D/\tau_T$, where $\tau_T = 75$ days is the restoring time scale.

These are defined (in nondimensional quantities) as

$$\int_{\phi_W}^{\phi_E} v \cos\theta \, d\phi = \frac{\partial \Psi_M}{\partial z} \quad ; \quad \int_{\phi_W}^{\phi_E} w \cos\theta \, d\phi = -\frac{\partial \Psi_M}{\partial \theta} \qquad (6.55)$$

and

$$\int_{\theta_S}^{\theta_N} u \, d\theta = -\frac{\partial \Psi_Z}{\partial z} \quad ; \quad \int_{\theta_S}^{\theta_N} w \, d\theta = \frac{\partial \Psi_Z}{\partial \phi} \qquad (6.56)$$

With the scaling used in section 6.6.1, the dimensional volume transport Ψ_* is given by $\Psi_* = r_0 U D \; \Psi$ (for both zonal and meridional overturning).

The steady state at standard values of the parameters has a maximum meridional overturning of 20 Sv (Fig. 6.32a) and shows the typical unicellular structure with sinking confined to the northernmost part of the domain. In Fig. 6.32c and Fig. 6.32e vector plots of the horizontal circulation for certain sections are shown, superposed on contour plots of the vertical velocity. The surface circulation is anti-cyclonic (Fig. 6.32c) with upward vertical velocities at the western part of the basin. A reversed flow occurs near the bottom (Fig. 6.32e), consistent with the overturning flow. A section of temperature in a north-south vertical plane shows a 'thermocline' in the upper 1000 m, with slight static instabilities in the northern part of the domain (Fig. 6.32b). Surface temperatures show small advective departures (Fig. 6.32d) from the zonally uniform state, while at depth there is only very little variation. The surface heat flux Q_T of this steady state is shown in Fig. 6.32f and has a maximum amplitude of $45 \; Wm^{-2}$. The heat flux is negative (positive) in the northern (southern) half of the basin with a slight signature of the western intensification of the ocean flow. The particular state in Fig. 6.32 is also a solution of the steady equations when the flow is forced by the prescribed heat flux Q_T. In other words, this heat flux is needed to maintain the circulation with a surface temperature which closely matches the imposed temperature T_S.

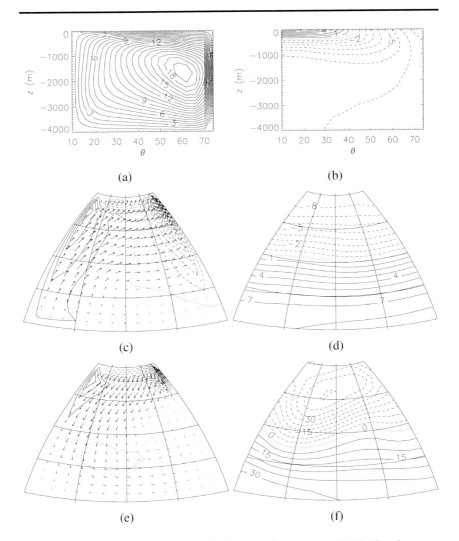

Figure 6.32.   Steady-state solution at standard values of the parameters. (a) Meridional overturning streamfunction (in Sverdrups). (b) Temperature (dimensionless) for a north-south vertical plane through the middle of the basin ($\phi = 318°$). The dimensional temperature can be obtained from $T_* = 15.0 + T$. (c) Velocity (dimensionless) near the surface (at 41 m depth). In this plot, vectors indicate the horizontal velocity, (u, v) and contours represent the dimensionless vertical velocity, w. Solid lines represent upwelling (flow out of the plane), dashed lines downwelling (flow into the plane). The maximum dimensional horizontal velocity is $1.7 \cdot 10^{-2}$ ms^{-1}, the maximum amplitude of the vertical velocity is $1.8 \cdot 10^{-6}$ ms^{-1} (downwelling). (d) Temperature near the surface. (e) Velocity at $z = -3200$ m. Maxima are $5.5 \cdot 10^{-3}$ ms^{-1} for the horizontal and $2.1 \cdot 10^{-5}$ ms^{-1} (downwelling) for the vertical velocity. (f) Surface heat flux Q_T (in Wm^{-2}), diagnosed from the solution obtained under restoring boundary conditions. Solid lines represent heat gain from the atmosphere, dashed lines heat loss to the atmosphere.

If one considers the linear stability of this steady state under restoring boundary conditions, such that temperature perturbations are considerably damped at the surface, it turns out that this state is linearly stable, because all eigenvalues have negative real part. The least damped mode has a centennial time scale of about 450 yr. One can also consider the stability of the steady state under the prescribed heat flux forcing Q_T. In this way, the temperature anomalies are not damped at the surface (Greatbatch and Zhang, 1995). Under this heat flux forcing condition, the state in Fig. 6.32 is unstable to a mode with an oscillation period $\mathcal{P} \approx 65$ years; this is the *multidecadal mode*.

The subsurface vertical velocities (Fig. 6.33a,c,e,g) and subsurface temperature anomalies (Fig. 6.33b,d,f,h) of this mode are plotted over nearly half of the oscillation period where in each plot the time scale in the figure caption is dimensional. The vertical velocity anomalies have their largest amplitudes near the northern boundary and propagate westwards (Fig. 6.33). The positive temperature anomaly at the surface, present at $t = 0$, follows the same propagation as the vertical velocities near the northern boundary (Fig. 6.33). Along the southern boundary, the anomalies are relatively weak and propagate eastwards.

The dependence of the growth rate and period of the multidecadal mode on the horizontal mixing of heat, the coefficient K_H, was determined. For five different states, the surface heat flux was diagnosed and the stability of the steady state determined under prescribed flux conditions. The growth rate (dashed) and period (drawn) corresponding to the multidecadal mode for the five different values of K_H are shown in Fig. 6.34a. The growth rate crosses the zero-axis at $K_H = 1670 \ m^2s^{-1}$. This indicates that a supercritical Hopf bifurcation occurs with decreasing K_H; the period at criticality is about 69 yr. For $K_H > 1670 \ m^2s^{-1}$, the steady state is (linearly) stable, but for $K_H < 1670 \ m^2s^{-1}$ it is unstable. The growth rate increases for smaller K_H and the period shortens slightly, being about 50 years at $K_H = 800 \ m^2s^{-1}$.

The location of the Hopf bifurcation defines the parameter value of K_H at fixed $K_V = 2.3 \times 10^{-4}$ (see Table 6.3) bounding a steady flow regime and an oscillatory regime. By following the path of this Hopf bifurcation in another parameter, a regime diagram in a two-parameter plane is obtained. In the (K_V, K_H) parameter plane such a diagram (based on only a limited number of points) is plotted in Fig. 6.34b. The regime below the curve, marked with the open squares, is the oscillatory regime. Slightly below this curve, periodic orbits of multidecadal period are expected. Increasing K_H damps the multidecadal mode, while increasing K_V has a destabilizing effect. The effect of K_V is mainly through the changes of the steady-state overturning, which increases with increasing K_V. The standard values of parameters are also indicated in Fig. 6.34b. The point labelled with a diamond is the standard parameter case, which is in the oscillatory regime. The period of the oscillation increases with decreasing K_V but remains in the multidecadal range.

The physical mechanism of oscillation of the multidecadal mode has been analysed in Colin de Verdière and Huck (1999) and Te Raa and Dijkstra (2002). In the latter paper, a detailed analysis is presented and the original paper should

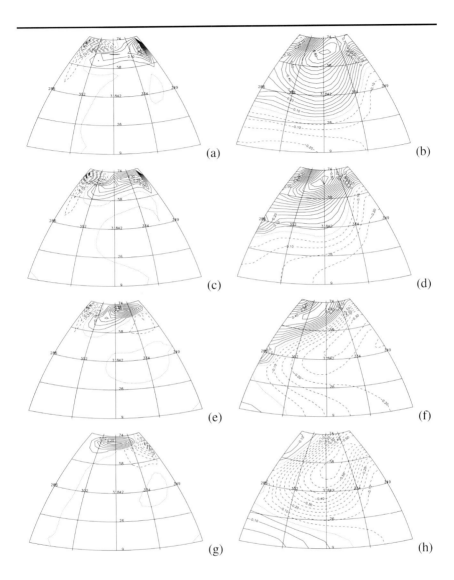

Figure 6.33. Patterns of the multidecadal mode. Vertical velocity perturbations $\tilde{w}$ (a,c,e,g) and temperature perturbations $\tilde{T}$ (b,d,f,h) both at $z = -19$ m at (a-b) $t = 0$ yr, (c-d) $t = 8.1$ yr, (e-f) $t = 16.3$ yr, (g-h) $t = 24.4$ yr. The plots are 1/8th period apart.

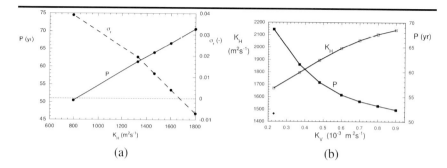

(a)   (b)

Figure 6.34.   (a) Period (in years, solid line) and growth rate (dimensionless, dashed line) as a function of the horizontal diffusivity K_H. The left vertical axis is for $\mathcal{P}$, the right vertical axis is for σ_r. (b) Regime diagram in the K_V-K_H plane for the interdecadal oscillation under prescribed-flux conditions. Open squares denote the values of K_H for which the Hopf bifurcation occurs. Below this curve , the steady state is unstable and oscillatory behavior is found; above this curve it is stable. The filled squares give the oscillation period at Hopf bifurcation. The point labelled with a diamond indicates the standard values of K_V and K_H.

be consulted for a full understanding of the energetics. There is a phase difference between the two terms in the anomalous buoyancy production, $<\bar{w}\tilde{T}>$ and $<\tilde{w}\bar{T}>$, where the quantities with a bar (tilde) refer to the steady state (perturbations). This, in turn, originates from the westward propagation of the temperature anomalies and the interplay of changing zonal and meridional temperature gradients with subsequent responses of the zonal and meridional overturning.

The physical mechanism can be summarized with help of Fig. 6.35. A warm

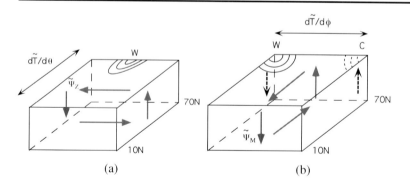

(a)   (b)

Figure 6.35.   Schematic diagram of the oscillation mechanism associated with the unstable multidecadal mode. The quantities $\tilde{\Psi}_M$ and $\tilde{\Psi}_Z$ are the meridional and zonal overturning anomalies, respectively. The phase difference between (a) and (b) is $\frac{\pi}{2}$; from Te Raa and Dijkstra (2002).

anomaly in the north-central part of the basin causes a positive meridional perturbation temperature gradient, which induces – via the thermal wind balance – a negative zonal overturning perturbation Ψ'_Z (Fig. 6.35a). The anomalous anticyclonic circulation around the warm anomaly causes cold water advection to the east and warm water advection to the west of the anomaly, resulting in westward phase propagation of the warm anomaly. The anomalous downwelling associated with the zonal overturning perturbation is consistent with this westward propagation, and the anomalous upwelling leads to a cold anomaly in the east. Due to the westward propagation of the warm anomaly, the zonal perturbation temperature gradient becomes negative, inducing a negative meridional overturning perturbation Ψ'_M (Fig. 6.35b). The resulting upwelling (downwelling) perturbations along the northern (southern) boundary cause a negative meridional perturbation temperature gradient, inducing a positive zonal overturning perturbation, and the second half of the oscillation starts. The crucial elements in this oscillation mechanism are the westward propagation of the temperature anomalies and the phase difference between the zonal and meridional overturning perturbations. The time scale of the oscillation depends on the basin crossing time, while the growth is related to the correlation of density anomalies — generated by the circulation anomalies — and those that cause them (Colin de Verdière and Huck, 1999; Te Raa and Dijkstra, 2002).

6.7.1.2 Finite-amplitude flows

The Hopf bifurcation associated with the destabilization of the steady thermal flows is very robust and must lead to finite-amplitude flows with multidecadal variability. Several of these flows have been obtained in so-called planetary, or frictional geostrophic models (FGM). In an FGM, the inertia terms and the local accelerations are put to zero, and a rigid-lid surface condition is used. In the momentum equations, also simplifications are usually made in the form of linear friction (Salmon, 1986; Colin de Verdière, 1988; Colin de Verdière and Huck, 1999) and no details of the continental geometry and bottom topography are included.

Using an FGM, where the flow is forced only by a zonally independent stationary heat flux (no freshwater flux), Greatbatch and Zhang (1995) found an oscillation with a period of about 50 years. The flow is characterized by changes in the MOC of about 7 Sv around a mean of 15 Sv. During the oscillation, the SST anomalies have largest amplitude in the small sinking region (of the mean state) and smaller amplitude over the larger upwelling regions. Robustness of these multidecadal oscillations to many physical processes was demonstrated in Colin de Verdière and Huck (1999). For example, they showed that both the β-effect and convective adjustment were not essential for the multidecadal variability to occur.

Specific transient solutions of the GFDL Modular Ocean Model (MOM), strongly connected to the results in Te Raa and Dijkstra (2002), were presented in Te Raa *et al.* (2004). MOM implements the full primitive equations on a specified domain on the sphere and solves these in time using an explicit time-marching scheme. A full description of this model, its capabilities and post-processing features is given in Pacanowski (1996) and serves also as a user manual of the model (see http://www.gfdl.gov/MOM/MOM.html). For a more detailed description on the numerics of the model, see Griffies (2004). The many configurations used in Te Raa *et al.* (2004) mainly differ in the choice of domain, mixing coefficients, resolution and surface forcing.

Finite-amplitude multidecadal flows in the simplest case (without continental and bottom topography, without wind forcing and with constant salinity) were computed for $K_H = 700 \ \mathrm{m^2 s^{-1}}$ and other parameters as in Table 6.3. The resolution used was the same as in Te Raa and Dijkstra (2002). The steady-state MOC under restoring conditions has an amplitude of 21 Sv with the velocity and temperature fields similar to those in Fig. 6.32. The heat flux through the surface at $t = 3000$ yr was diagnosed and then prescribed as a boundary condition instead of the restoring condition. The model is integrated for another 3000 yr under this prescribed heat-flux forcing. About 1000 yr after the switch in boundary condition, a multidecadal oscillation with a period of 45 yr appears in the MOC. The oscillation finally equilibrates with a peak-to-peak amplitude of about 10 Sv. The same results were obtained at a higher resolution of $2° \times 2°$ in the horizontal and 24 levels in the vertical.

The oscillation is characterized by large-scale temperature anomalies, having maximum amplitude near the surface, which propagate in a northwestward direction. The spatial patterns and propagation characteristics of these temperature anomalies strongly resemble those of the multidecadal mode in Fig. 6.33. To investigate whether the finite-amplitude multidecadal oscillaton found is indeed caused by the multidecadal mode, phase differences between meridional Ψ'_M and zonal Ψ'_Z overturning anomalies and between north-south and east-west temperature differences were computed. The north-south temperature difference ΔT_{N-S} was defined as the zonal average of $\Delta \hat{T}_{N-S}$, with

$$\Delta \hat{T}_{N-S}(\phi) = \int_{z=-1500/D}^{0} \int_{\theta_S}^{\theta_N} \frac{1}{\sin \theta} \frac{\partial T'}{\partial \theta} \, d\theta \, dz, \qquad (6.57)$$

where T' the difference between the total temperature and the time-mean temperature field. Similarly, ΔT_{E-W} is the meridional average of $\Delta \hat{T}_{E-W}$, with

$$\Delta \hat{T}_{E-W}(\theta) = \int_{z=-1500/D}^{0} \int_{\phi_W}^{\phi_E} \frac{1}{\sin \theta} \frac{\partial T'}{\partial \phi} \, d\phi \, dz. \qquad (6.58)$$

A plot of ΔT_{N-S} and ΔT_{E-W} over two oscillation periods at the end of the simulation is presented in Fig. 6.36a. The spatially-averaged meridional and zonal overturning anomalies for the same time interval are given in Fig. 6.36b. As can be seen, ΔT_{E-W} leads ΔT_{N-S} by about 15 yr, which is one-third of an oscillation period. The spatially averaged meridional overturning lags ΔT_{E-W} by about 2 yr, and the spatially averaged zonal overturning by about 11 yr. The features that characterized the multidecadal mode, namely the westward propagation of temperature anomalies in the north and the phase difference between zonal and meridional overturning, are thus also found in the finite-amplitude oscillation in the MOM model. Therefore, we can conclude that this finite-amplitude

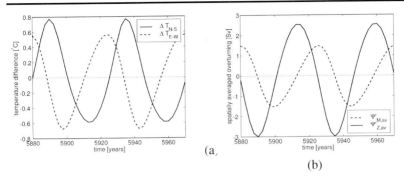

(a)

(b)

Figure 6.36. (a) *Zonally averaged north-south temperature difference* ΔT_{N-S} *(solid line) and meridionally averaged east-west temperature difference* ΔT_{E-W} *(dashed line), averaged over the upper* 1438 *m, during the last part of the integration for the MOM simulation with* $A_H = 1.6 \times 10^7$ $m^2 s^{-1}$ *and* $K_H = 7.0 \times 10^2\ m^2 s^{-1}$. (b) *Time series of the spatially averaged meridional overturning perturbation* Ψ'_M *(dashed line) and the spatially averaged zonal overturning perturbation* Ψ'_Z *(solid line).*

multidecadal oscillation is caused by the destabilization of the steady flow by the multidecadal mode; it is the periodic orbit coming from the Hopf bifurcation.

Te Raa *et al.* (2004) showed that the oscillatory flow due to multidecadal mode can be followed from the idealized case above — where it appears in a relatively viscous flow in a basin with a simple geometry under only thermal forcing — towards a complex flow in a more realistic domain. In a series of MOM simulations, the horizontal eddy viscosity was decreased by two orders of magnitude to a value commonly used in coarse-resolution ocean models ($A_H = 1.6 \cdot 10^5$ m^{2}s^{-1}, compare the value in Table 6.3). Next, a 'realistic' North-Atlantic continental geometry was added and finally also bottom topography and wind forcing were included. Along this model path, the main propagation direction of temperature anomalies in the oscillation remains (north)westward and phase differences remain such that Ψ'_M slightly lags ΔT_{E-W}, whereas Ψ'_Z lags ΔT_{N-S} by slightly more than half an oscillation period; Ψ'_M always lags Ψ'_Z.

6.7.2. The SH configuration: thermohaline flows

The immediate extension of the results in the previous subsection is the representation of the salt transport. In most cases, a mixed boundary formulation has been used (Cai, 1995) or the ocean is coupled to an energy balance atmospheric model with a prescribed freshwater flux (Chen and Ghil, 1996). Te Raa and Dijkstra (2003b) showed that the multidecadal mode also persists in a such a coupled ocean-atmosphere model and its growth rate is affected by the freshwater-flux pattern and strength. In Chen and Ghil (1996), the corresponding sustained decadal-to-multidecadal variability was found. In the next subsection, we will focus on the main new elements that are introduced through the presence of the salinity field: these are multiple equilibria and a new class of centennial modes.

6.7.2.1 Multiple equilibria and new internal modes

In Dijkstra *et al.* (2001), the bifurcation diagram for the single-hemispheric thermohaline flows under mixed boundary conditions was presented. The temperature forcing was of restoring type with T_S as in (6.54) and the surface freshwater flux F_S was taken as

$$F_S(\theta) = -\frac{1}{\cos\theta} \cos\left(\pi\frac{\theta - \theta_S}{\theta_N - \theta_S}\right) \tag{6.59}$$

where $\theta_S = 10°\text{N}$ and $\theta_S = 74°\text{N}$ and there is no wind forcing.

When the dimensionless strength of the freshwater flux, σ (section 6.6.1), is increased, a saddle-node bifurcation occurs on the branch of the TH solution. The flow changes from a northern-sinking (TH) solution — through a two-cell solution — to a southern-sinking (SA) solution. The bifurcation diagram (Fig. 6.37) consists of two back-to-back saddle-node bifurcations, qualitatively similar to that of the two-box Stommel (1961) model (section 3.1).

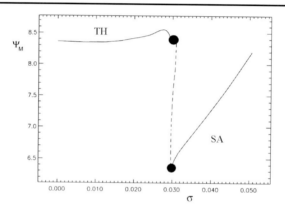

Figure 6.37.   Bifurcation diagram showing the absolute maximum meridional overturning Ψ_M of the flow in a single-hemispheric basin with the nondimensional strength of the freshwater flux σ as control parameter (Dijkstra et al., 2001). The drawn branches represent stable steady states, while the steady states on the dashed branch are unstable. The branches are separated by saddle-node bifurcations which are indicated by the filled circles. In the case shown here, the multiple equilibria regime in σ is very small.

By investigating the linear stability of the steady states for values of σ smaller than those at the saddle-node bifurcation on the TH branch, two types of damped centennial modes were found (Te Raa and Dijkstra, 2003a). One type of modes, say $\mathcal{C}_1$, appears also in pure thermal flows, but in the second type of modes, say

C_2, also salinity perturbations play a role. The imaginary and real parts of mode C_2 are shown in Fig. 6.38. To characterize this mode, the meridional overturning streamfunction and the zonally averaged temperature and salinity field are plotted. There is a strong positive meridional overturning anomaly in most of the basin

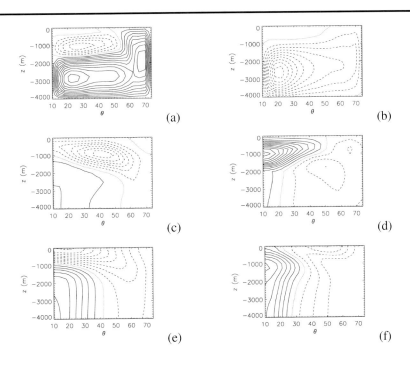

Figure 6.38. Imaginary (left) and real (right) parts of the eigenvector corresponding to the centennial mode C_2 (Te Raa and Dijkstra, 2003a). (a) and (b) Meridional overturning streamfunction. (c) and (d) Zonally averaged temperature. (e) and (f) Zonally averaged salinity.

(Fig. 6.38a), except in the southern part, where there is a weak negative anomaly near the surface. Half a period later, the negative overturning anomaly occupies the whole basin (Fig. 6.38b).

For the existence of both type of centennial modes, the advection of density anomalies by the steady state THC is essential. For mode C_1, the density perturbation is completely determined by temperature, even if salinity perturbations are allowed, whereas for mode C_2, both temperature and salinity determine the density. It was shown that the oscillation period is mainly determined by the overturning time scale of the steady-state flow and that the modes can be followed to a two-dimensional nonrotational situation (Te Raa and Dijkstra, 2003a). The latter suggests that the oscillations are caused by the same physics as the overturning oscillations in section 6.4.2 (Dijkstra and Molemaker, 1997). Although propagation characteristics are different in the three-dimensional version of the mode, these changes are not essential for the existence of the mode. The latter does, for example, not hold for the multidecadal mode, which is absent in a two-dimensional situation.

The centennial modes and multidecadal mode have their origin in the interaction of so-called stationary SST modes (Dijkstra, 2005). The latter modes are the eigenfunctions of the linear stability problem of the no-flow solution in the single-hemispheric basin. If one considers only a thermal forcing field, these modes can be ordered (n, m, l) where the numbers refer to the number of zeroes of the eigenmode in either zonal, meridional and vertical direction, respectively. At small thermal forcing, mergers between nonoscillatory SST modes occur to give rise to oscillatory modes. These mergers are similar to those which occurred in the barotropic QG model of wind-driven ocean circulation, where they lead to the gyre mode (section 5.5). With an increasing meridional temperature difference over the basin, say ΔT, first a merger occurs between the $(0, 0, 1)$ mode and the $(1, 0, 0)$ mode which leads to the multidecadal mode. For a slightly larger ΔT, the $(0, 1, 0)$ SST mode and the $(0, 1, 1)$ SST mode merge to give rise to the least-damped centennial mode. While the growth factor of the centennial mode (CM) remains negative for larger ΔT, that of the multidecadal mode (MM) eventually becomes positive.

Note that the type of merger explains why the CM is found in 2D models (Dijkstra and Molemaker, 1997; Te Raa and Dijkstra, 2003a), but the MM is not. The CM is a merger between two modes which have no zonal structure (the $(0, 1, 0)$ and $(0, 1, 1)$ modes). These SST modes will also be present in 2D models and hence a merger can occur. The MM is a merger between one mode with has zonal structure and one which has not; the $(1, 0, 0)$ SST mode is certainly absent in 2D models. Hence, a merger needed to obtain the MM cannot occur in these models and the MM mode is essentially three-dimensional.

Based on the origin of the oscillatory modes as mergers of the SST modes one can guess the classes of oscillatory modes present when salinity is included. Note that in the zero forcing limit, the classes of SST modes and SSS modes (solutions to the diffusion equation for salinity) are totally decoupled and the algebraic multiplicity of the eigenvalues is two. When the background flow is only thermally forced and its stability is considered under prescribed-flux conditions, mergers between the SST modes give rise to the MM and CM modes as above. However, mergers of SSS modes and between SSS and SST modes are also possible giving rise to the additional class of CMs found in Te Raa and Dijkstra (2003a). Because of the absence of a mean salinity gradient in Te Raa and Dijkstra (2003a), this new class of modes remained damped.

6.7.2.2 Finite-amplitude flows

The single-hemispheric thermohaline flows have been extensively analysed in a series of papers (Marotzke, 1991; Weaver and Sarachik, 1991) employing the MOM model and the different results were compared and further analysed in Weaver *et al.* (1993). In the latter study, three different zonally independent surface salinity profiles are used in the spin-up of the model under restoring boundary conditions, labelled A, B and C in Fig. 6.39a. For each of these profiles a (quasi-) steady state is found and the freshwater flux is diagnosed. The zonally averaged part of this flux is shown in Fig. 6.39b. For each situation A-C, two trajectories

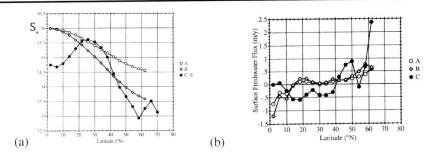

Figure 6.39.   (a) Surface salinity profiles (which are zonally independent) used in simulations under restoring boundary conditions in Weaver et al. (1993). (b) Zonally averaged freshwater flux of the steady state calculated with the restoring conditions in (a).

of the model were computed under mixed boundary conditions. One trajectory starts from the state at the end of the spin-up and the other starts from the ocean at rest. For case A, both trajectories end up in the same equilibrium state which is also the equilibrium state after spin-up. The flow consists of one overturning cell with sinking in the north and with a volume transport of about 12 Sv. For case B, both trajectories give different states (shown in Fig. 6.40) with different overturning strengths (about 9 and 12 Sv). The surface circulation is quite similar over most of the domain, showing the wind-driven transport in the form of Ekman cells. The solutions differ most in the extent of the sinking region at the northern boundary being much smaller in the case of stronger overturning (Fig. 6.40b). For case C, the restoring solution (Fig.6.41a) is unstable under mixed boundary conditions and both trajectories have a complicated temporal dependence but end up (after about 10,000 years) in a (quasi-) steady state shown in Fig.6.41b. For the latter solution, now a second cell has developed near the northern boundary and the strong downwelling region has been moved southward. Along the northern boundary 9 Sv of water is transported upwards.

These studies show the sensitivity to the shape and magnitude of the freshwater-flux strength on the type of equilibria found. In view of the bifurcation diagram in Fig. 6.37, the different equilibria can only be interpreted as being caused by small differences in convection sites as explained in section 6.6 (Vellinga, 1998). If the flows would be in the multiple equilibria regime, the only transition which could happen is that from a TH to a SA solution.

The temporal behavior in case C above has been investigated in Weaver and Sarachik (1991) and Weaver et al. (1993). The time series of the net basin-averaged surface heat flux is presented in Fig. 6.42a, starting from an initial ocean at rest and ending in the same state as presented in Fig. 6.41b. A Fourier spec-

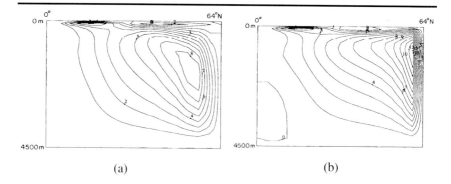

Figure 6.40.   *Two equilibrium solutions found for case B of Weaver* et al. *(1993) under the same mixed boundary conditions, but different initial conditions. (a) Equilibrium state found with the spin-up (restoring) solution as initial condition. (b) Equilibrium state found from an initially motionless ocean.*

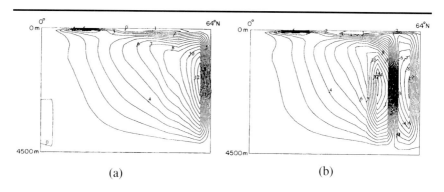

Figure 6.41.   *(a) Equilibrium state found after restoring conditions in case C of Weaver* et al. *(1993). This solution is unstable under mixed boundary conditions and the equilibrium state found after 10,000 years of integration is shown in panel (b).*

trum of this time series (Fig. 6.42b) shows significant energy in the decadal-to-multidecadal range. Weaver *et al.* (1993) demonstrated that the temporal behavior towards the steady state (Fig.6.41b) remains robust when the vertical viscosity is increased although in that case higher frequency motions are more damped. Seasonally varying forcing has no profound effect on the decadal variability and also with stochastic forcing in the freshwater flux, the decadal variability remains a dominant mode of variability.

Signatures of multidecadal variability are subsequently studied in Weaver and Hughes (1994). A single hemispheric basin, low resolution MOM-type model is used which captures features of the continental boundaries surrounding the Atlantic ocean. The model is forced by observed surface temperature (Levitus,

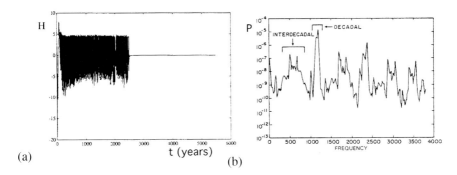

Figure 6.42. (a) Trajectory computed in case C of Weaver et al. (1993) starting from an initial state at rest; displayed is the net basin averaged surface heat flux H. (b) Power spectrum (with power spectral density P versus frequency) of the time-series in (a).

1982), observed freshwater flux (Schmitt *et al.*, 1989) and wind forcing (Hellerman and Rosenstein, 1983). After an initial spin-up, the model solution settled down into a periodic orbit with a period of about 22 years. There appears to be a slow phase of the oscillation associated with the decrease of the basin averaged surface heat flux followed by a fast phase of increase of this quantity. The mechanism of this variability is associated with large changes in convective activity in the Labrador Sea region of the model, leading to changes in the northward heat transport. In the slow phase of the oscillation, the Labrador Sea region is cooling while weak overturning occurs at the southwest boundary of Greenland. During the strong phase of the oscillation the deep water formation region moves to the Labrador Sea, once the cooling process and zonal overturning have set-up a sufficiently strong zonal pressure gradient. Patterns of the SST-anomaly show largest amplitude in the western North Atlantic with maximum amplitude of about 1.2 °C. It was shown that this variability is insensitive to the freshwater-flux forcing and wind-stress forcing and seems to be caused by processes involving the surface heat flux, convection and the overturning circulation, but the precise details are unclear.

In the FGM study of Winton and Sarachik (1993), the mean overturning is thermally direct with sinking in the north and upwelling at the southern boundary. An oscillation with a period in the order of the overturning time scale is found, with a period of about 250 year, and is termed a loop oscillation. The mechanism of the loop oscillation is related to advection of salinity anomalies along the mean overturning flow similar to that in Welander (1986). The presence of a positive salinity anomaly at low latitudes will decrease the overturning circulation. As a consequence, surface water is exposed longer to the freshwater forcing and becomes more saline at low latitudes. However, the salinity anomaly propagates with the flow, it reaches higher latitudes and eventually accelerates the circulation. It then escapes the freshening effect of the freshwater flux at higher latitudes. Simultaneously, a negative salinity anomaly develops at lower latitudes and the cycle starts all over again. It appears that the occurrence of these oscillations is very sensitive to the shape of the freshwater forcing profile. Moreover, the occurrence of the oscillation appears sensitive to the nonlinear nature of the equation of state used.

In simulations with a very strong freshwater flux at high latitudes, short periods of very strong overturning are found, which succeed long periods of collapsed overturning. These events are called flushes and the mechanism which has been proposed is essentially diffusive in nature. In Weaver *et al.* (1993), the circulation under restoring conditions (Fig. 6.43a) is unstable under mixed boundary conditions and eventually collapses (Fig. 6.43b). This collapse is referred to as the Polar Halocline Catastrophe (PHC) and is associated with the spreading of a tongue of low-salinity water equatorward. The circulation stays in the collapsed

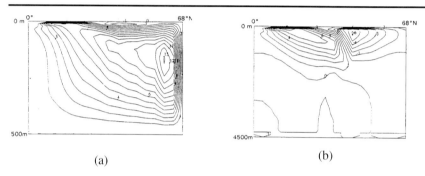

(a)   (b)

Figure 6.43.   (a) Steady state after restoring conditions in a slightly northward extended basin. (b) Collapsed state which results under mixed boundary conditions (Weaver et al., 1993).

state for about 2000 years and the low-latitude ocean absorbs an enormous amount of heat. Through vertical diffusion this heat reaches the low latitude deeper layers and through horizontal diffusion it reaches the deep layers at high latitudes. Meanwhile, the high-latitudes surface water remains cold through the atmospheric forcing. If the deep water becomes sufficiently warm at high latitudes, the water column becomes statically unstable and the induced overturning is so violent that the ocean looses all the heat stored over the period of collapse in only a few decades. After this flush, the freshening effect at high latitudes dominates again and the circulation collapses once more, explaining the oscillatory nature of these flushes.

In Marotzke (1991), flushes do not appear when wind forcing is included because the wind-driven poleward salt transport is sufficient to create dense enough polar water, such that convection is maintained. They do occur when the wind forcing is absent. However, in Weaver and Sarachik (1991), flushes do occur even with wind forcing since the high-latitude freshening is not compensated by the wind-driven salt transport. Winton and Sarachik (1993) find similar variability, which they call deep-decoupling oscillations, in an FGM. Each oscillation is comprised of three stages: (i) a decoupling period with a weak MOC and similar properties as the collapsed state discussed above, (ii) a flush, and (iii) a stage where the MOC recovers. With increasing strength of the freshwater flux, the decoupled phase becomes longer and the flush becomes more energetic.

6.7.3. The DH configuration: thermohaline flows

As will be explained in section 6.7.3.2, the discovery of multiple equilibria in the equatorially symmetric set-up was the! result (Bryan, 1986) which has stimulated nearly of the research on the stability of the THC. In this section, we have chosen to present these important results after the relevant bifurcation diagrams, which are presented next.

6.7.3.1 Bifurcation diagrams

Bifurcation diagrams for the equatorially-symmetric double-hemispheric configuration were computed in Weijer and Dijkstra (2001). The patterns of surface temperature T_S and surface freshwater flux F_S are prescribed as

$$T_S(\theta) \quad = \quad \cos \pi \, \frac{\theta}{\theta_N} \tag{6.60a}$$

$$F_S(\theta) \quad = \quad \frac{1}{\cos \theta} \cos \pi \, \frac{\theta}{\theta_N} \tag{6.60b}$$

over the domain $(\phi, \theta) \in [\phi_W = 286°, \phi_E = 350°] \times [\theta_S = 60°\text{S}, \theta_N = 60°\text{N}]$. The forcing is such that the meridional equator-to-pole temperature difference is equal to 20°C. The wind-stress forcing considered is an idealized profile for the Atlantic mimicking a double-gyre wind stress in each hemisphere, i.e., in dimensionless form

$$\tau^\phi(\theta) \quad = \quad -\tau \cos 4\pi \, \frac{\theta - \theta_S}{\theta_N - \theta_S} \tag{6.61a}$$

$$\tau^\theta \quad = \quad 0 \tag{6.61b}$$

The dimensional temperature profile T_S, zonal wind stress τ^ϕ and the freshwater flux F_S are shown in Fig. 6.44.

In the model used in Weijer and Dijkstra (2001), the horizontal resolution in 4° and the number of equidistant vertical levels is 16. In the case wind forcing is absent ($\tau = 0$) and other parameters similar to those in Table 6.3, (part of) the bifurcation diagram is plotted in Fig. 6.45. Here, the absolute maximum of the meridional overturning streamfunction Ψ_m is plotted versus σ. Again, the TH solution becomes unstable at the pitchfork bifurcation P and branches of asymmetric solutions appear. Note that, because of the norm chosen, both the NPP and SPP solutions have the same value of Ψ_m. At the saddle-node bifurcation L_1, the asymmetric branches become unstable but they regain stability through a second saddle-node bifurcation L_2. The MOC increases strongly with σ on this branch.

From the linear stability analysis of the steady states, the pattern of the mode destabilizing the TH state at the symmetry-breaking pitchfork bifurcation is determined from the eigenvector at P. Properties of the steady state flow at P are

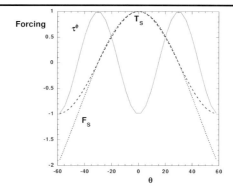

Figure 6.44. Plot of the pattern of the forcing functions for wind stress τ^ϕ, the atmospheric temperature $T_S/10$ and the freshwater flux F_S.

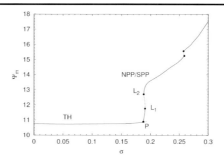

Figure 6.45. Bifurcation diagram in the control parameter σ. The asymmetric (NPP/SPP) solutions become unstable at the saddle-node bifurcation L_1 but regain stability at L_2. The saddle-node bifurcations close to $\sigma = 0.25$ reflect only minor rearrangements of the steady-state flow patterns.

plotted in Fig. 6.46. In addition to zonally averaged profiles of T, S and ρ (panels a-c), the overturning streamfunction (panel d) and the velocity fields at 100 m and 3000 m depth (panels e and f) are shown. All fields are equatorially symmetric and the density field is stably stratified except at high latitudes. The main downwelling is confined to small areas in the north- and south-eastern corners of the domain.

The same fields for the most unstable eigenvector at the pitchfork bifurcation P (in Fig. 6.45) are shown in Fig. 6.47. The eigenvector plotted here will favor a transition to a SPP state. Despite the zonal structure that results from the presence of rotation, the zonally averaged structures of the destabilizing perturbation are strikingly similar to those in purely two-dimensional models (Dijkstra and Molemaker, 1997) or zonally averaged models (Vellinga, 1996). Both the

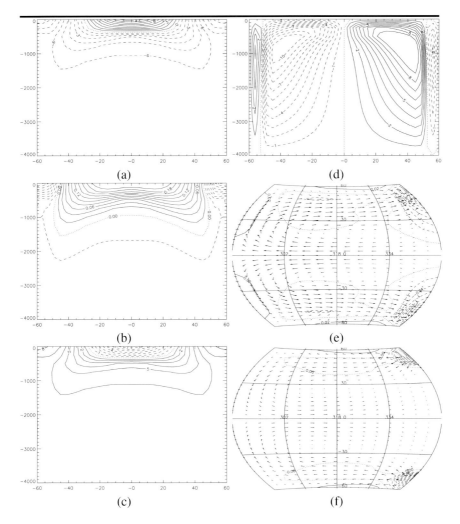

Figure 6.46.   Steady state at the pitchfork bifurcation P in Fig. 6.45. Shown are the zonally av-eraged fields of (a) temperature T, (b) salinity S, and (c) density, as well as (d) the meridional overturning streamfunction Ψ, and the velocity fields at (e) 100 m and (f) 3000 m depth. Dimen-sional temperature and salinity can be computed from T* − 15 = T and S* − 35 = S.

salinity and the temperature perturbations have a bi-polar structure (Fig. 6.47a-b), which is positive in the Southern Hemisphere, and negative in the Northern Hemisphere. This gives rise to an equatorially anti-symmetric density perturba-tion (Fig. 6.47c). Its mainly positive sign in the Southern Hemisphere shows that it is dominated by salinity. This bi-polarity of the anomalous density field sets up an inter-hemispheric pressure difference at depth. This pressure gradient gener-

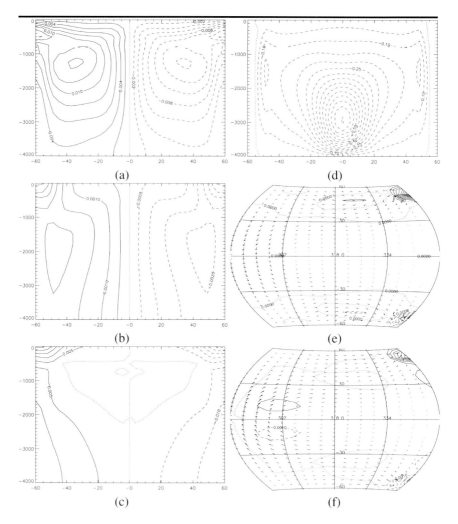

Figure 6.47. Most unstable eigenvector at the pitchfork bifurcation P in Fig. 6.45, destabilizing the steady TH state (Fig. 6.46). Same fields as in Fig. 6.46. Although the amplitude of the eigenvector is undetermined, the amplitudes of the individual fields are mutually consistent. The perturbation shown will favor a transition to an SPP state, but note that also the sign-reversed version of this perturbation is an eigenvector, and this will favor a transition to an NPP state.

ates deep cross-equatorial flow that is confined to the viscous western boundary layer (Fig. 6.47f). A return flow at shallower levels is established by continuity (Fig. 6.47e). Furthermore, the density at the surface and at depth increases southward and, through the thermal wind balance, this sets up a zonal flow. In the

Southern Hemisphere, this flow is eastward at the surface and westward at depth, in the north it is the other way around (Fig. 6.47e-f).

The MOC perturbation advects heat and salt from (sub)tropical regions southward, and enhances the thermal and saline anomalies on that hemisphere. When the meridional salinity gradient is strong enough, the density perturbation that initially generated the overturning anomaly, is amplified. The surface salt flux amplitude σ must therefore exceed a critical value for the TH state to become unstable.

6.7.3.2 Finite-amplitude flows

The issue of multiple equilibria in ocean models was addressed in a double-hemispheric set-up in Bryan (1986) using the MOM model and has been reproduced in Weaver and Sarachik (1991). Bryan (1986) first used the freshwater flux and solution of a single-hemispheric version of the model (over the domain 0-90°N) which was obtained under restoring conditions with observed salinity. This solution and the surface forcing were reflected through the equator to provide a full basin solution as initial condition (Fig. 6.48a). The freshwater flux of this state was diagnosed and used in subsequent runs under mixed boundary conditions.

When a negative salinity anomaly of 1 psu is (instantaneously) added poleward of 45°S, the deep convection in the Southern Hemisphere is interrupted. The residence time of water parcels in the surface layer increases and leads through the convective feedback to a collapse of the circulation in the Southern Hemisphere and within 50 years a pole-to-pole circulation is reached (Fig. 6.48b) with sinking in the north. Adding a positive salinity anomaly of 2 psu in the same region (poleward of 45°S) induces an intensification of the meridional overturning, which leads to a southern-sinking solution through the advective feedback mechanism in about 200 years (Fig. 6.48c). A similar simulation using an initial condition of a 2 psu salinity anomaly in the northern region gives a northern-sinking solution (Fig. 6.48d).

More recent studies looked more systematically to the structure of solutions of the MOM model. In Klinger and Marotzke (1999), a clever way is found to determine asymmetric states under equatorially symmetric conditions by varying the temperature differences over the northern and southern part independently. Several equilibria are found in a double-hemispheric configuration and for the case when the equator-to-pole temperature difference is the same in both hemispheres, the structure of equilibria again appears to arise through a (subcritical) pitchfork bifurcation. A case with a smaller temperature difference in the Northern Hemisphere leads to several different asymmetric states and the bifurcation diagram is more complicated. However, precise statements on the bifurcation structure and symmetry breaking cannot be obtained through a multiple of time-dependent simulations, since the unstable steady states will always remain hidden.

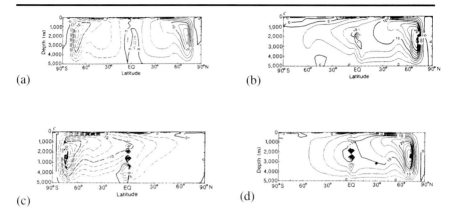

Figure 6.48. Meridional overturning streamfunction of equilibrium states under mixed boundary conditions in Bryan (1986), obtained from different initial conditions. (a) Reference state under restoring boundary conditions. (b) Circulation after a negative 1 psu salinity perturbation southward of 45°S. (c) Circulation after a positive 2 psu salinity perturbation southward of 45°S. (d) Circulation after a positive 2 psu salinity perturbation northward of 45°N.

6.7.4. Multi-basin and global models

As a next step in the model hierarchy, we proceed to models where aspects of interbasin transport are represented. Again, bifurcation diagrams are presented first (section 6.7.4.1) and these results are followed by those of low-resolution ocean models (section 6.7.4.2).

6.7.4.1 Bifurcation diagrams

As a first step towards a multi-basin set-up, the bifurcation diagram of the Atlantic configuration with an open southern channel (Fig. 6.31d), referred to as the 'open' case, was presented in Dijkstra *et al.* (2003). This is an imperfection of the equatorially double-hemispheric case discussed in the previous section (referred to as the 'closed' case). The equatorial symmetry is broken as soon as the southern channel is opened and the different branches of equilibria disconnect. Branches of solutions are again determined versus the dimensionless amplitude of the freshwater forcing σ, with the pattern as in (6.32). The main results are schematically summarized in Fig. 6.49.

The results for the 'closed' case (thermohaline forcing, no wind) are shown in Fig. 6.49a with the pitchfork bifurcation in the relevant range of σ. Note that this bifurcation is now subcritical in accordance with the results in section 6.5.1. It is the shape of the freshwater flux which determines this property of the pitchfork and a physical explanation is given in Dijkstra *et al.* (2003). The changes in bifurcation diagram between the idealized 'closed' and 'open' cases demonstrates the impact of opening the Southern Ocean (compare Fig. 6.49a and Fig. 6.49b). Even without wind, the open southern channel gives a preference for the northern-sinking solution as the southern-sinking solution becomes an isolated branch. The physical reason for this preference is that the southern passage reduces the zonal pressure gradient that can be sustained and thus decreases the meridional overturning circulation of the southern-sinking solution. In the idealized 'open' case, the

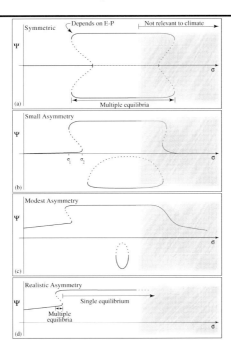

Figure 6.49. Sketch of different bifurcation diagrams relating the symmetric 'closed' case to the strongly-asymmetric, realistic Atlantic basin case. For each, a measure of the equatorially-asymmetric component of the overturning is shown as a function of the freshwater-flux parameter σ. (a) For latitudinally symmetric boundary conditions. The subcritical pitchfork requires fairly realistic latitude structure of the freshwater flux F_S as in (6.32). (b) With a slight latitudinal asymmetry introduced, such as opening the Southern Ocean. The back-to-back saddle node bifurcation diagram appears in the interval from σ_1 to σ_2. (c) With larger latitudinal asymmetry, such as including winds (but at smaller than observed strength) of the Southern Ocean. (d) Including all sources of asymmetry. The region of the multiple equilibria regime and unique regime are indicated.

effect of including wind stress, driving an ACC and associated Ekman circulation in the southern ocean, is dramatic. With increasing wind stress, the southern-sinking branch shrinks through an isola (Fig. 6.49c) into a single point and finally disappears completely (Fig. 6.49d). This indicates that the open Southern ocean with wind forcing is an important factor in the structure of THC equilibria.

The bifurcation diagram of a global ocean model with full bottom topography, realistic continental geometry and coupled to an energy balance model of the atmosphere, was presented in Weijer *et al.* (2003). This model has a 4° horizontal resolution and 12 vertical levels and after discretization a dynamical system with 280,000 degrees of freedom appears. With a diagnosed freshwater forcing F_S^1 determined from a steady state under restoring conditions to the Levitus salinity climatology (Levitus, 1994), the equilibrium state is unique. The strength of the Atlantic meridional overturning of this state is that of the point labelled A in Fig. 6.50a.

An anomalous freshwater flux is added over a local region near Newfoundland (the domain $P = (\phi, \theta) \in [300°, 336°] \times [54°, 66°]$) with an amplitude γ_p (in Sv) and the freshwater flux forcing is given by

$$F_S = F_S^1(\phi, \theta) + \gamma_p F_S^2(\phi, \theta) \qquad (6.62)$$

with $F_S^2 = 1$ in P and zero elsewhere. For this case, the bifurcation diagram as in Fig. 6.50a results. It is represented as a plot of the maximum value of the Atlantic meridional overturning streamfunction (Ψ_{atl}) of the steady solutions of the model versus the strength γ_p of the perturbation freshwater flux. For clarity, each point on the curve represents a steady global ocean circulation pattern which is a solution of the full three-dimensional primitive equations. By solving for a giant generalized eigenvalue problem, also the linear stability of each steady state is determined. Linearly stable states are indicated by a solid linestyle, whereas unstable steady states have a dashed linestyle. Three steady states exist for $\gamma_p = -0.15$ Sv (states B, C and D in Fig. 6.50a) of which B and D are stable and state C is unstable. When γ_p is smaller than -0.19 Sv, another unique regime appears with a reverse Atlantic MOC state (state E in Fig. 6.50a), the latter having strong southern sinking (Weijer *et al.*, 2003).

When there is only a single steady state for the values of the parameters chosen, say indicated by A, this state is globally stable. When this flow is subjected temporarily to a freshwater-flux perturbation, the MOC will decrease. As soon as the forcing has disappeared, however, the MOC will recover to that of the original state A. However, when there are more stable equilibria, say states B and D, the situation is different. When the anomalous forcing is applied to the state B, the state which is reached at the time the forcing is removed may evolve towards the state D. In this way, a transition takes place which can be viewed as a finite-amplitude instability (section 2.4) of state B (Dijkstra *et al.*, 2004).

The stability of the global ocean circulation was studied in a course-resolution (5° horizontally) ocean-only model by Weijer and Dijkstra (2003). Several stable modes with a a millennial time scale were found. The pattern of the buoyancy anomaly propagates over the time-mean global ocean flow. Such internal oscillatory modes may be at the origin of the Dansgaard-Oeschger oscillations, as was suggested by Sakai and Peltier (1997) and Sakai and Peltier (1999) and more recently by Timmermann *et al.* (2003).

6.7.4.2 Finite-amplitude flows

Marotzke and Willebrand (1991) used an idealized configuration of the MOM model consisting of two similar ocean basins (from 48°S-64°N), mimicking the Atlantic and Pacific basin. These basins are connected in the south by a channel

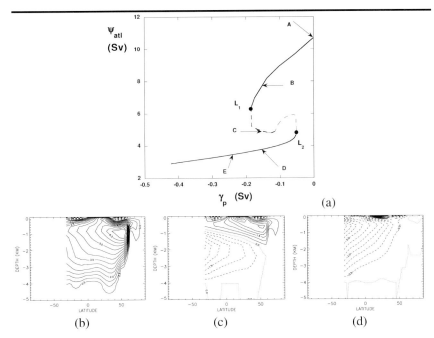

Figure 6.50.   (a) Bifurcation diagram of the global ocean circulation as a plot of the maximum meridional Atlantic meridional overturning of the steady states versus the anomalous freshwater-flux strength γ_p in Sv. (b) Pattern of the meridional overturning streamfunction in the Atlantic for the state B in (a); contour values are in Sv. (c) Similar as (b) but for state C. (d) Similar as (b) but for state D.

(from 64°S-48°S) with specified transport, representing the ACC, the latter inducing a north-south asymmetry. Similar simulations have been done as in Bryan (1986), using the solution obtained under restoring conditions (in an equatorially symmetric version of the model) as reference. The diagnosed freshwater flux of this solution is equatorially asymmetric and the model has been run under mixed boundary conditions using different initial conditions.

Under mixed boundary conditions, four different types of equilibria have been found. There is a solution with northern sinking in both basins, termed the northern-sinking solution (Fig. 6.51a), with a global meridional overturning strength of 36 Sv. By inducing negative (positive) buoyancy perturbations in the Atlantic (Pacific), a conveyor type of circulation was found; the Atlantic MOC of the solution is shown in Fig. 6.51b and the Pacific MOC in Fig. 6.51c. Also a so-called 'inverse conveyor' could be found; this is a conveyor solution with the roles of Pacific and Atlantic interchanged. Finally, a southern-sinking state could be found (Fig. 6.51d), with sinking in the south in both ocean basins. When a strong asymmetric freshwater-flux forcing is considered, the spin-up solution is unstable under mixed boundary conditions, a polar halocline catastrophe occurs, and eventually a southern-sinking solution is obtained. The diagnosed freshwater flux shows high precipitation over the Northern Hemisphere such that a northern-sinking state cannot exist in this case.

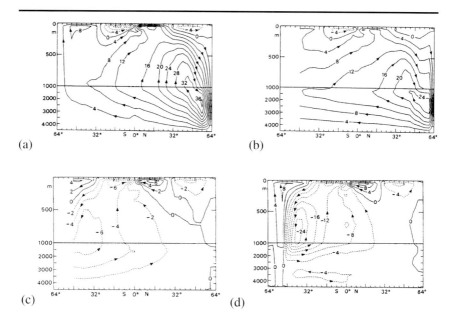

Figure 6.51.   Equilibrium states under mixed boundary conditions in Marotzke and Willebrand (1991), obtained from different initial conditions. (a) Global MOC of the northern-sinking solution. (b) Conveyor belt, Atlantic MOC. (c) Conveyor belt, Pacific MOC. (d) Global MOC of the southern-sinking solution.

A similar MOM configuration was used in Weaver *et al.* (1994) but the extent of Pacific and Atlantic was different and a fully prognostic ACC-channel flow was included. The procedure of computation was similar to that in Marotzke and Willebrand (1991) and three different solutions of the Atlantic MOC were found. A 'normal' conveyor solution (Fig. 6.52a), a weak conveyor (Fig. 6.52b) and a strong conveyor (Fig. 6.52c). For these three equilibrium flows, the Pacific circulation is quite the same. They did not find northern sinking and inverse conveyor solutions, likely because of a more limited northward extension of the Pacific basin. When stochastic noise is added to the freshwater flux, the 'normal' conveyor solution displays variability on decadal-to-centennial time scales. With increasing amplitude, this state collapses and also flushes appear, similar to those in the single-hemispheric models discussed in section 6.7.2.

    The effect of slightly more realistic surface boundary conditions for the temperature was considered in Rahmstorf (1995a) using a configuration similar to that in Marotzke and Willebrand (1991). With this new boundary condition for the heat flux, which is derived from an energy balance model, there is a stronger scale

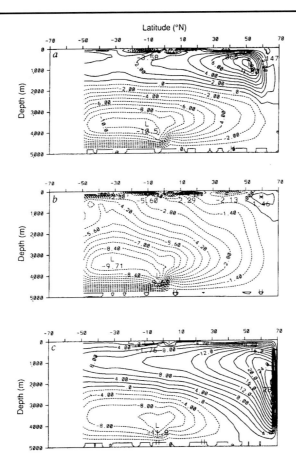

Figure 6.52. Atlantic meridional overturning streamfunction of three different equilibrium solutions in Hughes and Weaver (1994).

selectivity of the atmospheric response to SST anomalies than in the traditional restoring boundary condition. With this new boundary condition, the conveyor solution is more stable than under the more traditional conditions. In this configuration, also different equilibria associated with the convective feedback have been found (Rahmstorf, 1994). The Atlantic MOC patterns of the different states are plotted in Fig. 6.53, and differ by small details of the flow in the northern part of the basin. The actual differences come from changes in the pattern of sites where mixing occurs due to the convective adjustment procedure.

The global version of the MOM, incorporating fairly realistic geometry, has been used by England (1993) and later by Rahmstorf (1995b) to study the stability

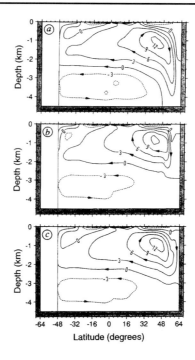

Figure 6.53. Atlantic meridional overturning streamfunction of three different 'conveyor' solutions due to different convection patterns (Rahmstorf, 1994).

of THC. The circulation was driven by a prescribed freshwater flux, wind-stress forcing and instead of a prescribed temperature, a zeroth order model of ocean-atmosphere interaction was used. The freshwater flux forcing was changed by adding slowly varying perturbations at different locations and the behavior of the THC was monitored. When the freshwater flux was added in the northern North Atlantic with a magnitude of 0.05 Sv per 1,000 years, a response as shown in Fig. 6.54 was obtained. On the vertical axis, the amount of North Atlantic Deep Water is plotted while on the horizontal axis, the freshwater forcing is shown (in Sv). With increasing freshwater forcing, the strength of the MOC decreases. At some point, the overturning rapidly decreases, while convective transitions cause the wrinkles in the curve. When the freshwater input is reversed, hysteresis occurs and it takes a negative freshwater (i.e., salt) input obtain a MOC state with strong northern sinking. A nice overview of the role of the hysteresis behavior of the THC in paleoclimate variability and its relevance to climate change is provided in Rahmstorf (2000). Transitions between the THC, involved in a so-called stochastic resonance, are suggested as a mechanism for the Dansgaard-Oeschger oscillations during the last glacial (Ganopolsky *et al.*, 2001).

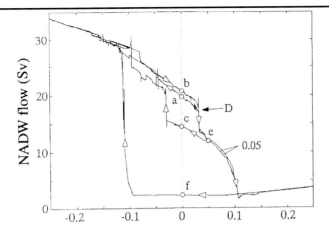

Figure 6.54. Trajectories computed with the global version of the MOM (Rahmstorf, 1995b). On the vertical axis the amount of NADW is plotted while on the horizontal axis, the strength of the freshwater forcing perturbation (in the northern North Atlantic) is plotted (in Sv).

The Large Scale Geostrophic (LSG) model was suggested by Hasselmann (1982) and subsequently developed, tested and used by Maier-Reimer *et al.* (1993). The idea behind the model is to filter the fast time scale phenomena (which are not relevant for the changes in the ocean on large space and time scales). This is done by retaining the time derivatives in the temperature and salinity equations and by diagnosing the velocity using planetary geostrophy. Inertia is completely neglected in the momentum balance as are the local accelerations. The free surface is treated prognostically, and the barotropic part of the equations is integrated with an implicit time-marching scheme to allow relatively large time steps. Prahl *et al.* (2003) use a version of the LSG model to investigate the width of the THC hysteresis versus the diapycnal mixing coefficient of heat and salt, K_V. They show that, under a constant surface freshwater flux, it decreases with decreasing K_V and eventually seems to disappear at very small K_V values.

Variability in the global LSG model has been investigated in Mikolajewicz and Maier-Reimer (1990). The model is spun-up using restoring conditions on the surface salinity and temperature and annual-mean wind stress. The barotropic streamfunction of the state obtained after 3800 year of integration is shown in Fig. 6.55a and the Atlantic MOC in Fig. 6.55b. The freshwater flux is derived from the situation at spin-up and the circulation is continued under mixed boundary conditions; it is found that the circulation is stable. Subsequently, a stochastic component is added to the freshwater flux (as seen in Fig. 6.56a) and the response of the NADW outflow at 30°S is plotted in Fig. 6.56b. Low-frequency variability is found and linked due to integration of high frequency surface forcing by the ocean (Hasselmann, 1976). However, at a time scale of about 300 years, more energy in the spectrum is found than can be expected from this mechanism. Analyses

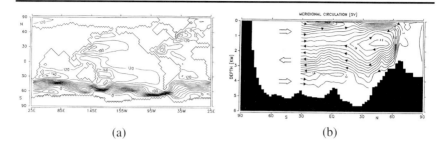

(a)   (b)

Figure 6.55.   (a) Time-mean barotropic streamfunction and (b) MOC in the LSG model (Mikola-jewicz and Maier-Reimer, 1990).

of the salinity anomalies associated with this variability shows a dipole pattern, which is advected and interacts with the Atlantic THC.

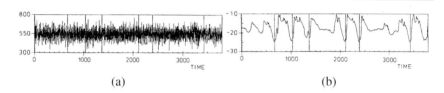

(a)   (b)

Figure 6.56.   (a) Time series (horizontal scale in years) of the net freshwater flux (in 10^3 ms^{-1}) in the Southern Ocean in the LSG model Mikolajewicz and Maier-Reimer (1990). (b) Outflow (in Sv) of NADW at $30°S$ for the same model.

6.8. Coupled ocean-atmosphere models

Low-resolution (globally) coupled ocean-atmosphere models have been used to study both the possibility of multiple flow patterns of the ocean circulation as well as its variability on multidecadal time scales. In so-called Earth system Models of Intermediate Complexity (EMICs), the atmosphere and/or ocean component are strongly simplified (Claussen *et al.*, 2002). Although the EMICSs are certainly useful for climate studies, since one can simulate the climate over longer time periods, they will not be considered here. The main reason is that it is not clear for most, if not all, of these models whether and how close their solutions approximate those of primitive equation models. Hence, unless robust qualitative information can be extracted (e.g., from a broad parameter study), not much long lasting knowledge is expected to arise from some individual trajectories of these models.

The other class of models are those which directly derive from the primitive equations through discretization. The best analysed coupled ocean-atmosphere model, which has been integrated over long times, is the GFDL-R15 model described in Manabe and Stouffer (1988), with improvements made later on (Manabe and Stouffer, 1993; Delworth *et al.*, 1993; Manabe and Stouffer, 1995). The horizontal resolution of the global ocean model is $4°$ and it has 12 vertical levels.

Technical box 6.4:
Flux correction

'Early' atmosphere models (see chapter 10 by J.T. Kiehl in Trenberth (1993)) were either coupled to a mixed layer ocean model or simply the sea surface temperature was prescribed. Similarly, many ocean models have been developed (and tuned) with prescribed wind stress and buoyancy forcing. The latter is usually of the form of so-called 'restoring' conditions

$$F_H = \lambda_T(T_S - T) \quad ; \quad \lambda_T = \frac{\rho_0 C_p H_m}{\tau_T}$$

$$F_S = \lambda_S(S_S - S) \quad ; \quad \lambda_S = \frac{H_m}{\tau_S}$$

where τ_T and τ_S are restoring time scales of the heat flux F_H and freshwater flux F_S forcing towards observed sea surface temperature T_S and sea surface salinity S_S profiles. In most models, these fluxes are applied as source terms in the upper (vertical) level of the flow domain having a thickness H_m. Suppose that both equilibrium states of a certain atmosphere model and a certain ocean model have been determined using the procedures above. Let $\overline{F}_H^o$ and $\overline{F}_S^o$ be the heat and freshwater flux computed from the ocean state at equilibrium and $\overline{F}_H^a$ and $\overline{F}_S^a$ the downward heat flux and freshwater flux computed from the atmosphere solution, then in general $\overline{F}_H^a \neq \overline{F}_H^o$ and $\overline{F}_S^a \neq \overline{F}_S^o$. When both models are coupled, the (coupled) initial state is no equilibrium state of the coupled model. Imagine that the atmospheric freshwater flux is much larger in the northern North Atlantic than that maintaining the overturning circulation in the ocean model, hence $\overline{F}_S^a > \overline{F}_S^o$. In the coupled simulation, the density of the upper ocean will decrease in time and as a consequence, the lateral buoyancy gradient decreases and convection is inhibited. If no other mechanisms compensate for the reduced lateral buoyancy gradient, the circulation may collapse completely. This troubles the simulation of a reasonable mean state. To prevent this 'drift', correction procedures have been devised to keep the modelled mean state near a realistic state usually referred to as 'flux-correction' or 'flux-adjustments' (Sausen *et al.*, 1988). In one of these procedures (Cubasch *et al.*, 1992) corrections $\overline{C}_H$ and $\overline{C}_S$ are computed from

$$\overline{C}_H = \overline{F}_H^o - \overline{F}_H^a \quad ; \quad \overline{C}_S = \overline{F}_S^o - \overline{F}_S^a$$

Within the coupled simulations, the heat flux prescribed to the ocean model is then taken as

$$F_H^o = F_H^a + \overline{C}_H \quad ; \quad F_S^o = F_S^a + \overline{C}_S$$

where F_H^a and F_S^a are the actual time dependent heat and freshwater fluxes computed from the atmosphere model. The effect of the time-independent corrections is that the equilibrium state of the uncoupled models is now by construction an equilibrium solution of the coupled model (if $F_H^a = \overline{F}_H^a$ then $F_H^o = \overline{F}_H^o$). Another method is to use the equilibrium fluxes of the atmospheric model during the construction of the equilibrium of the ocean model (during the so-called 'spin-up'). In this case,

$$F_H^o = \overline{F}_H^a + \lambda_T(T_S - T) \;\; ; \;\; F_S^o = \overline{F}_S^a + \lambda_S(S_S - S)$$

are prescribed as boundary conditions during spin-up. When the ocean reaches equilibrium, the fluxes $\overline{F}_H^o$ and $\overline{F}_S^o$ are determined from the final temperature and salinity profile. Flux adjustments are determined from the last hundred (or thousand) years of integration (which provide the solution as $(\overline{T}, \overline{S})$)

$$\overline{C}_H = \lambda_T(T_S - \overline{T}) \;\; ; \;\; \overline{C}_S = \lambda_S(S_S - \overline{S})$$

and these corrections are applied in the same way as the first procedure in the coupled simulation. Here too, when $F_H^a \approx \overline{F}_H^a$ and $F_S^a \approx \overline{F}_S^a$, a coupled mean state results which is close to the equilibrium of both uncoupled ocean and atmosphere models. Although there are several other variants of flux-correction (Weaver and Hughes, 1996) they all share the same goal: to correct systematic errors in the individual model components. The defence of the use of flux adjustments has been that the corrections are time-independent and hence should not influence the time-dependent behavior of the coupled system. That this is not true, will be shown below using two-dimensional models. Other studies, using box models (Marotzke and Stone, 1995), have indicated that stability properties can also change. With the improvement of the individual model components, also the need for flux correction will decrease and finally, the concept and practise will (must!) be totally abandoned.

Indications for the existence of multiple equilibria were already found in early versions of the GFDL-R15 model (Manabe and Stouffer, 1988). From a state of rest, both ocean and atmosphere were spun-up during a preliminary integration. The circulation obtained after 1000 'upper ocean' years had hardly any overturning and showed very low salinity North Atlantic surface waters. At this point, it was concluded that the atmosphere model could not provide the correct freshwater flux. So, one simulation was continued using restoring conditions on salinity for 1000 'upper ocean' years. From this simulation, the freshwater flux was diagnosed and used to correct the freshwater flux from the atmosphere model (see Technical Box 6.4) . Next, the simulation was continued and it ended up in an equilibrium state E_1. When the preliminary simulation was continued using the same flux correction, an equilibrium state E_2 was obtained. The two different meridional overturning streamfunction patterns of the states E_1 and E_2 are shown in Fig. 6.57; note that north is to the left. The states differ considerably in the amount of overturning, which is about 12 Sv in Fig. 6.57a, but nearly zero in Fig. 6.57b. As can be expected, the states display an enormous difference in surface temperature and salinity patterns with the weak overturning state having a smaller surface density in the northern North Atlantic.

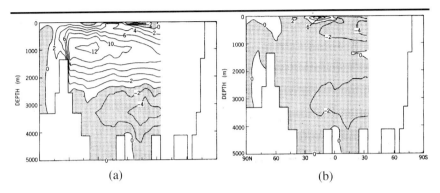

Figure 6.57. The meridional overturning streamfunction of the two equilibrium states obtained under similar forcing conditions in the coupled model of Manabe and Stouffer (1988). (a) State E_1. (b) State E_2; note that north is to the left.

The multiple equilibria structure appears to be a robust feature in coupled ocean-atmosphere models. Manabe and Stouffer (1999), for example, showed that a weak MOC state is reached as a result of a certain freshwater perturbation in case the ocean model has a small vertical mixing coefficient of heat and salt that varies with depth. They found that with a larger, but constant, vertical mixing coefficient, the MOC reduced substantially but recovered after some time. Tziperman (1997, 2000) demonstrates that the occurrence of a weak MOC ('collapsed') state depends strongly on the mean salinity field in the North Atlantic.

More recently, Vellinga *et al.* (2002) investigated the response of a coupled ocean-atmosphere model (HadCM3) to a sudden negative change in surface salinity in the northern North Atlantic. The overturning circulation is strongly reduced initially, but it recovers after about 120 years. The salt transport by the subtropical gyre appears a crucial factor in the recovery process since it is able to restore the salt deficit caused by the initial perturbation. The latter result suggests that the particular realization of the time-mean global ocean circulation in HadCM3 is in a unique regime.

The response of the THC to changes in the surface buoyancy forcing is also a central factor in the different climate change scenarios computed with current climate models that investigate the impacts of increased atmospheric greenhouse gas concentrations (IPCC, 2001). In chapter 8 of the IPCC (2001) report, a good overview can be obtained on the different coupled models, which are currently used for this task. The increased radiative forcing changes the hydrological cycle to induce more precipitation in the North Atlantic, which slows down the MOC in many, but not all, models. Apart from changes in the freshwater flux, also

changes in the ocean-atmosphere heat flux may affect the MOC. The degree of
the slowdown depends on the model configuration and on the properties of the
mean climate state (Clark *et al.*, 2002).

Variability on multidecadal time scales has been investigated in the GFDL-
R15 coupled model by Delworth *et al.* (1993). The 200 year time-mean of the
meridional overturning streamfunction (with a pattern similar to Fig. 6.57a) has
about 18 Sv overturning. The variability of the MOC is monitored by plotting the
annual-mean maximum value of the meridional overturning streamfunction; the
so-called THC index. Pronounced variability (Fig. 6.58a) is found with a period of
about 50 years. The difference in annual-mean model SST between four decades

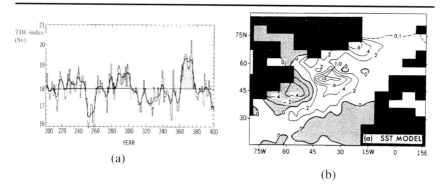

(a)   (b)

*Figure 6.58.   (a) Time series of annual-mean maximal value of the meridional overturning
streamfunction (THC index) over 200 year of integration of the GFDL coupled model. (b) Pattern obtained by Delworth* et al. *(1993) as the difference of four decades of high THC index states
and low THC index states.*

of high THC index states and low THC index states is shown in Fig. 6.58b. This
pattern has a dipole like appearance, with action centers in the western part of
the basin. It clearly resembles the pattern shown in section 1.4 obtained from
observations (Kushnir, 1994) as the difference of SST patterns of relatively cold
years (1970-1984) and relatively warm (1950-1964) years (Fig. 1.20b).

Analysis of relations between the different fields and heat and salt budgets (Delworth *et al.*, 1993) have shown that the oscillation is mainly an ocean-only phenomenon. It appears driven by density anomalies in the sinking region of the
North Atlantic combined with smaller density anomalies of different sign in the
broad upwelling region. The density variations are induced by fluctuations of the
overturning circulation and the gyre circulation.

A nice set of sensitivity experiments with the GFDL R15 model (Delworth and
Greatbatch, 2000) showed that coupled feedbacks between the ocean and atmosphere are not involved in the multidecadal variability. When the ocean model is
forced by the climatological seasonal cycle of surface fluxes, no multidecadal variability appears. When the ocean model is forced by annual-mean surface fluxes,
the multidecadal variability is similar to that of the coupled model. The surface

heat flux is shown to be the essential component of the surface fluxes causing the variability. It is concluded that the multidecadal variability can be attributed to "a damped mode in the ocean system, which is continuously excited by low-frequency atmospheric forcing". Although this is an important result, it provides neither a mechanistic explanation for the physical processes setting the time scale nor for its spatial pattern.

An analysis of the multidecadal variability in a 900-year simulation (Delworth *et al.*, 2002) with the GFDL-R30 model was performed in Dijkstra *et al.* (2005). The main difference between this version the GFDL-R15 model is the higher spatial resolution in both the atmospheric and oceanic components. The horizontal resolution of the ocean model in the GFDL-R30 model is $1.875°$ in longitude by $2.25°$ in latitude, with 18 unevenly spaced levels in the vertical. As in previous versions of the GFDL-climate model, 'flux adjustments' are used.

The simulated maximum time-mean meridional overturning of 25 Sv is quite large compared to observations; the sinking mainly occurs between $60°N$ and $65°N$. An M-SSA analysis reveals a statistical mode of variability of 44 years in the annual-mean Atlantic meridional overturning streamfunction. The maximum of the meridional overturning streamfunction anomaly, ψ_M, is plotted in Fig. 6.59a over a 200-year interval. The same time scale of variability is found in potential temperature, salinity, horizontal velocities and potential density at several vertical model levels. The anomaly patterns of the meridional overturning streamfunction ψ of this statistical mode are shown in Fig. 6.59b-e for four phases during the oscillation. The starting time was chosen to be year 600, when the amplitude ψ_M of the statistical mode is near maximum, and each subsequent panel is 6 years later. Together, the plots show nearly half of the cycle of the oscillation and the other half-cycle is similar but with anomalies of reversed sign. In Fig. 6.59b, the overturning is about 2.5 Sv stronger than its average value, with a single-cell anomaly pattern. After 6 years, the overturning anomaly has slightly decreased (Fig. 6.59c) while keeping the same pattern. The anomaly is nearly zero after about 12 years (Fig. 6.59d) and after 18 years in the oscillation, the overturning is weaker than normal (Fig. 6.59e), again with the same anomaly pattern.

In Fig. 6.60, the anomaly patterns of the potential density and the horizontal velocity field (at 680 m depth) of the statistical mode are plotted for four phases during the oscillation. There is a clear westward propagation of the anomalies at this depth over the 44-year oscillation. The horizontal velocity patterns in Fig. 6.60 are caused by the density anomalies. At year 612, there is an anomalous anti-cyclonic circulation east of Newfoundland, mainly induced by the positive temperature anomaly (i.e., the negative density anomaly in Fig. 6.60c). This anomalous circulation is present in the upper 1000 m of the central North Atlantic. The intensity of this circulation lags ψ_M by approximately 10 years (Fig. 6.60b).

6.9. Synthesis

In this section, an attempt is made to interpret the behavior of the solutions of ocean-only and coupled ocean-atmosphere models, as presented in the previous

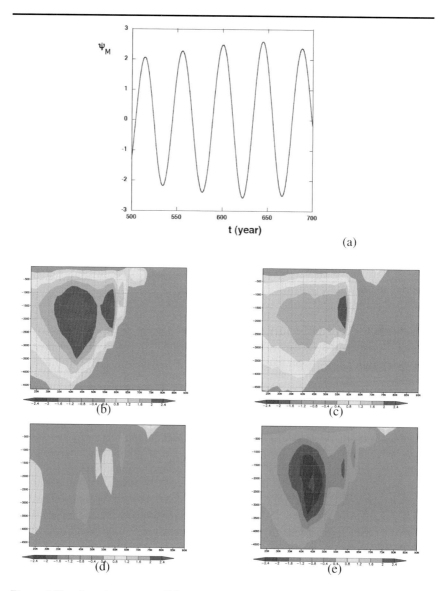

Figure 6.59.   (in color on page 525).   Properties of the statistical mode, having a 44-year timescale, in the GFDL-R30 model. (a) Maximum of the meridional overturning streamfunction anomaly (Sv). (b-e) Patterns of meridional overturning streamfunction anomaly (Sv) in the North Atlantic region. The patterns are shown at a 6-yearly interval, starting in model year 600, over about one half-cycle of the oscillation; the other half-cycle is similar but with anomalies of reversed sign.

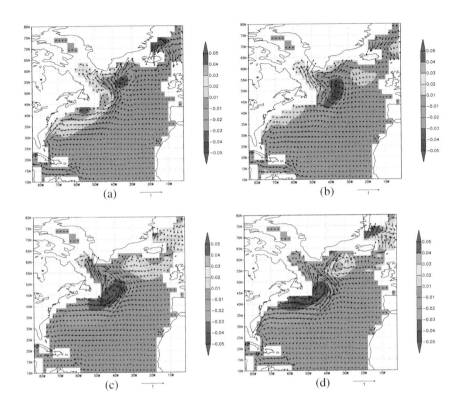

Figure 6.60. (in color on page 526). Properties of the statistical mode, having a 44-year timescale, in the GFDL-R30 model. Potential density (kg m⁻³) and horizontal velocity (cm s⁻¹) anomalies at 680 m depth in the North Atlantic region. The patterns are shown at a 6-yearly interval, starting in model year 600, over about one half-cycle of the oscillation; the other half-cycle is similar but with anomalies of reversed sign.

sections, with help of the bifurcation diagrams presented in the sections 6.3 to 6.7 for a hierarchy of simpler models. We focus on two specific questions:

(i) Can the transition between a pattern of a strong and a weak Atlantic meridional overturning, as suggested in the hysteresis behavior in Fig. 6.54 and the results in Manabe and Stouffer (1988), be understood in terms of a trajectory between two stable steady states and, if so, between which states?

(ii) Can the Atlantic variability on decadal-to-multidecadal time scales, in particular the pattern and time scale of the Atlantic Multidecadal Oscillation (AMO, see section 1.3.2) be understood from the excitation of an internal mode of variability of the Atlantic THC?

6.9.1.   Different mean thermohaline flows?

Having computed bifurcation diagrams for a hierarchy of models of the THC, quite a definite answer can be given to question (i) above: *the hysteresis behavior of the THC arises through a transition between the pole-to-pole NPP and SPP solutions which is possible once the freshwater flux becomes strongly equatorially asymmetric.*

Just as for the wind-driven circulation in chapter 5, where one could take either the single- or double gyre flow as a starting point, here one can take either the SH or DH configuration. Again, the situation with the highest symmetry, the DH configuration, is the most easy starting point. In the lowest level of the model hierarchy, this would be the Welander three-box model (Welander, 1986; Thual and McWilliams, 1992) as discussed in section 3.1. Here, as in two-dimensional Boussinesq models (Cessi and Young, 1992; Quon and Ghil, 1992; Dijkstra and Molemaker, 1997), in zonally averaged models (Wright and Stocker, 1991; Vellinga, 1996) as well as in 3D primitive equation models (Weijer and Dijkstra, 2001), a symmetry-breaking pitchfork bifurcation exists. The pole-to-pole NPP and SPP solutions arise through symmetry breaking and the important characteristic of these solutions is that the salt-advection feedback is active to maintain their circulation. Hence, the overturning increases with increasing strength of the equatorially-symmetric freshwater flux (Dijkstra and Molemaker, 1997; Weijer and Dijkstra, 2001). The character of the pitchfork bifurcation, being either subcritical or supercritical, depends on the shape of the freshwater flux. With a 'realistic' representation of this flux, which takes the equatorial precipitation regions into account, the pitchfork bifurcation is subcritical (Klinger and Marotzke, 1999; Dijkstra *et al.*, 2003).

As soon as asymmetry is present, the pitchfork bifurcation ceases to exist and isolated branches appear. The degree of asymmetry of forcing and boundary conditions (e.g., continental geometry) subsequently becomes central to the location of the different branches of steady states in parameter space. The dominant asymmetries strongly favor the northern-sinking pole-to-pole (NPP) solution for the following reasons:

- The slightly asymmetric surface freshwater flux favors the northern-sinking (NPP) solution, because the salinity is relatively larger in the northern part of the basin (Dijkstra and Neelin, 2000).

- The asymmetric continental geometry of the Atlantic also favors the NPP solution, through the asymmetric air-sea interaction (Dijkstra and Neelin, 2000).

- The lateral salt and heat input into the Atlantic basin due to interbasin exchanges also favor the NPP solution (Weijer *et al.*, 1999, 2001) because of the relative decrease in surface density in the southern part of the basin.

■ The presence of the opening in the Southern Ocean also strongly favors the NPP solution and this effect is exacerbated by the presence of the asymmetric wind-stress field (Dijkstra *et al.*, 2003). This is due to the lack of a zonal pressure gradient, needed to sustain a southward geostrophic flow in the Southern Ocean, and the northward Ekman transport.

The fact that all asymmetries are qualitatively preferring the NPP state may be a good reason why we see this pattern in the observations today.

Now look at the effect of a regional anomalous freshwater flux, say near New Foundland, with an amplitude represented by γ_p. At large γ_p, the asymmetry due to the additional freshwater flux is so strong that it opposes and overrules the preference of the northern-sinking solution due to the southern channel and winds. Therefore, when the variations of the equilibria in γ_p are considered, one smoothly connects the NPP branch with the SPP branch, creating a large hysteresis between northern sinking and southern-sinking solutions. This appears like a bifurcation diagram in a Stommel two-box model (with two back-to-back saddle-node bifurcations), but the situation is essentially different because the salt-advection feedback maintains both NPP and SPP states (see Dijkstra and Weijer (2003) for a more detailed discussion).

6.9.2. Temporal variability through internal modes?

The results of the analysis of the GFDL-R30 climate model in section 6.8 show the existence of a dominant statistical oscillatory pattern of variability with a time scale of 44 years. From observations, the only characteristics which are available of the AMO are its time scale (~ 50 year) and its spatial pattern (Fig. 1.20) from Kushnir (1994).

To connect the finite-amplitude results in Te Raa *et al.* (2004) to the analysis of the GFDL-R30 climate model, additional computations of variability with the MOM were presented in Dijkstra *et al.* (2005) for the SH configuration with continental geometry. Nearly all parameters (except A_H and K_H) are chosen similar as in Table 6.3 and only thermal (no wind and freshwater) forcing is used (as in section 6.7.1). The resolution used is $2° \times 2°$ horizontally and 16 layers in the vertical; values of $A_H = 1.6 \times 10^5$ m^{2}s^{-1} and $K_H = 700$ m^{2}s^{-1} are used for the mixing coefficients. After the spin-up under restoring conditions, the surface heat flux is diagnosed and thereafter prescribed as a flux condition. An oscillation with a period of about 45 year develops in the flow, as can be seen from a plot of the maximum of the meridional overturning streamfunction (Fig. 6.61a). A plot of the zonal and meridional overturning anomalies (Fig. 6.61b) shows a signature of the characteristic phase difference between the zonal and meridional overturning associated with the multidecadal mode.

To establish a connection between these results and those of the GFDL-30 model (and observations), one has to realize that many of the idealized models have been using prescribed-flux conditions and hence the atmospheric damping of SST anomalies is absent. In a coupled model such as the GFDL-30 model, as well as in observations, there is substantial atmospheric damping. This damping

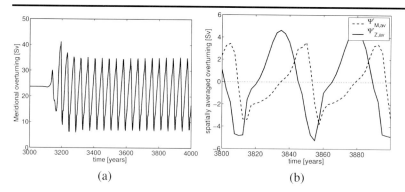

Figure 6.61.   (a) Periodic orbit in the MOM SH configuration with continental boundaries and a resolution of $2° \times 2°$ *and 16 layers;* $K_H = 700\ m^2 s^{-1}$ *and* $A_H = 1.6 \times 10^5\ m^2 s^{-1}$. *At* $t = 3000$ *yr, the boundary condition was changed to a prescribed heat flux. (b) Spatially-averaged zonal and meridional overturning anomalies.*

will decrease the amplitude of the multidecadal signal in Fig. 6.61a. In fact, it will decrease the growth factor of the multidecadal mode in the idealized configurations (Te Raa and Dijkstra, 2003b) and it may actually stabilize the THC with respect to this mode. Stabilization can also occur due to bottom topography (Winton, 1997; Huck *et al.*, 2001), which can also be responsible for substantial changes in the mean THC flow (Spall and Pickart, 2001). Even if the multidecadal mode is damped, still the results are relevant because in the coupled model there is a strong component of 'noise'. Under this 'noise', the multidecadal mode can still be excited; in fact, this is a case of a stochastic Hopf bifurcation (Gardiner, 2002). The properties of such as noise-driven signal were studied in Griffies and Tziperman (1995) and Rivin and Tziperman (1997) using simple box models.

For the oscillation in Fig. 6.61, the SST difference field between Atlantic warm (corresponding to maximum MOC) and Atlantic cold (corresponding to minimum MOC) states is plotted in Fig. 6.62a. The same field is plotted in Fig. 6.62b for the 44-year statistical mode in the GFDL-R30 model. Considering also the corresponding field from observations (Fig. 1.20b) there is a striking qualitative correspondence between the patterns. More specifically, with respect to: (i) the secondary maximum in the warm anomaly near 70°W, 30°N, (ii) the area of negative anomalies near the North American coast, and (iii) the 90-degrees anti-clockwise rotated V-pattern of the positive anomalies. There are of course also discrepancies, in particular in the eastern basin where the pattern of the observations shows positive anomalies and the GFDL-R30 pattern has slightly negative anomalies. In addition, both amplitudes of the MOM and GFDL-30 pattern are a factor 2 to 5 larger than those in the observations.

The results provide support for the following answer to question (ii) above: *the Atlantic Multidecadal Oscillation is caused by the multidecadal mode of variability of the North Atlantic THC.* The time scale of the variability is set by the

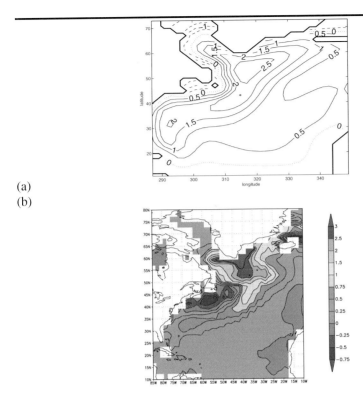

(a)
(b)

Figure 6.62.   *(in color on page 527). (a) Difference in SST fields between maximum and minimum strength of the MOC in the MOM3.1 model. (b) As in (a), but for the statistical 44-year mode in the GFDL-R30 model.*

basin crossing time of density anomalies in a background density field according to Fig. 6.35. The pattern of variability arises through a deformation, due to the continental boundaries, of the pattern of the multidecadal mode found in the SH configuration. As there is no sustained multidecadal oscillation in the GFDL-R30 model, the multidecadal mode is likely to be damped in this model. In the latter case, it is the atmospheric noise that is able to excite the spatial pattern of this mode to sufficient amplitude that it is detected as a statistical mode by a technique such as M-SSA.

This view is consistent with many, if not all, model results on multidecadal variability. For example, it was discussed at length in Te Raa and Dijkstra (2002) that the westward propagation of the temperature anomalies within an oscillation cycle is seen in most sustained multidecadal oscillations in ocean-only models (Greatbatch and Zhang, 1995; Chen and Ghil, 1995; Huck *et al.*, 1999). The view is also not in contradiction with the results in Delworth and Greatbatch (2000). Their CLIM simulation (the ocean model forced with only climatological atmospheric fluxes) nicely demonstrates that the flow regime is not supercritical, i.e., there is no sustained oscillation. The TOTAL (ocean forced by total fluxes of the coupled model run) and RANDOM (ocean forced by only the annual-mean atmospheric fluxes chosen at random) simulations demonstrate that coupled feedbacks and also atmospheric noise on time scales < 1 year are not needed to generate the multidecadal variability.

The simulations HEAT_LP (only low-frequency part (> 20 year) of the atmospheric fluxes) and HEAT_HP (only high-frequency part (< 20 year) of the atmospheric fluxes) seem at first sight puzzling. As the multidecadal variability has a much smaller amplitude under HEAT_HP, it looks like the low-frequency component of the atmospheric variability is driving the multidecadal variability as is also the interpretation in Delworth and Greatbatch (2000). However, as shown in Roulston and Neelin (2000), the result that high-frequency noise is not able to excite the multidecadal signal is due to the fact the underlying ocean model is only weakly nonlinear. In this case, there is no effective pathway to channel energy from the smaller to the larger scales. Consequently, the low-frequency part of the noise forcing is essential to excite the multidecadal mode. This does, however, not mean that the variability is 'driven' by the low-frequency part of the atmospheric forcing. In reality, the ocean is highly nonlinear, there is a mechanism for channeling energy over a range of scales and hence high-frequency atmospheric noise is capable of exciting the multidecadal mode.

To summarize, the dynamical systems analysis of a hierarchy of models has lead to an interpretation framework for results of a large number of ocean-only and coupled ocean-atmosphere models. In addition, it has lead to detailed physical mechanisms of climate variability. The idea to isolate these mechanisms in a most elementary context and then trace their characteristics through the model hierarchy provides an approach that can also be followed with respect to other phenomena of climate variability. In fact, a similar approach has already been shown to be very successful in determining the physical mechanism of propagation of anomalies in ENSO as we will see in the next chapter.

6.10. Exercises on Chapter 6

(E6.1) *Diagnostic model*

Since the momentum equations are linear when inertia is neglected, approximate expressions of the perturbation geostrophic velocities can be obtained in terms of a prescribed temperature (or more general buoyancy) field $\bar{T}$ (in the absence of wind and freshwater forcing). Let the domain be $(\phi, \theta, z) \in [\phi_w, \phi_e] \times [\theta_s, \theta_n] \times [-1, 0]$. Away from boundaries, the velocity field is pure geostrophic and horizontal and vertical mixing of momentum can be neglected. In that case, consider the set of dimensionless diagnostic equations for the velocities and pressure as

$$-v \sin \theta = -\frac{1}{\cos \theta}\frac{\partial p}{\partial \phi}$$

$$u \sin \theta = -\frac{\partial p}{\partial \theta}$$

$$\frac{\partial p}{\partial z} = Ra\,\bar{T}$$

$$\cos \theta \frac{\partial w}{\partial z} + \frac{\partial u}{\partial \phi} + \frac{\partial(v \cos \theta)}{\partial \theta} = 0$$

a. Determine the pressure explicitly from the temperature field and show that the result is

$$p = Ra\left[\int_{-1}^{z}\bar{T}\,dz' - \int_{-1}^{0}\left(\int_{-1}^{z}\bar{T}\,dz'\right)dz\right]$$

b. Show that for the meridional overturning streamfunction ψ, it follows that

$$\psi(\theta, z, t) = -\frac{Ra}{\sin \theta}\int_{-1}^{z}\left[\int_{\phi_w}^{\phi_e}f(\phi, \theta, z', t)d\phi\right]dz'$$

$$f(\phi, \theta, z, t) = \int_{-1}^{z}\frac{\partial \bar{T}}{\partial \phi}\,dz' - \int_{-1}^{0}\left(\int_{-1}^{z}\frac{\partial \bar{T}}{\partial \phi}\,dz'\right)dz$$

c. Calculate the horizontal velocities and meridional overturning streamfunction for the case

$$\bar{T} = e^{\kappa z}\,\cos \pi\left(\frac{\phi - \phi_w}{\phi_e - \phi_w}\right)$$

Further reading: Te Raa and Dijkstra (2002).

(E6.2) *Ekman horizontal boundary layers*

Consider the steady linearized momentum equations in Cartesian coordinates on an f-plane, i.e.,

$$-fv = -\frac{1}{\rho_0}\frac{\partial p}{\partial x} + A_H \nabla^2 u + A_V \frac{\partial^2 u}{\partial z^2}$$

$$fv = -\frac{1}{\rho_0}\frac{\partial p}{\partial y} + A_H \nabla^2 v + A_V \frac{\partial^2 v}{\partial z^2}$$

$$\frac{\partial p}{\partial z} = -g\rho$$

$$\frac{\partial u}{\partial x} + \frac{\partial v}{\partial y} + \frac{\partial w}{\partial z} = 0$$

Let the density ρ be given as $\rho(y) = \alpha y$, with $\alpha > 0$ such that higher-density water is at northern latitudes. The flow is within a basin of dimensions $L \times L \times D$ and no-slip conditions are imposed on the lateral boundaries.

a. Where do you expect lateral boundary layers to appear in the flow?

b. Show that the thickness of the boundary layer at the eastern boundary scales with $\sqrt{A_H/f}$.

c. What will happen in numerical ocean models when this boundary layer is not resolved ?

Further reading: Winton (1996).

(E6.3) *Surface boundary conditions*

Consider the salt balance for a surface box at the ocean-atmosphere interface (Fig. 6.63). The salinity of the ocean changes through advection and differences between evaporation (E) and precipitation (P). The units of the latter fluxes are $m^3 s^{-1}$ per m^2 and hence ms^{-1}.

Consider the two-dimensional salt balance in the surface box in rectangular coordinates, i.e.,

$$\frac{\partial S}{\partial t} + \frac{\partial (uS)}{\partial x} + \frac{\partial (wS)}{\partial z} = K_V \frac{\partial^2 S}{\partial z^2}$$

where horizontal mixing is neglected and K_V is the vertical mixing coefficient.

a. Show that the discrete salinity balance over the box can be written as

$$\frac{\partial S}{\partial t} + \frac{1}{\Delta x}[(uS)^+ - (uS)^-] + \frac{1}{\Delta z}[(wS)^+ - (wS)^-] = K_V[(\frac{\partial S}{\partial z})^+ - (\frac{\partial S}{\partial z})^-]$$

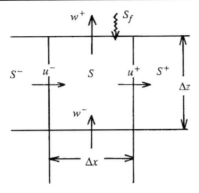

Figure 6.63.   A finite-difference grid box in the $x - z$-plane.

The virtual salt flux S_f is formally defined as

$$S_f = (E - P)S$$

b. Argue why the S in the equation for S_f should be substituted by a mean reference salinity S_0. (*Hint: consider the integral of S_f over the surface*).

In the mixed boundary formulation (section 6.2), one puts $w^+ = 0$ and $S_f = (E - P)S_0$.

c. Determine the discrete salinity balance over the box in the mixed boundary condition formulation.

In the natural boundary formulation, one puts $w^+ = E - P$ and

$$(wS)^+ - K_V(\frac{\partial S}{\partial z})^+ = 0$$

d. Determine the discrete salinity balance over the box in the natural boundary condition formulation.

e. Argue why natural boundary conditions represent the physics of the air-sea exchange in a better way.

Further reading: Huang (1993).

(E6.4) *Diffusive solution*

Assume that the dominant balance in the two-dimensional temperature equation is between meridional and vertical diffusion of heat. In this case, analytical progress can be made when the temperature at the top is assumed to be prescribed. With a restoring surface forcing $T_S = 1 + \cos \pi y$, on the domain $(y, z) \in [0, 1] \times [0, 1]$, the temperature is determined from the following problem

$$K_H \frac{\partial^2 T}{\partial y^2} + K_V \frac{\partial^2 T}{\partial z^2} = 0$$

$$y = 0, 1 : \frac{\partial T}{\partial y} = 0$$

$$z = 0 : \frac{\partial T}{\partial z} = 0 \quad ; \quad z = 1 : T = 1 + \cos \pi y$$

a. Show that the solution of this problem is given by

$$T(y, z) = 1 + \cos \pi y \frac{\cosh \pi \gamma z}{\cosh \pi \gamma}$$

with $\gamma = \sqrt{K_H/K_V}$.

b. Determine the diagnostic geostrophic zonal velocity arising through this temperature field.

Further reading: Wright and Stocker (1991).

(E6.5) *Scaling*

One of the important issues in the dynamics of the meridional overturning circulation concerns the scaling of its maximum value ψ_M with the meridional surface density difference $\Delta \rho$.

a. Scale the horizontal velocity U according to the geostrophic balance and use the hydrostatic pressure scale Δp to derive that

$$U = \frac{Dg\Delta\rho}{fL}$$

where f is the local Coriolis parameter, g is the gravitational acceleration, and D is a measure of the depth of the thermocline (large vertical gradient in temperature).

b. Use the advective-diffusive balance in the thermocline

$$w\frac{\partial \rho}{\partial z} = K_V \frac{\partial^2 \rho}{\partial z^2}$$

to show that

$$U = \frac{K_V L}{D^2}$$

c. Show that in this case $\psi_M \approx (\Delta\rho)^{\frac{1}{3}}$.

Further reading: Rahmstorf (1996).

(P6.1) *Bifurcation analysis of a Cahn-Hilliard type equation*

As was shown in Technical Box 6.1, in the large-aspect ratio case, the equilibria of the diffusive thermohaline flow can be modeled by the boundary value problem

$$\delta^2\chi'' = r\sin x + \mu^2\chi(\chi + \sin x)^2 + \chi$$
$$\chi(-\pi) = \chi(\pi) = 0$$

for a function $\chi = \chi(x)$.

a. Write this equation as a first order problem.

The solutions to this boundary value can be numerically computed with the AUTO software (see the beginning of chapter 4 how to obtain the software). Be aware that the boundary value option (IPS = 4) has been turned on in AUTO and that the appropriate boundary conditions are defined.

b. We want to determine the bifurcation diagram for $\delta = 0.1$ and $\mu^2 = 7$ using r as a parameter. Design a continuation path to compute this bifurcation diagram efficiently.

c. Determine this bifurcation diagram numerically.

d. Follow the paths of both saddle-node bifurcations to smaller values of δ and locate the value of the cusp bifurcation for $\delta = 0.05$ (see also Fig. 6.21).

Further reading: Cessi and Young (1992) and Dijkstra and Molemaker (1997).

(P6.2) *Critical thresholds*

When stable multiple equilibria are present, we would like to determine how we have to perturb one state to induce a transition to the other state. We consider this problem here for the Stommel (1961) two-box model as discussed in section 3.1. As shown there, the dimensionless equations become

$$\frac{dT}{dt} = \eta_1 - T(1 + M(T - S))$$
$$\frac{dS}{dt} = \eta_2 - S(\eta_3 + M(T - S))$$

where $T = T_e - T_p$, $S = S_e - S_p$ are the scaled temperature and salinity differences between the equatorial and polar box and $\Psi = T - S$ is the dimensionless flow rate. The function M indicates the modulus function which is smoothed as in (3.1.3). Three parameters appear in these equations above: the parameter η_1 measures the strength of the thermal forcing, η_2 that of the freshwater forcing and η_3 is the ratio of the relaxation times of temperature and salinity.

a. Implement this box model into AUTO (see the beginning of chapter 4 how to obtain the software) and compute the bifurcation diagram for $\eta_1 = 3.0$, $\eta_3 = 0.4$ with η_2 as control parameter.

b. Compute also the bifurcation diagrams in η_2 for $\eta_3 = 0.3$ and $\eta_3 = 0.5$.

For $\eta_3 = 0.4$ and $\eta_2 = 1.3$, there are two stable steady states. The state A_1 is called a thermally-driven state (or TH state) with $\Psi > 0$ and the state A_2 is called a salinity-driven state (or SA state) with $\Psi < 0$.

c. What happens to the system when it is initially in state A_1 for $\eta_3 = 0.3$ and suddenly η_3 is changed to $\eta_3 = 0.5$? What if it is changed suddenly to $\eta_3 = 0.3$? In which case will the meridional overturning drastically change?

Now, look at the time-dependent problem with state A_1 as initial conditions. The value of η_3 is changed over the time interval $t \in [0, t_m]$ to 0.3 and after t_m it is switched back to 0.4.

d. Which qualitatively different trajectories, depending on t_m, are possible?

e. In the previous problem, how can you determine the critical value of t_m that is needed to induce a transition from the state A_1 (with $\Psi > 0$) to one with $\Psi < 0$?

Further reading: Dijkstra et al. (2004).

Chapter 7

THE DYNAMICS AND PHYSICS OF ENSO

Dancing on equatorial waves.
Danza Española No. 5, E. Granados.

In chapter 1, the El Niño /Southern Oscillation (ENSO) phenomenon was introduced as an interannual climate variation in the Tropical Pacific. Sea surface temperature (SST) anomalies of up to a few degrees occur in the eastern part of the Pacific (El Niño /La Niña) and are accompanied by a weakening and strengthening of the trade winds (Southern Oscillation). ENSO is the most prominent example of interannual variability in the climate system. Because it evolves on relatively short time scales, it is one of the best studied climate phenomena, both observational and theoretical. ENSO is caused by processes both in the tropical ocean and atmosphere with a central role for the SST. The observed spatial structures involved, their temporal development and the relationship between the oceanic and atmospheric variables are now fairly well known (Rasmusson and Carpenter, 1982; Wallace *et al.*, 1998). A historical overview of key research leading to this knowledge is given in Philander (1990) and Wallace *et al.* (1998).

In this chapter, focus is on the dynamical understanding of El Niño which has been obtained over the past decades through mechanistic studies with intermediate complexity coupled ocean-atmosphere models (ICMs). Because bifurcation analysis has not been performed on a hierarchy of models such as that for the ocean circulation, this chapter has a slightly different setup than the previous two chapters. We will mainly focus on one class of ICMs and dynamical systems methods will help interpret its solutions.

The chapter starts with a short description of the phenomena under study in section 7.1, which ends with the central questions posed. Modeling of the equatorial ocean is subject of section 7.2 where the relevant equatorial waves and adjustment processes are discussed. In section 7.3, the physics of coupled processes between the equatorial ocean and atmosphere is addressed while simultaneously the additional ingredients for an ICM are introduced. An overview of results of the first ICM, which was able to successfully simulate ENSO-like behavior (Zebiak and Cane, 1987) is given in section 7.4. In the next section 7.5, the development is sketched towards a conceptual framework, the delayed oscillator, to understand the results in this ICM. It is here that a dynamical systems approach turns out to be useful. The involvement of coupled processes in the annual-mean state and its consequences for ENSO are subject of section 7.6 and 7.7, while the interaction of the seasonal cycle and ENSO is dealt with in section 7.8. In the section 7.9, ENSO variability is addressed in a broader context together with an overview of results from Coupled General Circulation Models (CGCMs). A synthesis of the results in this chapter follows in section 7.10.

7.1. Basic Phenomena

The spatial and temporal structures of the annual-mean state, the seasonal cycle and of ENSO are described below, but only those features which easily relate to those computed from the ICMs introduced later on. Hence, this description has a very limited scope and other sources should be consulted (for example Philander (1990), Horel (1982) and Wallace *et al.* (1998)) for a more complete view of the phenomena involved.

7.1.1. The annual-mean state

The annual-mean wind stress over the Tropics is shown in Fig. 7.1. Clearly, the trade winds over the Pacific are mainly zonal and directed from east to west. The maximum amplitude of the zonal wind stress is about 0.2 Pa. At the equator, there is a small component of the meridional wind stress with an amplitude of about 0.05 Pa. The structure of the winds is not symmetric with respect to the equator since the convergence of the South Pacific trade winds and North Pacific trade winds is located slightly north of the equator. This is associated with the fact that the Intertropical Convergence Zone (ITCZ) is on average located north of the equator.

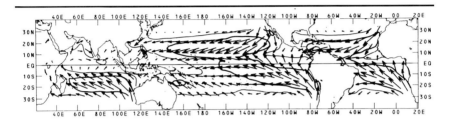

Figure 7.1.   Annual-mean wind stress in the Tropical Pacific (from Ropelewski and Halpert (1987)). The maximum amplitude of the zonal component of the wind stress over the Pacific is about 0.2 Pa.

The annual-mean SST over the same region (as plotted in Fig. 7.2) indicates that there is a strong asymmetry between the relatively warm western part of the basin (the so-called warm pool) and the cooler eastern basin, (the so-called cold tongue). The thick curve is the 25°C isotherm, which indicates that the cold tongue has a mean temperature of about 24°C while the warm pool temperature is about 29°C . The equatorial zonal temperature difference over the basin is about 5°C. There is also a north-south asymmetry about the equator, with more warm water situated north of the equator.

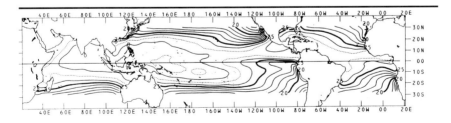

Figure 7.2.   Annual-mean SST in the Tropical Pacific (from Ropelewski and Halpert (1987)). The thick contour line is the 25°C isotherm; contour levels are in degrees.

Physical processes up to a few hundred meters depth in the ocean play an important role in the Pacific climate system. During the Tropical Ocean Global Atmosphere (TOGA) program (1985-1995), a whole array of measurement devices has been set-up in the Tropical Pacific (McPhaden and coauthors, 1998). Hence, only over the last decade, the temperature at these depths have been measured routinely[1] through the TAO-buoy network (see http://www.pmel.noaa.gov/toga-tao). In Fig. 7.3, a longitude-depth section of the equatorial temperature (from 2°S to 2°N) is shown for November 1996. This situation is close to annual-mean conditions as used by the TAO-project Office[2] for the Tropical Pacific. At each longitude, there is a strong vertical gradient in the temperature distribution; this transition region is the equatorial thermocline. The depth of the 20°C isotherm is a reasonable measure of the location of the thermocline. This depth changes from about 200 m at the western part of the basin to about 50 m at the eastern boundary. Hence, in the annual-mean state, the colder water is much closer to the surface in the east than in the west.

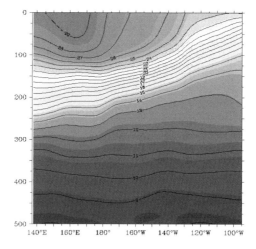

Figure 7.3. (in color on page 528). Depth-longitude section of the near-equatorial temperature in the Pacific monthly averaged over November 1996. This situation is close to annual-mean conditions for the Tropical Pacific Kessler and McCreary (1995). The crosses in the figure indicate the measurement positions of the TAO-buoys.

From this brief description of the spatial patterns of the annual-mean state, the most important feature is the strong zonal asymmetry in the equatorial ocean,

---

[1] The figures which follow have all been plotted through the graphics software and data made available through the TAO realtime data-access site at http://www.pmel.noaa.gov/toga-tao/realtime.html.
[2] for further details, see Kessler and McCreary (1995) and http://www.pmel.noaa.gov/toga-tao/clim.html.

shown here for both the SST field and the thermocline field. In the sections 7.6 and 7.7, the physics of this asymmetry is described and explained using results of ICMs.

7.1.2. The seasonal cycle

Within a calendar year, the trade winds, the pattern of the SST and that of the thermocline all change as the Sun moves two times a year over the equator (in March and in September). Fig. 7.4 provides snapshots of the seasonal cycle of SST and surface winds. In April (Fig. 7.4a), the east-west contrast in SST along the equator is minimal, which coincides with relatively weak trade winds. In October (Fig. 7.4b), the equatorial zonal SST gradient is strongest, which coincides with stronger trade winds.

Maximum precipitation areas are associated with the position of the ITCZ where the trade winds converge (Fig. 7.4). In Augustus/September, this point is farthest north and the strength of the winds is maximal (Horel, 1982). In March/April, the convergence is located closest to the equator and the trade winds are relatively weak. Connected to this is the migration of the ITCZ which, for example at 120°W, moves from about 2°N in March to about 12°N in September. A strong coupling also exists between the movements of the ITCZ and those of other convection zones in the Tropics.

The equatorial pattern and amplitude of the SST and zonal wind field are shown for 1993 in Fig. 7.5; note that time increases downwards. Within the seasonal cycle of equatorial SST in the Pacific (Fig. 7.5, left panel), the Pacific cold tongue is coldest during September/October and warmest during March/April. Peculiar is the fact that this cold tongue has a strong annual variation, whereas the forcing is semi-annual (Mitchell and Wallace, 1992). In the western part of the basin (the warm pool) a weak semi-annual cycle with warmest temperatures in February/March and August/September can be observed. Maximum amplitudes of the easterly zonal wind (Fig. 7.5, right panel) occur around November/December near about 140°W and the equatorial trade winds are weakest from March to May. In both fields, one notices the slight westward propagation of the signal in the second half of the year.

Much more detail on the description of the seasonal cycle is given in Horel (1982), Philander (1990) and Yu and McPhaden (1999). Although in section 7.8, some attention is given on the physics of the seasonal variations, this discussion will be quite superficial. Reason is that different ICMs are needed to understand the seasonal cycle than those by which ENSO can be understood. Moreover, the Pacific seasonal cycle is probably not as well understood as ENSO.

7.1.3. Interannual variability

What makes El Niño unique among other interesting phenomena of natural climate variability is that it has both a well-defined pattern in space and a relatively well-defined time scale. No sophisticated statistical tools are needed to isolate the El Niño pattern in SST variability. For example, consider the SST anomaly for

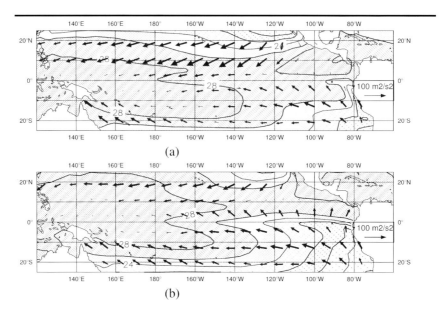

(a)

(b)

Figure 7.4.   (in color on page 529).   SST and wind-stress climatologies for (a) April and (b) October. The contours give the 1961–1990 SST climatology (contour interval $2°C$) from NCEP (National Centers for Environmental Prediction), the arrows the 1961-1992 pseudo-wind stress climatology (in $m^2 s^{-2}$) from FSU (Stricherz et al., 1997). Pseudo-wind stress has the direction of the surface wind and the magnitude of the wind speed squared.

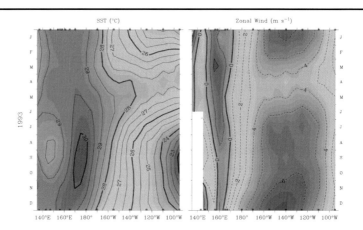

Figure 7.5.   (in color on page 529).   Seasonal cycle in 1993 of the monthly averaged equatorial SST (left panel) and zonal wind (right panel).   The figure is plotted through the graphics software and data made available through the TAO realtime data-access site at http://www.pmel.noaa.gov/toga-tao/realtime.html. Note that time is downwards.

December 1997 (Fig. 7.1.3), obtained by subtracting the average December SST of a long reference period from the actual December 1997 SST. The pattern, which 'springs to the eye', is characterized by higher than usual temperatures east of the date line in the equatorial Pacific.

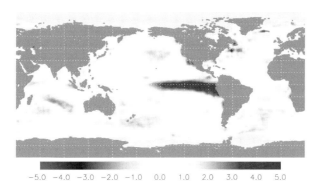

Sea-surface temperature anomaly field of December 1997, at the height of the 1997/1998 El Niño. Data from NCEP.

In section 1.3, two indices monitoring the state of the Tropical Pacific system, the SOI and the NINO3 index were shown to have strong variability on interannual time scales and to be well anti-correlated. The correlation between 12-month means of the monthly NINO3 index and the SOI was close to -0.9 over the last 50 years. The spectrum of both the NINO3 index and the SOI is dominated by interannual frequencies (Fig. 7.6), but the SOI contains more high-frequency variability and has a white spectral tail while the NINO3 index has a red tail. It appears that the occurrence of El Niño 's (positive NINO3) and La Niña 's (negative NINO3) is quite irregular. Strong El Niño 's such as in 1982-1983 are rare (as are strong La Niña 's) and long periods exist with either weak warm or weak cold conditions. El Niño is to some extent phase locked to the seasonal cycle as most El Niño's and La Niña's peak around December. The root mean square of the NINO3 index is almost twice as large in December than in April.

A long enough accurate data set is now available of Tropical Pacific SST fields, such that dominant patterns of variability at interannual time scales can be extracted. When the seasonal signal is filtered out, the equatorial SST anomalies over the years 1986-1999 show a fairly standing irregularly oscillating signal (Fig. 7.7, middle panel) with maximum amplitudes in the cold tongue region. The maximum temperature anomaly during the 1997-1998 El Niño was about 5°C. There was slight eastward propagation of the signal in the western part of the basin.

In Fig. 7.7, also time-longitude diagrams of zonal wind and thermocline depth anomalies are shown. The wind-stress response associated with a NINO3 anomaly is closely related to the sea-level pressure (SLP) anomaly pattern. The wind response (Fig. 7.7, left panel) is concentrated around the equator in an area

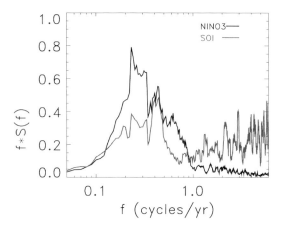

Figure 7.6. Spectra of monthly mean SOI and NINO3 index. Shown are normalized periodograms smoothed over 11 bins, that is over 0.11 cycles per year. Note that x-axis is logarithmic, and $f*S(f)$ rather than $S(f)$ is shown, in order that equal areas make equal contributions to the variance.

around the date line west of the NINO3 area. The westerly wind response to the SST anomaly indicates that the trade winds are weakened (and sometimes even reversed) during El Niño's. Sea-surface height and thermocline depth anomalies roughly vary in phase with El Niño in the eastern Pacific, and are in anti-phase in the western Pacific (Fig. 7.7, right panel). On closer inspection, thermocline depth anomalies in the western Pacific seem to precede SST anomalies in the eastern Pacific.

As a measure of thermocline anomalies, the upper ocean heat content H_c in 10^{10} J/m^2 is often used. It is defined as defined as

$$H_c = \rho_0 C_p \int_{-H_f}^{0} T \, dz \tag{7.1}$$

where H_f is a fixed depth, chosen usually as 300 m. If the thermocline (e.g. the depth of the 20°C isotherm) is depressed (elevated), then the upper layer is warmer (colder) leading to a larger (smaller) heat content. Just as the thermocline anomalies, the anomalies in heat content show a clear propagation eastward with west leading east.

Together with the variations of the SOI and NINO3 indices, the pictures above suggest an oscillatory signal which can be characterized by several patterns at different phases of the oscillation. During an El Niño , the SST anomaly is positive over the eastern equatorial Pacific basin having a maximum amplitude in the Pacific cold tongue. This positive SST anomaly is accompanied by a west-

Monthly Zonal Wind, SST, and 20°C Isotherm Depth Anomalies 2°S to 2°N Average

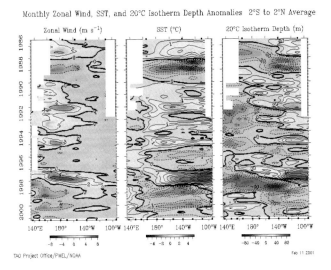

TAO Project Office/PMEL/NOAA

Feb 11 2001

Figure 7.7. Time-longitude diagrams of equatorial anomalies of zonal wind (left panel), SST (middle panel) and in the right panel the depth of the 20° isotherm (a measure for the thermocline depth). Data are for the period 1986-2000, and measured by the TAO/TRITON array. The plot is made through data and software at http://www.pmel.noaa.gov/tao

erly wind-stress anomaly, with a maximum amplitude west of the maximum SST anomaly. In the western part of the basin, the thermocline depth anomaly, and hence the heat content anomaly is negative. This is consistent with the equatorial anomalies in Fig. 7.7. Positive easterly SST anomalies occur simultaneously with an anomalously low western Pacific heat content. In this oscillatory view, the cold phase of the oscillation (in this case a weak La Niña -phase), occurred at the end of 1995. Between the cold and warm phase of the oscillation is a 'transition phase', which occurred at the beginning of 1997 (Fig. 7.7). During this 'transition phase', SST anomalies are nearly zero over the basin as are the wind-stress anomalies. Simultaneously, the equatorial thermocline depth and heat content anomaly in the western and central Pacific are positive.

The pictures above give an impression of the propagating features of anomalies associated with El Niño , but again this description is far from complete and other sources should be consulted to get a better view of the full complexity involved in ENSO (Rasmusson and Carpenter, 1982; Wallace *et al.*, 1998). Much information has been obtained during the recent strong 1997-1998 El Niño and a thorough description of the latter event can be found in McPhaden (1999).

7.1.4. Low-frequency variability of ENSO

From proxy data, more and more information becomes available on the behavior of ENSO under different global climate conditions such as different global mean temperatures. Based on the analysis of annually-banded corals from Papua New Guinea, Tudhope *et al.* (2001) show that ENSO has existed for the past 130,000 years. However, there have been substantial changes in its strength through time. Based on analysis of Ecuadorian varved lake sediments, Rodbell *et al.* (1999) find that ENSO periods were > 15 year from about 15 ka to 7 ka and that modern periodicities of 2 - 8 years appeared afterwards. Data from microfossils show that ENSO events were less intense around 3 ka but more pronounced around 1.5 ka (Woodroffe *et al.*, 2003).

When one considers the spectrum of the NINO3 index in Fig. 7.6b, also energy is found in lower frequencies, in particular in the decadal-to-interdecadal range (Fedorov and Philander, 2000). In Fig. 7.8, the interannual variability of the SST averaged over the box $[120°W, 80°W] \times [5°S - 5°N]$ in the Eastern Pacific is shown on a background of decadal variability. The strength of El Niño variability before the mid-1970s seems to be smaller than that after this period. According

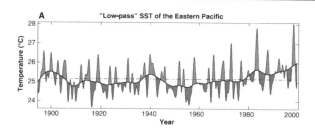

Figure 7.8.   (in color on page 531). Plot of the SST averaged over the box $[120°W, 80°W] \times [5°S - 5°N]$ in the Eastern Pacific on a background of decadal variability, the latter obtained through a low-pass filter (from Fedorov and Philander (2000)).

to NCEP data, the standard deviations of the SOI (NINO3) for 1951-1975 is 1.64 (0.81), to be compared for 1976-2000 where it is 1.84 (1.00). The spatial patterns of this decadal change are fairly similar to that of the interannual variability, but the SST anomalies at the eastern side of the basin extend more from the equator to midlatitudes (Zhang *et al.*, 1997).

7.1.5. Central questions and Approach

A theory of the variability of the Tropical Pacific climate system through which the above phenomena, in particular El Niño, can be understood from elementary physical principles, must contain a description of the physical processes and balances involved in the mean state, the seasonal cycle and the interannual and longer time scale variability. In particular, the questions related to the patterns and time scales, as shown above, of this system are

(i) Why is there a strong spatial asymmetry in the annual-mean state, with a warm pool in the west and a cold tongue in the east ?

(ii) Why is the ITCZ mostly located north of the equator ?

(iii) Why is there an annual period in the eastern Pacific seasonal cycle although the Sun moves twice over the equator ?

(iv) What determines the time scale and the patterns associated with ENSO variability ?

(v) What are the processes controlling the decadal-to-interdecadal variability of ENSO?

In the following sections, an attempt is made to present a theoretical framework from which partial answers to these questions can be given. Many of these results are scattered over the literature and the set-up chosen is to give the reader a good entrance into this literature. The text has been guided by the reviews of Neelin *et al.* (1994) and Neelin *et al.* (1998), but much more details and derivations are provided. Notational consistency with the literature (where possible) has been maximized and derivations have been kept as transparent as possible.

7.2. Models of the Equatorial Ocean

From the description of the phenomena, it appears that that their characteristic zonal scale is the basin length, but that the meridional scale is much smaller. Furthermore, the vertical scale is only a few hundred meters. In fact, most phenomena of interest are present only in a relatively small zone around the equator. This motivates the use of the $\beta-$ plane models which were introduced in section 5.3. Again, the homogeneous case is considered first and later extended to a layer-type model.

7.2.1. Constant density ocean model

For the case of constant density ρ, the starting equations are the dimensional equations (5.23). The only thing to change for the equatorial case is the value of the Coriolis parameter at the central latitude, which is the equator and hence $f_0 = 0$. In this way, the dimensional equations become

$$\frac{Du_*}{dt_*} - \beta_0 y_* v_* = -\frac{1}{\rho}\frac{\partial p_*}{\partial x_*} +$$
$$+A_H\left[\frac{\partial^2 u_*}{\partial x_*^2} + \frac{\partial^2 u_*}{\partial y_*^2}\right] + A_V\frac{\partial^2 u_*}{\partial z_*^2} \tag{7.2a}$$

$$\frac{Dv_*}{dt_*} + \beta_0 y_* u_* = -\frac{1}{\rho}\frac{\partial p_*}{\partial y_*} +$$
$$+A_H\left[\frac{\partial^2 v_*}{\partial x_*^2} + \frac{\partial^2 v_*}{\partial y_*^2}\right] + A_V\frac{\partial^2 v_*}{\partial z_*^2} \tag{7.2b}$$

$$\frac{\partial p_*}{\partial z_*} = -\rho g \tag{7.2c}$$

$$\frac{\partial w_*}{\partial z_*} + \frac{\partial v_*}{\partial y_*} + \frac{\partial u_*}{\partial x_*} = 0 \tag{7.2d}$$

$$\frac{D}{dt} = \frac{\partial}{\partial t} + u\frac{\partial}{\partial x} + v\frac{\partial}{\partial y} + w\frac{\partial}{\partial z} \tag{7.2e}$$

with dimensional boundary conditions at the ocean-atmosphere interface, described by $z_* = \eta_*$, of the form

$$p_* = p_{a*} \tag{7.3a}$$

$$\rho A_V \frac{\partial u_*}{\partial z_*} = \tau_0 \tau^x \tag{7.3b}$$

$$\rho A_V \frac{\partial v_*}{\partial z_*} = \tau_0 \tau^y \tag{7.3c}$$

$$\frac{D}{dt_*}(z_* - \eta_*) = 0 \tag{7.3d}$$

In these equations, u_* and v_* are the horizontal velocities, w_* is the vertical velocity and p_* is the pressure. The quantities g, τ_0, A_H, and A_V are the acceleration due to gravity, a typical amplitude of the wind stress and the horizontal and vertical mixing coefficients of momentum; the quantity p_{a*} is a background atmospheric pressure. Other boundary conditions for the flow, for example at the continental boundaries, will be specified later on.

7.2.2. The reduced gravity model

A slight extension of the previous model is the flow in a two-layer ocean in which the bottom layer is assumed to be motionless (Fig. 7.9). In this case, the equations (7.2- 7.3) hold for the top layer (with density ρ and equilibrium depth H) and also for the second layer (with slightly larger density $\rho + \Delta\rho$). The horizontal pressure gradient is zero in the second layer and hence only the hydrostatic pressure equation applies, i.e.

$$\frac{\partial p_{2*}}{\partial z_*} = -(\rho + \Delta\rho)g \tag{7.4}$$

Let the interface between the layers be prescribed through $z_* = -H + \zeta_*$, as seen in Fig. 7.9, then at the interface, the continuity of pressure and the kinematic condition become

$$p_{1*} = p_{2*} \tag{7.5a}$$

$$\frac{D}{dt_*}(z_* + H - \zeta_*) = 0 \tag{7.5b}$$

where the material derivative can be taken in both layers, since the vertical velocity is continuous. The equatorial reduced gravity ocean model is obtained by

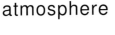

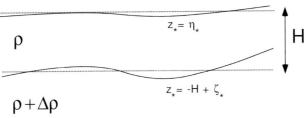

Figure 7.9.   Sketch of the reduced gravity ocean model. The upper active layer has a density ρ and equilibrium depth H. The bottom layer has a density $\rho + \Delta\rho$ and is motionless.

integrating over the upper layer, with total thickness $h_* = \eta_* + H - \zeta_*$. The equations become

$$\frac{\partial u_*}{\partial t_*} + u_* \frac{\partial u_*}{\partial x_*} + v_* \frac{\partial u_*}{\partial y_*} - \beta_0 y_* v_* =$$

$$= -g' \frac{\partial h_*}{\partial x_*} + \frac{\tau_0 \tau^x}{h_* \rho} + A_H \left[\frac{\partial^2 u_*}{\partial x_*^2} + \frac{\partial^2 u_*}{\partial y_*^2} \right] \qquad (7.6a)$$

$$\frac{\partial v_*}{\partial t_*} + u_* \frac{\partial v_*}{\partial x_*} + v_* \frac{\partial v_*}{\partial y_*} + \beta_0 y_* u_* =$$

$$= -g' \frac{\partial h_*}{\partial y_*} + \frac{\tau_0 \tau^y}{h_* \rho} + A_H \left[\frac{\partial^2 v_*}{\partial x_*^2} + \frac{\partial^2 v_*}{\partial y_*^2} \right] \qquad (7.6b)$$

$$\frac{\partial h_*}{\partial t} + \frac{\partial(u_* h_*)}{\partial x_*} + \frac{\partial(v_* h_*)}{\partial y_*} = 0 \qquad (7.6c)$$

where $g' = g\Delta\rho/\rho$ is the reduced gravity.

This equivalent barotropic (or reduced gravity) shallow water type model is the first cornerstone of the theory underlying the Pacific ocean dynamics relevant for the ENSO variability. In the following two sections, the spatial/temporal behavior of the linearized version of this model is considered.

7.2.3. Equatorial waves

Consider the motionless ($\bar{u}_* = \bar{v}_* = 0$) reference state with $\bar{h}_* = H$, which is a stationary solution of the unforced, nondissipative equations (7.6). The equations governing small amplitude motions are obtained by linearizing the equations (7.6) around this reference state and become

$$\frac{\partial u_*}{\partial t_*} - \beta_0 y_* v_* = -g' \frac{\partial h_*}{\partial x_*} \tag{7.7a}$$

$$\frac{\partial v_*}{\partial t_*} + \beta_0 y_* u_* = -g' \frac{\partial h_*}{\partial y_*} \tag{7.7b}$$

$$\frac{\partial h_*}{\partial t_*} + H\left(\frac{\partial u_*}{\partial x_*} + \frac{\partial v_*}{\partial y_*}\right) = 0 \tag{7.7c}$$

It is convenient to introduce nondimensional quantities by

$$t_* = \frac{L}{c_o} t \; ; \; x_* = Lx \; ; \; y_* = \lambda_o y \tag{7.8a}$$

$$h_* = Hh \; ; \; u_* = c_o u \; ; \; v_* = \frac{\lambda_o}{L} c_o v \tag{7.8b}$$

Here, L is the zonal basin length, c_o is a shallow water gravity wave speed and λ_o is a characteristic meridional length scale, the equatorial Rossby radius of deformation, given by

$$c_o = \sqrt{g'H} \; ; \; \lambda_o = \sqrt{\frac{c_o}{\beta_0}} \tag{7.9}$$

Using these scales, the dimensionless equations become

$$\frac{\partial u}{\partial t} - yv + \frac{\partial h}{\partial x} = 0 \tag{7.10a}$$

$$\zeta_o^2 \frac{\partial v}{\partial t} + yu + \frac{\partial h}{\partial y} = 0 \tag{7.10b}$$

$$\frac{\partial h}{\partial t} + \frac{\partial u}{\partial x} + \frac{\partial v}{\partial y} = 0 \tag{7.10c}$$

with $\zeta_o = \lambda_0/L$.

Travelling wave solutions are sought of the form

$$u(x, y, t) = \hat{u}(y)e^{i(kx - \sigma t)} \tag{7.11a}$$

$$v(x, y, t) = \hat{v}(y)e^{i(kx - \sigma t)} \tag{7.11b}$$

$$h(x, y, t) = \hat{h}(y)e^{i(kx - \sigma t)} \tag{7.11c}$$

with k being the nondimensional wavenumber and σ the angular frequency. The boundary conditions are

$$y \to \pm\infty : \hat{u}, \hat{v}, \hat{h} \to 0 \tag{7.12}$$

The solutions with $\hat{v} \equiv 0$ have a dispersion relation

$$\sigma^2 = k^2 \tag{7.13}$$

and the meridional structure of the wave is

$$\hat{u}(y) = \hat{u}(0)e^{\frac{-ky^2}{2\sigma}} \tag{7.14a}$$

$$\hat{h}(y) = \frac{\sigma}{k}\hat{u}(y) \tag{7.14b}$$

with $\hat{u}(0)$ being an arbitrary amplitude. The solutions which are bounded for $y \to \pm\infty$ exist only when $\sigma = +k$. Hence, the phase velocity of these waves is positive and the waves only move eastward. These are the well-known Kelvin waves with a dimensional wavelength and phasespeed (σ/k) given by

$$\lambda_* = \frac{2\pi L}{k} \; ; \; c_* = c_o \tag{7.15}$$

Patterns of the the thermocline field h of a Kelvin wave are plotted in Fig. 7.10 for four stages during the propagation. The dimensionless wavenumber is chosen $k = \pi$, corresponding to a wavelength of exactly twice the basin $\lambda_* = 2L$. For the Kelvin wave, the dimensionless period $\mathcal{P}$ is $2\pi/\sigma = 2$ and the pictures in Fig. 7.10 are at times $t = 0, t = 1/8, t = 1/4, t = 3/8$, which covers a quarter of a period. The maximum amplitude of the thermocline field for the Kelvin wave is located just at the equator.

Also free wave solutions with $\hat{v} \neq 0$ exist. In (7.10), $\hat{u}$ and $\hat{h}$ can be eliminated (see e.g., Pedlosky (1987), section 8.3) to give a scalar equation for $\hat{v}$, i.e.

$$\hat{v}'' + \hat{v} \left[\zeta_0^2(\sigma^2 - k^2) - \frac{k}{\sigma} - y^2 \right] = 0 \tag{7.16}$$

where the $'$ indicates the differentiation to y. Equation (7.16) has only bounded solutions when

$$\zeta_0^2(\sigma^2 - k^2) - \frac{k}{\sigma} = 2j + 1 \tag{7.17}$$

for integers $j = 0, 1, \cdots,$. These solutions are of the form

$$\hat{v}_j(\eta) = \psi_j(y) = \frac{e^{\frac{-y^2}{2}} H_j(y)}{(2^j j! \pi^{1/2})^{1/2}} \tag{7.18}$$

with H_j being the Hermite polynomials and the ψ_j are called the Hermite functions. The first couple of Hermite polynomials are

$$H_0(y) = 1 \quad ; \quad H_1(y) = 2y \tag{7.19a}$$
$$H_2(y) = 4y^2 - 2 \quad ; \quad H_3(y) = 8y^3 - 12y \tag{7.19b}$$

First, we consider the full spectrum by putting $\zeta_o = 1$, which is equivalent to use L also as a meridional length scale. The dispersion relation (7.17) can be written as

$$k = -\frac{1}{2\sigma} \pm \frac{1}{2} \left[(\frac{1}{\sigma} - 2\sigma)^2 - 8j \right]^{1/2} \tag{7.20}$$

For $j > 0$, two real roots exist provided $(1/\sigma - 2\sigma)^2 \geq 8j$ in which case σ satisfies

$$0 < \sigma < \frac{1}{\sqrt{2}}((j+1)^{1/2} - j^{1/2}) \tag{7.21a}$$

or

$$\sigma > \frac{1}{\sqrt{2}}(j^{1/2} + (j+1)^{1/2}) \tag{7.21b}$$

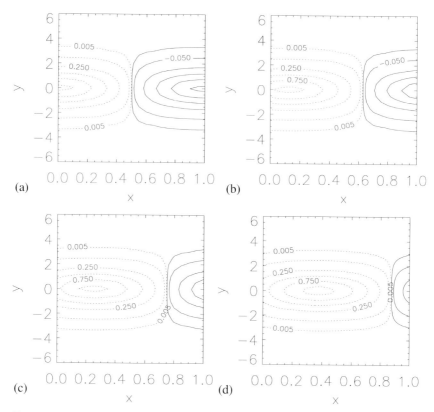

Figure 7.10.   Patterns of the dimensionless thermocline field for the Kelvin wave for four different times during one period $\mathcal{P} = 2$ of evolution (a) $t = 0$ (b) $t = \mathcal{P}/8$, (c) $t = \mathcal{P}/4$ and $t = 3\mathcal{P}/8$. The wavenumber $k = \pi$ and plotted is $\psi_0(y)\cos(\pi(x-t))/\sqrt{2}$, where ψ_0 is the Hermite function in (7.18). Note that x and y are scaled according to (7.8).

The first interval of σ is in the low frequency range and the waves are called equatorial Rossby waves. The second interval represents the high frequency so-called 'inertia-gravity' waves.

For the case $j = 0$, two roots are found from (7.20), the first one being $\sigma = -k$ which leads to a westward travelling Kelvin wave which becomes unbounded far from the equator. The second root is

$$k = -\frac{1}{\sigma} + \sigma \qquad (7.22)$$

which gives a bounded wave called the Yanai wave. For large σ, the character of the wave becomes Kelvin like, whereas for small σ it becomes Rossby like. A classical picture of the dispersion relation for the Kelvin wave, the Yanai wave and $j = 1$ Rossby and inertia-gravity waves is plotted in Fig. 7.11. The Yanai and

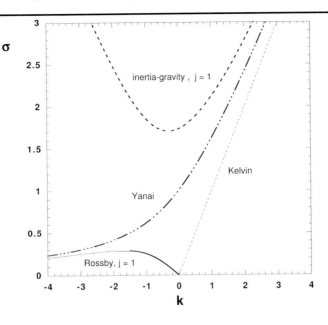

Figure 7.11.   *Dispersion relation of equatorial free waves. Shown are the Kelvin wave, the Yanai wave and the $j = 1$ Rossby wave.*

Kelvin waves have a positive group velocity and for inertia-gravity and Rossby waves, the group velocity c_g becomes

$$c_g = \frac{\partial \sigma}{\partial k} = \frac{1 + 2\sigma k}{2\sigma^2 + \frac{k}{\sigma}} \tag{7.23}$$

For long, low frequency Rossby waves (Fig. 7.11), the group velocity is negative and the approximate dispersion relation is (note that both $k^2 << 1$ and $\sigma^2 << 1$)

$$\sigma = -\frac{k}{2j + 1} \tag{7.24}$$

Their dimensional phase velocity is given by

$$c_* = -\frac{c_o}{2j + 1}$$

and depends the meridional wavenumber j. These long waves only remain as non-dispersive waves (in addition to the Kelvin wave) in the limit $\zeta_o \to 0$, which can be immediately concluded from (7.17). This limit is therefore called the long wave limit. The first long Rossby wave (j = 1) travels westward with a phase velocity which is 1/3 of that of the Kelvin wave. From the expressions of the Hermite functions in (7.18), one can see that the amplitude is restricted to a relatively small meridional interval around the equator; these waves are therefore called "equatorially trapped".

Patterns of the thermocline field for the $j = 1$ Rossby wave, with again a dimensionless wavenumber $k = \pi$, are plotted Fig. 7.12 for four stages during the propagation. The dimensionless period of the $j = 1$ Rossby wave is $\mathcal{P} = 6$, and the pictures are shown at $t = 0, t = 3/8, t = 3/4, t = 9/8$, which again covers a quarter of the period. The maximum amplitude of the j = 1 Rossby wave is

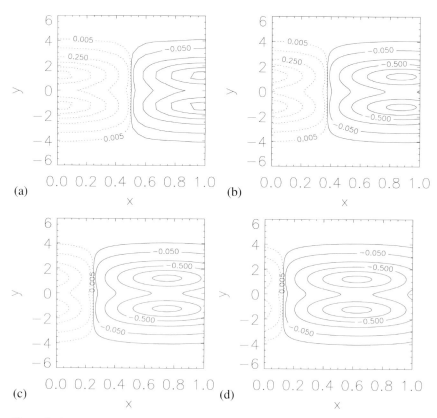

Figure 7.12.   Patterns of the thermocline field h of the $j = 1$ Rossby wave for four different times during one period $\mathcal{P} = 6$ of evolution (a) $t = 0$ (b) $t = 3/8$, (c) $t = 3/4$ and (d) $t = 9/8$. The wavenumber k is equal to π and plotted is $(\psi_0(y) + \psi_2(y)/\sqrt{2})cos(\pi(x - t)))/(2\sqrt{2})$, where ψ_0 and ψ_2 are Hermite functions as in (7.18).

off-equatorial and at about $1.33 \times \lambda_o$ from the equator and this distance increases (Fig. 7.13) for higher Rossby waves, i.e. larger j. For $c_o = 2$ m/s, the dimensional

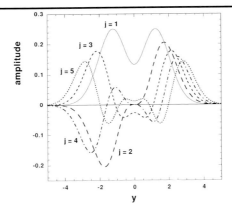

Figure 7.13. Meridional structure of the dimensionless thermocline field $\hat{h}_j$ in (7.31b) associated with the first 5 (long) free Rossby waves (j = 1, ..., 5).

values of the meridional locations at which Rossby wave thermocline amplitudes have their maximum are shown in Table 7.1 for the long waves with j = 1, ..., 5 together with dimensional crossing times for a basin of 15,000 km.

Wave type	y_{max}	θ_{max}	τ_c (days)
Kelvin	0.0	0.0	87
Rossby, $j = 1$	1.22	±3.31	260
Rossby, $j = 2$	1.75	±4.75	434
Rossby, $j = 3$	2.17	±5.88	608
Rossby, $j = 4$	2.50	±6.77	781
Rossby, $j = 5$	2.83	±7.67	955

Table 7.1. Typical quantities of free equatorial waves for $c_o = 2$ m/s, and $\beta_0 = 2.2 \times 10^{-11} (ms)^{-1}$, such that $\lambda_o = 301.5$ km. The dimensionless quantity y_{max} is the position of the maximum amplitude of the thermocline depth as seen in Fig. 7.13; θ_{max} is the latitude of this position. The travel time is based on the time it takes for the wave to cross a basin of 15,000 km. The Kelvin wave travels from west to east whereas all Rossby waves travel from east to west.

7.2.4. Forced response in a basin

Using the model derived in the previous section, the changes in the ocean circulation in a finite basin due to the presence of a prescribed wind stress are considered next. Under limitations of small amplitude forced motion, the shallow water model can be linearized around a motionless reference state with constant thermocline depth H. Small amplitude zonal winds are assumed to be present, while the meridional component of the wind is neglected. A further simplification arises by idealizing the horizontal friction to be linear rather than harmonic. This can be justified by recognizing that for equatorially trapped motions of which the zonal length scale L is much larger than the meridional scale λ_o,

$$A_H \left[\frac{\partial^2 u_*}{\partial x_*^2} + \frac{\partial^2 u_*}{\partial y_*^2} \right] \approx -\frac{2A_H}{\lambda_o^2} u_* = -a_m u_* \qquad (7.25)$$

which can, for example, be derived through central differences around the equator. With the scaling (7.8), the dimensionless problem to determine the small amplitude response to the wind stress is obtained from (7.6) and given by

$$\frac{\partial u}{\partial t} - yv + \frac{\partial h}{\partial x} + \epsilon_o u = F_0 \tau^x \qquad (7.26a)$$

$$\zeta_o^2 \frac{\partial v}{\partial t} + yu + \frac{\partial h}{\partial y} + \epsilon_o \zeta_o v = 0 \qquad (7.26b)$$

$$\frac{\partial h}{\partial t} + \frac{\partial u}{\partial x} + \frac{\partial v}{\partial y} + \epsilon_o h = 0 \qquad (7.26c)$$

where $F_0 = \tau_0 L / (c_o^2 \rho H)$ is the dimensionless amplitude of the zonal wind stress and $\epsilon_o = a_m L / c_o$ is the dimensionless linear damping coefficient. In a finite basin on the equatorial β-plane, the boundary conditions are

$$x = 0, 1 \quad : \quad u = 0 \qquad (7.27a)$$

$$y \to \pm\infty \quad : \quad u, v, h \to 0 \qquad (7.27b)$$

With $\mathbf{F} = (\tau^x, 0, 0)$ and $\mathbf{u} = (u, v, h)$, this system of equations can be written as

$$\mathcal{M} \frac{\partial \mathbf{u}}{\partial t} + \mathcal{L}\mathbf{u} = \mathbf{F} \qquad (7.28a)$$

$$\mathcal{L} = \begin{pmatrix} \epsilon_o & -y & \frac{\partial}{\partial x} \\ y & \zeta_o \epsilon_o & \frac{\partial}{\partial y} \\ \frac{\partial}{\partial x} & \frac{\partial}{\partial y} & \epsilon_o \end{pmatrix} \quad ; \quad \mathcal{M} = \begin{pmatrix} 1 & 0 & 0 \\ 0 & \zeta_o^2 & 0 \\ 0 & 0 & 1 \end{pmatrix} \qquad (7.28b)$$

Applying Fourier transformation in x, according to

$$\hat{\mathbf{u}}(k, y, t) = \int_{-\infty}^{\infty} \mathbf{u}(x, y, t) e^{-ikx} \, dx \qquad (7.29a)$$

$$\hat{\mathbf{F}}(k, y, t) = \int_{-\infty}^{\infty} \mathbf{F}(x, y, t) e^{-ikx} \, dx \qquad (7.29b)$$

then all x-derivatives in $\mathcal{L}$ will transform to ik in $\hat{\mathcal{L}}$ and $\hat{\mathcal{M}} = \mathcal{M}$. All free wave solutions of the previous section, say written as $\hat{\mathbf{U}}$, are solutions of the eigenvalue problem (for $\epsilon_o = 0$)

$$\hat{\mathcal{L}}\hat{\mathbf{U}} = i\sigma \hat{\mathcal{M}}\hat{\mathbf{U}} \qquad (7.30)$$

where σ is given through the dispersion relation (7.17).

In the limit $\zeta_o \to 0$, only the long (small k), low frequency modes (small σ) Rossby waves remain, having a dispersion relation and eigenfunctions for j = 1,2,...,

$$\sigma_j = \frac{-k}{2j+1} \tag{7.31a}$$

$$\hat{u}_j(y) = \frac{1}{2\sqrt{2}}\left(\frac{\psi_{j+1}(y)}{\sqrt{j+1}} - \frac{\psi_{j-1}(y)}{\sqrt{j}}\right) \tag{7.31b}$$

$$\hat{h}_j(y) = \frac{1}{2\sqrt{2}}\left(\frac{\psi_{j+1}(y)}{\sqrt{j+1}} + \frac{\psi_{j-1}(y)}{\sqrt{j}}\right) \tag{7.31c}$$

$$\hat{v}_j(y) = \psi_j(y) \tag{7.31d}$$

To that, the Kelvin waves with dispersion relation and eigenfunction

$$\sigma_0 = k \tag{7.32a}$$

$$\hat{u}_0(y) = \frac{1}{\sqrt{2}}\psi_0(y) \tag{7.32b}$$

$$\hat{h}_0(y) = \frac{1}{\sqrt{2}}\psi_0(y) \tag{7.32c}$$

$$\hat{v}_0(y) = 0 \tag{7.32d}$$

have to be added to get a complete system of basis functions for the meridional structure of the solutions (Cane, 1979a,b) of the problem (7.26). The vector eigenfunctions (7.31) and (7.32) will below be indicated with $\mathbf{\Phi}_j$ and $\mathbf{\Phi}_0$, respectively. A consequence of the elimination of the small waves in the limit $\zeta_o \to 0$ is that one can no longer satisfy the kinematic boundary condition ($u = 0$) at the western boundary of the basin. A consistent boundary condition is to balance the incoming and outgoing zonal mass flux (Cane and Sarachik, 1977), which gives

$$x = 0 \ : \ \int_{-\infty}^{\infty} u \, dy = 0 \tag{7.33}$$

As a next step, the zonal wind stress is assumed to have the particular form

$$\tau^x(x, y, t) = \delta(x - x_0)g(y)e^{i\omega t} \tag{7.34}$$

where δ is the delta distribution, x_0 a point in the basin and $g(y)$ a prescribed function. The time dependence is assumed periodic with frequency ω; since the system of equations (7.26) is separable in time, also the solutions $\mathbf{u}$ have the same time dependence, i.e. $\mathbf{u}(x, y, t) = e^{i\omega t}\tilde{\mathbf{u}}(x, y)$. If the solution $\tilde{\mathbf{u}}(x, y)$ for the wind stress shape (7.34) is determined and is indicated by $\mathbf{G}(x, y; x_0)$ then the solution for every wind stress with spatial dependence $\tau^x(x, y, t) = f(x)g(y)e^{i\omega t}$, is given by

$$\mathbf{u}(x, y, t) = e^{i\omega t}\int_0^1 \mathbf{G}(x, y; x_0)f(x_0) \, dx_0 \tag{7.35}$$

which is easily verified by substitution of (7.35) into the equations (7.28). Hence the solution $\mathbf{G}$ acts as a Green's function and it is worthwhile to determine this solution explicitly. The problem of the determination of the Green's function is addressed in Technical Box 7.1.

Technical box 7.1:
Green's function

In this technical box, the derivation of the Green's function $\mathbf{G}$ in (7.35) is provided. First the (particular) solution $\mathbf{G}_f$ to the inhomogeneous problem is derived followed by the total solution $\mathbf{G}$ which satisfies the boundary conditions. After Fourier transformation of the equations (7.28) for $\mathbf{u} = \tilde{\mathbf{u}} e^{i\omega t}$, the following system of equations results

$$\phi \hat{u} - y \hat{v} + ik\hat{h} = e^{-ikx_0} g(y) \tag{7.36a}$$

$$y \hat{u} + \frac{\partial \hat{h}}{\partial y} = 0 \tag{7.36b}$$

$$\phi \hat{h} + ik\hat{u} + \frac{\partial \hat{v}}{\partial y} = 0 \tag{7.36c}$$

with $\phi = \epsilon_o + i\omega$.

The forcing function $g(y)$ and the dependent quantities $\hat{u}, \hat{v}, \hat{h}$ are expanded into the free wave solutions as follows

$$g(y) = r_0 \hat{u}_0(y) + \sum_{j=1}^{\infty} r_j \hat{u}_j(y) \tag{7.37a}$$

$$\hat{u} = a_0 \boldsymbol{\Phi}_0(y) + \sum_{j=1}^{\infty} a_j \boldsymbol{\Phi}_j(y) \tag{7.37b}$$

where the $\boldsymbol{\Phi}_j$ satisfy (7.30). Equating term by term, the coefficients a_j are solved in terms of the r_j as

$$a_j = \frac{r_j e^{-ikx_0}}{\phi + i\sigma_j} \tag{7.38}$$

where σ_j is the frequency of eigenmode j. The inverse Fourier transform now gives the formal solution as

$$\mathbf{G}_f(x, y, \phi; x_0) = \frac{1}{2\pi i} \int_{-\infty}^{\infty} e^{ik(x-x_0)} \times$$

$$\times \left[\frac{r_0}{k - i\phi} \boldsymbol{\Phi}_0(y) - \sum_{j=1}^{\infty} \frac{r_j(2j+1)}{k + i\phi(2j+1)} \boldsymbol{\Phi}_j(y) \right] dk \tag{7.39}$$

The integrals can be evaluated through the residue theorem and one gets

$$\mathbf{G}_f(x, y, \phi; x_0) = r_0 \mathbf{\Phi}_0(y) e^{-\phi(x-x_0)} \mathcal{H}(x - x_0) +$$
$$+ \sum_{j=1}^{\infty} (2j + 1)\, r_j\, e^{\phi(2j+1)(x-x_0)} \mathbf{\Phi}_j(y) \mathcal{H}(x_0 - x) \qquad (7.40)$$

where $\mathcal{H}$ is the Heaviside function. The physics of this forced response is easy to understand. If a pulse wind stress forcing is applied at $x = x_0$, then to the west ($x < x_0$) only a Rossby wave response ($\mathbf{\Phi}_j$) is found whereas to the east ($x > x_0$), a Kelvin wave response ($\mathbf{\Phi}_0$) is found.

This solution does not satisfy the boundary conditions at the eastern and western boundaries and solutions of the homogeneous problem have to be added to accomplish this. The latter are the actual eigenfunctions $\mathbf{\Phi}_j$, the free wave solutions, say with up to now undetermined amplitudes b_j. One obtains for $x > x_0$,

$$\mathbf{G}(x, y, \phi; x_0) = (r_0 + b_0) \mathbf{\Phi}_0(y) e^{-\phi(x-x_0)} +$$
$$+ \sum_{j=1}^{\infty} b_j e^{\phi(2j+1)(x-x_0)} \mathbf{\Phi}_j(y) \qquad (7.41a)$$

while for $x < x_0$, one obtains

$$\mathbf{G}(x, y, \phi; x_0) = b_0 \mathbf{\Phi}_0(y) e^{-\phi(x-x_0)} +$$
$$+ \sum_{j=1}^{\infty} ((2j + 1) r_j + b_j) e^{\phi(2j+1)(x-x_0)} \mathbf{\Phi}_j(y) \qquad (7.41b)$$

From the condition at the eastern boundary ($u = 0$), it follows from (7.26b) that

$$\partial h(1, y, \phi; x_0)/\partial y = 0 \Rightarrow h_E^G(\phi; x_0) = h(1, y, \phi; x_0) \qquad (7.42)$$

where the superscript G refers to the Green's function. To determine the coefficients b_j, the identities (Cane and Sarachik, 1977)

$$0 = \lim_{M \to \infty} \left[\hat{u}_0 + 2 \sum_{j=0}^{M} \alpha_{2j+1} \hat{u}_{2j+1} \right] \qquad (7.43a)$$

$$\pi^{-\frac{1}{4}} = \lim_{M \to \infty} \left[\hat{h}_0 + 2 \sum_{j=0}^{M} \alpha_{2j+1} \hat{h}_{2j+1} \right] \qquad (7.43b)$$

$$\alpha_{2j+1} = \frac{\sqrt{(2j + 1)!}}{2^j j!} \qquad (7.43c)$$

are used. Note that the convergence with M in these identities is very poor for the zonal velocity component and for both zonal velocity and thermocline off the equator. Convergence is best for the thermocline deviation on the equator.

Application of the eastern boundary condition and equating term by term to zero gives

$$\begin{aligned}
(r_0 + b_0)e^{-\phi(1-x_0)} &= \pi^{\frac{1}{4}} h_E^G \\
b_{2j+1} e^{\phi(4j+3)(1-x_0)} &= 2\alpha_{2j+1}\, \pi^{\frac{1}{4}} h_E^G \\
b_{2j} &= 0
\end{aligned}$$

from which the coefficients b_j can be solved. Eventually, the complete solution to the pulse forcing at $x = x_0$, i.e. the Green's function for the problem, is found as

$$\mathbf{G}(x, y, \phi; x_0) = \pi^{\frac{1}{4}} h_E^G \mathbf{K}(\phi(1-x), y) - \mathbf{L}(\phi(x_0 - x), y)\mathcal{H}(x_0 - x) \quad (7.45)$$

where vector functions $\mathbf{K}$ and $\mathbf{L}$ are defined as

$$\mathbf{K}(\eta, y) = e^{\eta}\mathbf{\Phi}_0(y) + 2\sum_{j=0}^{\infty} \alpha_{2j+1} e^{-\eta(4j+3)} \mathbf{\Phi}_{2j+1}(y) \quad (7.46a)$$

$$\mathbf{L}(\eta, y) = r_0 e^{\eta}\mathbf{\Phi}_0(y) - \sum_{j=0}^{\infty} (2j+1) r_j e^{-\eta(2j+1)} \mathbf{\Phi}_j(y) \quad (7.46b)$$

Up to this point, only the eastern boundary amplitude of the thermocline h_E^G is still unknown, but it can be determined from the western boundary condition (7.33) and becomes

$$\pi^{\frac{1}{4}} h_E^G(\phi; x_0) = \frac{\int_{-\infty}^{\infty} L_u(\phi x_0, y) dy}{\int_{-\infty}^{\infty} K_u(\phi, y) dy} \quad (7.47)$$

where K_u and L_u are the first components of $\mathbf{K}$ and $\mathbf{L}$, respectively. This completes the basic machinery needed in the next sections to understand the response of the ocean to varying wind stress forcing.

7.3. Physics of Coupling

Anomalies in SST somehow manage to change the winds, and in the first subsection a model is sketched how to compute the low level wind response due to SST anomalies. Next, wind stress anomalies induce changes in the ocean circulation and examples are shown in 7.3.2, using the results of Technical box 7.1. Finally, a model is considered in 7.3.3 to determine how changes in ocean circulation induce SST anomalies.

7.3.1. Atmospheric response to diabatic heating

A class of simple models to analyze the low level wind response due to heating anomalies in the tropics was proposed by Matsuno (1966). These models are also of shallow water type, following the same approach as in section 7.2. The steady response of one of these models was analyzed in detail in Gill (1980) and since then, this type of model is referred to as a Gill model. The equations are

$$\frac{\partial U_*}{\partial t_*} - \beta_0 y_* V_* - \frac{\partial \Theta_*}{\partial x_*} + a_M U_* = 0 \tag{7.48a}$$

$$\frac{\partial V_*}{\partial t_*} + \beta_0 y_* U_* - \frac{\partial \Theta_*}{\partial y_*} + a_M V_* = 0 \tag{7.48b}$$

$$\frac{\partial \Theta_*}{\partial t_*} - c_a^2 \left(\frac{\partial U_*}{\partial x_*} + \frac{\partial V_*}{\partial y_*} \right) + a_M \Theta_* = Q_* \tag{7.48c}$$

where (U_*, V_*) are the low level winds, Θ_* the geopotential height (with dimension m^2/s^2), a_M is a damping coefficient and c_a is the phase speed of the first baroclinic Kelvin wave in the atmosphere. The flow is forced by a representation of the adiabatic heating term Q_* (having dimension m^2/s^3). Note the similarities with the reduced gravity ocean model with the difference being in the forcing terms. More accurate derivations of these type of models can be found in Holton (1992).

To study the response, it is convenient to scale the equations with

$$t_* = \frac{L}{c_o} t \; ; \; x_* = L x \; ; \; y_* = \lambda_a y \tag{7.49a}$$

$$\Theta_* = c_a^2 \Theta \; ; \; U_* = c_a U \; ; \; V_* = \frac{\lambda_a}{L} c_a V \tag{7.49b}$$

$$Q_* = q_0 Q \; ; \; \lambda_a = \sqrt{\frac{c_a}{2\beta_0}} \tag{7.49c}$$

Note that the factor 2 in the definition of λ_a is different from the scaling of the ocean model. On the other hand, already anticipating coupling, the time is scaled with the advective time scale in the ocean. The dimensionless equations become

$$c \frac{\partial U}{\partial t} - \frac{y}{2} V - \frac{\partial \Theta}{\partial x} + \epsilon_a U = 0 \tag{7.50a}$$

$$c \zeta_a^2 \frac{\partial V}{\partial t} + \frac{y}{2} U - \frac{\partial \Theta}{\partial y} + \zeta_a \epsilon_a V = 0 \tag{7.50b}$$

$$c \frac{\partial \Theta}{\partial t} - \left(\frac{\partial U}{\partial x} + \frac{\partial V}{\partial y} \right) + \epsilon_a \Theta = \mu_0 Q \tag{7.50c}$$

with $\epsilon_a = a_M L / c_a$, $\mu_0 = q_0 L / c_a^3$, $c = c_o / c_a$ and $\zeta_a = \lambda_a / L$. Of these parameters, both c and ζ_a are small and to a good approximation, the atmospheric time derivatives can be neglected, as well as the damping in the meridional momentum balance. All fields must be bounded far from the equator. The solution of this linear problem is provided in Technical box 7.2.

Technical box 7.2: Solution of the Gill model

With $c \to 0$, $\zeta_a \to 0$ and by introducing new independent variables $S = \Theta + U$ and $R = \Theta - U$ the problem (7.50) becomes

$$-(\frac{y}{2}V + \frac{\partial V}{\partial y}) - \frac{\partial S}{\partial x} + \epsilon_a S = \mu_0 Q \qquad (7.51a)$$

$$\frac{y}{2}(S - R) - (\frac{\partial S}{\partial y} + \frac{\partial R}{\partial y}) = 0 \qquad (7.51b)$$

$$\frac{y}{2}V - \frac{\partial V}{\partial y} + \frac{\partial R}{\partial x} + \epsilon_a R = \mu_0 Q \qquad (7.51c)$$

Subsequently, the variables S, R and the forcing Q are expanded into parabolic cylinderfunctions $D_n(y)$ with coefficients depending on x,

$$Q(x, y, t) = \sum_{n=0}^{\infty} Q_n(x) D_n(y) e^{i\omega t} \qquad (7.52a)$$

$$R(x, y, t) = \sum_{n=0}^{\infty} R_n(x) D_n(y) e^{i\omega t} \qquad (7.52b)$$

$$S(x, y, t) = \sum_{n=0}^{\infty} S_n(x) D_n(y) e^{i\omega t} \qquad (7.52c)$$

where a periodic time dependence in the forcing has been assumed with frequency ω. The parabolic cylinderfunctions $D_n(y)$ are related to the Hermite polynomials through

$$D_n(y) = 2^{\frac{-n}{2}} e^{-\frac{y^2}{4}} H_n(\frac{y}{\sqrt{2}}) \qquad (7.53)$$

For all n, the relations

$$\frac{y}{2}D_n + D_n' = nD_{n-1} \; ; \; \frac{y}{2}D_n - D_n' = D_{n+1} \qquad (7.54)$$

are valid. Substitution of the expansions (7.52) into the equations (7.51) gives a system of ordinary differential equations for the coefficient functions S_n and R_n. For $n = 0$,

$$\epsilon_a R_0 + R_0' - \mu_0 Q_0 = 0 \qquad (7.55a)$$

$$R_1 = 0 \qquad (7.55b)$$

$$\epsilon_a S_0 - S_0' - V_1 - \mu_0 Q_0 = 0 \qquad (7.55c)$$

from which R_0 and R_1 are directly determined. For $n = 1$, one obtains

$$\epsilon_a S_1 - S_1' - 2V_2 - \mu_0 Q_1 = 0 \qquad (7.56a)$$
$$2R_2 = S_0 \qquad (7.56b)$$
$$V_0 - \mu_0 Q_1 = 0 \qquad (7.56c)$$

from which V_0 directly follows. For $n > 1$, the equations become

$$\epsilon_a S_n - S_n' - (n+1)V_{n+1} - \mu_0 Q_n = 0 \qquad (7.57a)$$
$$\epsilon_a R_n + R_n' + V_{n-1} - \mu_0 Q_n = 0 \qquad (7.57b)$$
$$(n+1)R_{n+1} - S_{n-1} = 0 \qquad (7.57c)$$

Using (7.57c) to eliminate the terms involving S_n in (7.57a) and adding the results to (7.57b) for $n \to n+2$ gives a single equation for R_{n+2}, $n > 0$, i.e.

$$(2n+3)\epsilon_a R_{n+2} - R_{n+2}' - \mu_0(Q_n + (n+1)Q_{n+2}) = 0 \qquad (7.58)$$

from which R_{n+2} and eventually the total solution for U, V and Θ can be calculated. The results for U and Θ are

$$U(x,y,t) = \frac{e^{i\omega t}}{2}[(2R_2(x) - R_0(x))D_0(y) + 3R_3(x)D_1(y)]$$
$$+ \frac{e^{i\omega t}}{2}\left[\sum_{n=2}^{\infty}((n+2)R_{n+2}(x) - R_n(x))D_n(y)\right] \qquad (7.59a)$$

$$\Theta(x,y,t) = \frac{e^{i\omega t}}{2}[(R_0(x) + 2R_2(x))D_0(y) + 3R_3(x)D_1(y)]$$
$$+ \frac{e^{i\omega t}}{2}\left[\sum_{n=2}^{\infty}((n+2)R_{n+2}(x) + R_n(x))D_n(y)\right] \qquad (7.59b)$$

where

$$R_0(x) = \mu_0\int_0^x e^{-\epsilon_a(x-s)}Q_0(s)\,ds$$
$$R_1(x) = 0$$
$$R_{n+2}(x) = \mu_0\int_x^1 e^{(2n+3)\epsilon_a(x-s)}((n+1)Q_{n+2}(s) + Q_n(s))\,ds$$

for $n = 0, 1, \cdots$,. This completes the full solution of the Gill model.

An example of the steady response of the Gill model is considered with the forcing described by

$$Q(x,y) = \psi_0(\frac{\lambda_a}{\lambda_o}y)\,\sin \pi x\,; \; \omega = 0 \qquad (7.61)$$

with ψ_0 the Hermite function defined in (7.18). Note that since $y = y_*/\lambda_a$, the argument in the Hermite function is y_*/λ_o and hence the meridional scale of the forcing is the Rossby radius of deformation of the ocean. With $c_a = 30 \ m/s$, $c_o = 2 \ m/s$, the ratio of the Rossby deformation radii of atmosphere and ocean is about 3 ($\lambda_a \approx 826 \ km$).

The forcing (7.61) is shown Fig. 7.14a and in subsequent panels, the stationary ($\omega = 0$) zonal wind response (7.59a) is plotted for $\epsilon_a = 0, 2.5$ and 5.0. For each value of ϵ_a, there are westerly (easterly) winds to the west (east) of the maximum heating. The signal west of the heating maximum is mainly due to Rossby waves, while that to the east is due to the Kelvin wave. The zonal wind response becomes more local as the value of ϵ_a increases. This is also clear physically, since ϵ_a is a ratio of the basin length L and an atmospheric damping length scale c_a/a_M. When the damping a_M increases, the length scale over which anomalies are damped decreases and hence the response is more localized to the forcing.

An approximation to the equatorial zonal wind response U is obtained by truncating the solution (7.59a) for only the first three parabolic cylinderfunctions, i.e. with $D_0(0) = 1, D_1(0) = 0, D_2(0) = -1$, the equation (7.59a) gives

$$U(x, 0, t) = e^{i\omega t}(\frac{3}{2}R_2(x) - \frac{1}{2}R_0(x)) \qquad (7.62)$$

where the R_4 contribution is also neglected. As we will show later on, this expression turns out to be useful when considering reduced models which only take the equatorial response into account.

When the diabatic heating structure is known, the low level wind response can be computed from the Gill model. However, this leaves the problem to relate the diabatic heating structure and the SST anomalies. The simplest connection (Zebiak, 1982) is that convection mostly occurs over the warmest water which leads to a direct coupling with SST anomalies $\tilde{T}_*$ and those in latent heat $\tilde{Q}_*$ through

$$\tilde{Q}_* = \alpha_T \tilde{T}_* \qquad (7.63)$$

with some constant coefficient α_T (with dimension $m^2/(s^3 K)$). If a typical scale of the temperature anomaly is ΔT, then $q_0 = \alpha_T \Delta T$. The dimensionless parameter measuring the amount of heating per SST anomaly is then given by

$$\mu_0 = \frac{\alpha_T \Delta T L}{c_a^3} \qquad (7.64)$$

which will be part of the main coupling parameter introduced in the ocean-atmosphere model in subsequent sections. The relation between SST anomalies and diabatic forcing above is far from perfect and many improvements based on detailed atmospheric modelling have been suggested (see e.g., Neelin *et al.* (1998) and references therein).

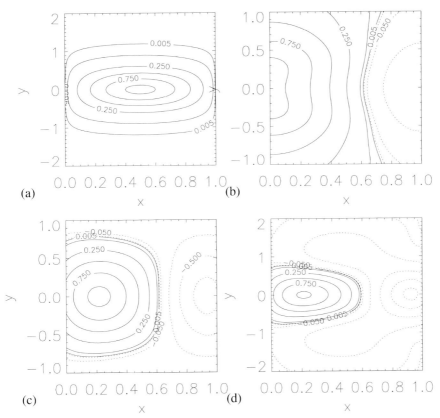

Figure 7.14.   (a) Pattern of the diabatic forcing $Q(x,y)$ given by (7.61). The zonal wind response (7.59a) is plotted for three different values of ϵ_a in subsequent panels, (b) $\epsilon_a = 0$, (c) $\epsilon_a = 2.5$ and (d) $\epsilon_a = 5.0$. Note that y is scaled with $\lambda_a = 826$ km and that x is scaled with the basin length $L = 1.5 \times 10^4$ km. The zonal velocity is scaled with $c_a = 30$ m/s and the factor $\mu_0 = 1$.

7.3.2. Adjustment of the ocean

The low level surface winds exert a wind stress on the ocean surface according to the bulk formula

$$(\tau^x, \tau^y) = C_d \rho_a |\,\mathbf{U}\,|\mathbf{U} \tag{7.65}$$

where C_d is the drag coefficient, ρ_a the density of air and $\mathbf{U} = (U, V)$. Considering perturbations $\tilde{\mathbf{U}}$ from some reference state $\bar{\mathbf{U}}$, the perturbation wind stress can be taken proportional to the perturbation velocity in the lower atmospheric layer, i.e.

$$\frac{\tilde{\tau}^x_*}{\rho H} = \gamma \tilde{U}_* \;;\; \frac{\tilde{\tau}^y_*}{\rho H} = \gamma \tilde{V}_* \tag{7.66}$$

where γ is a constant (having dimension s^{-1}).

Reason for starting with a detailed analysis of the ocean response to a simplified wind structure in section 7.2.4, is that the solutions can be used to illustrate the response of the shallow water model to changes in the equatorial winds. When the zonal wind stress does not depend on the meridional coordinate, i.e. $g(y) = 1$ in (7.37a), then the solution to periodic wind forcing with frequency ω can be explicitly calculated (Technical box 7.3).

Technical box 7.3:
Explicit solution
to periodic forcing

In this technical box, an explicit solution of the forced ocean response to a zonal forcing $\tau^x = f(x)e^{i\omega t}$ is constructed, using the Green's function calculated in Technical box 7.1. The basic identity used is the explicit summation (Cane and Sarachik, 1977)

$$e^{-iz} \begin{pmatrix} \hat{u}_0 \\ \hat{h}_0 \end{pmatrix} + 2\sum_{j=0}^{\infty} \alpha_{2j+1} e^{iz(4j+3)} \begin{pmatrix} \hat{u}_{2j+1} \\ \hat{h}_{2j+1} \end{pmatrix} =$$

$$= \pi^{-\frac{1}{4}} e^{\frac{i}{2}y^2 \tan 2z} \frac{1}{\sqrt{\cos 2z}} \begin{pmatrix} -i\sin 2z \\ \cos 2z \end{pmatrix} \tag{7.67}$$

for complex z with $\Im(z) \geq 0$. Note that for $z = 0$, the identities reduce to (7.43).

Computation of the values of r_j in (7.37a) gives (for $g(y) = 1$)

$$r_0 = \frac{1}{\sqrt{2}} \int_{-\infty}^{\infty} \psi_0(y)\,dy \;\; = \;\; \pi^{\frac{1}{4}}$$

$$r_{2j+1} = \frac{1}{2\sqrt{2}} \int_{-\infty}^{\infty} \left(\frac{\psi_{2j+2}}{\sqrt{2j+2}} - \frac{\psi_{2j}}{\sqrt{2j+1}}\right) dy \;\; = \;\; -\pi^{\frac{1}{4}} \frac{2\alpha_{2j+1}}{4j+3}$$

$$r_{2j} \;\; = \;\; 0$$

and hence one finds from (7.46b) that

$$\mathbf{L}(\eta, y) = \pi^{\frac{1}{4}} \mathbf{K}(\eta, y) \tag{7.69}$$

Using the integral

$$\int_{-\infty}^{\infty} e^{i\frac{y^2}{2}\tan(2z)}\,dy = \sqrt{\frac{2\pi}{i\tan 2z}}$$

one obtains as a solution for h_E^G in (7.47), using the identity (7.67) for $z = -i\phi$, as

$$h_E^G(\tilde{\phi}; x_0) = \sqrt{\frac{\sin(2\tilde{\phi}x_0)}{\sin(2\tilde{\phi})}}$$

where $\tilde{\phi} = -i\phi = \omega - i\epsilon_o$. The Green's function **G** is then completely known and the total response to a wind stress with zonal dependence $f(x)$ can be computed from (7.35) as

$$u(x,y;\tilde{\phi}) = h_E(\tilde{\phi})\frac{\sin 2\tilde{\phi}(x-1)}{\sqrt{\cos 2\tilde{\phi}(x-1)}}e^{i\frac{y^2}{2}\tan(2\tilde{\phi}(x-1))} -$$

$$- \int_x^1 f(x_0)\frac{\sin 2\tilde{\phi}(x-x_0)}{\sqrt{\cos 2\tilde{\phi}(x-x_0)}}e^{i\frac{y^2}{2}\tan 2\tilde{\phi}(x-x_0)}\,dx_0 \qquad (7.70a)$$

$$h(x,y;\tilde{\phi}) = h_E(\tilde{\phi})\sqrt{\cos 2\tilde{\phi}(x-1)}e^{i\frac{y^2}{2}\tan(2\tilde{\phi}(x-1))} -$$

$$- \int_x^1 f(x_0)e^{i\frac{y^2}{2}\tan(2\tilde{\phi}(x-x_0))}\sqrt{\cos 2\tilde{\phi}(x-x_0)}\,dx_0 \qquad (7.70b)$$

where

$$h_E(\tilde{\phi}) = \int_0^1 \sqrt{\frac{\sin(2\tilde{\phi}x_0)}{\sin(2\tilde{\phi})}}f(x_0)\,dx_0 \qquad (7.71)$$

is the thermocline amplitude at the east coast.

The thermocline response to a periodic wind stress with the spatial structure

$$\tau^x = 0.6(0.12 - \cos^2\frac{\pi(x - 0.57)}{1.14})\cos\frac{2\pi t}{\mathcal{P}} \qquad (7.72)$$

is shown in Fig. 7.15 with a period $\mathcal{P}$ corresponding to 3 years (Neelin *et al.*, 1998). In the panels, time $t = 0$ the indicates the phase of maximum westerly winds and no winds are present at $t = -\mathcal{P}/4$. At times when the wind stress is present, the thermocline response is nearly in steady balance with the wind stress. The ocean does not only react to the instantaneous wind structure but also to previous winds through propagation of waves. The structures off the equator to the west of the wind are partly free Rossby waves which are still adjusting to the wind but part of this response is just a forced response in steady balance with the wind stress. It is the departure of this steady balance, which is crucial to further evolution of the flow and provides the ocean with a memory.

A measure of this memory was considered in Neelin *et al.* (1998) to be the difference between the actual response and that which would be at every time in steady balance with the wind stress. This difference is plotted in Fig. 7.15b from which it can be seen that the 'ocean memory' is largest near the western boundary. It acts like as a reservoir of unadjusted heat content that is fed down the western boundary to an equatorial boundary layer. The importance of this memory component for ENSO dynamics will become clear in later sections.

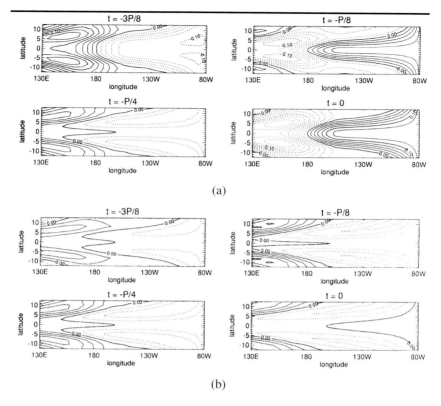

Figure 7.15. (a) Response of the thermocline to a periodic wind fluctuation having a spatial structure (7.72). Plotted are the spatial patterns of the oscillation for several phases of the oscillation. (b) Memory of the ocean, defined as the difference between the actual thermocline response and that which would be obtained when the thermocline response at every time would be in steady balance with the wind stress (Neelin et al., 1998).

7.3.3. Processes determining the SST

Once the ocean fields are known, next step is to determine the changes in SST. The upper layers of the ocean are generally well-mixed up to a depth of $50\ m$ and the temperature is fairly vertically homogeneous.   Consider such a mixed layer in Fig. 7.16, having a constant depth H_m. The temperature in the mixed layer changes due to air-sea interaction, processes at the bottom of the mixed layer and advection. The net heat flux from the atmosphere into the ocean is denoted by Q_{oa} (positive when heat is transferred from atmosphere into the mixed layer) and the heat flux at the bottom of the mixed layer by Q_b (positive when heat leaves the mixed layer). The general temperature equation is given by

$$\frac{\partial T_*}{\partial t_*} + \mathbf{u}_*.\nabla T_* = K_H \nabla_H^2 T_* + K_V \frac{\partial^2 T_*}{\partial z_*{}^2} \qquad (7.73)$$

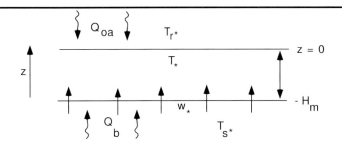

Figure 7.16. Sketch of the mixed layer ocean model. The heat flux Q_{oa} is taken positive when heat is transferred from the atmosphere to the ocean and the heat flux Q_b is taken positive when heat leaves the mixed layer.

where ∇_H^2 is the horizontal Laplace operator. The boundary conditions are

$$z_* = 0 \quad : \quad \rho C_p K_V \frac{\partial T_*}{\partial z_*} = Q_{oa} \qquad (7.74a)$$

$$z_* = -H_m \quad : \quad \rho C_p K_V \frac{\partial T_*}{\partial z_*} = Q_b \qquad (7.74b)$$

In the approximation that the temperature is vertically homogeneous over the layer, one can integrate (7.73) over the layer which results in

$$\frac{\partial T_*}{\partial t} + u_* \frac{\partial T_*}{\partial x_*} + v_* \frac{\partial T_*}{\partial y_*} = K_H \nabla_H^2 T_* + \frac{Q_{oa} - Q_b}{\rho C_p H_m} \qquad (7.75)$$

The ocean/atmosphere heat flux is composed of four contributions, i.e. short wave, longwave, latent and sensible heat fluxes. The net heat flux Q_{oa} can be parameterized into a simple form as, for example, given by Haney (1971)

$$\frac{Q_{oa}}{\rho C_p H_m} = -a_1 (T_* - T_{r*}) \qquad (7.76)$$

where T_{r*} is a reference atmospheric equilibrium temperature. At the lower boundary, the heat flux is composed of diffusive and advective contributions, with the latter dominating, i.e.

$$\frac{Q_b}{\rho C_p H_m} = w_* \frac{T_* - T_{s*}}{H_u} \qquad (7.77)$$

where w_* is a typical vertical velocity at the bottom of the mixed layer, H_u a vertical distance such that the temperature gradient between the mixed layer and the subsurface temperature T_{s*} is well approximated. The subsurface temperature will depend on the vertical temperature distribution and hence on the position of the thermocline. If the thermocline depth increases (decreases), then T_{s*} will increase (decrease). More detailed dependencies are considered in later sections.

There are two ways in which the vertical velocities w_* are generated. Changes in the thermocline (for example through the propagation of waves) are accompanied by vertical velocities. In addition, vertical velocities are caused by horizontal convergences/divergences in the Ekman layer where frictional processes in the upper layer provide the momentum transfer from the surface down to the interior. With a linear frictional damping coefficient a_s, a balance between the frictional processes, the Coriolis effect and wind stress leads to

$$a_s u_{E*} - \beta_0 y_* v_{E*} = \frac{\tau_0 \tau^x}{\rho H_m} \tag{7.78a}$$

$$a_s v_{E*} + \beta_0 y_* u_{E*} = \frac{\tau_0 \tau^y}{\rho H_m} \tag{7.78b}$$

where the subscript E refers to the Ekman layer velocities. The vertical velocities due to the Ekman dynamics are given by

$$w_{E*} = H_m \left(\frac{\partial u_{E*}}{\partial x_*} + \frac{\partial v_{E*}}{\partial y_*} \right) \tag{7.79}$$

Using the scaling (7.8) and a vertical velocity scale $H_m c_0 / L$, the dimensionless equations (7.78) become

$$\epsilon_s u_E - y v_E = \frac{H}{H_m} F_0 \tau^x \tag{7.80a}$$

$$\epsilon_s \zeta_o^2 v_E + y u_E = \frac{H}{H_m} F_0 \tau^y \tag{7.80b}$$

$$w_E = \frac{\partial u_E}{\partial x} + \frac{\partial v_E}{\partial y} \tag{7.80c}$$

with $\epsilon_s = a_s L / c_0$ is the surface layer damping and $F_0 = \tau_0 L / (\rho H c_o^2)$ as before. These equations can be easily solved for given wind stress and the expressions are

$$u_E = \frac{H}{H_m} F_0 \frac{\epsilon_s \zeta_o^2 \tau^x + y \zeta_0 \tau^y}{\epsilon_s^2 \zeta_o^2 + y^2} \tag{7.81a}$$

$$v_E = \frac{H}{H_m} F_0 \frac{-y \tau^x + \epsilon_s \zeta_0 \tau^y}{\epsilon_s^2 \zeta_o^2 + y^2} \tag{7.81b}$$

such that the dimensionless upwelling at the equator (with $\tau^y = 0$) is given by

$$w_E = \frac{H}{H_m} F_0 \left[\frac{1}{\epsilon_s} \frac{\partial \tau^x}{\partial x} - \frac{\tau^x}{\epsilon_s^2 \zeta_o^2} \right] \tag{7.82}$$

For a constant zonal wind stress $\tau^x = -1$, the dimensional upwelling w_{E*} is constant along the equator and given by

$$w_{E*} = \frac{\tau_0 \beta_0}{\rho a_s^2} \tag{7.83}$$

and with a wind stress amplitude $\tau_0 = 0.05 \, Pa$ and $a_s = 5.0 \times 10^{-6} \, s^{-1}$, its dimensional value is a few meters per day.

7.3.4. Feedbacks

In the previous sections, the elementary physical processes have been discussed and there is enough background now to discuss feedbacks in the coupled system. There are three important feedbacks, which are called the thermocline, the upwelling and the zonal advection feedback. The first two are associated with changes in the vertical heat transport modelled by (7.77), and the third is associated with horizontal heat transport. The feedbacks are sketched in this section in their most elementary form.

7.3.4.1 Thermocline feedback

This feedback is best explained by looking at a sloping thermocline in a constant upwelling ocean as sketched in Fig. 7.17. The sloping thermocline is brought about by background winds. The formulation in Technical box 7.3 provides analytic expressions for the thermocline shape, if the stationary forcing ($\omega = 0$) is only zonally dependent ($\tau^x = f(x)$, $\tau^y = 0$). In in the limit $\epsilon_o \to 0$ and hence $\tilde{\phi} \to 0$), the expression (7.70b) becomes

$$\bar{h}(x) = \int_0^1 f(x_0)\sqrt{x_0}\,dx_0 - \int_x^1 f(x_0)\,dx_0 \qquad (7.84)$$

For example, when the zonal easterly wind stress is constant, $f(x) = -f_0 < 0$ it follows that $\bar{h}(x) = f_0(1/3 - x)$. The thermocline slope is constant for this case and the thermocline anomaly is positive over one third of the basin. This background wind also gives upwelling at the equator, according to (7.82) as

$$\bar{w}_E = \frac{H}{H_m}\frac{F_0 f_0}{\epsilon_s^2 \zeta_o^2} \qquad (7.85)$$

which is constant and positive.

Now assume that a positive SST perturbation $\tilde{T}$ is present at some location, for example in the eastern part of the basin (Fig. 7.17). This leads to a perturbation in the low level zonal wind which is westerly with a maximum located west of the maximum of the SST anomaly according to the Gill model response shown in the Figs. 7.14. Since the background winds are weakened locally (f_0 smaller), the slope of the thermocline decreases and it becomes more flat. In this case, the colder water will be closer to the surface in the west but it will be farther down in the east. In other words, in the east the thermocline is deeper and hence the subsurface temperature is higher. Hence, the subsurface temperature effectively increases at the level of upwelling, giving a positive heat flux perturbation at the bottom of the mixed layer according to (7.75) and (7.77), i.e.

$$\frac{\partial \tilde{T}_*}{\partial t_*} \approx -\bar{w}_*\frac{\tilde{T}_* - \tilde{T}_{s*}}{H_u} \qquad (7.86)$$

As $\bar{w}_*$ is positive, the first term in the right hand side represent the local damping of SST anomalies. However, when $\tilde{T}_{s*} > 0$, then the second term on the right hand side is positive and the original disturbance may be amplified. Due to west-ward shift of the wind response with respect to the maximum SST anomaly, the maximum response of the thermocline is eastward of the maximum SST anomaly. Hence, this induces eastward propagation tendencies to the perturbations.

7.3.4.2 Upwelling feedback

As another prototype situation, consider that the thermocline is fixed with a cer-tain slope related to the background winds, the latter similar to the previous case. Again a positive SST-anomaly is present in the east which generates the same changes in the wind as before (Fig. 7.17). However, now the changes only influ-ence the upwelling, mainly through the Ekman layer dynamics. Weaker easterly winds imply less upwelling and hence less colder water enters the mixed layer. This can also be seen from (7.77), i.e.

$$\frac{\partial \tilde{T}_*}{\partial t_*} \approx -\tilde{w}_* \frac{\bar{T}_* - \bar{T}_{s*}}{H_u} \tag{7.87}$$

If $\tilde{w} < 0$ and the background vertical temperature gradient is stably stratified ($\bar{T}_* > \bar{T}_{s*}$), then the surface temperature perturbation is amplified. The maximum downwelling anomaly occurs west of the SST anomaly and hence the upwelling feedback introduces westward propagating tendencies to the SST anomalies.

7.3.4.3 Zonal advection feedback

The zonal advection feedback arises through zonal advection of heat induced by zonal velocity anomalies driven by wind anomalies. Imagine a region with a strong annual-mean SST gradient, say $\partial \bar{T}_*/\partial x_* < 0$. Such a region occurs, for example, at the eastern side of the warm pool. Suppose a positive SST anomaly ($\tilde{T}_* > 0$) occurs, which leads again to westerly wind anomalies. Consequently, the zonal surface ocean current ($\tilde{u}_* > 0$) is intensified (Fig. 7.17) leading to am-plification of the positive temperature perturbation, according to (7.75), i.e.,

$$\frac{\partial \tilde{T}_*}{\partial t_*} \approx -\tilde{u}_* \frac{\partial \bar{T}_*}{\partial x_*} \tag{7.88}$$

Part of the mixed layer zonal velocity is due to equatorial wave dynamics and part is due to Ekman dynamics.

7.3.4.4 Strength of the feedbacks

From the description of the coupled processes of the ocean and atmosphere, a measure of the strength of the feedbacks can be obtained. An SST anomaly $\tilde{T}$ with an amplitude of ΔT gives an atmospheric forcing according to the Gill model of amplitude $q_0 = \alpha_T \Delta T$. From technical box 7.2, the amplitude of the zonal wind response due to this anomaly is given by

$$c_a \mu_0 = \frac{\alpha_T \Delta T L}{c_a^2} \tag{7.89}$$

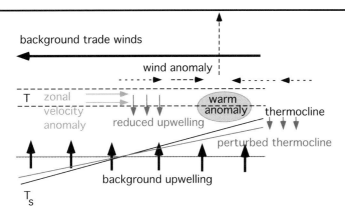

Figure 7.17.   (in color on page 532). Sketch to illustrate the thermocline feedback, the upwelling feedback and the zonal advection feedback. In each case, a warm SST anomaly induces wind-stress anomalies. (i) Thermocline feedback: the wind anomaly leads to changes in the thermocline slope, which induces — with constant background upwelling — an amplification of the SST anomaly. (ii) Upwelling feedback: the wind anomaly leads to changes in the upwelling which induces — in a background stably stratified temperature field — an amplification of the SST anomaly. (iii) Zonal advection feedback: the wind anomaly induces stronger zonal advection which, if the annual-mean zonal SST gradient is negative, leads to amplification of the SST anomaly.

The surface wind anomaly creates a wind stress anomaly of amplitude, according to (7.66) of amplitude

$$\frac{\tilde{\tau}^x_*}{\rho H} = \frac{\gamma \alpha_T \Delta T L}{c_a^2} \tag{7.90}$$

Hence, the main parameter controlling the strength of the feedback is the amount of wind stress per SST anomaly, which is indicated here by μ_* (with dimension $N/(m^2 K)$),

$$\mu_* = \frac{\rho H \gamma \alpha_T L}{c_a^2} \tag{7.91}$$

7.4. The Zebiak-Cane Model

One of the first models that was able to reasonably simulate ENSO was that of Zebiak and Cane (1987). In its original version, an annual-mean state and seasonal cycle of both ocean and atmosphere is obtained from observations and within the model the evolution of anomalies with respect to this reference state are computed. The model produces recurring warm events that are irregular in both amplitude and spacing, but favor a 3 to 4 year period. A summary of the model set-up is given in section 7.4.1 using elements of model development already presented in the previous sections. An overview of the main results of the model is given in section 7.4.2.

7.4.1. Formulation

The model captures the evolution of large scale motions in the tropical ocean and atmosphere in a domain of infinite extent in the meridional direction. The ocean is bounded by meridional walls at the west ($x_* = 0$) and east ($x_* = L$) coast. The ocean component of the model consists of a well-mixed layer of mean depth H_1 embedded in a shallow water layer of mean depth $H = H_1 + H_2$ having a constant density ρ (Fig. 7.18). Only long wave motions above the thermocline are considered and the deep ocean (having a constant density $\rho + \Delta\rho$) is assumed to be at rest.

As a first step, mean horizontal velocities over the layers are defined as

$$\mathbf{u}_{1*} = \frac{1}{H_1} \int_{-H_1}^{0} \mathbf{u}_* \, dz_* \tag{7.92a}$$

$$\mathbf{u}_{2*} = \frac{1}{H_2} \int_{-H}^{-H_1} \mathbf{u}_* \, dz_* \tag{7.92b}$$

$$\mathbf{u}_{m*} = \frac{1}{H} \int_{-H}^{0} \mathbf{u}_* \, dz_* = \frac{1}{H}(H_1\mathbf{u}_{1*} + H_2\mathbf{u}_{2*}) \tag{7.92c}$$

It is then assumed that the difference velocity between the first layer and the total mean velocity is exactly the component of the velocity which is induced by frictional processes in the Ekman layer. Hence,

$$\mathbf{u}_{s*} = \mathbf{u}_{1*} - \mathbf{u}_{m*} = \frac{H_2}{H}(\mathbf{u}_{1*} - \mathbf{u}_{2*}) = \frac{H_2}{H}\mathbf{u}_{E*} \tag{7.93}$$

The mean velocity $\mathbf{u}_{*m}$ satisfies the reduced gravity model with a thermocline h_*, having an equilibrium depth H.

The evolution of the mixed layer temperature T_* is governed by the equation

$$\frac{\partial T_*}{\partial t_*} + a_T(T_* - T_0) \quad + \quad \frac{w_{1*}}{H_u}\mathcal{H}(w_{1*})(T_* - T_{s*}(h_*))$$

$$+ \quad u_{1*}\frac{\partial T_*}{\partial x_*} + v_{*1}\frac{\partial T_*}{\partial y_*} = 0 \tag{7.94}$$

where $\mathcal{H}$ is a continuous approximation of the Heaviside function. The horizontal mixing in (7.75) has been simplified by approximating the dissipative processes as linear damping,

$$K_H \nabla_H^2 T_* \approx -a_2 T_* \tag{7.95}$$

with a damping coefficient incorporated in a_T. The approximations of the surface and bottom fluxes (7.76) and (7.77) have also been used. The second term in (7.94) is usually referred to as the Newtonian cooling term, with inverse damping time a_T, representing all processes as horizontal mixing, sensible and latent heat surface fluxes, and long wave and shortwave radiation. T_0 is the temperature of radiative equilibrium which is realized in the absence of large-scale horizontal motion in the upper ocean and atmosphere. The next term models the heat flux due

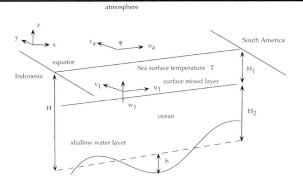

Figure 7.18.   Schematic representation of the Zebiak-Cane model showing both surface layer and shallow water layer. the latter bounded below by the thermocline.

to upwelling through the total velocity w_{*1} and the approximate vertical temperature gradient $(T_* - T_{s*}(h_*))/H_u$. The subsurface temperature (T_{s*}) depends on the thermocline deviations and models the effect that heat is transported upwards (if $w_{*1} > 0$) when the cold water is further from the surface. Explicit expressions will be given below. The last two terms in (7.94) represent horizontal advection. The vertical velocity component, the upwelling, is determined from continuity

$$w_{1*} = H_1 \left(\frac{\partial u_{1*}}{\partial x_*} + \frac{\partial v_{1*}}{\partial y_*} \right) \tag{7.96}$$

For readers who have not gone through the previous subsections, the full dimensional equations of the simplest ZC-type model are given. In addition to (7.94), the ocean model is the reduced gravity model (section 7.2)

$$\frac{\partial u_*}{\partial t_*} + a_m u_* - \beta_0 y_* v_* + \frac{\partial h_*}{\partial x_*} = \frac{\tau_*^x}{\rho H} \tag{7.97a}$$

$$\beta_0 y u_* + g' \frac{\partial h_*}{\partial y_*} = \frac{\tau_*^y}{\rho H} \tag{7.97b}$$

$$\frac{\partial h_*}{\partial t_*} + a_m h_* + c_o^2 \left(\frac{\partial u_*}{\partial x_*} + \frac{\partial v_*}{\partial y_*} \right) = 0 \tag{7.97c}$$

with boundary conditions

$$\int_{-\infty}^{\infty} u_*(0, y_*, t_*) dy = 0 \; , \; u_*(L, y_*, t_*) = 0 \tag{7.98}$$

The equations for the surface layer velocities are

$$a_s u_{s*} - \beta_0 y_* v_{s*} = \frac{H_2}{H} \frac{\tau_*^x}{\rho H_1} \tag{7.99a}$$

$$a_s v_{s*} + \beta_0 y_* u_{s*} = \frac{H_2}{H} \frac{\tau_*^y}{\rho H_1} \tag{7.99b}$$

The atmospheric zonal and meridional boundary-layer velocities (U_*, V_*) and geopotential Θ_* satisfy the steady balances in the Gill model and hence,

$$\frac{\partial U_*}{\partial t_*} + a_M U_* - \beta_0 y_* V_* - \frac{\partial \Theta_*}{\partial x_*} = 0 \qquad (7.100a)$$

$$\frac{\partial V_*}{\partial t_*} + a_M V_* + \beta_0 y_* U_* - \frac{\partial \Theta_*}{\partial y_*} = 0 \qquad (7.100b)$$

$$\frac{\partial \Theta_*}{\partial t_*} + a_M \Theta_* - c_a^2 \left(\frac{\partial U_*}{\partial x_*} + \frac{\partial V_*}{\partial y_*} \right) = \alpha_T (T_* - T_{r*}) \qquad (7.100c)$$

The right hand side of (7.100c) is the approximation of the heat flux according to (7.76). The model is closed when explicit expressions are provided for the function T_{s*} and T_0. Moreover, the reference temperature T_{r*} has to be chosen. In addition, the model contains quite a set of parameters for which 'best' values have to be provided. In Zebiak and Cane (1987), more details of the model set-up are provided and the atmospheric model is slightly more complicated since the convergence feedback scheme of Zebiak (1982) is implemented.

7.4.2. Results

In Zebiak and Cane (1987), the model is used in 'anomaly mode' where only the evolution of quantities with respect to some reference state are computed. The latter state is derived from observations and can be an annual-mean state or a state varying with annual period. If the reference state is indicated by $\bar{\mathbf{u}}_* = (\bar{u}_*, \bar{v}_*), \bar{h}_*, \bar{T}_*$, then equation for the SST anomalies $\tilde{T}$ used is

$$\frac{\partial \tilde{T}_*}{\partial t_*} = -a_T \tilde{T}_* - (M(\bar{w}_{1*} + \tilde{w}_{1*}) - M(\bar{w}_{*1})) \frac{\partial \bar{T}_*}{\partial z_*}$$

$$-M(\bar{w}_{1*} + \tilde{w}_{1*}) \frac{(\tilde{T}_* - T_{s*}(\tilde{h}_*))}{H_1} - \bar{\mathbf{u}}_{1*}.\nabla \tilde{T}_* - \tilde{\mathbf{u}}_{1*}.\nabla (\bar{T}_* + \tilde{T}_*) \qquad (7.101)$$

with $\partial \bar{T}_*/\partial z_*$ a prescribed vertical temperature gradient and $M(x) = x$ if $x > 0$ and zero elsewhere. The function T_{s*} is chosen in such a way that saturation to a temperature T_1 and T_2 occur when h_* becomes very shallow and deep, respectively. Extra parameters are introduced to control the transition between these temperatures at intermediate values of h_*. The reference temperature T_{r*} is related to $\bar{T}_*$ although a more complicated diabatic heating scheme is used.

In Zebiak and Cane (1987), a 90-year simulation is described which is initialized by a four month (from December to April of the first year) duration westerly wind anomaly in the region 145°E - 170°W with an meridional exponential decay scale of 20° and an amplitude of 2 m/s. After this initial disturbance, the model computes the evolution of the anomalous fields with respect to the prescribed annual cycle. A time series of the SST anomalies (represented by the NINO3 and NINO4 indices) for this simulation is shown in Fig. 7.19, where the NINO4 index is defined in the caption. Sustained oscillations are found with a period of about 3-4 years, with recurrent warm and cold events, each of about 14-18 month du-

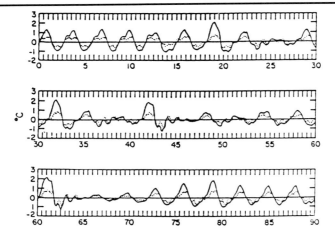

Figure 7.19.   Time series (time in years) of the NINO3 (drawn) index (area 5°N - 5°S × 90°W - 150°W and the NINO4 (dotted) index (area 5°N - 5°S × 150°W - 160°W of the Zebiak-Cane model (Zebiak and Cane, 1987)

ration. The peaks of the warm events, with an amplitude of about 2-3°C, closely follow the annual cycle and tend to occur either in June or at the end of the year.

Equatorial anomalies of thermocline depth and zonal wind stress are plotted in Fig. 7.20. The major warm events (for example, year 32 and 42) are preceded by anomalously high equatorial heat content. An example is provided at the beginning of year 31 (Fig. 7.20a), where the zonally averaged thermocline depth anomaly is positive. Following warm events, the equatorial heat content is low over the whole basin. Strong westerly wind anomalies occur during warm events (Fig. 7.20b) with maximum amplitude in the central Pacific. The equatorial anomalies show the characteristic propagation of the heat content anomalies, with west leading east, and a nearly stationary wind response similar to that observed (section 7.1). The effect of the nonzero zonally averaged heat content is important, since when its effect is neglected in the model, through modification in T_{s*}, the interannual oscillation disappears and an annual oscillation on either a too warm or too cold mean state is found. If its effect is only partially taking into account, an interannual oscillation with period longer than 4 year found, whereas when its effect is exaggerated the period is much shorter than 4 year.

The occurrence of the oscillation appears to be quite robust to variation of parameters, although it disappears when the coupling, i.e. the amount of wind stress per SST anomaly, becomes too weak. There is substantial influence of the annual cycle. When the annual cycle is fixed at some stage in the evolution of a warm event, the development of anomalies is very different. Fixation of April conditions before the peak of the event in December, results in slower growth but longer persistence of the SST anomalies. Fixation of August conditions in the same year

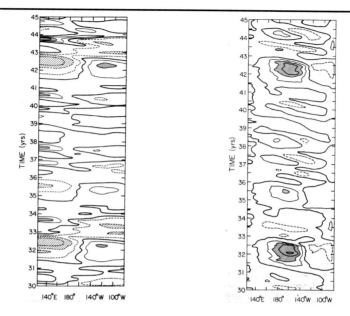

Figure 7.20.   Equatorial anomalies for (a) thermocline and (b) zonal wind stress. In both panels, positive anomalies are indicated with solid lines and negative by dashed lines. In (a) the contour interval is 10 m and anomalies greater (smaller) than 20 (-20) m are stippled (hatched). In (b) anomalies greater than 0.015 Pa are stippled (Zebiak and Cane, 1987).

causes a substantial increase in growth rate while fixation in December that year (or July next year) hardly has impact on the development. The (northern) Summer appears to be most favorable for rapid growth of the anomalies and Spring seems less favorable. This is directly related to the annual cycle development, with large east-west gradients in the (northern) summer, stronger trade winds and stronger upwelling (section 7.1). Even without an annual cycle present, for example by fixing background conditions at July during the whole simulation, an interannual oscillation is found, but it is much more regular (nearly periodic). This suggests that the annual cycle contributes to the irregularity of the interannual oscillation.

Although this model has been a major breakthrough in ENSO research, it has taken a while before the interannual response of this model was understood. In Zebiak and Cane (1987), the major physical processes at work in the interannual signal are already identified. First, a positive feedback between the large scale ocean and atmosphere leads to amplification of SST anomalies. Next, equilibration of temperature anomalies occurs through nonlinear effects, in the model mainly in the SST equation, which set the finite amplitude of the anomalies. The period of the cycle is determined by a systematic time delay between dynamical changes in the eastern ocean and associated large-scale fluctuations in the equatorial wind stress. In the following sections this descriptive view will be made more precise through more detailed analysis of the solutions of the model.

7.5. Towards the Delayed Oscillator

As deduced from the simulations in ZC-model, it appears that when the amount of wind stress per SST anomaly is too small, no ENSO like oscillations occur (Zebiak and Cane, 1987; Battisti, 1988). This indicates that an amplification mechanism is at work only above some critical coupling strength and has motivated studies on the stability of simple annual-mean states in Zebiak-Cane type models.

7.5.1. Coupled modes: periodic ocean basin

Philander *et al.* (1984) consider a substantial simplification of the ZC-model, by neglecting the surface layer physics and taking a basin of infinite zonal extend. As annual-mean state a flat thermocline, with a zonal linear temperature profile is taken and the mean state is motionless. The heat flux anomaly forcing the atmospheric wind anomalies is taken proportional to the thermocline depth anomaly, i.e.

$$\tilde{Q}_* = \alpha \, \tilde{h}_* \tag{7.102}$$

and wind stress anomalies are taken proportional to the surface wind anomalies. In this way, SST anomalies are linearly related to thermocline anomalies and an SST equation is not needed. Using the steady Gill model and the reduced gravity ocean model leads to a linear system of equations governing the evolution of perturbations on the mean state. A few special cases are considered such as the non-rotating and constant rotating (no β-effect) case, and the particular limit $c_o = c_a$. Indications of unstable modes are found, but the eigenvalue problem associated with linear stability of the particular mean state is not solved.

The latter was done in Hirst (1986), with the same simplifications as (no surface layer, horizontally unbounded domain, motionless basic state) in Philander *et al.* (1984). When (7.94) is linearized around such a basic state a more general SST equation results,

$$\frac{\partial \tilde{T}_*}{\partial t_*} + a_T \tilde{T}_* - \tilde{u}_* \frac{d\bar{T}_*}{dx_*} - K_T \tilde{h}_* = 0 \tag{7.103}$$

where $d\bar{T}_*/dx_*$ is the prescribed basic state zonal temperature gradient and K_T arises through linearization of the vertical heat flux around the basic state. The adiabatic heat flux forcing of the atmosphere is taken proportional to the SST anomalies, with proportionality constant K_Q, and the wind stress is proportional to the lower level wind with constant factor K_S. Within this SST evolution equation, three different simplifications are considered. One of these simplifications (model I in Hirst (1986)) is the limit of large K_T and a_T. In this case, one obtains from (7.103) that $\tilde{T}_* = \kappa \tilde{h}_*$, where $\kappa = K_T/a_T$ and hence the heat flux anomaly in the atmosphere is proportional to the thermocline anomaly as in Philander *et al.* (1984).

A consequence of the zonal extent of the domain is that the equations can be separated in x_*. Hence, in a normal mode approach, solutions are sought of the form

$$\phi_*(x_*, y_*, t_*) = e^{i(k_* x_* - \sigma_* t_*)} \hat{\phi}_*(y_*) \tag{7.104}$$

for all quantities ϕ_* of the model. The evolution of the perturbation quantities on the mean state then reduces to an eigenvalue problem which in dimensional quantities becomes

$$a_M \hat{U}_* - \beta_0 y_* \hat{V}_* - i k_* \hat{\Theta}_* = i \sigma_* \hat{U}_* \tag{7.105a}$$

$$a_M \hat{V}_* + \beta_0 y_* \hat{U}_* - \frac{\partial \hat{\Theta}_*}{\partial y_*} = i \sigma_* \hat{V}_* \tag{7.105b}$$

$$a_M \hat{\Theta}_* - c_a^2 (i k_* \hat{U}_* + \frac{\partial \hat{V}_*}{\partial y_*}) + K_Q \hat{T}_* = i \sigma_* \hat{\Theta}_* \tag{7.105c}$$

$$a_T T_* - \hat{u}_* \frac{d\bar{T}_*}{dx_*} - K_T \hat{h}_* = i \sigma_* \hat{T}_* \tag{7.105d}$$

$$a_m \hat{u}_* - \beta_0 y_* \hat{v}_* + i k_* \hat{h}_* - K_S \hat{U}_* = i \sigma_* \hat{u}_* \tag{7.105e}$$

$$a_m \hat{v}_* + \beta_0 y_* \hat{u}_* + \frac{\partial \hat{h}_*}{\partial y_*} - K_S \hat{V}_* = i \sigma_* \hat{v}_* \tag{7.105f}$$

$$a_m \hat{h}_* + c_o^2 (i k_* \hat{u}_* + \frac{\partial \hat{v}_*}{\partial y_*}) = i \sigma_* \hat{h}_* \tag{7.105g}$$

together with boundary conditions that all perturbation quantities approach zero with $y_* \to \pm\infty$. The corresponding eigenvalue problem for σ_* can only be solved numerically and was done in Hirst (1986) using a finite difference and a spectral approach.

Parameter	Value		Parameter	Value	
a_M	5.0×10^{-6}	s^{-1}	a_T	9.2×10^{-8}	s^{-1}
c_a	30	ms^{-1}	K_T	$2.0 \ 10^{-11}$	$K(ms)^{-1}$
$\bar{T}_x$	5.0×10^{-7}	Km^{-1}	c_o	2.0	ms^{-1}
β_0	$2.2 \ 10^{-11}$	$(ms)^{-1}$	K_S	8.0×10^{-8}	s^{-1}
a_m	1.0×10^{-7}	s^{-1}	K_Q	$7 \qquad 10^{-3}$	$m^2 s^{-3} K^{-1}$

Table 7.2. Typical values of dimensional parameters used in Hirst (1986) to solve the stability of a motionless state of the equatorial ocean-atmosphere with constant depth thermocline and a constant zonal temperature gradient $\bar{T}_x$.

For typical values of the parameters, shown in Table 7.2, a result is shown in Fig. 7.21 for the model where anomalies in SST are taken proportional to the thermocline anomalies, i.e.

$$\hat{T}_* = \kappa \hat{h}_* \tag{7.106}$$

instead of (7.105d). In Fig. 7.21a, the growth rate of each mode is plotted as

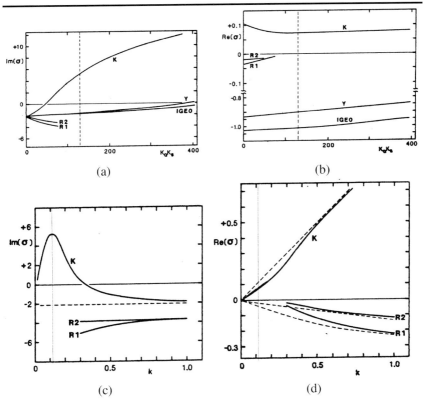

Figure 7.21. (a) Growth rate $(Im(\sigma))$ and (b) angular frequency $(Re(\sigma))$ of the most unstable mode in the model of Hirst (1986) for a particular value of the dimensionless wavenumber k (indicated by the vertical line in (b)) as a function of the coupling strength $K_Q K_S$. Length is scaled with the oceanic Rossby radius of deformation $\lambda_o = 250$ km. (c) and (d) Similar as (a) and (b), but now coupling strength $K_Q K_S$ is fixed at a value (indicated by the vertical line) in (a) and the wavenumber is varied.

a function of the coupling parameter $K_Q K_S$ for fixed $k = 0.106$, which corresponds to a dimensional wavelength $\lambda_* = 2\pi\lambda_o/0.106 \approx 15{,}000$ km. As coupling increases, the free oceanic Kelvin wave (labelled K) becomes unstable. In Fig. 7.21b, it is shown that the frequency of the wave slightly decreases giving a (slightly) larger travel time of the wave. Coupling affects also other free modes [3] but all are more stable than the coupled Kelvin wave. The same is found when coupling is fixed and the wavenumber is varied as in Fig. 7.21c. There is a band of wavenumbers for which each Kelvin wave is unstable. The shorter Kelvin waves

---

[3]Here the Rossby modes with index j are labelled R_j, the Yanai mode is labelled Y and an inertia-gravity mode is labelled $IGEO$.

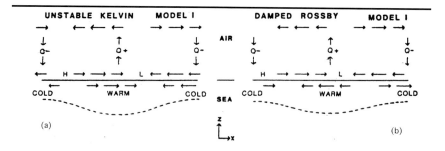

Figure 7.22.   Sketch of the mechanism of (a) destabilization of the oceanic Kelvin mode through coupling and (b) stabilization of the first baroclinic Rossby wave. The anomalies are considered at the equator with Q indicating the heat flux into/out of the atmosphere.

stabilize again, while the frequency increases substantially with k, but not very different from the uncoupled modes (Fig. 7.21d).

Within this simple model, the effect of coupled processes on the free oceanic waves can be easily determined. The two different cases of Kelvin and Rossby modes are considered in Fig. 7.22. Consider a positive equatorial thermocline anomaly $\tilde{h}$ as in Fig. 7.22a (a deeper thermocline). As the SST anomaly $\tilde{T}$ is in phase with $\tilde{h}$, air rises above the warmer sea surface leading to the heat flux indicated by Q_+. According to the response of the Gill model, westerly surface wind anomalies appear to the west of the heating anomalies with easterlies to the east. The effect on the zonal velocities in the upper ocean creates a convergence in the upper ocean layer with a maximum slightly east of the initial thermocline anomaly. Hence, the anomaly is amplified and tends to propagate eastwards. For the westward travelling Rossby waves, the wind induced zonal velocities create a divergence which damps the initial positive thermocline anomaly (Fig. 7.22b).

In Hirst (1986), the sensitivity of other forms of the SST - equation is considered. It appears that the type of mode actually destabilized is sensitive to the processes which control SST. In the advective limit (model II in Hirst (1986)), with $K_T = 0$, zonal advection is able to destabilize the Rossby modes, whereas the Kelvin mode is stable and the coupled mode has westward propagating tendencies. Within the full SST equation (model III in Hirst (1986)), a slow westward propagating mode becomes unstable which seems unrelated to free waves in the uncoupled situation, but has its origin in the adjustment of the SST. The latter mode has features of modes found in earlier work by Anderson and McCreary (1985).

In Neelin (1991b), a next step is taken by considering the effect of surface layer processes in more detail. A simplification of the full SST equation to an equatorial strip is introduced by meridional differencing. More specific, the SST equation

(7.94) is simplified as

$$\frac{\partial T_*}{\partial t_*} + a_T(T_* - T_0) + \frac{w_{1*}}{H_u}\mathcal{H}(w_{1*})(T_* - T_{s*}(h_*))$$

$$+u_{1*}\frac{\partial T_*}{\partial x_*} - \mathcal{H}(-v_{N*})\frac{2v_{N*}}{\lambda_o}(T_* - T_{N*}) = 0 \qquad (7.107)$$

where v_{N*} is the meridional velocity at the northern boundary of the strip and T_{N*} a fixed off-equatorial temperature. In the strip, the surface layer physics can also be simplified since the wind stress is mainly zonal ($\tau_*^y \approx 0$). From the surface layer model (7.99) one obtains

$$u_{s*} \approx \frac{H_2}{a_s H_1}\frac{\tau_*^x}{\rho H} = b_u\frac{\tau_*^x}{\rho H} \qquad (7.108\text{a})$$

$$w_{s*} \approx H_1 b_u \frac{\partial}{\partial x_*}\frac{\tau_*^x}{\rho H} - b_w\frac{\tau_*^x}{\rho H} \qquad (7.108\text{b})$$

$$v_{N*} \approx -\frac{\lambda_o}{2H_1}b_w\frac{\tau_*^x}{\rho H} \qquad (7.108\text{c})$$

with $b_u = H_2/(a_s H_1), b_w = (H_1/\lambda_o)b_u$ and where the meridional velocity v_{N*} is determined from the continuity equation through discretization over the strip.

In addition to the unstable Kelvin and Rossby modes found in the periodic basin, the so-called slow SST mode is found which is related to the adjustment of SST. This mode is characterized by amplification of temperature anomalies, which give wind stress anomalies, which by surface layer processes give tendencies in SST. It can have interannual time scale and the mode may exist, even if the time derivative terms in the momentum equations of the reduced gravity model are put to zero. Hence, ocean wave dynamics does not play a role in the destabilization of the mode. Although analytic approximations were obtained in Neelin (1991b) for this mode in the horizontally unbounded basin, we will not address these here, since this mode will show up in the closed basin case discussed in the following sections.

7.5.2. Coupled modes: bounded basin

A natural next step is to consider the stability of more realistic basic states than those considered in Hirst (1986) in a bounded basin. In Hirst (1988), this problem is solved for the simple motionless, flat thermocline basic state using the same model as in Hirst (1986) for an ocean basin of zonal extent L. The unstable modes, as discussed for the periodic basin case are hardly affected by the basin dimensions except that the zonal wavenumbers become quantized.

The problem has been solved in much more detail and for general basic states in a series of papers by Jin and Neelin, both numerically and also analytically in reduced models. To understand the structure of the eigenmodes, for didactical reasons, a simpler configuration is considered: the equatorial strip approximation for the SST equation (7.107) is made.

Using the same scaling of atmosphere and ocean models as in (7.8) and neglecting the meridional winds, the dimensionless equations of the model become (from (7.50), (7.107) and (7.26))

$$-\frac{y}{2}V - \frac{\partial\Theta}{\partial x} + \epsilon_a U = 0 \qquad (7.109a)$$

$$\frac{y}{2}U - \frac{\partial\Theta}{\partial y} = 0 \qquad (7.109b)$$

$$-(\frac{\partial U}{\partial x} + \frac{\partial V}{\partial y}) + \epsilon_a \Theta - \mu_0(T - T_{r*}) = 0 \qquad (7.109c)$$

$$\frac{\partial T}{\partial t} + (w + w_s)\mathcal{H}(w + w_s)(T - T_s(h)) + \epsilon_T(T - T_0)$$

$$+(u + u_s)\frac{\partial T}{\partial x} - \mathcal{H}(-v_N)v_N(T - T_N) = 0 \qquad (7.109d)$$

$$\delta\frac{\partial u}{\partial t} - yv + \frac{\partial h}{\partial x} + \epsilon_o u = \frac{L}{c_0^2}\frac{\tau_*^x}{\rho H} \qquad (7.109e)$$

$$yu + \frac{\partial h}{\partial y} = 0 \qquad (7.109f)$$

$$\delta\frac{\partial h}{\partial t} + \frac{\partial u}{\partial x} + \frac{\partial v}{\partial y} + \epsilon_o h = 0 \qquad (7.109g)$$

with new damping parameter $\epsilon_T = a_T L/c_0$ and appropriate boundary conditions for the ocean as before. The mean velocities in the ocean layer, satisfying the shallow water equations are indicated by u, v and w. The expressions for the dimensionless surface layer quantities follow from (7.108) when divided by $c_0, H_1 c_0/L$ and $\lambda_o c_0/(2L)$, respectively. A factor δ has been included before the time-derivatives of the zonal momentum and continuity equation in the ocean to be able to continuously change between the situation where adjustment processes in the ocean are fast compared to the adjustment of SST (the fast wave limit, $\delta = 0$) and the case where SST adjustment is much faster (the fast SST limit, δ). The dimensionless parameter δ is therefore called the ratio of adjustment times of ocean wave dynamics and SST.

7.5.2.1 The near equatorial behavior

The first task is to construct a more general basic steady state (i.e., the annual-mean state) of which the stability is determined; this can be done in several ways. One can simply compute a realistic mean state from observations, say T_{o*} for SST. In general, this is not a solution of the SST equation but one simply neglects the residue with the argument that the model is too simple to model the mean state realistically. In this way, with $T_{r*} = T_{o*}$, the ocean model is only forced with wind stress anomalies which are caused by the atmospheric anomalous heat flux forcing $\mu_0(T_* - T_{o*})$. This follows the original approach in Zebiak and Cane (1987).

A second possibility is to construct a basic state for which the temperature is a solution of the SST equation. This can be done by prescribing a wind stress field

$$\frac{\tau_*^x}{\rho H} = \frac{\tau_0}{\rho H}\tau_z^x(x)\tau_m^x(y) \qquad (7.110)$$

with some fixed zonal and meridional structure $\tau_z^x(x)$ and $\tau_m^x(y)$. Using this wind stress forcing, the steady state of the ocean model can be computed. Once this state is known, the solution to the SST equation, say $\bar{T}_*$, is obtained. One then takes $T_{r*} = \bar{T}_*$, such that wind anomalies are computed from $T_* - \bar{T}_*$ and hence the wind stress to the ocean model is prescribed as

$$\frac{\tau_*^x}{\rho H} = \frac{\tau_0}{\rho H}\tau_z^x(x)\tau_m^x(y) + \gamma U_* \tag{7.111}$$

where U_* is the atmospheric zonal surface wind and γ is the wind stress coefficient as in (7.66). This is the approach taken in Hao *et al.* (1993) and Dijkstra and Neelin (1995b); the latter paper is followed.

For the Gill atmosphere model response, the equatorial zonal wind response to a temperature anomaly $\tilde{T}$ is approximated, using (7.62), through $U_* = c_a\mu_0(3R_2/2 - R_0/2)$ which can be written as

$$U_* = c_a\mu_0\mathcal{A}(\tilde{T}) \tag{7.112}$$

$$\mathcal{A}(T) = \frac{3}{2}\exp[3\epsilon_a x]\int_x^1 \exp[-3\epsilon_a s]T(s)ds$$

$$-\frac{1}{2}\exp[-\epsilon_a x]\int_0^x \exp[\epsilon_a s]T(s)ds \tag{7.113}$$

The dimensionless wind stress forcing can then be written as

$$\frac{L}{c_0^2}\frac{\tau_*^x}{\rho H} = F_0\tau_z^x(x)\tau_m^x(y) + \mu\mathcal{A}(T - T_r) \tag{7.114}$$

where again $F_0 = (L/c_0^2)\times\tau_0/(\rho H)$ and μ is the central dimensionless coupling parameter given by

$$\mu = \frac{L}{\rho Hc_0^2}\times\mu_* = \frac{\gamma\alpha_T\Delta T L^2}{c_0^2 c_a^2} \tag{7.115}$$

where μ_* was defined in (7.91). This parameter can be seen as the product as the amount of heat flux per SST anomaly times the amount of wind stress produced by the atmospheric surface wind field as a response to this heating anomaly. It indeed reflects the notion of the coupled processes operating in the tropical ocean-atmosphere system. With this choice of the wind stress forcing, the surface layer velocities can be written as

$$u_s = \delta_u(F_0\tau_z^x(x)\tau_m^x(y) + \mu\mathcal{A}(T - T_r)) \tag{7.116a}$$

$$w_s = -(\delta_{Fs}\tau_z^x(x)\tau_m^x(y) + \mu\delta_s\mathcal{A}(T - T_r)) + \frac{\partial u_s}{\partial x}$$

$$v_N = \delta_{Fs}\tau_z^x(x)\tau_m^x(y) + \mu\delta_s\mathcal{A}(T - T_r) \tag{7.116b}$$

where new parameters are introduced as

$$\delta_u = \frac{b_u c_0}{L} \; ; \; \delta_s = \frac{b_w c_0}{H_1} \; ; \; \delta_{Fs} = \delta_s F_0 \tag{7.117}$$

Of these, the parameter δ_s is most important since it measures the amount of up-welling generated through surface layer processes; this parameter will be referred to as the surface layer feedback parameter. The terms involving δ_u turn out to be small and the effect of zonal advection in the SST equation is neglected. The quantity δ_{Fs} sets the climatological upwelling and is fixed throughout as in Table 7.3.

Finally, a simple parameterization of T_{s*} is introduced through

$$T_{s*}(h_*) = T_{s0} + (T_0 - T_{s0}) \tanh \left[\frac{h_* + h_0}{h_1} \right] \tag{7.118}$$

where h_0 is an offset value, T_{s0} is the subsurface temperature for $h = -h_0$ and h_1 controls the steepness of the transition as h_* passes through $-h_0$. In this way, the range of the subsurface temperature is given by $[2T_{s0} - T_0, T_0]$. In dimensionless variables, this becomes

$$T_s(h) = T_{so} + (T_0 - T_{so}) \tanh(\eta_1 h + \eta_2) \tag{7.119}$$

with $\eta_1 = H/h_1, \eta_2 = h_0/h_1$.

If the wind stress has no meridional dependence, the equatorial ocean response (which one only needs within the equatorial strip) can be (easily) explicitly computed using the Green's function derived in Technical box 7.1 and the solution in Technical box 7.3. For any forcing in the right hand side of (7.109e) with time dependence $e^{\sigma t}$, and a zonal dependence of $f(x)$ this solution is given by, using (7.70) with $\omega = -i\sigma$,

$$
\begin{aligned}
u(x; \phi) &= -i\Big(\int_0^1 g_1(\xi; \tilde{\phi}) f(\xi) \, d\xi \, g_2(x - 1; \tilde{\phi}) \\
&\quad - \int_x^1 f(\xi) g_2(x - \xi; \tilde{\phi}) \, d\xi \Big) \tag{7.120a}
\end{aligned}
$$

$$
\begin{aligned}
h(x; \tilde{\phi}) &= \int_0^1 g_1(\xi; \tilde{\phi}) f(\xi) \, d\xi \, g_3(x - 1; \tilde{\phi}) \\
&\quad - \int_x^1 f(\xi) \, g_3(x - \xi; \tilde{\phi}) \, d\xi \tag{7.120b}
\end{aligned}
$$

$$g_1(z; \tilde{\phi}) = \sqrt{\frac{\sin 2\tilde{\phi}z}{\sin 2\tilde{\phi}}} \; ; \; g_2(z; \tilde{\phi}) = \frac{\sin 2\tilde{\phi}z}{\sqrt{\cos 2\tilde{\phi}z}}$$

$$g_3(z; \tilde{\phi}) = \sqrt{\cos 2\tilde{\phi}z}$$

with $\tilde{\phi} = -i(\delta\sigma + \epsilon_o)$.

To define the oceanic basic state, the wind stress is prescribed ($\mu = 0$) with meridional dependence $\tau_m^x(y) = 1$. The oceanic solution fields $(\bar{u}, \bar{h})$ can be computed for $\sigma = 0$, $f(x) = \tau_z^x$ and with

$$-i \sin ix = \sinh x \; ; \; \cos ix = \cosh x$$

these become

$$\bar{u}(x) = \int_0^1 \sqrt{\frac{\sinh 2\epsilon_o \xi}{\sinh 2\epsilon_o}} \, \tau_z^x(\xi) \, d\xi \, \frac{\sinh 2\epsilon_o(x-1)}{\sqrt{\cosh 2\epsilon_o(x-1)}}$$
$$- \int_x^1 \tau_z^x(\xi) \, \frac{\sinh 2\epsilon_o(x-\xi)}{\sqrt{\cosh 2\epsilon_o(x-\xi)}} \, d\xi \qquad (7.121a)$$

$$\bar{h}(x) = \int_0^1 \sqrt{\frac{\sinh 2\epsilon_o \xi}{\sinh 2\epsilon_o}} \, \tau_z^x(\xi) \, d\xi \, \sqrt{\cosh \, 2\epsilon_o(x-1)}$$
$$- \int_x^1 \tau_z^x(\xi) \sqrt{\cosh \, 2\epsilon_o(x-\xi)} \, d\xi \qquad (7.121b)$$

The latter expression reduces to (7.84) in the limit $\epsilon_o \to 0$. The upwelling velocity $\bar{w}_s$ can be obtained from (7.116) and subsequently with these ocean fields, the SST equation (7.109d) can be solved to obtain $\bar{T}$.

Parameter	Value			Parameter	Value	
a_M	5.0×10^{-6}	s^{-1}		a_T	9.2×10^{-8}	s^{-1}
c_a	30	ms^{-1}		T_0	30	$^\circ C$
H	200	m		c_o	2.0	ms^{-1}
β_0	$2.2 \; 10^{-11}$	$(ms)^{-1}$		T_N	30	$^\circ C$
a_m	1.3×10^{-8}	s^{-1}		$H_1 = H_u$	50	m
b_w	1.0×10^2	s		τ_0	5×10^{-2}	Pa
b_u	6.0×10^5	s		T_{s0}	22	$^\circ C$
L	1.5×10^7	m		h_0	25	m
h_1	30	m		a_s	5.0×10^{-6}	s^{-1}
Parameter	Value			Parameter	Value	
ϵ_a	2.5			ϵ_T	0.694	
η_1	6.667			η_2	0.833	
δ_s	1.0			δ_{Fs}	4.104	
F_0	1.0					

Table 7.3.   Standard values of dimensional and non-dimensional parameters used in Dijkstra and Neelin (1995b).

For the parameters as in Table 7.3, the solutions for $\bar{h}$ and $\bar{T}$ are plotted in Fig. 7.23 for

$$\tau_z^x(x) = 0.6 \, (0.12 - \cos^2\left[\frac{\pi(x-x_0)}{2x_0}\right]); \; x_0 = 0.57 \qquad (7.122)$$

Upwelling occurs over most of the basin except within a small interval near the western boundary. The thermocline is shallow in the east part of the basin and deep in the west. A cold tongue is present in the east having a temperature of about $23^\circ C$.

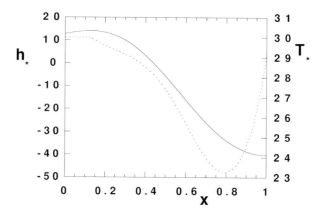

Figure 7.23.   Dimensional values of the basic state $\bar{T}_*$ (dotted) and $\bar{h}_*$ (drawn) computed with the prescribed wind forcing (7.122) using the standard values of the parameters. The zonal coordinate is the dimensionless length $x = x_*/L$.

Now choosing $T_r = \bar{T}$, the atmospheric response is computed for heating anomalies with respect to $\bar{T}$. The consequence of this way of basic state construction is that one obtains a family of basic states (depending on parameters) each of which, by construction, is a solution of the coupled model for every value of the coupling parameter μ.

If one considers now the stability of these basic states and limits to the perturbations of having the form

$$T(x, y, t) = \left[\bar{T}(x) + \hat{T}(x)e^{\sigma t}\right]\tau_m^x(y) \qquad (7.123)$$

then again with $\tau_m^x(y) = 1$ the Green's function can be used to express $\hat{u}$ and $\hat{h}$ into $\hat{T}$, using (7.120). In this way, an integro-differential eigenvalue problem arises through the SST equation. The latter problem can be written in general form as

$$\sigma\hat{T} = C_u(x)\hat{u} + C_h(x)\hat{h} - C_T(x)\hat{T} + \mu\delta_s\mathcal{A}(\hat{T}; x) \qquad (7.124)$$

with coefficients C_u, C_h and C_T given explicitly in Dijkstra and Neelin (1995b).

Both problems of the steady state determination and the linear stability of this basic state can be effectively tackled with the continuation methods discussed in chapter 4. To compute steady solutions as function of the parameters (for general μ), an integro-differential equation for the steady temperature field $\bar{T}(x)$ has to be solved. On an equidistant grid in x, $x_j = (j - 1)\Delta x$, $j = 1, \cdots, J$, let $z_j = \bar{T}(x_j)$. Since the wind stress is expressed into T by (7.114), and the ocean response to this wind stress through the Green's function, the SST equation provides a nonlinear system of algebraic equations for the vector **z**. This system

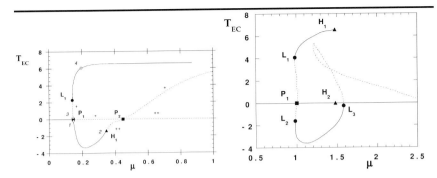

Figure 7.24. Bifurcation diagram based on a model for the near equatorial response for $\delta_s = 1.0$. (a) $\epsilon_a = 0.0$, (b) $\epsilon_a = 2.5$. On the vertical axis, the temperature deviation of the steady state (with respect to the reference state $\bar{T}$) is plotted at $x = 0.7$.

has many parameters, but there are three main control parameters ϵ_a, δ_s and μ. Once a steady state is known, the linear stability problem with the coefficients in (7.124) depending on the steady state can be solved. In Dijkstra and Neelin (1995b), branches of eigensolutions were traced as a function of parameters by the following procedure. Define

$$\hat{\mathbf{z}} = (\hat{T}_1^R, ..., \hat{T}_J^R, \hat{T}_1^I, ..., \hat{T}_J^I, \sigma^R, \sigma^I) \qquad (7.125)$$

where the superscripts R and I refer to real and imaginary parts of the eigenvector and eigenvalue, respectively. Then (7.124) provides also a nonlinear system of algebraic equations of order $2J$. The additional two equations are obtained through normalization of the eigenvector (real and imaginary parts). Then also a system of algebraic equations is obtained, which can be solved using pseudo-arclength continuation.

7.5.2.2 The fast wave limit

Results are simplest in the so-called fast wave limit where ocean wave processes are assumed much faster than the adjustment of SST. In the fast wave limit, $\delta \to 0$, the ocean response is in equilibrium to the actual wind stress. This regime was explored for the near equatorial response in Hao *et al.* (1993) using traditional methods and in Dijkstra and Neelin (1995b) with continuation methods. The bifurcation diagrams for both cases $\epsilon_a = 0.0$ and $\epsilon_a = 2.5$ and fixed $\delta_s = 1.0$, using μ as control parameter are plotted in Fig. 7.24. Plotted in the vertical axis is the deviation from the cold tongue temperature (at $x = 0.7$) from the value of the basic state ($\approx 23°C$). For $\epsilon_a = 0$ (Fig. 7.24a), the basic state becomes unstable at P_1 through a transcritical bifurcation. This type of bifurcation is expected here, since there is constructed solution for all values of μ (no saddle nodes) and there is no symmetry (no pitchforks).

For supercritical values of μ, a steady state with a cooler cold tongue stabilizes (Fig. 7.25a). Along this branch a Hopf bifurcation (H_1) occurs, destabilizing

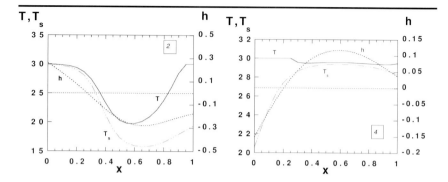

Figure 7.25.   Spatial structure of the total steady temperature field T, the subsurface temperature T_s and the thermocline field h for the two solutions along the branches of Fig. 7.24a. (a) point 2. (b) point 4.

this cold state through an oscillatory mode. The time longitude structure of the SST fields associated with this mode is plotted in Fig. 7.26a, showing a westward propagating orbit. At subcritical values of μ with respect to P_1, a steady state with a warm cold tongue appears. This state is unstable down to the value of the limit point (L_1) and then stabilizes again and approaches a state where the basin is nearly everywhere 30°C (Fig. 7.25b).

In the bifurcation diagram for $\epsilon_a = 2.5$ (Fig. 7.24b), also the basic state desta-bilizes through a transcritical bifurcation. The Hopf bifurcation H_1 is now lo-cated on the warm branch and a second Hopf bifurcation H_2 appears on the basic state branch; both oscillatory modes have westward propagating equatorial SST anomalies, similar to Fig. 7.26a. Contrary to what one might expect, in this very stripped version of the Zebiak-Cane model, a quite realistically looking basic state does not become directly unstable to oscillatory instabilities, but to a stationary in-stability. The immediate consequence is that multiple equilibria appear due to this bifurcation over quite a range of μ; one of these corresponds to an enhanced cold tongue, whereas the other consists of a warm state. Although oscillatory instabili-ties are around with reasonable periods, they do not occur as first bifurcations and therefore their relevance is at this point questionable.

The nice feature of continuation methods is that one easily follows these bifur-cation points in parameter space. For both values of ϵ_a, the path of the several of these bifurcations is plotted in the (δ_s, μ) space (Fig. 7.27). One can compare the location of the bifurcation points for $\delta_s = 1.0$ with those in Fig. 7.24. For $\epsilon_a = 0$ and large values of δ_s, the Hopf bifurcation H_1 becomes the primary instability (Fig. 7.27a). However, since the limit point L_1 still exists, multiple equilibria survive through an isolated branch (Dijkstra and Neelin, 1995b). The Hopf bifur-cation H_2 also becomes the primary instability for $\epsilon_a = 2.5$ (Fig. 7.27b) at large δ_s. At small δ_s, another Hopf bifurcation becomes the primary instability (point A). This instability is associated with an eastward propagating orbit of which the

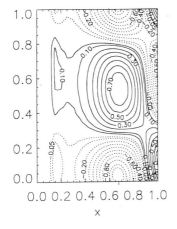
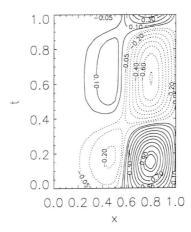

Figure 7.26. *(a) Time longitude plot of the SST anomaly corresponding to the oscillatory mode H_1 in Fig. 7.24a. (b) Same as (a) but for the Hopf bifurcation at point A in Fig. 7.27b. Only one period of the oscillation is plotted.*

pattern is shown in Fig. 7.26b. What the bifurcation diagrams neatly show, in par-

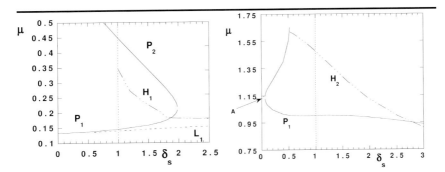

Figure 7.27. *Path of particular bifurcation points shown in Fig. 7.24 in the second control parameter δ_s for two values of ϵ_a. (a) $\epsilon_a = 0.0$; here the positions of L_1, H_1, P_1 and P_2 for $\delta_s = 1.0$ correspond to those in Fig. 7.24a. (b) $\epsilon_a = 2.5$; here the positions of H_2 and P_1 correspond to those in Fig. 7.24b for $\delta_s = 1.0$.*

ticular Fig. 7.27b, is that there is a continuous exchange of eastward propagating orbits as primary instabilities at small δ_s through stationary instabilities to westward propagating orbits at large δ_s. All oscillatory instabilities can be viewed as SST modes, because ocean wave dynamics is fully excluded in this model. When surface layer processes are absent, the upwelling feedback is not operating which leads to eastward propagating tendencies through the thermocline feedback. Similarly, when surface layer processes dominate, westward propagating tendencies are introduced through the upwelling feedback. In intermediate regimes, both tendencies compensate to give rise to stationary modes.

The results of this bifurcation analysis complement the results in Hao *et al.* (1993), where multiple equilibria were found together with both types of eastward and westward propagating oscillatory modes. Here, the view was taken of modified modes of the periodic ocean basin case as in Neelin (1991b). Propagation tendencies are indeed introduced in the same way in the periodic basin case and the presence of the lateral boundaries and the spatially varying basic state cause east-basin trapping. More complicated spatial-temporal behavior, such as relaxation oscillations, found in Hao *et al.* (1993), could also be explained dynamically in Dijkstra and Neelin (1995b) as being caused by mode interaction, in particular through interaction of a periodic orbit and a nearby stationary point.

The SST modes in the fast-wave limit are explored analytically detail for the case the coefficients in the eigenvalue problem (7.124) are constant in Jin and Neelin (1993b). Also for this case, the most unstable modes are stationary for small ϵ_a. The east-basin trapping is shown to occur through the east-west asymmetry associated with the β-effect. At small δ, the modes remain stationary and their growth rate increases. For larger ϵ_a, oscillatory modes become first unstable and remain present when wave time scales are included. Hence, the picture sketched in the fast wave limit remains valid in this regime.

7.5.2.3 The weak-coupling limit

Before going to the instabilities of a spatially dependent basic state in the full model (7.109), one special case is considered in more detail, because it adds greatly to the understanding of the total structure of the eigenmodes. When surface layer processes are neglected ($\delta_s = 0$), the eigenvalue problem (7.124) becomes

$$\hat{u}(x; \tilde{\phi}) = -i\mu \left(\int_0^1 g_1(\xi; \tilde{\phi}) \mathcal{A}(\hat{T}; \xi) \, d\xi \, g_2(x - 1; \tilde{\phi}) \right.$$
$$\left. - \int_x^1 \mathcal{A}(\hat{T}; \xi) g_2(x - \xi; \tilde{\phi}) \, d\xi \right) \tag{7.126a}$$

$$\sigma \hat{T} = C_u(x)\hat{u} + C_h(x)\hat{h} - C_T(x)\hat{T} \tag{7.126b}$$

$$\hat{h}(x; \tilde{\phi}) = \mu \left(\int_0^1 g_1(\xi; \tilde{\phi}) \mathcal{A}(\hat{T}; \xi) \, d\xi \, g_3(x - 1; \tilde{\phi}) \right.$$
$$\left. - \int_x^1 \mathcal{A}(\hat{T}; \xi) g_3(x - \xi; \tilde{\phi}) \, d\xi \right) \tag{7.126c}$$

with $\tilde{\phi} = -i(\delta\sigma + \epsilon_o)$ and the functions g_i as in (7.120).

If the limit of small coupling coefficient ($\mu \to 0$), the SST equation and the ocean dynamics equations become decoupled and each gives their own eigenmodes, i.e. SST modes and ocean dynamics modes, respectively. Nontrivial solutions for the zonal velocity and thermocline perturbation only arise if the denominator in the function $g_1(z)$ becomes zero. This leads to

$$\sin 2\tilde{\phi} = 0 \Rightarrow \tilde{\phi} = \frac{k\pi}{2} \Rightarrow \sigma = \sigma_r + i\sigma_i = \frac{-\epsilon_o + i\frac{k\pi}{2}}{\delta} \qquad (7.127)$$

for $k = 1, 2, \dots$. The spatial structure of these modes is given by

$$\hat{h}(x) = \sqrt{\cos k\pi(x-1)} \qquad (7.128a)$$

$$\hat{u}(x) = -i\frac{\sin k\pi(x-1)}{\sqrt{\cos k\pi(x-1)}} \qquad (7.128b)$$

since the term before this functional form is just a scalar and the atmospheric response is zero. Exactly these modes were shown to be eigenfunctions of the shallow water equations by Cane and Moore (1981) in the long wave approximation and they will be referred to as ocean basin modes. The gravest ocean basin mode ($k = 1$) has an angular frequency of $\pi/(2\delta)$, which for $\delta = 1$ corresponds to a period of $4L/c_o$, i.e. four time the basin crossing time of the free oceanic Kelvin wave. The zonal velocity of this mode is singular (for $\epsilon_o = 0$) at $x = 0.5$, but the thermocline field is smooth.

The uncoupled SST modes are just constants for constant C_T, with decay rate $\sigma_R = -C_T$ (if some Laplace diffusion is introduced they become simple cosine functions). In Neelin and Jin (1993), the correction to the growth rate and angular

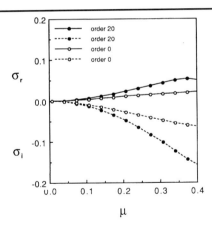

Figure 7.28. Modification of growth factors σ_r (drawn) and angular frequencies σ_i (dotted) in the weakly coupled limit for different orders of asymptotic approximation (Neelin and Jin, 1993).

frequency of the ocean basin modes and SST modes is analyzed using asymptotic expansions for constant coefficients in the SST equation. The correction of the eigenvalues due to coupling and hence as a function of μ is shown in Fig. 7.28.

The growth rate increases with coupling while the frequency decreases rapidly with respect to the uncoupled mode. For μ small, the structure of the mode still looks very much the same as the uncoupled basin mode, with zonal advection dominating in the SST structure (Fig. 7.29a). However, for larger μ the structure changes dramatically (Fig. 7.29b) with slow eastward propagation starting to appear in both fields. Both stationary and propagating SST modes (which occur when $\delta_s \neq 0$) do not modify strongly from the uncoupled case.

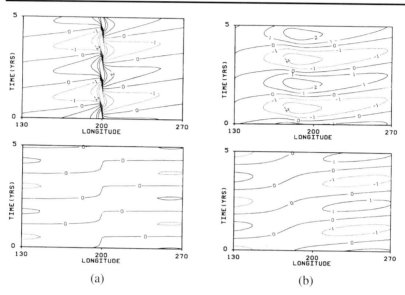

(a)   (b)

Figure 7.29.   Patterns of SST (left panel) and thermocline (right panel) anomalies for (a) $\mu = 0.1$ and (b) $\mu = 0.3$ of a (modified) ocean basin mode.

7.5.3. Modes in the full problem

Even when the equatorial strip approximation for the SST equation is kept, but a more general meridional structure of the forcing is assumed and more detail of the off-equatorial response is taken into account, the exact summation in the Green's function can no longer be carried out. The problem has to be tackled numerically with spectral or a finite difference approach similar to that in Hirst (1986); this problem was solved in Jin and Neelin (1993a). First a basic state is constructed which is a solution for every value of the coupling strength as in the previous section. This state has a prescribed Gaussian meridional structure with λ_o as the spatial decay scale. The zonal dependence at the equator of the fields $\bar{T}_*, \bar{h}_*$ and $\bar{w}_{s*}$ is plotted in Fig. 7.30. These fields show a cold tongue in the eastern part of the basin, a warm pool in the western part of the basin, a slope in the thermocline and strongest upwelling in the center of the basin.

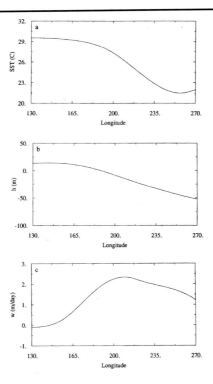

Figure 7.30.   Zonal section at the equator of the the dimensional basic state as used in Jin and Neelin (1993a). (a) SST ($\bar{T}_*$),   (b) thermocline depth ($\bar{h}_*$), and (c) upwelling ($\bar{w}_{s*}$).

The coupled model is considered as an anomaly model with respect to this basic state. In other words, the atmospheric heating anomaly is computed from SST anomalies with respect to $\bar{T}$. For the instabilities of this basic state, the parameter regime formed by the (μ, δ_s) - plane is investigated. The behavior of the modes in the weakly coupled regime were already explored above for a reduced version of the model, i.e. for the near equatorial response.

For the full model, there is quite a complication in the weakly coupled limit. The structure of the spectrum of the full ocean shallow water model in a bounded basin on an equatorial β-plane is continuous (Moore, 1968). When the equations are discretized, a resolution dependent scatter spectrum appears. How this spectrum looks like can again by determined from the Green's function analysis and is subject of Technical box 7.4.

Technical box 7.4:
Scatter spectrum

For the unforced, nondissipative, shallow water model on the equatorial β plane in the long wave limit, the spectrum of free modes, which satisfy boundary conditions is considered. Nontrivial solutions are only obtained for h_E^G in (7.47) if the denominator is zero and hence

$$\int_{-\infty}^{\infty} K_u(\phi, y) dy = 0$$

with

$$K_u(\phi, y) = \lim_{M \to \infty} \left[e^{\phi} \psi_0(y) + \frac{1}{\sqrt{2}} \sum_{j=0}^{M} \alpha_{2j+1} e^{-\phi(4j+3)} \left(\frac{\psi_{2j+2}(y)}{\sqrt{j+1}} - \frac{\psi_{2j}(y)}{\sqrt{j}} \right) \right]$$

for $\phi = \sigma$. When the infinite sum is taken, it is now known that the ocean basin modes (7.128) are found. However, when the sum is truncated at order M, the condition leads to

$$1 - \sum_{j=1}^{M} \frac{(2j-3)!!}{(2j)!!} z^j = 0 \; ; \; z = e^{-4\sigma} \qquad (7.129)$$

where $(2j)!! = 2 \times 4 \times 6 \times \cdots \times 2j$, with $(-1)!! = 1$. The real and imaginary parts of the eigenvalues σ are shown in Fig. 7.31 for $M = 11, M = 31$ and $M = 51$ by solving for the polynomial equation for z and relating $-4\sigma_j = \ln z_j - 2k\pi i$; only the set $k = 0$ is plotted in Fig. 7.31. All these modes damp faster than the physical damping rate (which is zero), and only one approaches a growth factor zero for large M and hence this is the ocean basin mode. It corresponds to the mode with $z = 1$, or $4\sigma = 2\pi k i$, similar to (7.127). The scatter modes exist because incident energy leaks into the system along the western boundary, through the long-wave approximation made.

All the ingredients are now available to understand the unified picture provided in Jin and Neelin (1993a). In Fig. 7.32, the frequency and growth rate of the modes are plotted as a function of μ with a larger dot representing a larger value of μ. For $\mu = 0$, SST modes and ocean dynamics modes are uncoupled. The latter modes consist of the ocean basin modes (labelled 'B') and the ocean scatter modes (labelled 'S'). With increasing coupling, the stationary SST mode becomes more unstable while simultaneously the ocean dynamics modes are strongly modified. Eventually, mode merging occurs where one of the ocean dynamics mode

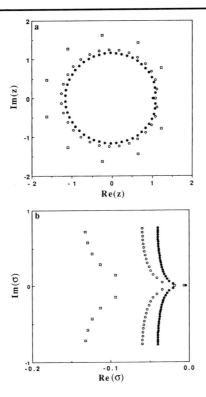

Figure 7.31. Spectrum of modes of the shallow water equations in the long wave approximation at three different truncations with (a) roots z_j and (b) eigenvalues σ_j. Truncation levels are $M = 11$ (open squares), $M = 31$ (open circles), $M = 51$ (filled circles).

merges with an SST mode. For small δ, the SST mode merges with scatter modes, whereas at large δ, the SST mode merges with and ocean basin mode. The resulting modes have been termed mixed SST/ocean dynamics modes. An example of such a merger is shown in Fig. 7.33a with patterns of the eigenmodes at location B in Fig. 7.33b. The resulting mode inherits the spatial pattern from the SST mode, while its frequency is determined from ocean adjustment processes providing a period in the interannual range. While this completes the investigation of the coupled modes within this intermediate complexity model context, it does not provide a simple physical view of the mechanism of the oscillation. The structure of the connection of the modes is complicated and asks for more simple models in which the dominant physics of the oscillatory modes can be understood. Therefore, attempts have been made to deduce more conceptual models related to these eigenmodes.

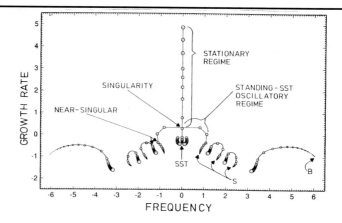

Figure 7.32.   Sketch of the connection between the different modes in parameter space. The modes labelled 'SST' are SST-modes, those labelled with 'S' are the scatter modes and those labelled 'B' are ocean basin modes.

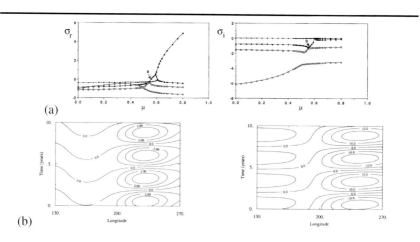

Figure 7.33.   (a) Typical curves of growth factors (left panel) and angular frequencies (for the case δ = 1.5), showing the connection between the stationary modes and a scatter mode near point B. (b) Equatorial patterns of the dimensional SST anomalies (left panel) and thermocline anomalies (right panel) for point B.

7.5.4. Conceptual models of the ENSO oscillation

Several attempts have been made to device more conceptual models capturing the essence of the oscillation as seen in intermediate coupled models (Zebiak and Cane, 1987). Although the earlier attempts by Battisti and Hirst (1989), Schopf and Suarez (1990) and Cane *et al.* (1990) have lead to successful models, the study of Jin (1997b) nicely follows the mixed SST/ocean dynamics mode framework discussed in the previous section. Hence, we first follow this approach here and come back to the earlier models later on.

7.5.4.1 The two-strip model

First step in obtaining a reduced model is a simplification of the shallow water response. For convenience, the dimensionless shallow water equations using the long wave approximation are again given below with the dimensionless zonal wind stress forcing indicated by τ, i.e.

$$\frac{\partial u}{\partial t} - yv + \frac{\partial h}{\partial x} + \epsilon_o u = \tau \tag{7.130a}$$

$$yu + \frac{\partial h}{\partial y} = 0 \tag{7.130b}$$

$$\frac{\partial h}{\partial t} + \frac{\partial u}{\partial x} + \frac{\partial v}{\partial y} + \epsilon_o h = 0 \tag{7.130c}$$

To obtain a single equation for h, (7.130a) is differentiated to y and the result is then multiplied by y. When (7.130a) is subtracted from this result one obtains

$$(yu_y - u)_t - y^2 v_y + yh_{xy} + \epsilon_o yu_y - h_x - \epsilon_o u = y\,\tau_y - \tau \tag{7.131}$$

where the subscripts now indicate differentiation. Next, (7.130b) is differentiated to x and the result multiplied by y. When also (7.130b) is differentiated to y, the two relations

$$y^2 u_x + yh_{xy} = 0 \tag{7.132a}$$

$$yu_y + u + h_{yy} = 0 \tag{7.132b}$$

are obtained. The terms yu_y and yh_{xy} are now eliminated from (7.131) using (7.132). When the relation $u = -h_y/y$ is used and the term with $u_x + v_y$ is eliminated using (7.130c), the final equation obtained is

$$y^2 \left(\frac{\partial h}{\partial t} + \epsilon_o h \right) + \left(\frac{2}{y}\frac{\partial}{\partial y} - \frac{\partial^2}{\partial y^2} \right)\left(\frac{\partial h}{\partial t} + \epsilon_o h \right) - \frac{\partial h}{\partial x} = y\frac{\partial \tau}{\partial y} - \tau \tag{7.133}$$

The boundary conditions then become

$$x = 0 \quad : \quad \int_{-\infty}^{\infty} \frac{1}{y}\frac{\partial h}{\partial y}\, dy = 0 \tag{7.134a}$$

$$x = 1 \quad : \quad \frac{\partial h}{\partial y} = 0 \tag{7.134b}$$

In Jin (1997b), it is assumed that h has a near parabolic dependence near the equator, a property which does not seem unreasonable, when looking at the thermocline structures of the free equatorial Rossby waves. Hence,

$$h(x, y, t) = h_e(x, t) + y^2 \Delta h(x, t) \tag{7.135}$$

If one takes $(h_n + h_e)/2 = h(x, 1, t)$, the latter being the thermocline deviation at a distance λ_o from the equator, then it follows that $\Delta h = (h_n - h_e)/2$. Note that the zonal velocity is given by $u = -h_y/y = h_e - h_n$. Now (7.135) is substituted into (7.133) and considered at $y = 0$ giving one equation relating h_e and h_n. A second equation is obtained by realizing that the second term in the left hand side of (7.133) is much smaller than the first at $y = y_n$ where $h \approx h_n$. This leads to the two-strip model

$$(\frac{\partial}{\partial t} + \epsilon_o)(h_e - h_n) + \frac{\partial h_e}{\partial x} = \tau_{|y=0} \tag{7.136a}$$

$$(\frac{\partial}{\partial t} + \epsilon_o)h_n - \frac{1}{y_n^2}\frac{\partial h_n}{\partial x} = \frac{\partial}{\partial y}(\frac{\tau}{y})_{|y=y_n} \tag{7.136b}$$

Note that the free wave solutions (in a zonally unbounded domain) with wavenumber k of (7.136b) have a frequency $-k/y_n^2$, and hence represent Rossby waves. For $y_n = 2$, these have a phase velocity $1/4$ of the free Kelvin wave signal (of the wave with the same wavenumber k) which is contained in (7.136a). The boundary conditions can be approximated by

$$h_n(1, t) = r_E h_e(1, t) \; ; \; h_e(0, t) = r_W h_n(0, t) \tag{7.137}$$

where r_E and r_W are a measure of the degree of zonal mass flux allowed at each boundary. For example, at the eastern boundary, the zonal velocity is given by $u_E(1, t) = h_e(1, t) - h_n(1, t) = (1 - r_E)h_e(1, t)$. Hence, if $r_E = 1$ the zonal mass flux is zero but for $r_E < 1$ a nonzero mass flux is allowed. In general, $r_W < 1$, since energy leaks through the western boundary under condition (7.134b) and a choice $r_W = 3/5$ is the appropriate value under the two-strip approximation with $h = 0$ at $y = 2y_n$ and beyond. Both r_W and r_E therefore monitor the degree of exchange of mass between the equatorial strip and off-equatorial regions.

The dispersion relation of the ocean adjustment modes (eigensolutions of the unforced problem (7.136)) which satisfy the boundary conditions becomes

$$\sigma_j = -\epsilon_o + \ln\frac{r_E r_W(1 + y_n^2) - r_E}{(1 + y_n^2) - r_E} + i\frac{2\pi j}{1 + y_n^2} \tag{7.138a}$$

$$h_n = H_n e^{\sigma_j t + (\sigma_j + \epsilon_o)xy_n^2} \tag{7.138b}$$

$$h_e = H_e e^{\sigma_j t - (\sigma_j + \epsilon_o)x} + \frac{h_n}{1 + y_n^2} \tag{7.138c}$$

for $j = 0, \pm 1, \pm 2, ...$. The second term in the right hand side of the first equation above has always a negative real part and represents damping through the energy loss at the boundaries. The mode for $j = 0$ is a purely stationary adjustment mode, which appears because the basinwide mean of h is a free mode solution of the two-strip model for $\epsilon_o = 0$. This mode can be seen as an approximation of the gravest scatter mode of the complete spectrum of ocean adjustment modes. The oscillatory modes for higher j are the equivalents of the ocean basin modes.

The equation governing the equatorial SST-perturbations follows from the general eigenvalue problem discussed in section 7.5.2, i.e.

$$\frac{\partial T_e}{\partial t} + C_T(x)T_e - C_h(x)h_e = 0 \qquad (7.139)$$

with C_T representing local damping and C_h the effect of thermocline variations (through background upwelling) on the temperature perturbations (thermocline feedback). As the SST perturbations change mostly in the eastern part of the basin, one can average the equation above over the eastern half of the basin, say from $x = 1/2$ to $x = 1$, to give

$$\frac{dT_{eE}}{dt} + C_{TE}T_{eE} - C_{hE}h_{eE} = 0 \qquad (7.140)$$

When, following the Gill atmosphere model, the wind stress anomaly is written as

$$\tau = \mu \mathcal{A}(T_e)e^{-(\alpha y)^2/2} \qquad (7.141)$$

where $\alpha = \lambda_o/\lambda_a$, the equatorial wind stress can be related to the temperature T_{eE}. A westerly wind response west of positive T_{eE} can be represented by

$$\tau_{|y=0} = \mu\, A_0\, T_{eE}\, f(x) \qquad (7.142)$$

with a fixed pattern $f(x)$ and amplitude A_0. The proportionality factor μ serves as coupling coefficient with $\mu = 1$ being a 'realistic' strength. The function $f(x)$ mimics the spatial pattern of the wind response and can be taken piecewise constant (Jin, 1997b), for example

$$f(x) = \frac{1}{x_2 - x_1} \quad \text{for } x_1 < x < x_2 \qquad (7.143)$$

and zero elsewhere. In this way, the forcing in the second equation of the two-strip model (7.136) can be approximated as

$$\frac{\partial}{\partial y}\left(\frac{\tau}{y}\right)_{|y=y_n} \approx -\mu\, A_0\, T_{eE}\, f(x)\frac{\theta}{y_n^2} \qquad (7.144a)$$

where θ is an $\mathcal{O}(1)$ coefficient.

A nice element in the coupled model developed in this way is that the two-strip equations can be integrated along the (Kelvin and Rossby) wave characteristics (for the general procedure, see any introductory text on partial differential equations, for example John (1986)), which are given by

$$x - x_0 \;=\; t - t_0 \qquad (7.145a)$$
$$x - x_0 \;=\; \frac{t_0 - t}{y_n^2}, \qquad (7.145b)$$

respectively, where (x_0, t_0) is any point in the domain. When damping is neglected, the solutions h_e and h_n can be obtained by first integrating (7.136b) along a Rossby wave characteristic starting at the eastern boundary and reaching the western boundary. Next, (7.136b) is integrated along a characteristic starting at the western boundary over the Kelvin crossing time, in which the wave has reached the eastern boundary. Using mean value approximations and the fact that $1 + y_n^2 >> 1$, this leads to delay equations of the form

$$
\begin{aligned}
h_{eW}(t) &= r_W r_E h_{eW}(t - 1 - y_n^2) + \\
+\mu A_0 r_W (r_E T_{eE}(t - 1 - x_P) &- \theta T_{eE}(t - y_n^2 x_P)) & (7.146a) \\
h_{eE}(t) &= r_W r_E h_{eE}(t - 1 - y_n^2) - \\
-\mu A_0 (\theta r_W T_{eE}(t - 1 - y_n^2 x_P) &- T_{eE}(t - 1 + x_P)) & (7.146b) \\
\frac{dT_{eE}}{dt} + C_{TE} T_{eE} - C_{hE} h_{eE} &= 0 & (7.146c)
\end{aligned}
$$

where x_P is a chosen fixed point within the area of wind response $[x_1, x_2]$.

7.5.4.2 The delayed oscillator

When the effect of the eastern boundary reflection is neglected ($r_E = 0$), then (7.146b) and (7.146c) give

$$
\begin{aligned}
\frac{dT_{eE}}{dt} &= -C_{TE} T_{eE} + \mu A_0 C_{hE} \\
(T_{eE}(t - 1 + x_P) &- \theta r_W T_{eE}(t - 1 - y_n^2 x_P)) & (7.147)
\end{aligned}
$$

which shows that the average eastern basin temperature is influenced by local damping and a remote signal due to propagation of Kelvin and Rossby waves. The delay time $1 - x_P$ is the effect due to the Kelvin wave and as this is fast, it can be neglected on long time scales. It provides the local amplification of temperature perturbations through the thermocline feedback through a forced Kelvin wave response. The delay $1 + y_n^2 x_P$ is the time taken for the Rossby wave to travel from the center of wind patch near x_P to the western boundary plus the time it takes the reflected Kelvin wave to cross the basin. When returned in the eastern part of the basin, it provides a delayed negative feedback to the temperature perturbation (since $r_W > 0$).

Based on other ad hoc approximations the picture of a delayed oscillator was derived by Battisti (1988) and Suarez and Schopf (1988). In these studies, a differential delay equation with local feedback was proposed of the form

$$
\frac{dT_*(t_*)}{dt_*} = a T_*(t_*) - b T_*(t_* - d) - c T_*^3(t_*) \qquad (7.148)
$$

Here a represents the growth rate of the temperature disturbance T in the eastern Pacific and would correspond to $\mu A_0 C_{hE} - C_{TE}$ in (7.147). The quantity d is the delay time due to the propagation of equatorial waves, corresponding to $1 + y_n^2 x_P$ in (7.147), and b measures its influence with respect to the local feedbacks. The nonlinear term in (7.148) is needed for equilibration of the temperature to finite amplitude.

Scaling time by $1/a$ and the temperature by $\sqrt{a/c}$, the dimensionless equation (7.148) becomes

$$\frac{dT(t)}{dt} = T(t) - \alpha T(t - \delta_T) - T^3(t) \tag{7.149}$$

with $\alpha = b/a$ and $\delta_T = ad$ is the dimensionless delay time. For $\alpha < 1$, this equation has three steady solutions, given by

$$\bar{T} = 0, \bar{T} = \pm\sqrt{1 - \alpha} \tag{7.150}$$

which correspond again to the multiple equilibria structure found in previous sections. Infinitesimal perturbations $\tilde{T}$ on each state satisfy

$$\frac{\partial \tilde{T}}{\partial t} = \tilde{T}(1 - 3\bar{T}^2) - \alpha \tilde{T}(t - \delta_T) \tag{7.151}$$

A normal mode analysis, with $\tilde{T} = \hat{T}e^{\sigma t}$, leads to the eigenvalue problem

$$\sigma = 1 - 3\bar{T}^2 - \alpha e^{-\sigma \delta_T} \tag{7.152a}$$
$$\sigma_r = 1 - 3\bar{T}^2 - \alpha e^{-\sigma_r \delta_T} \cos(\sigma_i \delta_T) \tag{7.152b}$$
$$\sigma_i = \alpha e^{-\sigma_r \delta_T} \sin(\sigma_i \delta_T) \tag{7.152c}$$

In Schopf and Suarez (1988), the regime $\alpha < 1$ is considered. In this regime, the trivial state is unstable to a stationary mode. For the solutions $\bar{T}^2 = 1 - \alpha$, the neutral curve is shown in the (δ_T, α) plane in Fig. 7.34a with the period of the oscillatory mode in Fig. 7.34b. The period of oscillation is larger than $2\delta_T$, which indicates that if the propagation time of the Rossby waves from the center of the basin is three times the Kelvin crossing time, the period is at least 18 months. In Schopf and Suarez (1988), it is shown that the same holds for the oscillation period of the fully nonlinear system. In Battisti (1988), the regime $\alpha > 1$ is

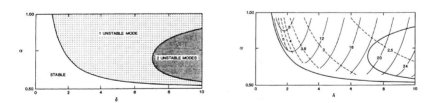

Figure 7.34. (a) Neutral curves of the nontrivial steady states in the delayed oscillator model (Suarez and Schopf, 1988). (b) Period the nonlinear oscillator in multiples of the dimensionless delay time $\delta = \delta_T$.

considered to be more realistic, and no additional unstable states appear. In this regime, the trivial state is unstable to one oscillatory solution.

This delayed oscillator model reflects the central elements of the ENSO cycle in terms of local growth due to instability and subsequent adjustment through (individual) Kelvin and Rossby waves. In the eastern part of the basin, strong feedback takes place which leads to amplification of disturbances: positive SST anomalies cause wind stress anomalies, which weaken the background trade winds, with maximum weakening west of the SST anomaly. This is turn creates a different slope in the thermocline which leads through the thermocline feedback to a larger SST anomaly. However, the ocean does not react instantaneously to the changing winds, but has a memory component which determines its long term evolution. The wind anomaly also generates westward travelling Rossby waves that make the thermocline shallower in off-equatorial regions in the western part of the basin. The Rossby waves reflect on the western boundary and cause a Kelvin wave reflection which causes an elevated thermocline. This Kelvin wave signal provides the delayed negative feedback, through which the SST anomaly reduces to zero and becomes slightly negative. Then the feedback start to operate with a different sign to amplify the negative temperature anomaly leading to a La Niña   state. Hence, the period of the oscillation is basically determined by the wave transit time to get the delayed feedback. Slightly different details have been suggested in several studies, but the basic mechanism of the delayed oscillator at work is that sketched above.

7.5.4.3 The coupled wave oscillator

Consider the situation of the fast SST limit where SST anomalies adapt nearly instantaneously to thermocline anomalies. This is modelled by putting the time derivative to zero in (7.146c) which gives $T_{Ee} = C_{hE}h_{eE}/C_{TE}$. When substituted in (7.146b) this gives the map

$$
\begin{aligned}
h_{eE}(t) \;=\; & r_W r_E h_{eE}(t-1-y_n^2) + \\
\mu A_0 \frac{C_{hE}}{C_{TE}} (h_{eE}(t-1+x_P) \;\; & - \;\; \theta r_W h_{eE}(t-1-y_n^2 x_P)) \qquad (7.153)
\end{aligned}
$$

This equation is the linear version of the coupled wave oscillator model introduced in Cane et al. (1990). A nonlinear version has been analyzed in Münnich et al. (1991), where A_0 is not constant but an operator (just as in the Gill model). Variants of these models were also considered in Schopf and Suarez (1990).

In this model, the wind stress response is (through the SST anomalies) directly related to thermocline depth anomalies. When the effect of the eastern boundary reflection is ignored, $r_E = 0$, forced Kelvin waves are generated which lead to amplification of thermocline anomalies. The delayed feedback through the forced Rossby waves which reflect to returning Kelvin waves is similar as in the delayed oscillator picture. In fact, using the Taylor-series expansion for the forced Kelvin wave response around time t, i.e.

$$
h_{eE}(t-1+x_P) = h_{eE}(t) - (1-x_P)\frac{dh_{eE}}{dt} + \cdots \qquad (7.154)
$$

leads to an equation of which the linear part is similar to (7.148). Since no SST adjustment is involved, the oscillatory phenomena in this model (Cane *et al.*, 1990) must be caused by a destabilized ocean adjustment mode. The oscillation mechanism is hence inherited from the uncoupled ocean basin modes (Jin, 1997a). The coefficients r_E and r_W control the growth rate of the mode and the propagating properties of the equatorial waves.

7.5.4.4 The recharge oscillator

Suppose that the timescale of the leading mode in the oscillation is much be larger than $1 + y_n^2$ (the period of the gravest ocean basin mode). Then similar Taylor series approximations as in (7.154) can be used to derive a system of two first order differential equations, i.e.

$$\frac{dh_{eW}}{dt} = -\frac{1 - r_W r_E}{1 + y_n^2} h_{eW} - \frac{\mu A_0 r_W (\theta - r_E)}{1 + y_n^2} T_{eE} \qquad (7.155a)$$

$$\frac{dT_{eE}}{dt} = -(C_{TE} + \mu A_0 C_{hE}) T_{eE} + C_{hE} h_{eW} \qquad (7.155b)$$

The first equation follows directly from (7.146a). The relation

$$h_{eE} = h_{eW} + \mu A_0 T_{eE}$$

the latter being an approximation of the 'quasi Sverdrup' steady balance over the basin, can be derived from (7.146b) and is used to obtain (7.155b) from (7.146c). The model above is the recharge oscillator model as presented in Jin (1997a) and Jin (1997b).

The right hand side of (7.155) can be considered as the Jacobian matrix associated with the stability problem of the trivial solution. Note that for $\mu = 0$, this state is stable, since $r_W r_E < 1$. A complex conjugate pair of eigenvalues crosses the imaginary axis with frequency ω_c at

$$\mu_c = \frac{C_{TE} + r}{A_0}$$

$$\omega_c = \sqrt{r_W (\theta - r_E) \frac{C_{TE} + r}{1 + y_n^2} - r^2} \qquad (7.156a)$$

where $r = (1 - r_W r_E)/(1 + y_n^2)$ is the decay rate of the zero frequency ocean adjustment mode. This oscillation only exists when $0 < r_W (\theta - r_E) \ll 1$ and in that case it has also a period longer than the intrinsic time scales in the model. Note also that when the time derivative of the SST-equation in (7.155) is put to zero, no oscillations will occur and hence adjustment processes of the SST are important.

The latter is reflected in the physics of the oscillation as described in Jin (1997a). Consider again a positive SST anomaly in the eastern part of the basin which induces a westerly wind response. Through the quasi-steady balance (7.5.4.4), this immediately changes the slope in the thermocline giving a deeper

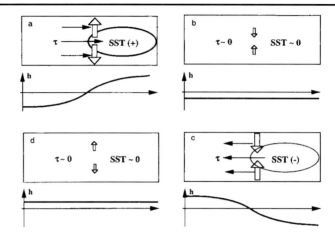

Figure 7.35.   Sketch of the different stages of the recharge oscillator (Jin, 1997a).

eastern thermocline. Hence, through the thermocline feedback, the SST anomaly
is amplified which brings the oscillation to the extreme warm phase (Fig. 7.35a).

Because of ocean adjustment, part of the equatorial mass is moved from the
equatorial strip to off-equatorial regions. Compare this with the picture provided
above where this mass exchange was caused by reflections of waves, while here it
is due to a nonzero divergence of the zonally integrated Sverdrup transport. This
effect is represented by the forcing, shown in the right hand side of (7.144). This
exchange, which can be seen as a discharge of equatorial heat content, causes
the equatorial thermocline to flatten and hence eventually reduces the temperature anomaly and consequently the wind stress anomaly (Fig. 7.35b). Eventually
a nonzero negative thermocline anomaly is generated, which allows cold water
to get into the surface layer by the background upwelling. This causes a negative SST anomaly leading through amplification to the cold phase of the cycle
(Fig. 7.35c). Through adjustment, the equatorial heat content is recharged (again
the zonally integrated Sverdrup transport is nonzero) and leads to a transition
phase with a positive zonally integrated equatorial thermocline anomaly.

7.6.   Coupled Processes and the Annual-Mean State

For the results so far, the annual-mean state has been prescribed or constructed,
but has certainly been independent of the nature of the coupled processes. This
has fixed also the mean atmospheric motions with trade winds along the equator,
rising motion in the west, eastward winds at height and sinking motion in the
eastern part of the Pacific. This cellular pattern is usually referred to as the Walker
circulation. In Bjerknes (1969) it is already noted that, "it seems reasonable to
assume that it is the gradient of SST along the equator which is the cause of (...)

the Walker circulation". Involvement of coupled processes in the spatial structure of the mean state will be considered next.

7.6.1. Constructed versus coupled mean states

As a first step, one might attempt to use the model of the near equatorial response as in section 7.5.2 to try to get prototype solutions for the annual-mean state. The simplified steady SST equation (7.109d), neglecting zonal advection of SST, is given by

$$\mathcal{H}(w_1)w_1(T - T_s(h)) - \mathcal{H}(-v_N)v_N(T - T_N) + \epsilon_T(T - T_0) = 0 \quad (7.157)$$

In the limit of negligible oceanic damping, $w_1 = w_s$ and the thermocline is given by

$$h(x) = \int_0^1 \sqrt{s}\, \tau^x(s)ds - \int_x^1 \tau^x(s)\, ds \quad (7.158)$$

where τ^x is the equatorial zonal wind stress. The velocities w_s and v_N follow from (7.116) and the simplified Gill atmosphere model, with the operator $\mathcal{A}$ defined by (7.113), is used. Standard values of oceanic and atmospheric model parameters were given in Table 7.3.

In the case considered in section 7.5.2.1 where the mean state was constructed, the ocean is forced with a mean wind stress derived from observations, τ^x_{obs} and the resulting steady state temperature was indicated by $\bar{T}$. In coupling the system, the wind stress τ^x fed into the ocean model then becomes

$$\tau^x = \tau^x_{obs} + \mu\mathcal{A}(T - \bar{T}) \quad (7.159)$$

where μ is the coupling parameter. Hence, there is a constructed solution, $T = \bar{T}$, $\tau^x = \tau^x_{obs}$ (referred to as the flux-corrected climatology in Dijkstra and Neelin (1995a)) which does not depend on coupling. The latter affects only the stability of this state to perturbations.

For models which attempt to simulate the full tropical coupled ocean-atmosphere system, the determination of the climatology is a major end in itself. This climatology depends on coupling strength and the values of many parameters in the model. For the case of a model with active ocean only in a single basin (here the Pacific) there will be a part of the wind stress which depends on the atmospheric response to the temperature pattern within the basin and a part, say τ^x_{ext}, which is determined externally. The part determined within the basin will depend on temperature departures, $T - T_0$, from surface heat flux equilibrium, thus

$$\tau^x = \tau^x_{ext} + \mu\mathcal{A}(T - T_0) \quad (7.160)$$

where $T = T_0$ is assumed outside the basin (as appropriate for continental conditions). For the external part τ^x_{ext}, a zonally constant component, for example, due to the zonally symmetric circulation can be considered. Since the latter obeys strong dynamical constraints, it will depend relatively little on the coupled dynamics within the basin, compared to the Walker type circulation $\mathcal{A}(T - T_0)$

driven by zonal gradients. A more extensive discussion of the form of τ_{ext}^x and T_0 is given in Dijkstra and Neelin (1995a). For (7.160), the stationary state is easily computed at $\mu = 0$, but depends on coupled feedbacks otherwise. Temperature solutions computed with (7.160) are referred to as "coupled climatologies".

7.6.2. Demise of multiple equilibria

How does the bifurcation structure of the constructed case change as we relax the restriction towards the more difficult coupled climatology case ? To this end, the observed wind stress is split into the part τ_{ext} and the remainder $\tau_{obs} - \tau_{ext}$. To be able to follow the deformation of the bifurcation diagram, a homotopy parameter α_F is introduced. The wind stress τ is then given by

$$\tau^x = \tau_{ext}^x + \alpha_F(\tau_{obs}^x - \tau_{ext}^x) + \mu\left[\alpha_F A(T - \overline{T}) + (1 - \alpha_F)A(T - T_0)\right] \quad (7.161)$$

where $\overline{T}$ still refers to the (constructed) equilibrium state with $\tau^x = \tau_{obs}^x$. It is clear that for $\alpha_F = 1$, the flux corrected problem is recovered. For $\alpha_F = 0$, there is no flux correction and we obtain the coupled climatology case. Equilibrium solutions will be determined by the zonally constant shear stress of magnitude τ_{ext}^x and the full feedback processes within the basin.

For the case $\tau_{ext}^x = -0.2$ (which corresponds to about 0.01 Pa) and the already used wind stress field (7.122) which roughly corresponds to observed mean equatorial winds (Hao et al., 1993), the bifurcation diagram for $\alpha_F = 1.0$ is plotted in Fig. 7.36a. The constructed state becomes unstable at a transcritical bifurcation near $\mu = 1.0$ and two new branches of steady states appear. Solutions on the lower branch have a relatively cold eastern basin while on the upper branch, there is a relatively warm eastern basin (Fig. 7.37a/b). In strong contrast to Fig. 7.36a, no bifurcations occur in the bifurcation diagram (Fig. 7.36c) for the coupled climatology case ($\alpha_F = 0$). There is only one solution over the range of μ considered and the spatial pattern of the total field at $\mu = 1.1$ (Fig. 7.37d) closely resembles that of the constructed case on the "cold" branch (Fig. 7.37a).

What is the connection between the bifurcation diagrams in Fig. 7.36a and Fig. 7.36c, for example, what happens to the "warm" branch present in Fig. 7.36a ? The answer can be deduced from Fig. 7.36b where bifurcation diagrams for intermediate α_F are presented. These pictures have no direct physical significance individually, but show clearly the transition in parameter space between the physically relevant cases $\alpha_F = 0$ and $\alpha_F = 1$. When α_F is decreased slightly to $\alpha_F = 0.99$, the transcritical bifurcation is broken. This breakup occurs according to the imperfection of the transcritical bifurcation, as discussed in chapter 3. The spatial structure of the solutions for $\alpha_F = 0.99$ at $\mu = 1.1$ remain nearly identical to those in Fig. 7.37a/b. With decreasing α_F, the upper branch quickly moves to larger μ (see the branch for $\alpha_F = 0.95$), indicating that the states on this branch (see Fig. 7.37c at the labelled point 1 in Fig. 7.36b) can only be maintained through large coupling. The "cold" branch, however, hardly changes with decreasing α_F ($\alpha_F = 0.95$). This indicates that for this state the upwelling and thermocline slope that were maintained by prescribed winds in the constructed case can still be maintained by coupled processes. With decreasing α_F only this

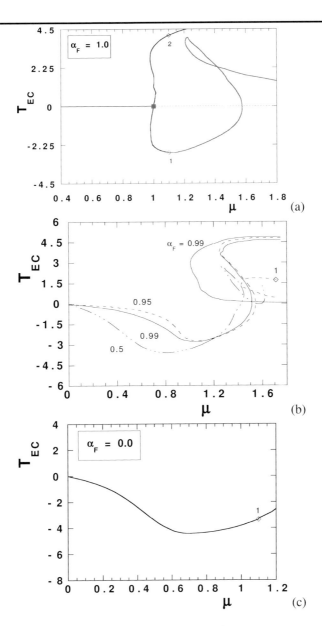

Figure 7.36.   Bifurcation diagrams for several values of the homotopy parameter α_F. In panel (a), the constructed case is shown ($\alpha_F = 1$), in panel (b) bifurcation diagrams for several intermediate values of α_F are plotted and in (c) the coupled climatology case is shown ($\alpha_F = 0$).

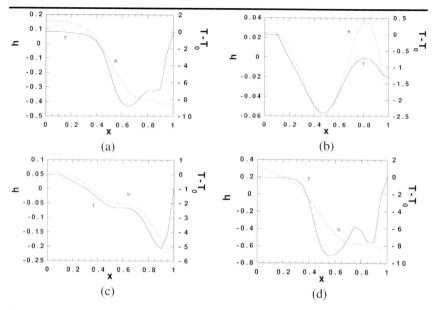

Figure 7.37. Solutions of the total fields of temperature and thermocline at labelled points in Fig. 7.36. (a) At the point labelled 1 in Fig. 7.36a. (b) At the point labelled 2 in Fig. 7.36a. (c) At the point labelled 1 in Fig. 7.36b. (d) At the point labelled 1 in Fig. 7.36c.

branch remains and is deformed smoothly into the coupled climatology branch of Fig. 7.36c.

To understand this behavior qualitatively, especially the shift of the warm branch to larger values of coupling, (7.161) is rewritten as

$$\tau^x = \tau^x_{ext} + \alpha_F(\tau^x_{obs} - \tau^x_{ext}) + \mu(\mathcal{A}(T - T_0) + \alpha_F \mathcal{A}(T_0 - \bar{T})) \qquad (7.162)$$

The first two terms represent a large scale easterly wind of which the magnitude does not vary much if α_F is close to 1 and which remains easterly, though smaller, as $\alpha_F \to 0$. Of the terms which multiply μ, the second term $\mathcal{A}(T_0 - T)$ corresponds to large scale westerlies. These westerlies, which are introduced by flux-correction, are essential in maintaining the warm branch, which has a relatively warm eastern basin. Their structure is fixed by the flux-corrected climatology and hence only the magnitude can be changed by coupling. On the contrary, the first term multiplying μ represents coupled easterly winds which can change according to a temperature variation. If α_F is decreased, the solution on the 'warm-branch' can only be maintained if μ increases to maintain the contribution of the westerlies essential to the balances on this branch. Eventually, the warm branch will disappear by moving to infinite μ in the limit $\alpha_F \to 0$. On the cold branch, in contrast, the coupled system is able to produce easterlies by internal feedbacks as the flux-correction is relaxed. This is possible because the temperature difference $(T - T_0)$ in the $\mu\mathcal{A}(T - T_0)$ term can be self-consistently maintained at finite negative values by a combination of upwelling and thermocline feedbacks. The temperature can thus adjust at fixed μ to a change in α_F to maintain the balances on the branch while the artificial westerly wind contribution vanishes (as $\alpha_F \to 0$).

Early discussion of the ENSO phenomenon was often phrased in terms of multiple stationary states, thought of as a warm "El Niño " state and the cold counterpart "La Niña ". Subsequent modeling (Zebiak and Cane, 1987; Battisti and Hirst, 1989) and observational work (Rasmusson *et al.*, 1990) has lead to a consensus view of ENSO as an essentially cyclic phenomenon. It has, however, been difficult to lay the notion of multiple stationary states entirely to rest because these were in fact found in some tropical coupled models in some flow regimes. Even when the first bifurcation from the climatology is a Hopf bifurcation with interannual period and ENSO-like characteristics, it is common for ENSO models to exhibit transcritical bifurcations from the constructed climate state at higher coupling. In many of the results of the previous section, these were encountered. Also the delayed oscillator model Schopf and Suarez (1988), which was intended to explain cyclic behavior, was originally used in a regime which has unstable stationary states in addition to the limit cycle. Although Battisti and Hirst (1989) pointed out that a better estimate of the parameters suggested a different regime, the presence of these states in this model nonetheless continued to attract attention (McCreary and Anderson, 1991; Wu *et al.*, 1993; Wakata and Sarachik, 1994).

The results from the bifurcation analyses indicate that the existence of multiple equilibria in ENSO models is an artifact of the way the annual-mean state is constructed and does not correspond to any structure which is likely to occur in the fully coupled ocean/atmosphere system. The construction places a restriction on the system such that the trivial solution, in terms of anomalous variables, exists independent of the model parameters. Modifications to this stationary state can thus occur only via codimension-1 bifurcations and additional stationary branches arise through transcritical bifurcations. Suarez and Schopf (1988) and Battisti and Hirst (1989) obtained pitchfork bifurcations in the delayed oscillator model because they excluded quadratic terms from the nonlinearity. Transcritical bifurcations are not robust when the conditions imposed by the construction are relaxed and the constructed solution ceases to exist. If the change to the system could be regarded as a small perturbation, multiple stationary points could continue to exist by having two unconnected solution branches, at least one of which has a saddle node bifurcation—this configuration can, of course, be seen when an artificial parameter is used to move gradually from the flux corrected case to a fully coupled corrected case. However, for the particular physical situation relevant to the tropical coupled climatology, one of the branches (the warm branch) moves away to infinity. This leaves a single stationary branch which evolves nonlinearly as a function of model parameters.

7.6.3. The position of the cold tongue

Although the discussion of the previous section has shown that multiple equilibria are not very likely in the tropical ocean/atmosphere system, it is not a priori guaranteed that the coupled processes represented in the model can explain the

east-west asymmetry of the observed mean state with the wind forcing taken as in
(7.161) for $\alpha_F = 0$. In order to distinguish coupled from uncoupled contributions
to the solution in the fully coupled case, it is convenient to define T_{ext} and h_{ext} to
be the response of the uncoupled ocean to the external wind stress τ_{ext}. Then the
contribution associated with coupling is $\hat{T} = T - T_{ext}$, etc.

In this section, the variation of the spatial structure of the coupled mean state
with respect to the internal coupled feedbacks is considered, while keeping a constant external wind stress, τ_{ext}^x. In this case, the structure of the steady states is
controlled by the parameters ϵ_a, δ_s and μ. For $\tau_{ext}^x = -0.2$ the externally forced
thermocline and SST, denoted h_{ext} and T_{ext}, are shown in Fig. 7.38; upwelling is
constant across the basin. The temperature decreases monotonically from west to
east and the thermocline profile is linear.

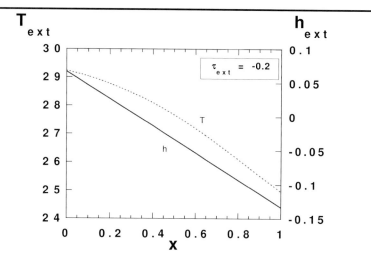

*Figure 7.38.   External state obtained with a constant external wind ($\mu = 0$) with amplitude
$\tau_{ext}^x = -0.2$ according to the fully coupled set-up.*

In Fig. 7.39a, bifurcation diagrams are plotted for three values of δ_s and show
how coupling influences the structure of the steady states. The case of $\epsilon_a = 1.25$
(i.e., a relatively long atmospheric damping scale, comparable to the basin length)
is chosen here for illustration of the behavior. In each bifurcation curve there is
a unique solution for the climatology at each μ, but the feedback between trade
winds, upwelling and thermocline slope within the basin rapidly modifies the solution as coupling is increased. Spatial structures of the solutions at four labelled
point indicated are plotted in the panels b-e of Fig. 7.39.

The balance between thermocline feedback and upwelling feedback strongly
affects the preferred spatial form of the cold tongue. Fig. 7.39 shows cases where
the surface-layer feedback is alternately shut off or increased (strong coupling is
used to highlight the differences). For the solution in Fig. 7.39b (point 1, $\delta_s = 0$),

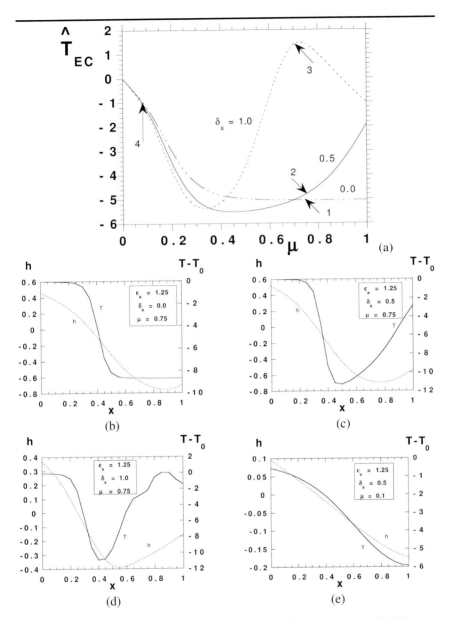

Figure 7.39. (a) Bifurcation diagrams for different values of δ_s with fixed $\epsilon_a = 1.25$. The quantity $\hat{T}_{EC}$ is the difference of the total temperature and the external temperature in the cold tongue (at $x = 0.7$). (b - e) Solutions of the total fields at the labelled points 1 - 4, respectively, in panel (a).

only the thermocline feedback is active. Since there is only externally induced up-welling, SST departures from T_0 are governed by the temperature of the upwelled water. The thermocline is deep in the west and very shallow in the east, giving temperatures nearly equal to the surface equilibrium temperature in the west. The cold tongue has constant temperature in the east due to saturation at minimum subsurface temperature. Fig. 7.39d shows the case of strong surface-layer feed-backs (point 3, $\delta_s = 1.0$). Because the effect of wind modifications on upwelling is stronger, the cold tongue shifts west toward the region of maximum easterlies, which in turn shift further west. The deep thermocline in the western Pacific limits this process. On the eastern side of the basin, westerly wind perturbations almost cancel τ_{ext}^x, giving greatly reduced upwelling and consequent warming: a narrow cold tongue in mid-basin results.

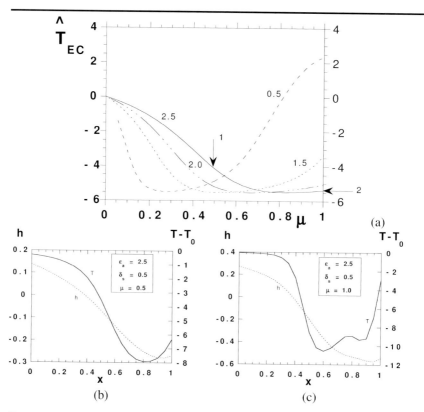

Figure 7.40.   (a) Bifurcation diagrams for different values of ϵ_a with fixed $\delta_s = 0.5$. The quantity $\hat{T}_{EC}$ is the difference of the total temperature and the external temperature in the cold tongue (at $x = 0.7$). (b - c) Solutions of the total fields at two labelled points in panel (a).

Results so far have used the case of relatively small ϵ_a so that it is clear that the decay scale in the atmosphere is not the primary length scale. Bifurcation diagrams for the standard value of $\epsilon_a = 2.5$ and an intermediate value of $\delta = 0.5$ are presented in Fig 7.40a with spatial patterns in the panels b-c. An increase of ϵ_a leads to a more localized response of the wind field. The most significant effect of ϵ_a is on the spatial pattern of the cold tongue and contrary to what might be expected, decreases in the atmospheric spatial scale (larger ϵ_a) are not associated with narrower spatial scale for the cold tongue in this range. Rather, the main effect is to move the cold tongue minimum eastward. This is because larger ϵ_a tends to weaken the easterly wind response to the west of a cold region relative to the response within the region (see Fig. 7.14). This reduces upwelling to the west and shifts the region of shallowest thermocline, with consequent feedbacks on SST.

At $\mu = 0.5$ (point 1, Fig. 7.40b), the cold tongue is situated at a reasonable position ($x = 0.8$) and the extent of the SST minimum and the slight increase in SST at the east coast are quite satisfactory. In fact, the climatology looks very much like that of a constructed climatology forced with an approximation to observed wind stress. If the coupling is increased to $\mu = 1.0$, again perturbation downwelling causes a rise in temperature at the east coast (point 2, Fig. 7.40c), and also the cold tongue becomes broader and the minimum is shifted back to the west. Slight changes in ϵ_a make only a modest difference but tend to shift the cold tongue minimum. The main point of the results is that a reasonable climatology (cold tongue position, shape, and amplitude) can be generated through coupled feedbacks and a relatively small constant external wind stress (τ_{ext}) in this simple model. The shape and size of the contribution by coupled feedbacks in (Fig. 7.40b) relative to the externally driven contribution in Fig. 7.38 neatly illustrates the importance of coupling. The balance of processes involved in the internal feedbacks is crucial to determining the shape of the cold tongue and the position of the warm pool margin. The thermocline feedback tends to move the cold tongue eastward, and favors broad spatial extent. The upwelling feedback tends to move the cold tongue westward. Depending on other factors, the upwelling feedback can narrow the cold tongue in the central basin, or can create longitudinal variations of shorter spatial scale within the cold tongue. In an extreme case, it is possible to develop a cold tongue which is strongest in the western basin.

The most important role of the externally forced contribution to the cold tongue is to create a basic cooling on which the coupled processes can then feed back to amplify the cold tongue. The size of this externally maintained cooling strongly affects the magnitude of the cold tongue that can be attained at reasonable values of the coupling. However, it has little effect on the shape of the cold tongue. In addition, the external wind is needed to provide a basic upwelling along the equator so that thermocline variations are communicated to the surface to affect SST. If this externally forced contribution to the upwelling is small, coupled processes alone can still generate a cold tongue since increased trades tend to increase upwelling which amplifies the cold tongue, etc. However, the shape of the resulting cold tongue differs from observed because the upwelling feedback dominates over the thermocline feedback.

It is possible within this model to obtain ENSO-like unstable modes on a climatology which has a reasonable warm-pool/cold-tongue structure. However, it requires some level of tuning, mainly to get the period of the oscillation in the correct regime; parameter values are given in Dijkstra and Neelin (1999). However, the simultaneous occurrence of desirable characteristics of the ENSO mode with the right period, and a coupled climatology that has an adequate cold tongue structure is not found over a large volume in parameter space within the model simulating only the near equatorial response.

7.7. Unifying Mean State and Variability

The results in the previous section have clearly supported the fact that the spatial form of the warm-pool/cold-tongue pattern depends on the nature of the coupled feedbacks, referred to as the "climatological version of the Bjerknes hypothesis" in Neelin and Dijkstra (1995). Since the model of the near equatorial response is not sufficient to study variations of the ENSO mode with changes in the mean state due to coupling, a next step is to consider the full model as described in section 7.5.2 to compute simultaneously coupled mean states and their oscillatory instabilities. This has been carried out in Van der Vaart *et al.* (2000), with the numerical details for solving the set of governing equations (7.109) provided in Van der Vaart (1998). Variables are expanded into spectral basis functions, with Chebyshev polynomials in zonal direction and Hermite functions in meridional direction. Using collocation techniques, a set of nonlinear algebraic equations is obtained for the steady states of the model. The analysis of the stability of these steady states leads to a generalized eigenvalue problem. Both steady states and their linear stability are traced through parameter space using the continuation techniques of chapter 4.

7.7.1. The warm pool/cold tongue state

At zero coupling ($\mu = 0$), the ocean circulation and consequently SST is determined by the external zonal wind stress τ_{ext}^x. This wind stress is assumed to have the form

$$\tau_{ext}^x = -F_0\, e^{-\frac{(\alpha y)^2}{2}} \; ; \; \alpha = \frac{\lambda_o}{\lambda_a} \tag{7.163}$$

where α controls the meridional extension and F_0 the amplitude of the external wind. In response to the external wind with $F_0 = 0.1$, the equatorial temperature T_{ext} increases monotonically from about 25.5 °C in the east to about 28.5 °C in the west. The thermocline is approximately linear at the equator, its depth is increasing westwards and it has slight off-equatorial maxima near the western boundary.

At small μ, the additional wind stress due to coupling is approximately the atmospheric response to the cooling $T_{ext} - T_0$. This enhances the easterly winds over most of the basin, leading to larger upwelling and a stronger thermocline slope, strengthening the cold tongue in the eastern part of the basin. The temperature $(T - T_0)_{EC}$ of the cold tongue and the vertical velocity just below the

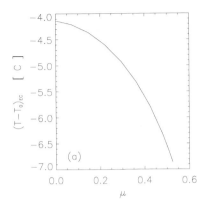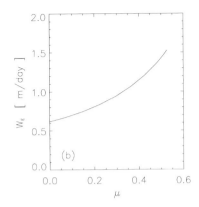

Figure 7.41.   Eastern Pacific ($x = 0.8$) equatorial SST deviation from $T_0 = 30\,^{\circ}C$ (a) and dimensional upwelling velocity (b) as a function of the coupling strength μ.

cold tongue w_E (Fig. 7.41) demonstrate that there is a unique steady solution as a function of μ, with more upwelling as coupling gets stronger. At $\mu = 0.5$, the spatial structure of the mean state is shown in Fig. 7.42. The zonal scale of the cold tongue (panel a) is set by a delicate balance of thermocline and surface layer feedbacks (Dijkstra and Neelin, 1995a). The meridional extent of the cold tongue is determined both by the Ekman spreading length (a_s/β_0) and by meridional advection. The thermocline field (Fig. 7.42b) displays the off-equatorial maxima and a deeper (shallower) equatorial thermocline in the west (east). This indicates that the reservoir of heat content lies off-equatorial in the central and western part of the basin. The zonal wind response U (Fig. 7.42c) shows the intensification of the westward winds, with a maximum west of the cold tongue. The vertical velocity structure (Fig. 7.42d) is clearly controlled by Ekman divergences. Upward velocities are restricted to an equatorial zone and the maximum amplitude occurs in the eastern part of the basin.

7.7.2.   The ENSO mode

Along the branch of steady states in Fig. 7.41 the linear stability is determined simultaneously, by writing the total solution vector ϕ, consisting of ocean, atmosphere quantities and SST, as

$$\phi(x, y, t) = \overline{\phi}(x, y) + \tilde{\phi}(x, y)\, e^{\sigma t}. \tag{7.164}$$

Here, the vector $\overline{\phi}$ represents the mean state, $\tilde{\phi}$ perturbations with respect to this mean state and σ is again the complex growth rate of the perturbation. In Fig. 7.43, the path of six modes – which become leading eigenmodes at high coupling – is plotted as a function of the coupling strength μ. In Fig. 7.43a, a larger dot size indicates a larger value of μ and both period and growth rate of the modes are

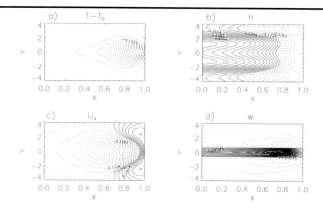

Figure 7.42.   The mean state at standard parameter values and $\mu = 0.5$. (a) $T_* - T_0$; maximum 6.6 °C. (b) Thermocline depth; maximum 82.3 m. (c) Zonal wind U_*; maximum 9.5 m/s. (d) Vertical velocity w_{1*}; maximum 1.44 m/day. In all panels, values are scaled with the maximum value of each field and the contour levels (with interval 0.069) are with respect to this maximum.

given in year^{-1}. In Fig. 7.43b, only the growth rate is plotted against μ. One oscillatory mode becomes unstable as μ is increased and a Hopf bifurcation occurs near $\mu = 0.5$.

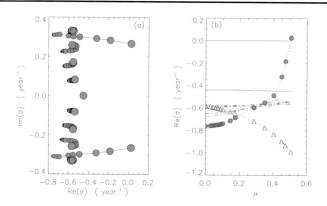

Figure 7.43.   (a) Plot of the eigenvalues for the six leading eigenmodes in the $(\mathrm{Re}(\sigma), \mathrm{Im}(\sigma))$-plane. Values of the coupling strength μ are represented by dot size (smallest dot is the uncoupled case ($\mu = 0$) for each mode, largest dot is the fully coupled case at the Hopf bifurcation ($\mu_c = 0.5$)). The Hopf-bifurcation that yields the ENSO mode occurs where the path of one eigenvalue first crosses $\mathrm{Re}(\sigma) = 0$. (b) The growth rate of the leading modes as a function of coupling strength.

At the critical value $\mu_c = 0.5$, for which the mean state was shown in Fig. 7.42, time-longitude diagrams of the equatorial thermocline, temperature and zonal wind anomalies of this oscillatory mode are shown in Fig. 7.44. The SST pattern (Fig. 7.44b) displays a nearly standing oscillation for which the spatial scale is confined to the cold tongue of the mean state. There is a slight eastward propagation of the SST anomaly in the central equatorial Pacific. The thermocline anomaly (Fig. 7.44a) shows western anomalies in heat content leading those with the same sign at the eastern boundary. These anomalies are out of phase with the SST anomalies with a lag of about 5 months. The wind response (Fig. 7.44c) is much broader zonally and is in phase with the SST anomaly.

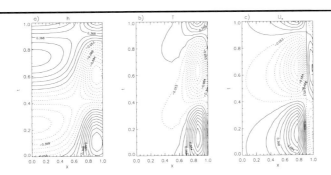

Figure 7.44.   *Time(t)-longitude diagram at the equator of the anomalies of (a) Thermocline depth* $\tilde{h}_*$ *(max = 9.5 m), and (b) Sea-surface temperature* $\tilde{T}_*$ *(max = 1.4° C) and (c) Zonal wind* $\tilde{U}_*$ *(max = 3.0 m/s). The period of the oscillation is 3.7 years. Note that the amplitude of the oscillation is not determined by the linear stability analysis. The maximum amplitudes are relative magnitudes of the different fields, i.e. the mode displays a thermocline deviation of about 10 m per degree SST anomaly.*

The phase relationships between wind, temperature and thermocline anomalies can be seen more clearly in Fig. 7.45. Here, the thermocline depth in the western Pacific h_W and the zonal mean equatorial thermocline displacement h_{ZM} are related to the SST anomaly in the eastern Pacific (at $x = 0.92$) over one cycle of the oscillation. Panel (a) shows the characteristic ENSO phase relationship between SST and thermocline anomalies with a relatively shallow (deep) western thermocline in case of a warm (cold) event. As the closed curve is traversed clockwise over one cycle of the oscillation, it is seen that a cold event is followed by an extreme positive western thermocline anomaly. As the SST anomaly becomes zero, the western thermocline anomaly is still positive and panel (b) shows that this also holds for the zonally averaged thermocline anomaly. Hence, the equatorial heat content is slowly built up after the cold event by the increase of the trade winds. This sets the stage for the following warm event in which the equatorial heat content is discharged. After the warm event, the zonally averaged thermo-

cline anomaly is negative as the SST anomaly goes through zero again and the
equatorial heat content is low, which causes the next cold event.

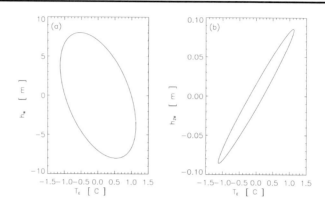

*Figure 7.45.   (a) Phase relation between the equatorial thermocline depth anomaly in the western
part of the basin and the SST anomaly in the east. (b) The phase relation between the zonally
averaged equatorial thermocline anomaly h_{ZM} and SST anomaly in the east (T_E). Note that the
amplitudes in both panels are arbitrary, but that their ratio is fixed. The direction of rotation with
time is clockwise in both cases.*

The meridional structure of the ENSO mode is shown by plotting the different
fields (Figs. 7.46 to 7.48) at several phases of the oscillation relative to the period,
i.e. phase $t = 1/2$ indicates the fields after half a period. The starting point of the
description is a positive SST-anomaly in the eastern Pacific (early El Niño phase),
as shown in Fig. 7.46 at $t = 0$. Eastward zonal wind anomalies to the west of the
maximum in SST-anomaly (Fig. 7.46) are present as can be seen in Fig. 7.48 at
$t = 0$. The wind response amplifies the positive SST-anomaly ($t = 1/16$ to $1/8$)
and the spatial scale of the SST anomaly is controlled by the shape of the cold
tongue (cf. Fig. 7.42). The equatorial thermocline response to the weaker surface
winds up to $t = 1/8$ results in a negative anomaly (i.e. negative heat content)
in the Western Pacific (Fig. 7.47, $t = 1/8$). This anomaly is at its minimum a
few months later than the maximum of equatorial SST. As long as the positive
thermocline/SST-anomaly in the eastern part of the basin does not weaken, this
negative anomaly cannot be discharged. However, due to ocean wave reflections
at the eastern boundary, the mass fed along the equator to the eastern basin is
transformed into a collective of long Rossby waves which propagates westward
(Fig. 7.47, $t = 0 - 1/8$).   At the equator, the eastern positive thermocline
anomaly and consequently the SST anomalies are weakened. This reduces the
east-west SST gradient, causing the anomalous eastward winds to weaken (Fig.
7.48, t = 1/8 to 5/16). Termination of the El Niño phase sets in, as the western
warm pool discharges its previously built up negative heat content (t = 3/8 through

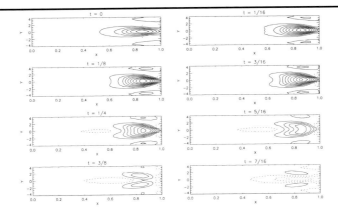

Figure 7.46.   Planforms of the SST anomaly during the oscillation in the x-y plane; times are with respect to the period of the oscillation. Drawn (dotted) lines represent warm (cold) anomalies.

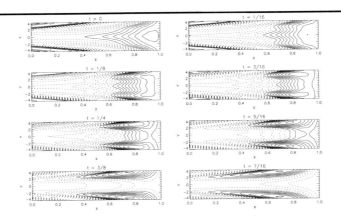

Figure 7.47.   Planforms of the thermocline anomaly during the oscillation as in Fig. 7.46. Drawn (dotted) lines represent positive (negative) anomalies.

7/16 in Fig. 7.47). As the thermocline rises in the east, the SST anomaly becomes negative and through coupled processes its amplitude increases. The trade winds recover (Fig. 7.48, t = 7/16) and the positive off-equatorial thermocline anomalies propagate westwards (Fig. 7.47, t = 3/8 - 7/16). Then the cycle starts over again but with the signs of the perturbations reversed.

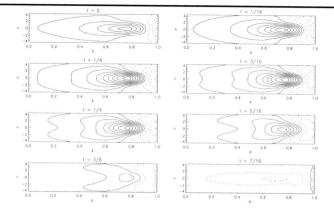

Figure 7.48. Planforms of the zonal wind anomaly during the oscillation as in Fig. 7.46. Drawn (dotted) lines represent anomalous eastward (westward) winds.

7.7.3. Model reduction

The results of the linear stability analysis do not provide information on the finite amplitude of the fields for supercritical conditions. In this section, the equilibration of the perturbations to finite amplitude is studied in a weakly nonlinear context, i.e. for coupling values of μ just above the Hopf-bifurcation (Van der Vaart, 1998). Let

$$\epsilon = \frac{\mu - \mu_c}{\mu_c} \ll 1 \,, \tag{7.165}$$

be a measure of the distance beyond critical conditions, i.e. $\mu = \mu_c$ at which $Re(\sigma) = 0$. For these values of the coupling strength μ, equilibration of the unstable perturbations will occur on a time-scale long compared to the time-scale of growth. Therefore a new time variable $\tau = \epsilon^2 t$ is introduced.

The coupling strength μ, time and the solution vector ϕ are expanded in terms of ϵ and the 'ENSO' mode $\tilde{\phi}$ with time-dependence $e^{i\omega_c t}$, where $\omega_c = \Im(\sigma)$ at $\mu = \mu_c$,

$$
\begin{aligned}
\phi &= \bar{\phi} + \epsilon A(\tau)\tilde{\phi}e^{i\omega_c t} + \epsilon^2(|A(\tau)|^2\tilde{\phi}_{02} \\
&+ A^2(\tau)\tilde{\phi}_{22}e^{2i\omega_c t}) + \epsilon^3\tilde{\phi}_{13}e^{i\omega_c t} + c.c. \tag{7.166a} \\
\frac{\partial}{\partial t} &\to i\omega_c + \epsilon^2\frac{\partial}{\partial \tau} \tag{7.166b} \\
\mu &= \mu_c(1 + \epsilon^2 m) \,, \qquad m = \mathcal{O}(1) \tag{7.166c}
\end{aligned}
$$

In these expansions, $c.c.$ denote complex conjugate, m is the new control parameter, and $A(\tau)$ is the (complex) amplitude of the initially unstable mode with spatial structure $\tilde{\phi}$.

By substitution of the expansions (7.166) into the governing equations and collecting terms of like orders in ϵ and $e^{i\omega_c t}$, one can reduce the full equations to a scalar equation for the amplitude $A(\tau)$. This becomes a Landau equation

$$\frac{\partial A}{\partial \tau} = m\frac{\partial \sigma}{\partial \mu}A - \Lambda A|A|^2 \qquad (7.167)$$

where the coefficients are evaluated at $\mu = \mu_c$ and which are calculated numerically within the pseudo spectral set-up (Van der Vaart, 1998). Solving (7.167) for A, the solution for the SST field becomes

$$T(x, y, t) = \bar{T}(x, y) + \epsilon A(\tau)\tilde{T}(x, y)e^{i\omega_c t} + \mathcal{O}(\epsilon^2) \qquad (7.168)$$

where the mean state is represented by $\bar{T}$ and the SST pattern of the critical mode by $\tilde{T}$. If the coefficients $\frac{\partial \sigma}{\partial \mu}$ and Λ satisfy the conditions for a supercritical Hopf bifurcation, $(\Re(\frac{\partial \sigma}{\partial \mu}) > 0$ and $\Re(\Lambda) > 0)$ finite amplitude solutions to (7.167) exist of the form

$$A(\tau) = \sqrt{\frac{\Re(\frac{\partial \sigma}{\partial \mu})}{\Re(\Lambda)}}e^{i\Omega\tau} \; ; \; \Omega = \Im(\frac{\partial \sigma}{\partial \mu}) - \frac{\Im(\Lambda)}{\Re(\Lambda)} \qquad (7.169)$$

Using these expressions one can derive the amplitude and total period of the stable periodic orbit for coupling values beyond the Hopf-bifurcation. For example, the period $\mathcal{P}$ is given by

$$\mathcal{P} = \frac{2\pi}{\omega_c + \epsilon^2\Omega}, \qquad (7.170)$$

In Fig. 7.49a the period of the orbit is plotted as a function of the coupling strength μ. The period of ENSO oscillation is set by the critical period (at the Hopf-bifurcation), rather than the frequency of the instability at supercritical coupling strength; this is also found in other studies (Battisti and Hirst, 1989; Neelin *et al.*, 1994; Jin, 1997b). The amplitude of the SST-anomaly in the cold tongue increases strongly with coupling, being about $3°C$ for $\epsilon = 0.1$ (Fig. 7.49b).

7.8. Presence of the Seasonal Cycle

As already described in the beginning of this chapter, another major time-dependent phenomenon in the Tropical Pacific is the seasonal cycle (Horel, 1982). Important feature is the annual march of the ITCZ with the corresponding movement of the trade winds as sketched in section 7.1. At the equator, eastern Pacific SST is warmest during April/March and is coldest during August/September when the southerly winds are maximal. In this section, the interaction between the seasonal cycle and the ENSO oscillation is discussed.

7.8.1. Coupled processes and the seasonal cycle

The annual component of the seasonal cycle in the Eastern Pacific is a strange phenomenon considering the semi-annual component of the solar forcing. There are strong indications that coupled processes, in particular those in the surface

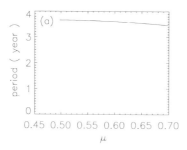
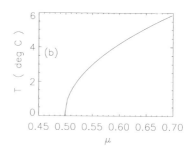

Figure 7.49.   *The frequency (a) and amplitude (b) of the finite amplitude limit cycle as obtained from the weakly nonlinear analysis within the fully coupled Zebiak-Cane model in a finite ocean basin. Shown is the maximum amplitude of the SST-anomaly near the position of the cold tongue.*

layer, are involved to get an annual response to a semi-annual forcing (Philander *et al.*, 1996; Mitchell and Wallace, 1992; Chang *et al.*, 1995). Slight westward propagation of equatorial SST anomalies is also seen in the eastern Pacific. Since the seasonal thermocline anomalies are much weaker than the seasonal SST anomalies, it is believed that surface layer processes play a dominant role. Hence, of the coupled modes, SST modes are potential candidates for the seasonal signal in the equatorial SST anomalies.

Do these coupled processes play a role to create such a strong equatorial asymmetry as to get the ITCZ mostly north of the equator? For example, there is north-south asymmetry due to the shape of the continents, but do coupled processes amplify this asymmetry? Are coupled processes involved to causes the annual period and the westward propagation of SST anomalies in the eastern Pacific? Possible answers have been given in Xie (1994) and Xie and Philander (1994) using zonally independent models, which include more detailed physics in the atmosphere and oceanic mixed layer. In these type of models, both evaporation (in combination with the amplification of the wind) and vertical mixing can break equatorial symmetry and both seem to be are important in more extensive models (Philander *et al.*, 1996). Chang and Philander (1994) perform a stability analysis exploring the hypotheses (Mitchell and Wallace, 1992) that meridional wind and its interaction with local SST gradient play a crucial role. They find that a positive feedback between SST and surface layer dynamics can give anti-symmetric and symmetric unstable SST modes. Based on a similar model, Liu and Xie (1994) suggest that the annual equatorial signal may come from the extratropics.

In all these studies a sufficiently strong southerly wind response at the equator is essential. This is too demanding for the type models introduced in these chapters. The Gill atmosphere model is known to give too small southerly winds at the equator and this problem is not easily fixed. One way out is just to prescribe the annual signal into these models. In this way, the meridional asymmetry is created through the observed seasonal heat flux and wind stress. Although this circumvents the interesting point about the role of coupled processes in the seasonal cycle, the interaction of the seasonal cycle with ENSO can be explored.

7.8.2. Interaction of seasonal cycle and ENSO

The results of Zebiak and Cane (1987), as described in section 7.4 showed that the warm events followed the seasonal cycle and peaks tend to occur either in June or at the end of year. Observations of NINO3 seem to show a favorable peak month at November-December (Rasmusson and Carpenter, 1982; Neelin *et al.*, 2000), but the standard deviation is also relatively large in this period.

First, the interaction of the annual cycle and ENSO is considered in the coupled wave oscillator model (Cane *et al.*, 1990; Münnich *et al.*, 1991). Using a particular variant of these point-coupling models, the interaction of the ENSO mode and the seasonal cycle has been studied (Tziperman *et al.*, 1994a) by looking at solutions of

$$\frac{dh_*}{dt_*} = a\,\mathcal{A}\left[h_*(t - \frac{L}{2c_K})\right] - b\,\mathcal{A}\left[h_*(t_* - (\frac{L}{2c_K} + \frac{L}{2c_R})\right] + c\,cos\,\omega_a t_* \quad (7.171)$$

The first term of the right hand side represents the direct feedback due to the Kelvin wave which is initiated at half the basin and therefore has a travel time $L/(2c_K)$. Similarly, the second term is the delayed feedback due to a Rossby wave initiated at the same position, traveling westward, reflecting and travelling back to the central part of the basin as a Kelvin wave. The last term is the seasonal forcing with frequency ω_a corresponding to a one-year period. An example of the shape of the function $\mathcal{A}$ is a tangent hyperbolic function which saturates both at large negative and positive thermocline anomalies and has a sharp transition near $h_* = 0$. For this function, κ is the slope at $h_* = 0$ and measures the strength of the coupled thermocline feedback; this parameter is used as a control parameter.

Results of the function $h(t) = h_*(t)/H$ of 1,024 years integrations and its spectrum are plotted in Fig. 7.50 for four values of κ. For small κ, there is a simple periodic orbit with annual period induced by the forcing. For $\kappa = 1.2$, the ENSO frequency appears which is incommensurate with the annual frequency and hence a quasi-periodic signal is obtained. For larger κ, locking to the frequency of the annual cycle occurs just as was seen for the circle map in section 3.4. The dominant frequency is exactly 4 years and hence the system is located in one of the 'Arnold' tongues'. For even larger nonlinearity, the spectrum becomes broad banded, which is an indication of chaotic behavior in the system.

Nearly simultaneously, similar results were presented in Jin *et al.* (1994) using the equatorial strip approximation for the SST-equation and these were more extensively presented in Jin *et al.* (1996). The coupled model (7.109) is extended with an off-equatorial strip having a temperature T_{A*} and the SST equations become

$$\frac{\partial T_*}{\partial t_*} = -a_T T_* + Q_E \quad - \quad u_{1*}\frac{\partial T_*}{\partial x_*} - \mathcal{H}(w_{1*})w_{1*}\frac{T_* - T_{s*}(h_*)}{H_u}$$

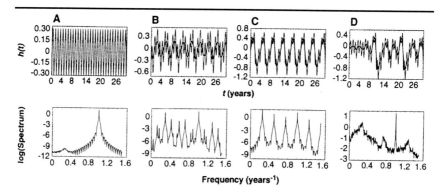

Figure 7.50. (a) Time series and (b) Power spectra of the wave oscillator model in Tziperman et al. (1994a) for four values of (A) $\kappa = 0.9$, (B) $\kappa = 1.2$, (C) $\kappa = 1.5$, (D) $\kappa = 2.0$.

$$-\frac{1}{4\lambda_o}(\mathcal{H}(v_{S*})v_{S*}(T_* + T_{A*}) \quad - \quad \mathcal{H}(-v_{N*})v_{N*}(T_* - T_{A*})) \qquad (7.172a)$$

$$\frac{\partial T_{A*}}{\partial t_*} \quad = \quad -a_T T_{A*} + Q_A \qquad (7.172b)$$

with obvious meaning for the meridional velocities north and south of the equatorial strip. The seasonal heat fluxes Q_E and Q_A are approximated from observations and are prescribed together with observed seasonal wind stress to construct a seasonal cycle with temperatures $(\bar{T}_*, \bar{T}_{A*})$. The time-longitude behavior of the equatorial temperature $\bar{T}_*$ (with annual period) within this model compares fairly well with observations (section 7.1) and the phase is about one month ahead, with minimum temperatures occurring in August rather than in September.

In the coupled model, the Gill atmosphere response is calculated from the forcing

$$Q_* = \alpha_T \left(T_* - \bar{T}_* + (T_{A*} - \bar{T}_{A*})\frac{y_*}{\lambda_a}\right)e^{-\frac{y_*^2}{2\lambda_a^2}} \qquad (7.173)$$

In this way, both zonal and meridional wind anomalies are obtained. Note that with the annual forcing in the wind stress and heat flux prescribed, the dynamical system corresponding to the coupled model becomes non-autonomous with a periodic basic state. Hence, Floquet theory (see chapter 3) must be applied to study the stability of the annual cycle. Again, the coupling strength μ, the surface layer feedback parameter δ_s and the dimensionless atmospheric damping length ϵ_a are main control parameters.

Results of the Floquet analysis were presented in Jin et al. (1996) and compared with the unstable modes of the annual-mean state. For two different values of δ_s, both eigenvalues (for the annual mean state) and Floquet exponents (for the annual cycle) are shown in Fig. 7.51. For small δ_s, the Floquet exponents show basically the same result as the eigenmodes for the annual-mean state (compare Fig. 7.51a

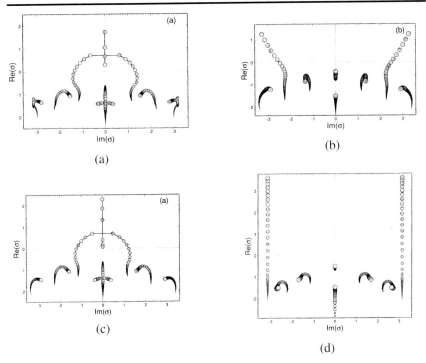

Figure 7.51. *(a) Collective plot (Jin et al., 1996) of the five leading eigenvalues for the annual-mean climatology for (a) $\delta_s = 0.25$ and (c) $\delta_s = 0.6$. The range of the coupling strength μ is from 0.1 to 1.0. (b) Same but now the Floquet exponents for the seasonal cycle for $\delta_s = 0.25$. (d) Same as (b) but for $\delta_s = 0.6$.*

and Fig. 7.51c). The mode which becomes unstable through a Hopf bifurcation on the annual-mean state is very similar to that destabilizing the annual cycle through a Naimark-Sacker bifurcation. However, for larger δ_s the frequency of the most unstable mode in the annual-mean case becomes modified and a Floquet exponent with frequency π appears. This gives a Floquet multiplier with $\rho = -1$ and hence a period doubling bifurcation occurs; near onset, a periodic orbit of 2 years is expected. The pattern of this coupled mode (see Technical box 3.3) is very similar (Fig. 7.52b) to that of the mode destabilizing the annual-mean state (Fig. 7.52a).

When the coupling strength is increased above criticality, the time series of the model become very complicated. For $\delta_s = 0.25$, power spectra for three values of μ are plotted in Fig. 7.53a. The annual cycle is stable for $\mu = 0.8$, for $\mu = 0.9$, the solution is frequency locked to a subharmonic oscillation having a 5-year period, with peaks at $1/5, 2/5, 3/5, 4/5$ and 1 year^{-1}. The relative size of peak at $4/5$ (which is the difference frequency of the annual cycle and the ENSO mode) shows that their interaction is stronger than that with the other

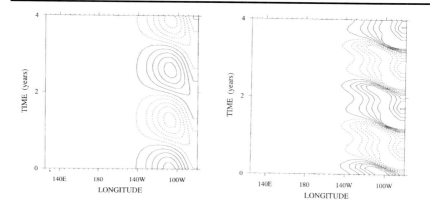

Figure 7.52.   (a) SST-pattern of the mode destabilizing the annual mean state. (b) SST-pattern of the periodic disturbance destabilizing the seasonal cycle; both (Jin et al., 1996) are for $\delta_s = 0.6$.

subharmonics. As μ increases further, a 4 year locked regime is encountered followed by a chaotic regime. The SST anomaly pattern for this case (Fig. 7.53b), shows that irregularity is present, but that the unstable mode of the linear stability analysis is still recognizable.

Similar behavior is found for other values of μ and δ_s and the behavior of the system in this parameter plane is captured (Jin *et al.*, 1996) in the Devil's terrace (Fig. 7.54). The colors indicate the frequency ratio of frequency locked solutions and the light areas indicate the chaotic solutions. Although chaotic solutions are nearly everywhere in parameter space over a fractal surface, the locked regime is broad, with the frequency ratio becoming smaller for smaller δ_s. For many solutions found, the phase locking to the annual cycle is near to that observed with January being the preferred month of the peak of the warm event.

The interaction between seasonal cycle and ENSO mode has been analyzed in Tziperman *et al.* (1995) within the Zebiak-Cane model. The coupling strength is varied by changing the drag coefficient in the bulk wind stress formula. When the amplitude of the annual cycle is zero, the annual-mean state is stable at standard value of the coupling. With perpetual July conditions as mean state, the system behaves very irregularly with a nearly continuous frequency spectrum; decreasing the coupling gives a nice periodic ENSO signal. This shows that even without the seasonal cycle, chaotic motion can appear in the ZC-model. This has later been attributed to mode interaction between the ENSO mode and another so-called 'mobile' mode (Mantua and Battisti, 1995). When the amplitude of the seasonal cycle is increased, frequency locking occurs in the same way as seen above, with chaotic regimes in between. Tziperman *et al.* (1997) analyze the factors which influence the interaction between seasonal cycle and ENSO in this model and point to the seasonality of the atmosphere as the primary effect. A similar sequence of transitions was also found in a more elaborate ICM by Chang *et al.* (1994), which

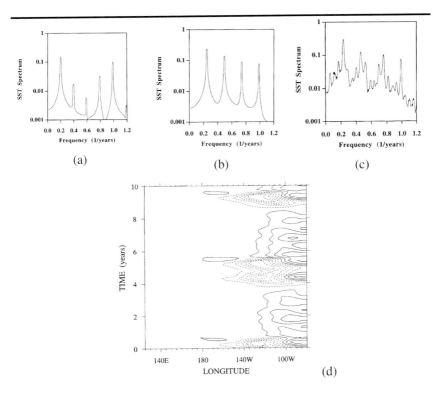

Figure 7.53. Power spectra of time series (Jin et al., 1996) for $\delta_s = 0.25$ and three supercritical values of μ, (a) $\mu = 0.9$, (b) $\mu = 1.0$ and (c) $\mu = 1.05$. (d) Pattern of SST anomaly for the time series for $\mu = 1.05$.

simulates both the seasonal cycle and ENSO. They stress the relative importance of the amplitude of the seasonal cycle and show that the ENSO frequency gets entrained (disappears from the signal) if the amplitude of the seasonal cycle becomes too large. Variations in phase locking behavior has been analyzed in Neelin et al. (2000).

7.8.3. The irregularity of ENSO

From the Devil's terrace picture, one observes that parameter regimes with chaotic behavior are relatively scarce with respect to frequency locked ones. Hence, although the interaction of the seasonal cycle and ENSO can give irregular behavior, it may not be the only source. Of course, there is strong variability due to various weather phenomena that have short lag-correlations of hours to days, related to processes which are unrelated to those controlling ENSO and are usually referred to as 'noise'.

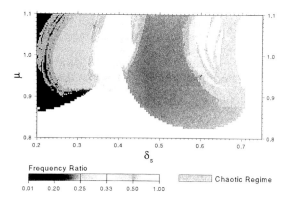

Figure 7.54.   (in color on page 532). The Devil's terrace as computed in Jin et al. (1996). Colors indicate the frequency locked regimes, with chaotic regimes in between. Along a section for fixed μ, a Devil's staircase, as discussed in section 3.4 appears.

In the model of Van der Vaart *et al.* (2000), for which results of the ENSO mode were shown in section 7.6, the applied noisy forcing incorporates no preferred spatial nor temporal scales. The stochastic forcing appears as an additional term in both the zonal wind stress and the upwelling.

$$\tau^x = F_0 \tau^x_{ext} + \mu \mathcal{A}(T - T_0) + \tau^x_{noise}, \qquad (7.174a)$$

$$w_1 = w + w_s + w_{noise}, \qquad (7.174b)$$

where $\tau^x_{noise}, w_{noise}$ have the actual form

$$\tau^x_{noise}(x, y, t) = a_{noise} R_\tau(x, y, t) e^{-y^2/4}, \qquad (7.175a)$$

$$w_{noise}(x, y, t) = a_{noise} R_w(x, y, t) e^{-y^2/4} \qquad (7.175b)$$

The coefficient a_{noise} controls the relative strength of the stochastic term. The functions R_w and R_τ are obtained through a pseudo random number generator and updated every time step. This gives a decorrelation time of about $1 - 2$ months, using a time step of 30 days (Blanke *et al.*, 1997). Note that this time step is possible, because the model is integrated using a fully implicit scheme. The seasonal cycle is represented through variations of the radiative equilibrium temperature T_0 with time and latitude, i.e.

$$T_0(y, t) = \bar{T}_0 \, \cos(\gamma_s(y + a_s \, \sin(\omega_a t))) \qquad (7.176)$$

The amplitude $\bar{T}_0$ is taken as 30 °C , the coefficient γ_s follows from a fit to the observed seasonal cycle and has the value 0.06. This accounts for an equilibrium temperature of 27 °C at 10 °N in the absence of temporal variations. The amplitude of the seasonal variations a_s is set to a value of 0.11, by the condition that T_0 has a maximum at 20°N at $t = \pi/2\omega_a$.

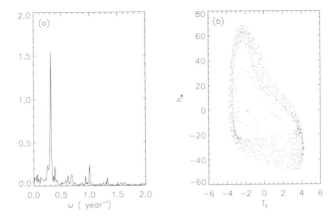

Figure 7.55. (a) Powerspectrum of equatorial SST at ($x = 0.8$), for the case of supercritical coupling strength ($\mu = 0.7$) with stochastic forcing and annual cycle included. (b) Phase diagram of eastern Pacific SST anomaly (T_{E*}) versus western Pacific thermocline depth anomaly h_{W*}.

With supercritical coupling strength ($\mu = 0.7$), and no stochastic and seasonal forcing, the model exhibits a regular periodic solution. When the seasonal cycle and noise are included, with $a_{noise} = 0.2$, at supercritical conditions the ENSO signal dominates over the seasonal cycle as can be seen in Fig. 7.55. The deterministic period of the supercritical solution retains its identity over the influence of stochastic forcing. The phase space view of eastern basin SST anomaly versus western Pacific thermocline depth, shows the effect of the noise as a fuzzy signature of the otherwise stable periodic orbit (Fig. 7.55b). At subcritical conditions ($\mu = 0.48$), ENSO variability is excited as is indicated by the low frequency signal in Fig. 7.56, but it remains of reasonably weak amplitude.

7.9. ENSO in General Circulation Models

From the previous sections, a framework emerges to understand the ENSO variability but as the ICMs contain only a limited part of relevant physical processes, the theory is far from complete to satisfactorily answer all questions as posed in section 7.1.5. Clearly, more work is needed to confirm that the physical mechanisms of ENSO variability as deduced from ICMs are actually dominant in these more detailed models and to answer the questions related to the seasonal cycle. Such a modelling effort has started since the early 1980s and the models are generally referred to as tropical General Circulation Models (GCMs). Many types of models are used but essentially two different classes can be distinguished: hybrid coupled models (HCMs) and coupled general circulation models (CGCMs). Both type of models consist of a state-of-the-art ocean model, but the atmospheric components differ substantially. In CGCMs, the atmosphere is

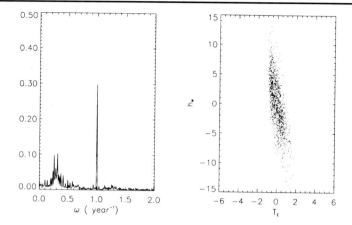

Figure 7.56.   As Fig. 7.55, for the case of subcritical coupling strength with stochastic forcing and annual cycle included.   Again in (a) the power spectrum of SST is shown and in (b) the phase space diagram.

a full physical-dynamical model whereas in HCMs, it is an empirically derived (statistical) model.

In the HCMs, the atmospheric response to variations in SST is captured by statistical relations inferred from data. The advantages of an HCM are a relatively low computational cost with respect to a CGCM. In addition, there is no oversimplification of nonlinear oceanic processes as in ICMs while still the level of complexity is lower than in a CGCM allowing an easier understanding of the phenomena simulated. HCMs have been quite successful in simulating many aspects of ENSO variability (Chang *et al.*, 1995; Syu *et al.*, 1995; Blanke *et al.*, 1997; Eckert and Latif, 1997; Syu and Neelin, 1995; Tang, 2002). Under steady forcing, the central Hopf bifurcation associated with the ENSO is found in HCMs as a transition from steady to periodic behavior. With seasonal forcing, a phase locking behavior similar to that in ICMs is found.

The CGCMs can be subdivided in so-called TOGA models and global models. In TOGA models, a high-resolution tropical Pacific ocean basin model is coupled to a global atmosphere model. In global CGCMs, a relatively low-resolution global ocean and atmosphere model are coupled. An extensive overview of the behavior of all these models is beyond the scope of this book and several reviews are available for this purpose (Neelin and coauthors, 1992; Mechoso and coauthors, 1995; Delecluse *et al.*, 1998; Latif *et al.*, 2001; AchutaRao and Sperber, 2002; Davey *et al.*, 2002). In the remainder of this section, however, a brief impression is given of the general capabilities (and remaining problems) of these models.

Many of the early models (in particular the coarse-resolution models) showed major errors in the annual-mean temperature and its zonal gradient, while interannual variability ranged from weak to moderate (Neelin and coauthors, 1992).

The errors in the annual-mean state have been an obstacle in coupled modelling for some time (Delecluse *et al.*, 1998) and were generally referred to as climate drift[4]. In some models, the cold tongue cut right across the basin or migrated to mid-basin. In others, the warm pool was displaced or warm water occurred in the eastern as well as in the western part of the basin. In many models, weak zonal SST gradients across the equator were common (Sperber *et al.*, 1987; Gordon, 1989; Meehl, 1990; Endoh *et al.*, 1991; Neelin and coauthors, 1992).

Since the early 1990s, relatively good seasonal cycles were obtained in TOGA-type CGCMs (Giese and Carton, 1994; Robertson *et al.*, 1995a,b; Terray *et al.*, 1995). In Mechoso and coauthors (1995), the capabilities of eleven different CGCMs in simulating the seasonal cycle were compared. Each CGCM was able to simulate a reasonable annual-mean state with sufficient zonal asymmetry along the equator. The development of the cold tongue over the year was more difficult to capture by the models and quite diverse behavior was found. Most CGCMs tended to develop a too narrow cold tongue around the equator, in some models warm water appeared south of the equator in April and most models had a too (equatorially) symmetric response. Some models generated a semi-annual component along the equator and some developed a convergence zone over the southeastern Pacific in March-April. Philander *et al.* (1996) conjectured that the movement of the ITCZ is caused by the equatorial asymmetry of the continents in the eastern Pacific. This small asymmetry is then amplified by the coupled feedbacks and leads to the strong asymmetry as seen in observations. The simulation of a correct seasonal cycle, however, poses (even today) a significant challenge to CGCMs.

ENSO-like behavior in early CGCMs was described in Sperber *et al.* (1987) and Philander *et al.* (1989), although significant drift occurred in the latter model. In Neelin and coauthors (1992), a multitude of behavior with both propagating and standing SST signals of interannual variability was found. It appeared that a correct structure of the annual-mean state was not an important factor to find interannual variability. Also in the Mechoso and coauthors (1995) intercomparison, a good simulation of the seasonal cycle did not seem to guarantee good interannual behavior.

Sometime ago, the Coupled Model Intercomparison Project (CMIP) was started with the aim to evaluate the behavior of the different models. More information on the models and the scientists participating in this intercomparison can be found on http://www-pcmdi.llnl.gov/modeldoc/cmip. The El Niño Simulation Intercomparison Project (ENSIP) has been one of the activities within the CMIP project (Latif *et al.*, 2001). The results show the capabilities of about 24 models in the simulation of the annual-mean state, the seasonal cycle and El Niño variability. For each of the models – in Latif *et al.* (2001), a short description of each model is provided – a 20-year simulation was used.

---

[4]The term 'drift' is used for both the equilibrium departure of the model climatology from observations and for the process of adjustment toward this equilibrium

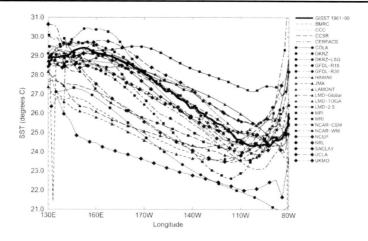

Figure 7.57.   Model simulated equatorial SSTs, averaged over 2°S - 2°N. The thick curve is the 1961-1990 mean from the GISST dataset (from Latif et al. (2001)).

In Fig. 7.57, the simulated annual-mean SST is shown for all of the models used (the acronyms of the models are explained in Latif *et al.* (2001)). The zonal temperature gradient in the central part of the basin is well simulated by most models which is a considerable improvement over earlier models (Neelin and coauthors, 1992). Many of the models have a cold bias in the central part of the basin and simulate too warm SSTs near the eastern boundary. Moreover, some models have large errors near the western boundary.

The performance of two of the models, the TOGA-type CERFACS model (Terray *et al.*, 1995) and the global GFDL-R30 model (Knutson and Manabe, 1998; Delworth *et al.*, 2002), in simulating the seasonal SST field is presented in Fig. 7.58. Both models do rather well in capturing the annual signal in the eastern Pacific and the semi-annual signal in the western Pacific. They also have a correct amplitude of the seasonal signal, but the CERFACS model shows a slightly better westward propagation than the GFDL-R30 model. Other models, however, have problems to simulate a correct amplitude and pattern of the seasonal cycle and many, in particular the coarse-resolution CGCMs, simulate a weak seasonal signal. Some models show a semi-annual signal in the eastern Pacific and some an annual signal in the western Pacific. It appears that a high-resolution ocean model component is necessary to simulate a correct seasonal cycle in the tropical Pacific.

The Hovmoeller diagrams of SST anomalies (with respect to the seasonal cycle) over the 20-year simulation period of both the CERFACS and GFDL-30 model are shown in Fig. 7.59. The CERFACS model (Fig. 7.59b) simulates very strong interannual variability during the first 10 years but weak variability over the next 10 years. The GFDL-R30 model displays (Fig. 7.59c), as many other models,

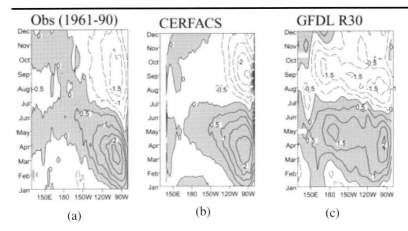

Figure 7.58. (a) Observed and (b-c) model simulated equatorial seasonal SSTs, averaged over 2°S - 2°N; (b) CERFACS model; (c) GFDL-R30 model. Results are from Latif et al. (2001).

too weak variability. According to Latif *et al.* (2001), the models can be separated into two classes with respect to interannual variability. There are models which simulate westward propagating SST anomalies (in particular the coarse-resolution models such as the GFDL-R30 model) and there are models which correctly simulate near standing SST anomalies. When the standard deviation of the NINO3 index is considered, there are only a few models which have a correct NINO3 amplitude and all of these models have deficiencies in simulating the seasonal cycle. The periods of the ENSO variability for the ensemble of models are in the range from 2 – 4 year.

There are only a few models which simulate the phase locking of ENSO to the seasonal cycle well and only one of these models (the CERFACS model) simulates a good seasonal cycle. This can be seen in Fig. 7.60, where the standard deviation of the NINO3 index is plotted versus calendar month. Many models do not capture the seasonal dependence of ENSO variability (weakest variability in April and strongest variability in December). There are models which show no seasonal dependence at all; these have also a weak seasonal signal. On the other hand, it is remarkable that there are models which do not simulate a good seasonal cycle, but still display a reasonable phase-locking behavior.

Also the regression patterns of SST and the heat content are analyzed in the models. There is a large range of patterns; many of the models do not capture an adequate wind response to SST anomalies. However, the high correlation of the NINO3 index and eastern upper ocean heat content at zero lag and the slow eastward propagation of the heat content is well captured in surprisingly many models. As concluded by Latif *et al.* (2001), there are unfortunately no CGCMs yet which simulate all aspects of the tropical climate and its variability well.

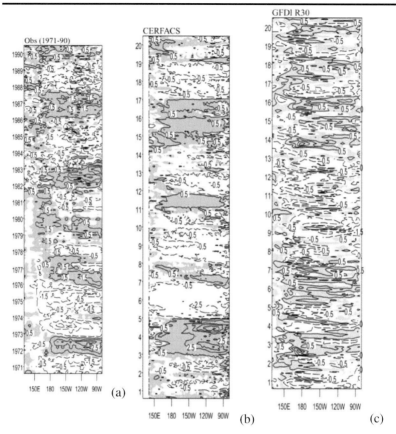

Figure 7.59. *(a) Observed and (b-c) model simulated equatorial seasonal SST anomalies, averaged over 2°S - 2°N; (b) CERFACS model; (c) GFDL-R30 model. Results are from Latif* et al. *(2001).*

7.10. Synthesis

The results from the previous section may appear as a disappointment having gone through the earlier sections where a clear mechanistic view of ENSO variability has been obtained through ICMs. One must realize, however, that the CGCMs aim to represent a multi-scale, multi-process 'virtual' reality and that in ICMs only a few of these processes and scales are represented. In this last section, we discuss the results of the ICMs in a broader context having the questions posed in section 7.1.5 in mind.

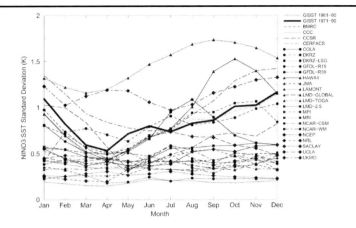

Figure 7.60. Model simulated standard deviation of the NINO3 index versus calendar month (Latif et al., 2001). The thick curves is from the GISST dataset.

7.10.1. A summary of ICM results

The coupled feedbacks discussed in section 7.3.4 shape the interesting and complicated tropical Pacific climate system. The strength of the feedbacks in the system is measured by the dimensionless coupling parameter (7.115)

$$\mu = \frac{\gamma \alpha_T \Delta T L^2}{c_o^2 c_a^2} \tag{7.177}$$

which scales with the square of the basin size. The zonal extent of the Pacific basin is a factor 3 larger than that of the Atlantic basin and consequently the coupling parameter is much larger. The feedbacks are thus stronger in the Pacific than in the Atlantic and hence more complicated behavior can be expected in the Pacific.

Section 7.6 provided strong arguments that the coupled feedbacks are involved to shape the zonal gradient of equatorial SST. If there is a weak background external wind stress τ_{ext}^x, the weak zonal SST gradient is amplified by coupled feedbacks to give a cold tongue – warm pool contrast. The analysis indicated that a subtle balance between the different feedbacks, thermocline and upwelling feedback, provides a correct spatial structure. A dominance of either of these feedbacks in ICMs may give a wrong zonal SST contrast in the annual-mean state (Dijkstra and Neelin, 1995a).

It would be strange if coupled processes would not be active in shaping the seasonal cycle. As mentioned in section 7.8.1, although there are many indications for this involvement, we do not have such a clear mechanistic view of how coupled processes operate in the seasonal cycle. This is a remaining fundamental problem — which is linked to the problem of the northern preference of the ITCZ (Philander *et al.*, 1996) — that still needs to be tackled with ICMs.

Linear stability analysis of annual-mean states, obtained under steady forcing in ICMs, convincingly show that coupled ocean-atmosphere internal modes are responsible for the interannual variability in these models. One can continuously follow these modes in the hierarchy of ICMs, starting with the Hirst (1986) periodic basin model in section 7.5.1 through the Jin and Neelin (1993a) model in section 7.5.2 to the Van der Vaart *et al.* (2000) model in section 7.7. If one starts with the motionless flow solution, there are two classes of modes. One class comes from the time derivative of the SST equation (the SST modes) and one class comes from the shallow-water dynamics (the ocean-dynamics modes). In the Hirst model, these modes are explicitly separated because of the simple structure of the steady state of which stability is analyzed. Both classes of modes interact in Jin and Neelin (1993a) to give the mixed SST-ocean dynamics modes (presented in section 7.5.3). The sole reason for the merging of these modes is the more complicated spatial structure of the annual-mean state and the presence of the zonal boundaries. A representation of these modes was presented in section 7.5.4. When variations of the coupling strength μ do not only influence the growth rate of the perturbations but also the underlying annual-mean state, the mixed SST-ocean dynamics modes as in the model of Van der Vaart *et al.* (2000) are obtained.

When the coupling parameter μ is increased, the annual-mean state becomes unstable through a supercritical Hopf bifurcation, say at μ_c. In several ICMs, the position of this Hopf bifurcation has been explicitly computed (for example, Van der Vaart *et al.* (2000)) and in others this Hopf bifurcation can be inferred from a transition from steady to periodic behavior. At the Hopf bifurcation, the growth rate of one of the mixed SST-ocean dynamics modes, i.e., the ENSO mode, becomes positive. The propagation mechanism is best described by the recharge-oscillator mechanism of Jin (1997a).

From the spatial patterns of the thermocline anomaly field in section 7.7.2. (Fig. 7.47), it can be seen that the ENSO mode is not simply a sum of a $j = 1$ Rossby mode (Fig. 7.12) and a Kelvin mode (Fig. 7.10), but that higher meridional Rossby modes are involved. Due to the presence of zonal boundaries, the free wave solutions do not exist and many of them combine to give a spatial structure more like the ocean-basin modes in Cane and Moore (1981). The consequence of this multimode structure is that the ocean adjustment time scale is much longer than that based on the propagation time scales of an individual Rossby and Kelvin wave. To obtain the correct spatial structure and time scale of the mode, the results in Van der Vaart *et al.* (2000) show that the spatial structure of the annual-mean state is important as the SST perturbations are mostly located in the cold tongue region.

For coupling values larger than that at the Hopf bifurcation (supercritical conditions), nonlinear processes become involved in the time-dependent behavior. The dependence of the period on μ is substantially different than that inferred from the angular frequency of the ENSO mode (see section 7.7.3). The precise mechanistic explanation still has to be given and this is also an important fundamental problem.

Under a seasonal cycle forcing, the periodic ENSO orbit nonlinearly interacts with the external seasonal forcing to give frequency and phase locking. Although the mathematical origin is the same as that causing the Arnold' tongues for the circle map (section 3.4.3) it is not easy to understand the physics of this synchronization phenomenon. As mentioned above, the interaction of the seasonal cycle and the ENSO mode has only been studied in ICMs under a prescribed seasonal cycle and not with a seasonal cycle which depends also on coupled processes.

In this framework, the time-dependent behavior under supercritical conditions can be modulated by slowly varying background changes, for example in the annual-mean state, to give decadal-to-inderdecadal variability of ENSO.

7.10.2. Multi-scale physics

The ICMs represent only part of the physical processes at work in the Pacific ocean-atmosphere system and one may ask which elements of this deterministic framework remain identifiable when placed in the context of a multi-scale physical system. Apart from observations, also the results from HCM and CGCM models are available to provide representations of this multi-scale system. In the next subsection, we first look whether we can recognize the 'building blocks' of ENSO dynamics provided by ICMs in the results of GCMs and observations. Next, we discuss the effect of the high-frequency variability ('noise') and in the last subsection, we discuss the effects of low-frequency background variability.

Recognition of 'building blocks'

Examining again the observations of the time-longitude diagrams of Fig. 7.7b, one sees that SST anomalies in the eastern and central part are associated with same-sign zonal wind anomalies further west. Thermocline anomalies have a tendency to propagate from the west to the east, while SST variations are dominated by a standing component. Most of these aspects can be reasonably captured by ICMs. However, in contrast to the atmosphere response in many ICMs, there are no clear systematic zonal wind anomalies west of SST anomalies.

The strong relation between SST anomalies and thermocline anomalies in the eastern Pacific (Fig. 7.7b) is an indication that the thermocline feedback is operating there. However, it is hard to make here a distinction between the thermocline and the zonal advection feedback because both have a similar signature (Picaut *et al.*, 1996; Jin and An, 1999). The central part of the Pacific exhibits the strongest correlations between local SST and wind. This is an indication that the zonal advection feedback and the upwelling feedback are probably more important in this region of larger mean zonal temperature gradient and wind anomalies. Probably, local-wind effects are important for the SST development in the central Pacific and wave-dynamic adjustment effects are dominating in the eastern Pacific.

Structures that correspond to the ocean adjustment waves, and reflection of these waves, have been observed by altimetry and TAO/TRITON measurements (Boulanger and Menkes, 1999; McPhaden and Yu, 1999). A complication in establishing properties of these waves, such as wave speeds and reflection coefficients, arises from the interaction of waves with the wind-stress variability, resulting in a huge uncertainty. In CGCMs, wave reflection is well-established, although wave dynamics is much more complicated than in ICMs, because there are several vertical modes instead of one, and because the background state is much more complicated. In Van Oldenborgh *et al.* (1999), the sensitivity of the NINO3 index to earlier sea-level fields and wind-stress fields is determined. Ocean adjustment waves can be seen in sea level influence fields that travel over the Pacific and reflect at the western boundary in agreement with the delayed-oscillator picture. A study of the energetics of El Niño variability (Goddard and Philander, 2000) within a CGCM also suggests a strong role for the delayed oscillator mechanism. In the half cycle after the El Niño peak, energy is fed into the ocean by the wind in the central and western Pacific, increasing the potential energy of the system. The gained energy adjusts via equatorial-wave dynamics and is released after the El Niño peak to the atmosphere.

Noise and its role in ENSO

From the perspective of El Niño, processes that evolve independently with smaller time and space scales can be considered as "noise". Noise has an important influence on the irregularity and predictability of El Niño, and probably is also important for sustaining El Niño variability.

In the atmosphere, an important phenomenon is the 30-60 day or Madden-Julian oscillation (Madden and Julian, 1994). This oscillation gives rise to westerly wind bursts, events of anomalous westerlies lasting typically a week, which are strongest somewhat west of the region of the El Niño wind response (Weller and Anderson, 1996; Vecchi and Harrison, 2000). They are stronger in boreal winter and spring, and are enhanced by local air-sea interaction. The Madden-Julian oscillation has substantial interannual variability. but El Niño events do not seem to affect westerly wind bursts although the location of the latter may shift slightly (Slingo *et al.*, 1999). The atmospheric variability not only causes wind stress anomalies, but also heating anomalies.

A comprehensive study of oceanic variability in the equatorial Pacific is presented in Kessler *et al.* (1996). Kelvin waves are excited very effectively by the Madden-Julian oscillation because they travel eastward with a similar speed. Also Rossby waves are excited by individual westerly wind bursts (Kessler *et al.*, 1995; Van Oldenborgh, 2000). Inertia-gravity waves, which have periods up to about a week (Philander, 1990), play a role as well. A striking feature of oceanic variability are tropical instability waves associated with the high shear of the equatorial currents at the northern boundary of the cold tongue (Legeckis, 1977). However, there is no evidence that these westward propagating structures are very relevant for El Niño variability.

The spectra of the NINO3 index and the SOI in Fig. 7.6 are moderately peaked compared to those of many deterministic ICMs. Hence, it is generally agreed that noise contributes substantially to the irregularity and predictability of El Niño. This is confirmed by many studies of the effect of explicit stochastic forcing in

ICMs and HCMs (e.g. Kleeman and Power (1994); Chang *et al.* (1996); Jin (1997a); Blanke *et al.* (1997); Eckert and Latif (1997)). Roulston and Neelin (2000) stress a sometimes overlooked basic fact: for linear systems, it is not the high-frequency part of the spectrum of the noise which influences the low-frequency part, but the low-frequency part. Case studies of the development of the 1997/1998 El Niño by McPhaden (1999) and Takayabu *et al.* (1999) and of events in the period 1986–1998 by Vecchi and Harrison (2000) point to a large role for variability associated with the Madden-Julian oscillation.

In CGCMs simulations, the noise is naturally present but its effect on the inter-annual variability has not been studied in detail. The HCM models, where noise can be added and which are much less expensive to run, have been used for this purpose. In Blanke *et al.* (1997) the noise forcing was estimated by removing variance associated with SST from the observed wind stress record in a way to get a white noise product preserving spatial correlations. The effect of realistic noise forcing applied to the HCM in a regime which otherwise would be periodic are sufficient to produce irregularity generally consistent with observations. In power spectra, the main spectral peak is broadened and rises modestly above the noise background, similar to that in ICMs. In the regime where the ENSO mode is damped, the stochastic forcing was able to get an ENSO signal above the noise level (Chang *et al.*, 1996).

A more theoretical approach to study how a perturbation $\mathbf{x}(t_0)$, given at an initial time t_0, develops at later times t is to linearize the system around a reference trajectory $\mathbf{x}_{ref}(t)$ which gives the tangent linear model for $\mathbf{x}(t)$, i.e. $\mathbf{x}(t) = \mathsf{A}(t, t_0)\mathbf{x}(t_0)$, where $\mathsf{A} = \mathsf{A}[\mathbf{x}_{ref}(t)]$. The adjoint A^T of the tangent model can be used to study the sensitivity of a function of the final state to perturbations at earlier times (Van Oldenborgh *et al.*, 1999). The singular vectors of A are the perturbation patterns that give the largest effects on the final state. Mathematically, due to the non-normality of A, transient error growth can be larger than that of the fastest growing mode (Blumenthal, 1991). Physically, this is related to the fact that disturbances propagate: perturbing the wind stress along the equator in the western Pacific has the largest effect on the NINO3 index a few months later. The above techniques can be applied both to unstable chaotic and stable noise driven systems. Transient error growth in a linearized version of the Cane-Zebiak model in a variety of parameter regimes is studied by Thompson and Battisti (2000). Moore and Kleeman (1999) extend the singular vector approach by considering the collective effect of noise at all earlier times. The stochastic optimals, noise patterns that give the largest contribution to interannual variability, have a large overlap with westerly wind burst variability. Van Oldenborgh (2000) uses the adjoint of an OGCM for studying what caused the development of the 1997/98 El Niño. He estimates that the initial-state wind anomaly of December 1996 and westerly wind bursts between December 1996 and May 1997 gave about equal contributions to the value of the NINO3 index in June 1997.

As was mentioned in the previous subsection, the ENSO mode has a positive growth rate when the coupling is larger than a critical value and the period and amplitude of ENSO variability are set by nonlinear effects. However, in the presence of noise, even if the system is linearly stable, it still exhibits oscillations. In this case of a so-called stochastic Hopf bifurcation (Gardiner, 2002), the amplitude is determined by the noise characteristics. The record of El Niño indices is too short to determine directly whether the El Niño system is stable or not, although it is clear that the stability properties depend on the season (Fedorov and Philander, 2000). Nonlinearity can show up in a broadening or even splitting of the central part of the probability density function of an index, such as the SOI. Originally, most research concentrated on the unstable regime, although the scenario of a stable oscillator excited by noise was already mentioned in the context of a HCM by Neelin (1991a).

Penland and Sardeshmukh (1995) concluded, through analysis of SST observations, that El Niño variability is noise driven and can be explained by the constructive interferences of three damped modes. Chang *et al.* (1996) add stochastic forcing to an ICM and to a HCM, and compare the behaviour to a CGCM. Their results indicate that El Niño in the CGCM was likely to be noise driven. Moore and Kleeman (1999) analyze the sensitivity of an intermediate model to stochastic perturbations and suggest that El Niño is a stochastically forced phenomenon, with westerly wind bursts being an important ingredient of the noise.

Burgers (1999) investigates how well a single "stochastic oscillator" can describe El Niño. The assumption of a complex damped mode leads to the autocorrelation function $\rho(t) = \exp(-\kappa t)\cos(\omega t + \alpha)/\cos\alpha$, which is very much like observed. A fit of the stochastic oscillator to observations of the NINO3 index yields $2\pi/\omega = 46 \pm 6$ months, $1/\kappa = 17 \pm 5$ months and $\alpha = 10° \pm 5°$. The fact that the spectra of the SOI and of the zonal wind fluctuations show a white background, and the NINO3 index hardly any, fits nicely with a picture of El Niño as stochastic oscillator driven by atmospheric noise. However, part of this white background could be due to observational noise, and the periods of rapid rise and decline of the 1997/1998 El Niño do not fit in this stochastic oscillator.

The picture that emerges from the above studies is that the sensitivity of the Tropical climate system to perturbations depends on the season, but that the system is close to neutral, with noise having a significant influence on the amplitude of the El Niño variability (Fedorov and Philander, 2000).

External processes and decadal variability

The decadal-to-interdecadal variability in the Pacific SST is currently an active field of research and it is impossible to review many of the ideas and studies here. A first mechanism that has been suggested is that white noise will inevitable lead to some decadal variability (Kirtman and Schopf, 1998). Latif (1998) have studied these decadal variations induced by stochastic forcing in a HCM and Flügel and Chang (1996) provide modeling evidence for stochastically induced shifts in El Niño behaviour. A second mechanism is that decadal variability arises due to complex nonlinear interactions between the basin-scale instabilities and smaller scale processes. After all, the deterministic system may be chaotic (Mantua and Battisti, 1995; Tziperman *et al.*, 1997) and thus contain a component of low-frequency variability. Finally, the annual-mean state on which the El Niño mode oscillates may be changed by external (to the Pacific basin) processes on a decadal time scale (Gu and Philander, 1997; S.-I. and Jin, 2000).

With respect to the latter possibility, it is recognized that the equatorial Pacific is part of the global climate system. The link between the equatorial Pacific system and the rest of the globe is both through the ocean and the atmosphere. The equatorial atmospheric circulation in the Pacific is intrinsically linked to extratropical circulation systems and the Asian monsoon system. The equatorial surface ocean current system is complicated and connects to the midlatitude circulation in the North- and South Pacific. Through transport of heat, both the thermocline shape and the sub-surface temperature may be changed on time scales which are controlled by processes other than those involved in El Niño .

In some simulations of global climate models, the dominant mechanism of low frequency variability of El Niño has its origin in the decadal variability in the extratratropical atmospheric winds (Barnett *et al.*, 1999; Pierce *et al.*, 2000). The large scale changes influence the trade winds and precondition the mean state of the thermocline in the equatorial ocean leading to prolonged periods of enhanced or reduced El Niño variability. Wang *et al.* (1998) stress the importance of the coupling between Walker and Hadley circulation to produce out-of-phase differences between tropical and extratropical SST anomalies.

There is also a clear linking of the Walker circulation with other tropical circulation systems, such as that over the Atlantic and the South-Asian monsoon (Webster and coauthors, 1996). Each year in northern Spring, the center of tropical convection migrates from the western Pacific warm pool to the northwest, announcing the arrival of the Asian monsoon. Normally, a weak Asian summer monsoon circulation is found during strong El Niño events. The normal low pressure system over the western Pacific shifts eastwards during El Niño events. Because of anomalous high pressure over the western Pacific/South Asian continent, precipitation is decreased. On the other hand, a strong monsoon with heavy rains and corresponding strong easterly winds tends to oppose El Niño conditions. Recently, this relation has weakened and no abnormal precipitation occurred during the last decades of increased El Niño occurrence (Kumar *et al.*, 1999).

Sub-surface propagation of midlatitude ocean temperature anomalies may also lead to interdecadal variability as demonstrated in a simple model (Gu and Philander, 1997). Temperature anomalies are formed in the winter mixed layer at midlatitudes, subduct when restratification occurs, propagate along the subtropical gyre and affect the tropical thermocline. Changes in tropical SST influence the midlatitude wind pattern which in turn affect the temperature in the midlatitude mixed layer, providing for an oscillation with an (inter)decadal time scale. In Zhang *et al.* (1998), observational evidence is presented of sub-surface temperature anomalies, which propagate over a period of about 10 years from the North Pacific to the low latitude Pacific. However, recent studies have indicated that these anomalies may not reach the equatorial region with sufficient amplitude to affect El Niño dynamics (Pierce *et al.*, 2000). Other oceanic connections have been suggested, such as (i) the signal of extratropical waves into the western equatorial Pacific (Lysne *et al.*, 1997) and (ii) the coupling of the meridional extent of the cold tongue to the shallow meridional overturning circulation connecting the equatorial upwelling region and the North Pacific subtropical gyre (Kleeman *et al.*, 1999).

7.10.3. The future of ENSO

Apart from a fundamental understanding of the basic physics of El Niño variability, most of the studies cited here were carried out because of (A) to predict the occurrence of next El Niño events and/or (B) to determine the changes of El Niño variability due to increased levels of greenhouse gases in the atmosphere.

The question (A) is intimately related to how well one is able to model the link between the interannual basin wide fluctuations associated with El Niño and the shorter time and smaller space scale variability in the Pacific. Different type of models are used to provide El Niño forecasts, ranging from statistical schemes to high-resolution coupled CCGMs. These can incorporate information from different data sources, e.g., in combining sub-surface data with wind-stress data and SST data. CGCMs can readily calculate the response to global SST patterns which is important for finding the teleconnections between equatorial variability and effects elsewhere in the world. However, the best statistical schemes are still competitive with the best CGCMs. In CGCMs, it is hard to find the 'right balance' of the coupled processes, since the way the various feedbacks and noise sources interact is quite subtle. None of the models predicted the onset of the 1997-1998 El Niño , but once this event was well underway, predictions of subsequent development were reasonable. Interpreting the past and future observations with the concepts of ICMs will make it possible to examine the performance of CGCMs in much more detail than is done so far, and allow CGCMs to be improved substantially.

The question (B) is related to how well one is able to model the link between El Niño variability and the global climate system. Whereas on the shorter time scale, there is immediate verification from available observations, this is not very well possible on the scale on which the global climate is changing. To properly capture El Niño variability in global models, these models should have enough resolution to represent (i) longwave ocean adjustment, (ii) the basic coupled feedbacks (iii) a representation of diabatic processes that provides the correct atmosphere response to equatorial SST anomalies (Newman *et al.*, 2000) and (iv) the influence of the annual-mean state on El Niño variability and vice versa.

The qualitative different results so far clearly show that modeling the effect of global warming on El Niño variability is still in its infancy. In some models (Knutson *et al.*, 1997) the mean zonal SST gradient decreases, with little change in 'El Niño-like' activity. In others models (Timmermann *et al.*, 1999; Timmermann, 2001), increasing greenhouse-gas concentrations induce a more "El Niño-like" mean state with more pronounced cold events than warm events. In the second Hadley Centre model (Collins, 2000a) changes in activity, frequency and phase-locking are found. However, in the third Hadley Centre model no change in El Niño statistics is found (Collins, 2000b). This is attributed to the difference in physical parameterization and not to the difference in resolution between the two versions. In one coupled model (Noda *et al.*, 1999), a change towards a more 'La Niña-like' mean state is found.

The latter problems indicate that still much work has to be done in the coming decades to get better solutions to the problems (A) and (B). Going through time, and considering that the connection between oceanic and atmospheric phenomena was only discovered in 1969 (Bjerknes, 1969), the subject has undergone a rapid transition to maturity. At the moment, the spatial and temporal patterns of interannual El Niño variability are well characterized, their dynamics is reasonable well understood and a diversity of models is used to predict the state of the Pacific a few months ahead. This work is continually stimulated by El Niño itself, which uncovers itself patiently every few years to have us learn a bit more of its identity.

7.11. Exercises on Chapter 7

(E7.1) *Reflection of equatorial waves*

Consider the reflection of an equatorial Rossby wave, with dispersion relation (7.17) for $j = M$, at the western boundary $x = 0$. The incoming wave has a wavenumber

$$k_I = -\frac{1}{2\sigma_I} + (\sigma^2 + \frac{1}{4\sigma^2} - (2M + 1))^{\frac{1}{2}}$$

and hence its group velocity is westward.

a. Show that the incoming velocity field of the wave is given by

$$v_I(x, y, t) = e^{i(k_I x - \sigma t)} \psi_M(y)$$

$$u_I(x, y, t) = \frac{i}{2} e^{i(k_I x - \sigma t)} \left(\frac{\sqrt{(2(M + 1))}}{\sigma - k_I} \psi_{M+1}(y) + \frac{\sqrt{(2M)}}{\sigma + k_I} \psi_{M-1}(y) \right)$$

$$h_I(x, y, t) = \frac{i}{2} e^{i(k_I x - \sigma t)} \left(\frac{\sqrt{(2(M + 1))}}{\sigma - k_I} \psi_{M+1}(y) + \frac{\sqrt{(2M)}}{\sigma - k_I} \psi_{M-1}(y) \right)$$

The reflected wave is a superposition of Rossby waves (with amplitude $B_m, m = 1, ..., M$) and a Kelvin wave (with amplitude B_K).

b.   Provide an expression for the zonal velocity field $u_R(x, y, t)$ of this superposition.

c. Determine the coefficients B_m and B_K.

d. Show that long Rossby waves reflect into a Kelvin wave.

Further reading: Philander (1990).

(E7.2) *Upwelling*

In the formulation of the Zebiak-Cane model, there is a Heaviside function in the vertical heat flux formulation (see e.g., (7.100)). Describe the physics motivating this representation.

Further reading: Zebiak and Cane (1987).

(E7.3) *Characteristics*

In section 7.5.4, the two-strip model is introduced. From this model, the delayed oscillator equations can be derived explicitly from the shallow-water model and SST evolution. Central in this derivation is the integration over characteristics defined by Kelvin and Rossby waves.

a. Determine the characteristics of a free Kelvin wave along the equator and sketch these in the $x - t$ plane.

b. Determine the characteristics of a free $j = 1$ Rossby wave along the y_{max} value in Table 7.1 and also sketch these in the $x - t$ plane.

c. Why can y_n in the two-strip model be identified with y_{max}?

d. Carry out the integration of the equations (7.135) along both characteristics under a. and b. and derive the equations (7.145).

Further reading: Jin (1997b).

(E7.4) *Exchange of heat and mass*

It is useful to ask how the mass and heat are transported during one ENSO cycle. The most transparant model to look at this transport is the recharge oscillator model in section 7.5.4.4.

a. Describe the exchange of mass between the equatorial strip to the off-equatorial region during one oscillation cycle.

b. Describe the exchange of heat between the equatorial strip to the off-equatorial region during one oscillation cycle.

Further reading: Philander (1990).

(E7.5) *ENSO Period*

In many CGCMs, the ENSO period is about 2-2.5 year and hence too short compared to that of the observed variability. One of the hypothesis of the cause for this short period is that the wind response of the atmosphere model is too meridionally confined to the equator.

a. With reference to the Gill model in section 7.3, formulate a few reasons why the wind response may be too localized.

b. Why does this localized wind response lead to a shorter ENSO period?

Further reading: Cane et al. *(1990) and Van der Vaart* et al. *(2000).*

(P7.1) *Pacific climate equilibria*

In this exercise, bifurcation analysis will be applied to a 'toy' model of the Tropical Pacific. In this model, the atmospheric wind response to SST anomalies is purely local,

$$A(T') = \mu T'$$

where T' is the perturbation from the appropriate reference temperature and μ is the coupling strength. As discussed in section 7.6, in the flux-corrected case we find that $T' = \hat{T} = T - \bar{T}$, where $\bar{T}$ is a solution for all values of μ. In the fully coupled case we find that $T' = T - T_0$, with T_0 being the fixed surface heat flux equilibrium temperature.

Thermocline depth h is parameterized to deepen locally under a westerly wind stress τ and the surface layer velocities are given by

$$h = \tau; \qquad w = -\delta_s \tau; \qquad -v_N = \delta_s \tau$$

with the scale of h suitably chosen. We introduce a homotopy parameter α_F and write the total wind stress τ as

$$\tau = \bar{\tau}_{ext} + \alpha_F(\bar{\tau}_{obs} - \bar{\tau}_{ext}) + \mu\{\alpha_F(T - \bar{T}) + (1 - \alpha_F)(T - T_0)\}$$

where $\bar{T}$ is the flux-corrected solution for $\alpha_F = 1$ (the response to 'observed' wind stress $\bar{\tau}_{obs}$) and $\bar{\tau}_{ext}$ is the external wind stress.

We use a piecewise continuous version of the T_s parameterization

$$
\begin{aligned}
T_s(h) &= T_1, \quad h \geq h_1; \\
T_s(h) &= T_2, \quad h \leq h_2; \\
T_s(h) &= T_2 + \gamma(h - h_2), \quad h_1 \geq h \geq h_2
\end{aligned}
$$

where $\gamma = (T_1 - T_2)/(h_1 - h_2) \geq 0$, $T_2 \leq T_1 \leq T_0$ and $h_1 \leq h_2$.

The steady temperature equation, due to the local approximations, reduces to the single scalar equation for T,

$$\mathcal{H}(w)w(T - T_s(h)) - \mathcal{H}(-v_N)v_N(T - T_N) + \epsilon_T(T - T_0) = 0$$

where $\mathcal{H}$ is the Heaviside function.

a. Show that h, w and v_N are given by

$$w = w_0 - \mu\delta_s\{\alpha_F(T - \bar{T}) + (1 - \alpha_F)(T - T_0)\}; \qquad v_N = w$$

with

$$w_0 = -\delta_s(\overline{\mathcal{T}}_{ext} + \alpha_F(\overline{\mathcal{T}}_{obs} - \overline{\mathcal{T}}_{ext}))$$

and

$$h = \overline{\mathcal{T}}_{ext} + \alpha_F(\overline{\mathcal{T}}_{obs} - \overline{\mathcal{T}}_{ext}) + \mu\{\alpha_F(T - \overline{T}) + (1 - \alpha_F)(T - T_0)\}$$

b. Implement the SST equation into the AUTO software (see the beginning of chapter 4 how to get this software) and determine the bifurcation diagram for $\alpha_F = 0$.

c. Follow the branches into α_F and determine the change in the bifurcation diagrams. What is the physical reason for the differences between the branches when changing α_F?

Further reading: Neelin and Dijkstra (1995).

(P7.2) *Frequency locking*

In this exercise, you will get more familiar with frequency locking as occurs between the ENSO mode and the seasonal cycle. Although there are relatively simple models where this can be demonstrated (Neelin *et al.*, 2000), it is bit much involved for an exercise to program these models and hence we turn to the familiar Van der Pol equation, which is now forced externally. The so-called forced Van der Pol equation is

$$\theta'' + \epsilon(\theta^2 - 1)\theta' + \theta = A\cos\omega t$$

a. With $x = \theta$ and $y = \theta'$, write this equation as a first order non-autonomous system.

You can either implement these equations yourself in a numerical solver for ordinary differential equations (such as available in MATLAB) or use a JAVA integrator on the web at the URL http://www.iro.umontreal.ca/~eckdoug/vibe/Relaxation/VanDerPolForced.html.

b. Consider first the case $A = 0$. Compute and plot trajectories for $\epsilon = 0.5, 1.0$ and 2.0.

c. Next, consider $A = 3.0$ and $\omega = 1.2$. Compute and plot trajectories for $\epsilon = 0.05, 0.25$ and 0.45. Do you observe any frequency locking?

d. Next, consider $A = 0.4$ and $\omega = 1.2$. Compute and plot trajectories for $\epsilon = 0.05, 0.25$ and 0.45. What type of behavior is found at $\epsilon = 0.25$?

e. Finally, consider $A = 5.0$ and $\omega = 1.788$. Compute and plot trajectories for $\epsilon = 1.0, 2.0$ and 3.0. What type of behavior is found at $\epsilon = 3.0$? How does this behavior arise?

Further reading: Neelin et al. *(2000) and Pikovsky* et al. *(2001).*

Bibliography

AchutaRao, K. and Sperber, K. B. (2002). Simulation of the El Niño Southern Oscillation: Results from the Coupled Model Intercomparison Project. *Clim. Dyn.*, **19**, 191–209.

Adcroft, A., Hill, C. N., and Marshall, J. C. (1999). A new treatment of the Coriolis terms in C-grid models at both high and low resolutions. *Monthly Weather Review*, **127**, 1928–1936.

Anderson, D. L. T. and McCreary, J. P. (1985). Slowly propagating disturbances in a coupled ocean-atmosphere model. *J. Atmos. Sci.*, **42**, 615–629.

Atkinson, K. E. (1976). *An Introduction to Numerical Analysis*. John Wiley and Sons.

Auer, S. J. (1987). Five-year climatological survey of the Gulf Stream system and its associated rings. *J. Geophys. Res.*, **92**, 11,709 – 11,726.

Bak, P., Bohr, T., and Høgh Jensen, M. (1985). Mode-locking and the transition to chaos in dissipative systems. *Physica Scripta*, **T9**, 50–58.

Bane, J. M. and Dewar, W. K. (1988). Gulf Stream bimodality and variability downstream of the Charleston Bump. *J. Geophys. Res.*, **93**, 6695–6710.

Baringer, M. O. and Larsen, J. C. (2001). Sixteen years of Florida Current transport at 27N. *Geophys. Res. Letters*, **28**, 3179–3182.

Baringer, M. O. and Molinari, R. (1999). Atlantic Ocean baroclinic heat flux at 24-26°N. *Geophys. Res. Letters*, **26**, 353–356.

Barnett, T. P., Pierce, D. W., Latif, M., Dommenget, D., and Saravanan, R. (1999). Interdecadal interactions between the tropics and midlatitudes in the Pacific basin. *Geophys. Res. Letters*, **26**, 615–618.

Barrett, R., Berry, M., Chan, T. F., Demmel, J., Donato, J., Dongarra, J., Eijkhout, V., Pozo, R., Romine, C., and der Vorst, H. V. (1994). *Templates for the Solution of Linear Systems: Building Blocks for Iterative Methods, 2nd Edition*. SIAM, Philadelphia, PA.

Batchelor, G. (1974). *Introduction to Fluid Dynamics*. Cambridge University Press, Cambridge, UK.

Batchelor, G. (2000). *Introduction to Fluid Dynamics*. Cambridge University Press, First Cambridge Mathematical Library Edition, Cambridge, UK.

Battisti, D. (1988). The dynamics and thermodynamics of a warming event in a coupled tropical ocean/atmosphere model. *J. Atmos. Sci.*, **45**, 2889–2919.

Battisti, D. and Hirst, A. (1989). Interannual variability in a tropical atmosphere-ocean model: Influence of the basic state, ocean geometry and nonlinearity. *J. Atmos. Sci.*, **46**, 1687–1712.

Baumgartner, A. and Reichel, E. (1975). *The World Water Balance*. Elsevier Publishers, Amsterdam, The Netherlands.

Beckmann, A. C. W., Böning, C. W., Köberle, J., and Willebrand, J. (1994). Effects of increased horizontal resolution in a simulation of the North-Atlantic Ocean. *J. Phys. Oceanogr.*, **24**, 326–344.

Berger, A. (1978). Long term variations of daily insolation and Quaternary climate change. *J. Atmos. Sci.*, **35**, 2362–2367.

Berger, A. and Loutre, M. F. (1991). Insolation values for the climate of the last 10 million years. *Quaternary Sciences Review*, **10**, 297–317.

Berger, W. H. and Jansen, E. (1994). *Younger Dryas episode: ice collapse and super-fjord heat pump*, pages 61–105. North Holland, Amsterdam, The Netherlands.

Berloff, P. S. and McWilliams, J. C. (1999a). Large-scale, low-frequency variability in wind-driven ocean gyres. *J. Phys. Oceanogr.*, **29**, 1925–1949.

Berloff, P. S. and McWilliams, J. C. (1999b). Quasi-geostrophic dynamics of the western boundary current. *J. Phys. Oceanogr.*, **29**, 2607–2634.

Berloff, P. S. and Meacham, S. P. (1997). The dynamics of an equivalent barotropic model of the wind-driven circulation. *J. Mar. Res.*, **55**, 407–451.

Berloff, P. S. and Meacham, S. P. (1998a). The dynamics of a simple baroclinic model of the wind-driven circulation. *J. Phys. Oceanogr.*, **28**, 361–388.

Berloff, P. S. and Meacham, S. P. (1998b). On the stability of the wind-driven circulation. *J. Mar. Res.*, **56**, 937–993.

Bjerknes, J. P. (1969). Atmospheric teleconnections from the equatorial Pacific. *Monthly Weather Review*, **97**, 163–172.

Blanke, B., Neelin, J. D., and Gutzler, D. (1997). Estimating the effect of stochastic wind stress forcing on ENSO irregularity. *J. Climate*, **10**, 1473–1486.

Bleck, R. and Chassignet, E. P. (1994). Simulating the ocean circulation with isopycnic coordinate models. *The Pennsylvania Academy of Science*, **XXX**, 17–39.

Blumenthal, M. (1991). Predictability of a coupled ocean-atmosphere model. *J. Climate*, **4**, 766–784.

Bond, G., Showers, W., Cheseby, M., Lotti, R., Almasi, P., deMenocal, P., Priori, P., Cullen, H., Hajdas, I., and Bonani, G. (1995). A pervasive millennial-scale cycle in the North Atlantic Holocene and glacial climates. *Science*, **278**, 1257–1265.

Botta, E. F. F. and Wubs, F. W. (1999). MRILU: An effective algebraic multi-level ILU-preconditioner for sparse matrices. *SIAM J. Matrix Anal. Appl.*, **20**, 1007–1026.

Boulanger, J.-P. and Menkes, C. (1999). Long equatorial wave reflection in the Pacific ocean from TOPEX/Poseidon data during the 1992-1998 period. *Clim. Dyn.*, **15**, 205–225.

Bradley, R. S. (1999). *Paleoclimatology: Reconstructing Climates of the Quaternary*. Academic Press, New York, U.S.A.

Briffa, K. R., Schweingruber, F. H., Jones, P. D., Osborn, T. J., Shiyatov, S. G., and Vaganov, E. A. (1998). Reduced sensitivity of recent tree-growth to temperature at high northern latitudes. *Nature*, **391**, 678–680.

Broecker, W. S. (1991). The great ocean conveyor. *Oceanography*, **4**, 79–89.

Broecker, W. S. (1995). *The Glacial World According to Wally*. Eldigio Press, Palisades New York, U.S.A.

Broecker, W. S. (2000). Abrupt climate change: causal constraints provided by the paleoclimate record. *Earth-Science Rev.*, **51**, 137–154.

Broecker, W. S., Peteet, D. M., and Rind, D. (1985). Does the ocean-atmosphere system have more than one stable mode of operation? *Nature*, **315**, 21–26.

Bryan, F. O. (1986). High-latitude salinity effects and interhemispheric thermohaline circulations. *Nature*, **323**, 301–304.

Bryan, F. O., Böning, C. W., and Holland, W. R. (1995). On the midlatitude circulation in a high-resolution model of the North Atlantic. *J. Phys. Oceanogr.*, **25**, 289–305.

Bryan, K., Manabe, S., and Pacanowski, R. C. (1974). A global ocean-atmosphere climate model. Part II. The oceanic circulation. *J. Phys. Oceanogr.*, **5**, 30–46.

Bryden, H. L. and Beal, L. M. (2001). Role of the Agulhas Current in Indian Ocean circulation and associated heat and freshwater fluxes. *Deep-Sea Research Part I*, **48**, 1821–1845.

Bryden, H. L., Roemmich, D., and Church, J. (1991). Heat transport across $24°N$ in the Pacific. *Deep-Sea Research*, **38**, 297–324.

Bryden, H. L., Griffiths, M. J., Lavin, A. M., Millard, R. C., Parilla, G., and Smethie, W. M. (1996). Decadal changes in water mass characteristics at $24°N$ in the subtropical North Atlantic Ocean. *J. Climate*, **9**, 3162–3186.

Burgers, G. (1999). The El Nino Stochastic Oscillator. *Clim. Dyn.*, **15**, 352–375.

Cai, W. (1995). Interdecadal variability driven by mismatch between surface flux forcing and oceanic freshwater/heat transport. *J. Phys. Oceanogr.*, **25**, 2643.

Cane, M. A. (1979a). The response of an equatorial ocean to simple wind stress patterns: I. Model formulation and analytic results. *J. Mar. Res.*, **37**, 233–252.

Cane, M. A. (1979b). The response of an equatorial ocean to simple wind stress patterns: II. Numerical results. *J. Mar. Res.*, **37**, 253–299.

Cane, M. A. and Moore, D. W. (1981). A note on low-frequency equatorial basin modes. *J. Phys. Oceanogr.*, **11**, 1578–1584.

Cane, M. A. and Sarachik, E. S. (1977). Forced baroclinic ocean motions: II. The linear equatorial bounded case. *J. Mar. Res.*, **35**, 395–432.

Cane, M. A., Münnich, M. M., and Zebiak, S. E. (1990). Study of self-excited oscillations of the tropical ocean-atmosphere system. Part I: Linear analysis. *J. Atmos. Sci.*, **47**, 1562–1577.

Cessi, P. (1996). Grid-scale instability of convective-adjustment schemes. *J. Mar. Res.*, **54**, 407–420.

Cessi, P. and Ierley, G. R. (1993). Nonlinear disturbances of western boundary currents. *J. Phys. Oceanogr.*, **23**, 1727–1735.

Cessi, P. and Ierley, G. R. (1995). Symmetry-breaking multiple equilibria in quasi-geostrophic, wind-driven flows. *J. Phys. Oceanogr.*, **25**, 1196–1205.

Cessi, P. and Primeau, F. (2001). Dissipative selection of low-frequency modes in a reduced gravity basin. *J. Phys. Oceanogr.*, **31**, 127–137.

Cessi, P. and Young, W. R. (1992). Multiple equilibria in two-dimensional thermohaline circulation. *J. Fluid Mech.*, **241**, 291–309.

Chandrasekhar, S. (1961). *Hydrodynamic and Hydromagnetic Stability.* Clarendon Press, Oxford, U.K.

Chang, K.-I., Ghil, M., Ide, K., and Lai, C.-C. A. (2001). Transition to aperiodic variability in a wind-driven double-gyre circulation model. *J. Phys. Oceanogr.*, **31**, 1260–1286.

Chang, P. and Philander, S. G. H. (1994). A coupled ocean-atmosphere instability of relevance to the seasonal cycle. *J. Atmos. Sci.*, **51**, 3628–3648.

Chang, P., Wang, B., Li, T., and Ji, L. (1994). Interactions between the seasonal cycle and the Southern Oscillation - frequency entrainment and chaos in a coupled ocean-atmosphere model. *Geophys. Res. Letters*, **21**, 2817–2820.

Chang, P., Ji, L., Wang, B., and Li, T. (1995). Interactions between the seasonal cycle and El Nino-Southern Oscillation in an intermediate coupled ocean-atmosphere model. *J. Atmos. Sci.*, **52**, 2353 – 2372.

Chang, P., Ji, L., Li, H., and Flügel, M. (1996). Chaotic dynamics versus stochastic processes in El Nino/ Southern Oscillation in coupled ocean-atmosphere models. *Physica D*, **98**, 301–320.

Chao, S.-Y. (1984). Bimodality of the Kuroshio. *J. Phys. Oceanogr.*, **14**, 92–103.

Chao, Y., Gangopadhyay, A., Bryan, F. O., and Holland, W. R. (1996). Modeling the Gulf Stream system: How far from reality ? *Geophys. Res. Letters*, **23**, 3155–3158.

Charney, J. G. and Flierl, G. (1981). Ocean analogues of large-scale atmospheric motions. In B. Warren and C. Wunsch, editors, *Evolution of Physical Ocenaography*, pages 504–548. MIT Press, Boston, USA.

Chassignet, E. P. and Bleck, R. (1993). The influence of layer outcropping on the separation of boundary currents: Part I: The wind-driven experiments. *J. Phys. Oceanogr.*, **23**, 1485–1507.

Chassignet, E. P. and Garaffo, Z. D. (2001). Viscosity parameterizations and the gulf stream separation. In P. Muller and D. Henderson, editors, *From stirring to mixing in a stratified ocean*, pages 37–41. Univ. of Hawaii, Honolulu, USA.

Chatfield, C. (2004). *The Analysis of Time Series: An Introduction, 6th edition.* CRC Press, Boca Raton, USA.

Chen, F. and Ghil, M. (1995). Interdecadal variability of the thermohaline circulation and high-latitude surface fluxes. *J. Phys. Oceanogr.*, **22**, 161–167.

Chen, F. and Ghil, M. (1996). Interdecadal variability in a hybrid coupled ocean-atmosphere model. *J. Phys. Oceanogr.*, **26**, 1561–1578.

Clark, P. U., Pisias, N. G., Stocker, T. F., and Weaver, A. J. (2002). The role of the thermohaline circulation in abrupt climate change. *Nature*, **415**, 863–869.

Claussen, M., Mysak, L., Weaver, A. J., Crucifix, M., Fichefet, T., Loutre, M.-F., Weber, S. L., Alcamo, J., Alexeev, V. A., Berger, A., Calov, R., Ganopolski, A., Goosse, H., Lohmann, G.,

Lunkeit, F., Mokhov, I. I., Petoukhov, V., Stone, P., and Wang, Z. (2002). Earth system models of intermediate complexity: closing the gap in the spectrum of climate system models. *Clim. Dyn.*, **18**, 579–586.

Colin de Verdière, A. (1988). Buoyancy driven planetary flows. *J. Mar. Res.*, **46**, 215–265.

Colin de Verdière, A. and Huck, T. (1999). Baroclinic instability: an oceanic wavemaker for inter-decadal variability. *J. Phys. Oceanogr.*, **29**, 893–910.

Collins, M. (2000a). The El Niño - Southern Oscillation in the second Hadley Centre Coupled model and its response to greenhouse warming. *J. Climate*, **13**, 1299–1312.

Collins, M. (2000b). Understanding uncertainties in the response of ENSO to greenhouse warming. *Geophys. Res. Letters*, **27**, 3509–3512.

Cox, M. (1984). A primitive equation, three-dimensional model of the ocean. Technical Report 1, GFDL Ocean Group.

Cox, M. (1987). An eddy-resolving model of the ventilated thermocline: Time-dependence. *J. Phys. Oceanogr.*, **17**, 1044–1056.

Cubasch, U., Hasselmann, K., Höck, H., E., M.-R., Mikolajewicz, U., Santner, B. D., and Sausen, R. (1992). Time-dependent greenhouse warming computations with a coupled ocean-atmosphere model. *Clim. Dyn.*, **8**, 55–69.

Cunningham, S. A., Alderson, S. G., King, B. A., and Brandon, M. A. (2003). Transport and variability of the Antarctic Circumpolar Current in Drake Passage. *J. Geophys. Res.*, **108**, 10.1029/2001JC001147.

Curry, R. G., McCartney, M. S., and Joyce, T. M. (1998). Oceanic transport of subpolar climate signals to mid-depth subtropical waters. *Nature*, **391**, 575–578.

Cushman-Roisin, B. (1994). *Introduction to geophysical fluid dynamics*. Prentice Hall.

Da Costa, E. D. and Colin de Verdiére, A. C. (2004). The 7.7 year North Atlantic oscillation. *Quart. J. Roy. Meteor. Soc.*, **128**.

Dansgaard, W., White, J. W. C., and Johnsen, S. J. (1989). The abrupt termination of the younger dryas climate event. *Nature*, **339**, 532–533.

Davey, M. K., Huddelston, M., Sperber, K. R., Braconnot, P., Bryan, F., Chen, D., Colman, R. A., Cooper, C., Cubasch, U., Delecluse, P., DeWitt, D., Fairhead, L., Flato, G., Gordon, C., Hogan, T., Ji, M., Kimoto, M., Kitoh, A., Knutson, T. R., Latif, M., Le Treut, H., Li, T., Manabe, S., Mechoso, C. R., Meehl, G., Power, S. B., Roeckner, E., Terray, L., Vintzileos, A., Voss, R., Wang, B., Washington, W., Yoshikawa, I., Yu, J., Yukimoto, S., and Zebiak, S. (2002). STOIC: A study of coupled model climatology and variability in tropical ocean regions. *Clim. Dyn.*, **18**, 403–420.

Delecluse, P., Davey, M., Kitamura, Y., Philander, S. G. H., Suarez, M., and Bengtsson, L. (1998). Coupled general circulation modeling of the tropical Pacific. *J. Geophys. Res.*, **103**, 14,357–14,374.

Delworth, T. L. and Greatbatch, R. G. (2000). Multidecadal thermohaline circulation variability driven by atmospheric surface flux forcing. *J. Climate*, **13**, 1481–1495.

Delworth, T. L. and Mann, M. E. (2000). Observed and simulated multidecadal variability in the Northern Hemisphere. *Clim. Dyn.*, **16**, 661–676.

Delworth, T. L., Manabe, S., and Stouffer, R. J. (1993). Interdecadal variations of the thermohaline circulation in a coupled ocean-atmosphere model. *J. Climate*, **6**, 1993–2011.

Delworth, T. L., Stouffer, R. J., Dixon, K. W., Spelman, M. J., Knutson, T. R., Broccoli, A. J., Kushner, P. J., and Wetherald, R. T. (2002). Review of simulations of climate variability and change with the GFDL R30 coupled climate model. *Clim. Dyn.*, **19**, 555–574.

Dengg, J., Beckmann, A., and Gerdes, R. (1996). The Gulf Stream separation problem. In W. A. Kraus, editor, *The Warmwatersphere of the North Atlantic Ocean*, pages 253–290. Borntraeger.

Deser, C. and Blackmon, M. L. (1993). Surface climate variations over the North Atlantic ocean during winter: 1900–1989. *J. Climate*, **6**, 1743–1753.

Dickson, R. and Brown, J. (1994). The production of North Atlantic Deep Water. *J. Geophys. Res.*, **99**, 12,319–12,341.

Dickson, R. R., Meincke, J., Malmberg, S., and Lee, A. (1988). The "Great Salinity Anomaly" in the northern North Atlantic 1968-1982. *Prog. Oceanogr.*, **20**, 103–151.

Dijkstra, H. A. (1992). On the structure of cellular solutions in Rayleigh-Bénard-Marangoni flows in small-aspect-ratio containers. *J. Fluid Mech.*, **243**, 73–102.

Dijkstra, H. A. (2005). The interaction of SST modes in the North Atlantic. *J. Phys. Oceanogr.* submitted.

Dijkstra, H. A. and Burgers, G. (2002). Fluid Dynamics of El Nino Variability. *Annual Review of Fluid Mechanics*, **34**, 531–558.

Dijkstra, H. A. and Katsman, C. A. (1997). Temporal variability of the wind-driven quasi-geostrophic double gyre ocean circulation: Basic bifurcation diagrams. *Geophys. Astrophys. Fluid Dyn.*, **85**, 195–232.

Dijkstra, H. A. and Molemaker, M. J. (1997). Symmetry breaking and overturning oscillations in thermohaline-driven flows. *J. Fluid Mech.*, **331**, 195–232.

Dijkstra, H. A. and Molemaker, M. J. (1999). Imperfections of the North-Atlantic wind-driven ocean circulation: Continental geometry and wind stress shape. *J. Mar. Res.*, **57**, 1–28.

Dijkstra, H. A. and Neelin, J. D. (1995a). Coupled ocean-atmosphere models and the tropical climatology. II: Why the cold tongue is in the east. *J. Climate*, **8**, 1343–1359.

Dijkstra, H. A. and Neelin, J. D. (1995b). On the attractors of an intermediate coupled ocean-atmosphere model. *Dyn. Atmos. Oceans*, **22**, 19–48.

Dijkstra, H. A. and Neelin, J. D. (1999). Coupled processes and the tropical climatology. III: Instabilities of fully coupled climatologies. *J. Climate*, **12**, 1630–1643.

Dijkstra, H. A. and Neelin, J. D. (2000). Imperfections of the thermohaline circulation: Latitudinal asymmetry versus asymmetric freshwater flux. *J. Climate*, **13**, 366–382.

Dijkstra, H. A. and Steen, P. H. (1991). Thermocapillary stabilization of the capillary break-up of an annular film of liquid. *J. Fluid Mech.*, **229**, 205–228.

Dijkstra, H. A. and Weijer, W. (2003). Stability of the global ocean circulation: The connection of equilibria in a hierarchy of models. *J. Mar. Res.*, **61**, 725–743.

Dijkstra, H. A., Molemaker, M. J., van der Ploeg, A., and Botta, E. F. F. (1995). An efficient code to compute nonparallel flows and their linear stability. *Comp. Fluids*, **24**, 415–434.

Dijkstra, H. A., Öksüzöglu, H., Wubs, F. W., and Botta, E. F. F. (2001). A fully implicit model of the three-dimensional thermohaline ocean circulation. *J. Comput. Phys.*, **173**, 685–715.

Dijkstra, H. A., Weijer, W., and Neelin, J. D. (2003). Imperfections of the three-dimensional thermohaline ocean circulation: Hysteresis and unique state regimes. *J. Phys. Oceanogr.*, **33**, 2796–2814.

Dijkstra, H. A., Te Raa, L. A., and Weijer, W. (2004). A systematic approach to determine thresholds of the ocean's thermohaline circulation. *Tellus*, **56A**, 362–370.

Dijkstra, H. A., Schmeits, M., Te Raa, L. A., and Gerrits, J. (2005). The physics of the Atlantic Multidecadal Oscillation. *Clim. Dyn.* submitted.

Doedel, E. J. (1980). AUTO: A program for the automatic bifurcation analysis of autonomous systems. In *Proc. 10th Manitoba Conf. on Numerical Math. and Comp.*, volume 30, pages 265–274.

Doedel, E. J. and Tuckermann, L. S. (2000). *Numerical Methods for Bifurcation Problems and Large-Scale Dynamical Systems*. Springer-Verlag, New York.

Drazin, P. G. and Reid, W. H. (2004). *Hydrodynamic Stability 2nd Edition*. Cambridge Univ. Press.

Duff, I. S., Erisman, E., and Reid, J. K. (1986). *Direct Methods for Sparse Matrices*. Clarendon Press, Oxford, U.K.

Dukowicz, J. K. and Smith, R. D. (1994). Implicit free-surface method for the Bryan-Cox-Semtner ocean model. *J. Geophys. Res.*, **99**, 7991–8014.

Eady, E. T. (1949). Long waves and cyclone waves. *Tellus*, **1**, 33–52.

Eckert, C. and Latif, M. (1997). Predictability of a stochastically forced hybrid coupled model of the Tropical Pacific ocean-atmosphere system. *J. Climate*, **10**, 1488–1504.

Endoh, M., Tokioka, T., and Nagai, T. (1991). Tropical Pacific sea surface temperature variations in a coupled atmosphere-ocean general circulation model. *J. Marine Systems*, **1**, 293–298.

Enfield, D. B., Mestas-Nunes, A. M., and Trimble, P. (2001). The Atlantic multidecadal oscillation and its relation to rainfall and river flows in the continental US. *Geophys. Res. Letters*, **28**, 2077–2080.

England, M. H. (1993). Representing the global-scale water masses in ocean general circulations models. *J. Phys. Oceanogr.*, **23**, 1523–1552.

Ertel, H. (1942). Ein neuer hydrodynamischer wirbesatz. *Meteorologisches Zeitschrift*, **59**, 277–281.

Fairbancks, R. (1990). The relationship between the deglacial record of sea level rise as recorded in Barbados corals and air temperature change as recorded in the Greenland ice cap. *Paleoceanography*, **5**, 937–948.

Farrell, B. F. and Ioannou, P. J. (1996). Generalized stability theory. I: Autonomous opertors. *J. Atmos. Sci.*, **53**, 2025–2040.

Fedorov, A. and Philander, S. (2000). Is El Nino Changing? *Science*, **288**, 1997–2002.

Fleury, L. and Thual, O. (1997). Stationary fronts of the thermohaline circulation in the low-aspect ratio limit. *J. Fluid Mech.*, **349**, 117–147.

Flierl, G. (1978). Models of vertical structure and calibration of two-layer models. *Dyn. Atmos. Oceans*, **2**, 341–381.

Flügel, M. and Chang, P. (1996). Stochastically induced climate shift of El Nino-Southern Oscillation. *Geophys. Res. Letters*, **26**, 2473–2476.

Fofonoff, N. P. (1954). Steady flow in a frictionless homogeneous ocean. *J. Mar. Res.*, **13**, 254–262.

Frankignoul, C. (1999). Sea surface temperature variability in the North Atlantic: monthly to decadal time scales. In A. Navarro, editor, *Beyond El Niño*, pages 25–48. Springer, Berlin-Heidelberg, Germany.

Fu, L.-L. (2001). Ocean Circulation and Variability from Satellite Altimetry. In G. Siedler, J. Church, and J. Gould, editors, *Ocean Circulation and Climate: Observing and Modeling the Global Ocean Ocean*, pages 141–172. Academic Press.

Ganachaud, A. and Wunsch, C. (2000). Improved estimates of global ocean circulation, heat transport and mixing from hydrographic data. *Nature*, **408**, 453–457.

Gangopadhyay, A., Cornillon, P., and Watts, D. (1992). A test of the Parsons-Veronis hypothesis on the separation of the Gulf Stream. *J. Phys. Oceanogr.*, **22**, 1286–1301.

Ganopolsky, A., Rahmstorf, S., Petoukhov, V., and Claussen, M. (1998). Simulations of modern and glacial climates with a coupled global model of intermediate complexity. *Nature*, **391**, 351–356.

Ganopolsky, A., Rahmstorf, S., Petoukhov, V., and Claussen, M. (2001). Rapid changes of glacial climate simulated in a coupled climate model. *Nature*, **409**, 153–158.

Gardiner, C. W. (2002). *Handbook of Stochastic Methods, 2nd edition*. Springer.

Gelfgat, A. Y. (1999). Different modes of Rayleigh-Bénard instability in two- and three dimensional rectangular enclosures. *J. Comp. Physics*, **156**, 300–324.

Gent, P. R. and McWilliams, J. C. (1990). Isopycnal mixing in ocean circulation models. *J. Phys. Oceanogr.*, **20**, 150–155.

Gent, P. R., Willebrand, J., McDougall, T. J., and McWilliams, J. C. (1995). Parameterizing eddy-induced tracer transports in ocean circulation models. *J. Phys. Oceanogr.*, **25**, 463–474.

Gerdes, R. and Köberle, C. (1995). On the influence of DSOW in a numerical model of the North Atlantic general circulation. *J. Phys. Oceanogr.*, **25**, 2624–2642.

Ghil, M. (1994). Cryothermodynamics: the chaotic dynamics of paleoclimate. *Physica D*, **77**, 130–159.

Giese, B. S. and Carton, J. A. (1994). The seasonal cycle in a coupled ocean-atmosphere model. *J. Climate*, **7**, 1208–1217.

Gill, A. E. (1980). Some simple solutions for heat induced tropical circulation. *Quart. J. Roy. Meteor. Soc.*, **106**, 447–462.

Gill, A. E. (1982). *Atmosphere-Ocean Dynamics*. Academic Press, New York, U.S.A.

Gnanadesikan, A. (1999). A simple predictive model of the structure of the oceanic pycnocline. *Science*, **283**, 2077–2081.

Goddard, L. and Philander, S. G. H. (2000). The energetics of El Nino and La Nina. *J. Climate*, **13**, 1496–1516.

Golub, G. H. and Van Loan, C. F. (1983). *Matrix Computations*. The Johns Hopkins University Press, Baltimore, U.S.A.

Golubitsky, M., Stewart, I., and Schaeffer, D. G. (1988). *Singularities and Groups in Bifurcation Theory, Vol. II*. Springer-Verlag, New York, U.S.A.

Gordon, A. L. (1986). Interocean exchange of thermocline water. *J. Geophys. Res.*, **91**, 5037–5046.

Gordon, C. (1989). Tropical ocean-atmosphere interactions in a coupled model. *Phil. Trans. Roy. Soc. London A*, **329**, 207–223.

Gray, W. M., Sheaffer, J. D., and Landsea, C. W. (1997). Variability of atlantic hurricane activity. In H. Diaz and D. Pulwarthy, editors, *Hurricanes*, pages 15–53. Springer, Germany.

Greatbatch, R. J. and Zhang, S. (1995). An interdecadal oscillation in an idealized ocean basin forced by constant heat flux. *J. Climate*, **8**, 82–91.

Griffies, S. M. (2004). *Fundamentals of ocean-climate models*. Princeton University Press, Princeton, USA.

Griffies, S. M. and Tziperman, E. (1995). A linear thermohaline oscillator driven by stochastic atmospheric forcing. *J. Climate*, **8**, 2440–2453.

Grove, J. M. (1988). *The Little Ice Age*. Methuen, London, U.K.

Gu, D. and Philander, S. G. H. (1997). Interdecadal climate fluctuations that depend on exchanges between the tropics and extratropics. *Science*, **275**, 805–807.

Guckenheimer, J. and Holmes, P. (1990). *Nonlinear Oscillations, Dynamical Systems and Bifurcations of Vector Fields, 2e edition*. Springer-Verlag, Berlin/Heidelberg.

Guckenheimer, J. and Kim, S. (1991). Computational environments for exploring dynamical systems. *Bifurcation and Chaos*, **1**, 269–276.

Haidvogel, D. B. and Beckmann, A. (1999). *Numerical Ocean Circulation Modelling*. Imperial College Press, London, U.K.

Haidvogel, D. B., McWilliams, J. C., and Gent, P. R. (1992). Boundary current separation in a quasi-geostrophic eddy-resolving ocean circulation model. *J. Phys. Oceanogr.*, **22**, 882–902.

Hall, M. and Bryden, H. (1982). Direct estimates of ocean heat transport. *Deep-Sea Research*, **29**, 339–359.

Haney, R. L. (1971). Surface thermal boundary conditions for ocean circulation models. *J. Phys. Oceanogr.*, **4**, 241–248.

Hao, Z., Neelin, J. D., and Jin, F.-F. (1993). Nonlinear tropical air-sea interaction in the fast-wave limit. *J. Climate*, **6**, 1523–1544.

Hartmann, D. L. (1994). *Global Physical Climatology*. Academic Press, San Diego, USA.

Hasselmann, K. (1976). Stochastic climate models. I: Theory. *Tellus*, **28**, 473–485.

Hasselmann, K. (1982). An ocean model for climate variability studies. *Progress in Oceanography*, **11**, 69–92.

Heinrich, H. (1988). Origin and Consequences of Cyclic Ice Rafting in the Northeast Atlantic Ocean during the Past 130,000 Years. *Quaternary Research*, **29**, 142–152.

Hellerman, S. and Rosenstein, M. (1983). Normal monthly wind stress over the world ocean with error estimates. *J. Phys. Oceanogr.*, **13**, 1093–1104.

Henry, D. and Bergeon, A., e. (2000). *Continuation Methods in Fluid Dynamics, volume 74 of Notes on Numerical Fluid Mechanics, Contributions to the ERCOFTAC/EUROMECH Colloquium 383, Aussois, France, 6-9 September 1998.* Vieweg, Germany.

Hirst, A. C. (1986). Unstable and damped equatorial modes in simple coupled ocean-atmosphere models. *J. Atmos. Sci.*, **43**, 606–630.

Hirst, A. C. (1988). Slow instabilities in tropical ocean basin-global atmosphere models. *J. Atmos. Sci.*, **45**, 830–852.

Hogg, N. G., Pickart, R. S., Hendry, R. M., and Smethie, W. J. (1986). The northern recirculation gyre of the Gulf Stream. *Deep-Sea Research*, **33**, 1139–1165.

Holland, W. R. and Bryan, F. O. (1994). Modelling the wind and thermohaline circulation in the North-Atlantic Ocean. In P. Malanotte-Rizzoli and A. Robinson, editors, *Ocean Processes in Climate Dynamics: Global and Mediterranean Examples*, pages 35–156. Kluwer Academic, Netherlands.

Holland, W. R. and Lin, L. B. (1975). On the generation of mesoscale eddies and their contribution to the ocean general circulation I. A preliminary numerical experiment. *J. Phys. Oceanogr.*, **5**, 642–657.

Holland, W. R. and Schmitz, W. J. (1985). Zonal penetration scale of model midlatitude jets. *J. Phys. Oceanogr.*, **15**, 1859–1875.

Holmes, P., Lumley, J. L., and Berkooz, G. (1996). *Turbulence, Coherent Structures, Dynamical Systems and Symmetry.* Cambridge University Press, Cambridge, U.K.

Holton, J. R. (1992). *An Introduction to Dynamic Meteorology.* Academic Press, New York, U.S.A.

Horel, J. D. (1982). The annual cycle in the Tropical Pacific atmosphere and ocean. *Monthly Weather Review*, **110**, 1863–1878.

Hsiung, J. (1985). Estimates of global, oceanic meridional heat transport. *J. Phys. Oceanogr.*, **15**, 1405–1413.

Huang, R. (1993). Real freshwater fluxes as a natural boundary condition for the salinity balance and thermohaline circulation forced by evaporation and precipitation. *J. Phys. Oceanogr.*, **23**, 2428–2446.

Huang, R. X. (1999). Mixing and energetics of the oceanic thermohaline circulation. *J. Phys. Oceanogr.*, **29**, 727–746.

Huang, R. X., Luyten, J. R., and Stommel, H. M. (1992). Multiple equilibrium states in combined thermal and saline circulation. *J. Phys. Oceanogr.*, **22**, 231–246.

Huck, T. and Vallis, G. (2001). Linear stability analysis of the three-dimensional thermally-driven ocean circulation: Application to interdecadal oscillations. *Tellus*, **53A**, 526–545.

Huck, T., Colin de Verdiére, A., and Weaver, A. J. (1999). Interdecadal Variability of the thermohaline circulation in box-ocean models forced by fixed surface fluxes. *J. Phys. Oceanogr.*, **29**, 865892.

Huck, T., Vallis, G., and de Verdiére, C. (2001). On the robustness of interdecadal oscillations of the thermohaline circulation. *J. Climate*, **14**, 940–963.

Hughes, T. M. and Weaver, A. J. (1994). Multiple equilibria of an asymmetric two-basin ocean model. *J. Phys. Oceanogr.*, **24**, 619–637.

Hurlburt, H. E. and Hogan, P. J. (2000). Impact of 1/8° to 1/64° resolution on Gulf Stream model-data comparisons in basin-scale subtropical Atlantic Ocean models. *Dyn. Atmos. Oceans*, **32**, 283–329.

Ierley, G. R. and Sheremet, V. A. (1995). Multiple solutions and advection-dominated flows in the wind-driven circulation. I: Slip. *J. Mar. Res.*, **53**, 703–737.

Imawaki, S., Uchida, H., Ichikawa, H., and Fukasawa (2001). Satellite altimeter monitoring the Kuroshio transport south of Japan. *Geophys. Res. Letters*, **28**, 17–20.

Imbrie, J. and Imbrie, J. Z. (1980). Model showing how Milankovitch cycles might produce the observed glacial cycles. *Science*, **207**, 943–953.

Iooss, G. and Joseph, D. D. (1997). *Elementary Stability and Bifurcation Theory, Second Edition*. Springer-Verlag.

IPCC (2001). *Climate Change 2001: The Scientific Basis. Contribution of Working Group I to the Third Assessment Report of the Intergovernmental Panel on Climate Change (IPCC) [Houghton, J.T., Y. Ding, D.J. Griggs, M. Noguer, P.J. van der Linden, X. Dai, K. Maskell and C.A. Johnson (eds)]*. Cambridge University Press, Cambridge, UK and New York, NY. Also available from www.ipcc.ch.

Jayne, S. R. and Marotzke, J. (2001). The dynamics of ocean heat transport variability. *Reviews of Geophysics*, **39**, 385–411.

Jiang, S., Jin, F.-F., and Ghil, M. (1995). Multiple equilibria and aperiodic solutions in a wind-driven double-gyre, shallow-water model. *J. Phys. Oceanogr.*, **25**, 764–786.

Jin, F.-F. (1997a). An equatorial recharge paradigm for ENSO. I: Conceptual Model. *J. Atmos. Sci.*, **54**, 811–829.

Jin, F.-F. (1997b). An equatorial recharge paradigm for ENSO. II: A stripped-down coupled model. *J. Atmos. Sci.*, **54**, 830–8847.

Jin, F.-F. and An, S.-I. (1999). Thermocline and zonal advective feedbacks within the equatorial ocean recharge oscillator model for ENSO. *Geophys. Res. Letters*, **26**, 2989–2992.

Jin, F.-F. and Neelin, J. D. (1993a). Modes of interannual tropical ocean-atmosphere interaction - a unified view. I: Numerical results. *J. Atmos. Sci.*, **50**, 3477–3503.

Jin, F.-F. and Neelin, J. D. (1993b). Modes of interannual tropical ocean-atmosphere interaction—a unified view. Part III: Analytical results in fully coupled cases. *J. Atmos. Sci.*, **50**, 3523–3540.

Jin, F.-F., Neelin, J. D., and Ghil, M. (1994). El Niño on the devil's staircase: Annual subharmonic steps to chaos. *Science*, **264**, 70–72.

Jin, F.-F., Neelin, J. D., and Ghil, M. (1996). El Niño/Southern Oscillation and the annual cycle: Subharmonic frequency-locking and aperiodicity. *Physica D*, **98**, 442–465.

John, F. (1986). *Partial Differential Equations*. Springer-Verlag, New York, U.S.A.

Johns, W. E., Shay, T. J., Bane, J. M., and Watts, D. R. (1995). Gulf Stream structure, transport and recirculation near 68° W. *J. Geophys. Res.*, **100**, 817–838.

Johns, W. E., Lee, T. N., Zhang, D., Zantopp, R., Liu, C. T., and Yang, Y. (2000). The Kuroshio east of Taiwan - Moored transport observations from the WOCE PCM-1 array. *J. Phys. Oceanogr.*, **31**, 1031–1053.

Johnsen, R. G. (1997). Climate control requires a dam at the Strait of Gibraltar. *EOS*, **78**, 280–281.

Johnsen, S. J., Clausen, H. B., Dansgaard, W., Gundestrup, N. S., Hammer, C. U., U., A., Andersen, K. K., Hvidberg, C. S., Dahl-Jensen, D., Steffensen, J. P., Shoji, H., Sveinbjörnsdottir, A. E., White, J. W. C., Jouzel, J., and Fisher, D. (1997). The $\delta^{18}O$ record along the Greenland Ice Core Project deep ice core and the problem of possible Eemian climatic instability. *J. Geophys. Res.*, **102**, 26,397–26,410.

Joseph, D. D. (1976). *Stability of Fluid Motions: Vol. I. and Vol. II.* Springer-Verlag, Berlin-Heidelberg, Germany.

Joyce, T. M. and Robbins, P. (1995). The long-term hydrographic record at Bermuda. *J. Climate*, **9**, 3122–3131.

Kaese, R. H. and Krauss, W. A. (1996). The Gulf Stream, the North Atlantic Current, and the origin of the Azores Current. In W. Krauss, editor, *The Warmwatersphere of the North Atlantic Ocean*, pages 291–337. Borntraeger, Berlin-Stuttgart, Germany.

Kamenkovich, V. M., Sheremet, V. A., Pastushkov, A. R., and Belotserkovsky, S. O. (1995). Analysis of barotropic model of the subtropical gyre in the ocean for finite Reynolds number. *J. Mar. Res.*, **53**, 959–994.

Kantz, H. and Schreiber, T. (1997). *Nonlinear Time Series Analysis.* Cambridge University Press, Cambridge, UK.

Katsman, C. A., Dijkstra, H. A., and Drijfhout, S. S. (1998). The rectification of the wind-driven circulation due to its instabilities. *J. Mar. Res.*, **56**, 559–587.

Katsman, C. A., Drijfhout, S. S., and Dijkstra, H. A. (2001). The interaction of the Deep Western Boundary Current and the wind-driven gyres as a casue for low-frequency variability. *J. Phys. Oceanogr.*, **31**, 2321–2339.

Kawabe, M. (1986). Transition processes between the three typical paths of the Kuroshio. *J. Oceanogr. Soc. Japan*, **42**, 174–191.

Kawabe, M. (1995). Variations of current path, velocity, and volume transport of the Kuroshio in relation with the large meander. *J. Phys. Oceanogr.*, **25**, 3103–3117.

Keller, H. B. (1977). Numerical solution of bifurcation and nonlinear eigenvalue problems. In P. H. Rabinowitz, editor, *Applications of Bifurcation Theory*. Academic Press, New York, U.S.A.

Kelly, K. A., Caruso, M. J., Singh, S., and Qiu, B. (1996). Observations of atmosphere-ocean coupling in midlatitude western boundary currents. *J. Geophys. Res.*, **101**, 6295–6312.

Kerr, R. A. (2000). A North Atlantic climate pacemaker for the centuries. *Science*, **288**, 1984–1986.

Kessler, W., Spillane, M., McPhaden, M., and Harrison, D. (1996). Scales of variability in the equatorial pacific inferred from the tropical atmosphere-ocean array. *J. Climate*, **9**, 2999–3024.

Kessler, W. S. and McCreary, J. P. (1995). The annual wind-driven Rossby wave in the sub-thermocline equatorial Pacific. *J. Phys. Oceanogr.*, **23**, 1192–1207.

Kessler, W. S., J., M. M., and Weickmann, K. M. (1995). Forcing of intraseasonal Kelvin waves in the equatorial Pacific. *J. Geophys. Res.*, **100**, 10,613–10,631.

Kirtman, B. and Schopf, P. (1998). Decadal Variability in ENSO Predictability and Prediction. *J. Climate*, **11**, 2804–2822.

Kiss, A. E. (2002). Potential vorticity "crisis", adverse pressure gradients, and western boundary current separation. *J. Mar. Res.*, **60**, 779–803.

Kleeman, R. and Power, S. (1994). Limits to the predictability in a coupled ocean-atmosphere model. *Tellus*, **46A**, 529–540.

Kleeman, R. P., McCreary, J. P., and Klinger, B. (1999). A mechanism for generating ENSO decadal variability. *Geophys. Res. Letters*, **26**, 1743–1746.

Klinger, B. A. and Marotzke, J. (1999). Behavior of double-hemispheric thermohaline flows in a single basin. *J. Phys. Oceanogr.*, **29**, 382–399.

Klinger, B. A., Drijfhout, S., Marotzke, J., and Scott, J. R. (2003). Sensitivity of basinwide meridional overturning to diapycnal difffusion and remote wind forcing in an idealized Atllantic-Southern Ocean geometry. *J. Phys. Oceanogr.*, **33**, 249–266.

Knutson, T. and Manabe, S. (1998). Model assesment of decadal variability and trends in the tropical Pacific Ocean. *J. Climate*, **11**, 2273–2296.

Knutson, T. R., Manabe, S., and Gu, D. (1997). Simulated ENSO in a global coupled ocean-atmosphere model: Multidecadal amplitude modulation and CO2 sensitivity. *J. Climate*, **10**, 131–161.

Koschmieder, E. L. (1993). *Bénard Cells and Taylor Vortices*. Cambridge University Press, Cambridge, UK.

Koschmieder, E. L. and Switzer, D. W. (1992). The wavenumbers of supercritical surface-tension-driven Bénard convection. *J. Fluid Mech.*, **240**, 533–548.

Kraus, W. A., editor (1996). *The Warmwatersphere of the North Atlantic Ocean*. Borntraeger, Berlin, Stuttgart, Germany.

Kumar, K. K., Rajagopalan, B., and Cane, M. A. (1999). On the weakening relationship between the Indian Monsoon and ENSO. *Science*, **284**, 2156–2159.

Kuo, H. L. (1951). The general circulation and the stability of zonal flow. *Tellus*, **3**, 268–284.

Kushnir, Y. (1994). Interdecadal variations in North Atlantic sea surface temperature and associated atmospheric conditions. *J. Phys. Oceanogr.*, **7**, 141–157.

Kuznetsov, Y. A. (1995). *Elements of Applied Bifurcation Theory*. Springer Verlag, New York, U.S.A.

Large, W. G., McWilliams, J. C., and Doney, S. C. (1994). Ocean vertical mixing: A review and a model with a nonlocal boundary layer parameterization. *Rev. Geophysics*, **32**, 363–403.

Large, W. G., Danabasogu, G., McWilliams, J. C., Gent, P. R., and Bryan, F. O. (2001). Equatorial circulation of a global ocean climate model with anisotropic horizontal viscosity. *J. Phys. Oceanogr.*, **31**, 518–536.

Latif, M. (1998). Dynamics of interdecadal variability in coupled ocean-atmosphere models. *J. Climate*, **11**, 602–624.

Latif, M., Sperber, K., Arblaster, J., Braconnot, P., Chen, D., Colman, A., Cubasch, U., Cooper, C., Delecluse, P., DeWitt, D., Fairhead, L., Flato, G., Hogan, T., Ji, M., Kimoto, M., Kitoh,

A., Knutson, T., Le Treut, H., Li, T., Manabe, S., Marti, O., Mechoso, C., Meehl, G., Power, S., Roeckner, E., Sirven, J., Terray, L., Vintzileos, A., Voss, R., Wang, B., Washington, W., Yoshikawa, I., Yu, J., and Zebiak, S. (2001). ENSIP: the El Nino simulation intercomparison project. *Clim. Dyn.*, **18**, 255–276.

Le Traon, P. Y., Nadal, F., and Ducet, N. (1998). An improved mapping method of multi-satellite altimeter data. *J. Atmos. Oceanic Technol.*, **15**, 522–534.

Lear, C. H., Elderfield, H., and Wilson, P. A. (2000). Cenozoic Deep-Sea temperatures and global ice volumes from Mg/Ca in Benthic Foraminiferal Calcite. *Nature*, **287**, 269–272.

Lee, D. and Cornillon, P. (1995). Temporal variation of meandering intensity and domain-wide lateral oscillations of the Gulf Stream. *J. Geophys. Res.*, **100**, 13,603–13,613.

Legeckis, R. (1977). Long waves in the eastern equatorial Pacific Ocean. *Science*, **197**, 1179–1181.

Lenderink, G. and Haarsma, H. (1994). Variability and multiple equilibria of the thermohaline circulation associated with deep-water formation. *J. Phys. Oceanogr.*, **24**, 1480–1493.

Levitus, S. (1982). Climatological Atlas of the World Ocean. *NOAA Professional Paper*, **13**, 1–150.

Levitus, S. (1994). World Ocean Atlas 1994, Volume 4: Temperature. *NOAA/NESDIS E/OC21, US Department of Commerce, Washington DC*, pages 1–117.

Levitus, S., Antonov, J., Boyer, T., and Stephens, C. (2000). Warming of the world ocean. *Science*, **287**, 2225–2229.

Liu, Z. and Xie, S. P. (1994). Equatorward propagation of coupled air-sea disturbances with application to the annual cycle of the eastern tropical Pacific. *J. Atmos. Sci.*, **51**, 3807–3822.

Lorenz, E. N. (1963). Deterministic nonperiodic flow. *J. Atmos. Sci.*, **20**, 130–141.

Lowe, J. J., Coope, G. R., Sheldrick, C., Harkness, D. D., and Walker, M. J. C. (1995). Direct comparison of UK temperatures and Greenland snow accumulation rates. *J. Quaternary Res.*, **10**, 175–180.

Lust, K. and Roose, D. (2000). Computation and bifurcation analysis of periodic solutions of large-scale systems. In E. Doedel and L. S. Tuckermann, editors, *Numerical Methods for Bifurcation Problems and Large-Scale Dynamical Systems*, pages 265–302. Springer.

Lust, K., Roose, D., Spence, A., and Champneys, A. R. (1998). An adaptive Newton-Picard algorithm with subspace iteration for computing periodic solutions of large-scale dynamical systems. *SIAM J. Scientific Computing*, **19**, 1188–1209.

Lysne, J., Chang, P., and Giese, B. (1997). Impact of extratropical Pacific on equatorial variability. *Geophys. Res. Letters*, **24**, 2589–2592.

Madden, R. and Julian, P. (1994). Observations of the 40-50-day tropical oscillation — A review. *Monthly Weather Review*, **122**, 814–835.

Maier-Reimer, E., Mikolajewicz, U., and Hasselman, K. (1993). Mean circulation of the Hamburg LSG OGCM and its sensitivity to the thermohaline surface forcing. *J. Phys. Oceanogr.*, **23**, 731–757.

Maltrud, M. E., R., S., and Malone, R. C. (1998). Global eddy-resolving ocean simulations driven by 1985–1995 atmospheric winds. *J. Geophys. Res.*, **103**, 30,825–30,853.

Manabe, S. and Stouffer, R. J. (1988). Two stable equilibria of a coupled ocean-atmosphere model. *J. Climate*, **1**, 841–866.

Manabe, S. and Stouffer, R. J. (1993). Century-scale effects of increased CO_2 on the ocean-atmosphere system. *Nature*, **364**, 215–220.

Manabe, S. and Stouffer, R. J. (1995). Simulation of abrupt climate change induced by freshwater input into the North Atlantic Ocean. *Nature*, **378**, 165–167.

Manabe, S. and Stouffer, R. J. (1999). Are two modes of thermohaline circulation stable? *Tellus*, **51A**, 400–411.

Mantua, N. J. and Battisti, D. S. (1995). Aperiodic variability in the Zebiak-Cane coupled ocean-atmosphere model: air-sea interaction in the western equatorial Pacific. *J. Climate*, **8**, 2897–2927.

Marotzke, J. (1991). Influence of convective adjustment on the stability of the thermohaline circulation. *J. Phys. Oceanogr.*, **21**, 903–907.

Marotzke, J. (2000). Abrupt climate change and thermohaline circulation: Mechanisms and predictability. *Proc. Natl. Acad. Sci.*, **97**, 1347–1350.

Marotzke, J. and Scott, J. R. (1997). Boundary mixing and the dynamics of the three-dimensional thermohaline circulation. *J. Phys. Oceanogr.*, **27**, 1713–1728.

Marotzke, J. and Scott, J. R. (1999a). Convective mixing and the thermohaline circulation. *J. Phys. Oceanogr.*, **29**, 2962–2970.

Marotzke, J. and Scott, J. R. (1999b). The location of diapycnal mixing and the meridional overturning circulation. *J. Phys. Oceanogr.*, **32**, 3578–3595.

Marotzke, J. and Stone, P. (1995). Atmospheric transports, the thermohaline circulation and flux adjustments in a simple coupled model. *J. Phys. Oceanogr.*, **25**, 1350–1364.

Marotzke, J. and Willebrand, P. (1991). Multiple equilibria of the global thermohaline circulation. *J. Phys. Oceanogr.*, **21**, 1372–1385.

Marotzke, J., Welander, P., and Willebrand, J. (1988). Instability and multiple steady states in a meridional-plane model of thermohaline circulation. *Tellus*, **40**, 162–172.

Masuda, A. (1982). An interpretation of the bimodal character of the stable kuroshio path. *Deep-Sea Research*, **29**, 471–484.

Mata, M. M., Tomczak, M., Wijffels, S., and Church, J. A. (2000). East australian current volume transports at 30s: estimates from the world ocean circulation experiment hydrographic sections pr11/p6 and the pcm3 current meter array. *J. Geophys. Res.*, **28**, 28,509–28,526.

Matsuno, T. (1966). Quasi-geostrophic motions in equatorial areas. *J. Meteorol. Soc. Jpn*, **2**, 25–43.

Matthes, F. E. (1940). Report on the Committee on glaciers. *Trans. American Geophysical Union*, **21**, 396–406.

McCalpin, J. D. and Haidvogel, D. B. (1996). Phenomenology of the low-frequency variabiliity in a reduced gravity quasi-geostrophic double-gyre model. *J. Phys. Oceanogr.*, **26**, 739–752.

McCreary, J. P. and Anderson, D. L. T. (1991). An overview of coupled ocean-atmosphere models of El Niño and the Southern Oscillation. *J. Geophys. Res.*, **96**, 3125–3150.

McManus, J. F., Bond, G. C., Broecker, W. S., and Johnson, S. (1994). High-resolution climate records from the North Atlantic during the last interglacial. *Nature*, **371**, 326–328.

McPhaden, M. (1999). Genesis and evolution of the 1997-98 El Niño. *Science*, **283**, 950–954.

McPhaden, M. and coauthors (1998). The Tropical Ocean-Global Atmosphere observing system: a decade of progress. *J. Geophys. Res.*, **103**, 14,169–14,240.

McPhaden, M. J. and Yu, X. (1999). Equatorial waves and the 1997-98 El Niño. *Geophys. Res. Letters*, **26**, 2961–2964.

McWilliams, J. C. (1977). A note on a consistent quasi-geostrophic model in a multiple connected domain. *Dyn. Atmos. Oceans*, **1**, 427–441.

McWilliams, J. C. (1996). Modeling the ocean general circulation. *Ann. Rev. Fluid Mechanics*, **28**, 215–248.

McWilliams, J. C. and Chow, J. H. S. (1981). Equilibrium geostrophic turbulence I: A reference solution in a β-plane channel. *J. Phys. Oceanogr.*, **11**, 921–949.

McWilliams, J. C. and Gent, P. R. (1980). Intermediate Models of Planetary Circulations in the Atmosphere and Ocean. *J. Atmos. Sci.*, **37**, 1657–1678.

Meacham, S. P. (2000). Low frequency variability of the wind-driven circulation. *J. Phys. Oceanogr.*, **30**, 269–293.

Meacham, S. P. and Berloff, P. S. (1997). Instabilities of a steady, barotropic, wind-driven circulation. *J. Mar. Res.*, **55**, 885–913.

Meacham, S. P. and Berloff, P. S. (1998). Barotropic, wind-driven circulation in a small basin. *J. Mar. Res.*, **55**, 523–563.

Mechoso, C. and coauthors (1995). The seasonal cycle over the tropical Pacific in coupled ocean-atmosphere general circulation models. *Monthly Weather Review*, **123**, 2825–2838.

Meehl, G. A. (1990). Development of global coupled ocean-atmosphere general circulation models. *Clim. Dyn.*, **5**, 19–33.

Mikolajewicz, U. and Maier-Reimer, E. (1990). Internal secular variability in an ocean general circulation model. *Clim. Dyn.*, **4**, 145–156.

Miller, A. J., Holland, W. R., and Hendershott, M. C. (1987). Open - ocean response and normal mode exitation in an eddy-resolving general circulation model. *Geophys. Astrophys. Fluid Dyn.*, **37**, 253–278.

Mitchell, T. P. and Wallace, J. M. (1992). The annual cycle in equatorial convection and sea surface temperature. *J. Climate*, **5**, 1140–1156.

Moore, A. M. and Kleeman, R. (1999). Stochastic Forcing of ENSO by the Intraseasonal Oscillation. *J. Climate*, **12**, 1199–1220.

Moore, D. W. (1968). *Planetary-Gravity Waves in an Equatorial Ocean.* Harvard University, Cambridge, MA, U.S.A.

Moron, V., Vautard, R., and Ghil, M. (1998). Trends, interdecadal and interannual oscillations in global sea-surace temperature. *Clim. Dyn.*, **14**, 545–569.

Mueller, T. J., Ikeda, Y., Zangenberg, N., and Nonato, L. V. (2000). Direct measurements of western boundary currents off brazil between 20s and 28s. *J. Geophys. Res.*, **28**, 28,509–28,526.

Munk, W. (1950). On the wind-driven ocean circulation. *J. Meteorol.*, **7**, 79–93.

Munk, W. and Wunsch, C. (1998). Abyssal recipes II: energetics of tidal and wind mixing. *Deep-Sea Research*, **45**, 1977–2010.

Münnich, M. M., Cane, M., and Zebiak, S. E. (1991). A study of self-excited oscillations of the tropical ocean-atmosphere system. II: Nonlinear cases. *J. Atmos. Sci.*, **48**, 1238–1248.

Mysak, L., Manak, D. K., and Marsden, R. F. (1990). Sea-ice anomalies observed in the Greenland and Labrador Seas during 1901-1984 and their relation to an interdecadal Arctic climate cycle. *Climate Dynamics*, **5**, 111–133.

Nadiga, B. T. and Luce, B. (2001). Global bifurcation of Shilnikov type in a double-gyre model. *J. Phys. Oceanogr.*, **31**, 2669–2690.

Nauw, J. and Dijkstra, H. A. (2001). The origin of low-frequency variability of double-gyre wind-driven flows. *J. Mar. Res.*, **59**, 567–597.

Nauw, J., Dijkstra, H. A., and Chassignet, E. (2004a). Frictionally induced asymmetries in wind-driven flows. *J. Phys. Oceanogr.*, **34**, 2057–2072.

Nauw, J., Dijkstra, H. A., and Simonnet, E. (2004b). Regimes of low-frequency variability in a three-layer quasi-geostrophic model. *J. Mar. Res.*, **62**, 684–719.

Nayfeh, A. H. and Balachandran, B. (1995). *Applied Nonlinear Dynamics*. John Wiley, New York, U.S.A.

Neelin, J. D. (1991a). A hybrid coupled general circulation model for El Nino studies. *J. Atmos. Sci.*, **47**, 674–693.

Neelin, J. D. (1991b). The slow sea surface temperature mode and the fast-wave limit: Analytic theory for tropical interannual oscillations and experiments in a hybrid coupled model. *J. Atmos. Sci.*, **48**, 584–606.

Neelin, J. D. and coauthors (1992). Tropical air-sea interaction in GCM's. *Clim. Dyn.*, **7**, 73–104.

Neelin, J. D. and Dijkstra, H. A. (1995). Coupled ocean-atmosphere models and the tropical climatology. I: The dangers of flux-correction. *J. Climate*, **8**, 1325–1342.

Neelin, J. D. and Jin, F.-F. (1993). Modes of interannual tropical ocean-atmosphere interaction–a unified view. II: Analytical results in the weak-coupling limit. *J. Atmos. Sci.*, **50**, 3504–3522.

Neelin, J. D., Latif, M., and Jin, F.-F. (1994). Dynamics of coupled ocean-atmosphere models: The tropical problem. *Ann. Rev. Fluid Mech.*, **26**, 617–659.

Neelin, J. D., Battisti, D. S., Hirst, A. C., Jin, F.-F., Wakata, Y., Yamagata, T., and Zebiak, S. E. (1998). ENSO Theory. *J. Geophys. Res.*, **103**, 14,261–14,290.

Neelin, J. D., Jin, F.-F., and Syu, H.-H. (2000). Variations of ENSO phase locking. *J. Climate*, **13**, 2570–2590.

New, A. L., Bleck, R., Jia, Y., Marsh, M., Huddleston, M., and Barnard, S. (1995). An isopycnic model study of the North Atlantic. I: Model experiment and water mass formation. *J. Phys. Oceanogr.*, **25**, 2667–2699.

Newman, M., Sardeshmukh, P., and Bergman, J. (2000). An assesment of the NCEP, NASA and ECMWF weanalyses over the Tropical west Pacific warm pool. *Bull. American Meteorological Society*, **81**, 41–48.

Nield, D. A. (1964). Surface tension and buoyancy effects in cellular convection. *J. Fluid Mech.*, **19**, 341–352.

Nilsson, J. and Walin, G. (2001). Freshwater forcing as a booster of thermohaline circulation. *Tellus*, **53A**, 629–641.

Noda, A., Yoshimatsu, K., Yukimoto, S., Yamaguchi, K., and Yamaki, S. (1999). *Relationship between natural variability and CO_2-induced warming pattern: MRI AOGCM Experiment.* *In* Proc. 10th Symposium on Global Change Studies, *10-15 January 1999, Dallas Texas*, volume 24. American Meteorological Society, Boston, USA.

North, G. R., Cahalan, R. F., and Coakley, J. A. (1981). Energy balance climate models. *Rev. Geophys. Space Phys.*, **19**, 19–121.

Oberhuber, J. M. (1988). *The Budget of Heat, Buoyancy and Turbulent Kinetic Energy at the Surface of the Global Ocean.* Max Planck Institute für Meteorologie Hamburg report nr. 15, Hamburg, Germany.

Oksendal, B. (1995). *Stochastic Differential Equations.* Springer-Verlag, Berlin, Germany.

Olson, D. B., Brown, O. B., and Emmerson, S. R. (1983). Gulf Stream frontal statistics from Florida Straits to Cape Hatteras derived from satellite and historical data. *J. Geophys. Res.*, **88**, 4569–4577.

Oort, A. H. (1983). *Global Atmospheric Circulation Statistics, 1975–1985.* NOAA Professional Paper nr. 14, U.S. Gov. Printing Office, Princeton, NJ, U.S.A.

OU-Staff (1989). *Ocean Circulation.* Pergamom Press, Oxford, U.K.

Ozawa, H., Ohmura, A., Lorenz, R. D., and Pujol, T. (2003). The second law of thermodynamics and the global climate system: a review of the maximum entropy production principle. *Reviews of Geophysics*, **41**, 4–1, doi:10.1029/2002RG000113.

Pacanowski, R. C. (1996). MOM 2 Documentation, User's Guide and Reference Manual. *GFDL Tech. Report*, **3.1**.

Paparella, F. and Young, W. R. (2002). Horizontal convection is non-turbulent. *J. Fluid Mech.*, **466**, 205–214.

Pedlosky, J. (1987). *Geophysical Fluid Dynamics. 2nd Edn.* Springer-Verlag, New York.

Pedlosky, J. (1996). *Ocean Circulation Theory.* Springer, New York.

Peixoto, J. P. and Oort, A. H. (1992). *Physics of Climate.* AIP Press, New York.

Penland, C. and Sardeshmukh, P. D. (1995). The optimal growth of tropical sea surface temperature anomalies. *J. Climate*, **8**, 1999–2024.

Philander, S. G. H. (1990). *El Niño and the Southern Oscillation.* Academic Press, New York.

Philander, S. G. H., Yamagata, T., and Pacanowski, R. C. (1984). Unstable air-sea interactions in the tropics. *J. Atmos. Sci.*, **41**, 604–613.

Philander, S. G. H., Pacanowski, R. C., Lau, N.-C., and Nath, M. J. (1989). Two different simulations of the Southern Oscillation and El Niño with coupled ocean-atmosphere circulation models. *Phil. Trans. R. Soc. London A*, **329**, 167–178.

Philander, S. G. H., Gu, D., Halpern, D., Lambert, G., Lau, N.-C., Li, T., and Pacanowski, R. C. (1996). Why the ITCZ is mostly north of the equator. *J. Climate*, **9**, 2958–2972.

Phillips, N. A. (1951). A simple three-dimensional model for the study of large-scale extratropical flow patterns. *J. Meteor.*, **8**, 381–394.

Piava, A. M., Chassignet, E. P., and Mariano, A. J. (2000). Numerical simulations of the North Atlantic subtropical gyre: sensitivity to boundary conditions. *Dyn. Atmos. Oceans*, **32**, 209–237.

Picaut, J., Masia, I., and du Penhoat, Y. (1996). An advective-reflective conceptual model for the oscillatory nature of ENSO. *Science*, **277**, 663–666.

Pierce, D. W., Barnett, T. P., and Latif, M. (2000). Connections between the Pacific Ocean tropics and midlatitudes on decadal timescales. *J. Climate*, **13**, 1173–1194.

Pikovsky, A., Rosenblum, M., and Kurths, J. (2001). *Synchronization*. Cambridge University Press.

Plaut, G., Ghil, M., and Vautard, R. (1995). Interannual and interdecadal variability in 335 years of Central England Temperature. *Science*, **268**, 710–713.

Prahl, M., Lohmann, G., and Paul, A. (2003). Influence of vertical mixing on the thermohaline hysteresis: analysis of an OGCM. *J. Phys. Oceanogr.*, **33**, 17071721.

Preisendorfer, R. W. (1988). *Principal Component Analysis in Meteorology and Oceanography*. Elsevier, Amsterdam, The Netherlands.

Primeau, F. W. (1998). *Multiple Equilibria and Low-Frequency Variability of Wind-Driven Ocean Models*. Ph.D. thesis, M.I.T. and Woods Hole, Boston, MA, U.S.A.

Primeau, F. W. (2002). Multiple equilibria and low-frequency variability of the wind-driven ocean circulation. *J. Phys. Oceanogr.*, **32**, 2236–2256.

Qiu, B. and Joyce, T. M. (1992). Interannual variability in the mid- and low-latitude western North Pacific. *J. Phys. Oceanogr.*, **22**, 1062–1079.

Qiu, B. and Miao, W. (2000). Kuroshio path variations south of Japan: bimodality as a self-sustained internal oscillation. *J. Phys. Oceanogr.*, **30**, 2124–2137.

Quon, C. and Ghil, M. (1992). Multiple equilibria in thermosolutal convection due to salt-flux boundary conditions. *J. Fluid Mech.*, **245**, 449–484.

Quon, C. and Ghil, M. (1995). Multiple equilibria and stable oscillations in thermosolutal convection at small aspect ratio. *J. Fluid Mech.*, **291**, 33–56.

Rahmstorf, S. (1994). Rapid climate transitions in a coupled ocean-atmosphere model. *Nature*, **372**, 82–84.

Rahmstorf, S. (1995a). Bifurcations of the Atlantic thermohaline circulation in response to changes in the hydrological cycle. *Nature*, **378**, 145–149.

Rahmstorf, S. (1995b). Multiple convection patterns and thermohaline flow in an idealized OGCM. *J. Climate*, **8**, 3028–3039.

Rahmstorf, S. (1996). On the freshwater forcing and transport of the Atlantic thermohaline circulation. *Clim. Dyn.*, **12**, 799–811.

Rahmstorf, S. (1998). Influence of Mediterranean Outflow on Climate. *EOS*, **79**.

Rahmstorf, S. (2000). The thermohaline circulation: a system with dangerous thresholds? *Climatic Change*, **46**, 247–256.

Rahmstorf, S. and England, M. (1997). Influence of Southern Hemisphere Winds on North Atlantic Deep Water Flow. *J. Phys. Oceanogr.*, **27**, 2040–2054.

Rasmusson, E. and Carpenter, T. H. (1982). Variations in tropical sea surface temperature and surface wind fields associated with the Southern Oscillation/El Niño. *Monthly Weather Review*, **110**, 354–384.

Rasmusson, E., Wang, X., and Ropelewski, C. (1990). The biennial component of ENSO variability. *J. Marine Systems*, **1**, 71–96.

Raymo, M. E., Ruddiman, W. F., Shackleton, N. J., and Oppo, D. W. (1990). Atlantic-Pacific carbon isotope differences over the last 2.5 million years. *Earth and Planetary Science Letters*, **97**, 353–368.

Reynolds, R. W. and Smith, T. M. (1994). Improved global sea surface temperature analysis using optimum interpolation. *J. Climate*, **7**, 929–948.

Richardson, P. (1980). Benjamin Franklin and Timothy Folger's first printed chart of the Gulf Stream. *Science*, **207**, 643–645.

Rintoul, S. R. (1991). South Atlantic interbasin exchange. *J. Geophys. Res.*, **96**, 2675–2692.

Rintoul, S. R., Hughes, C., and Olbers, D. (2001). The Antarctic Circumpolar Current System. In J. C. G. Siedler and J. Gould, editors, *Ocean Circulation and Climate*, pages 271–302. Academic Press, New York.

Rivin, I. and Tziperman, E. (1997). Linear versus self-sustained interdecadal thermohaline variability in a coupled box model. *J. Phys. Oceanogr.*, **27**, 1216–1232.

Roache, P. (1976). *Computational Fluid Dynamics*. Hermosa Publishing, Albequerque, NM, U.S.A.

Roberts, N. (1998). *The Holocene*. Blackwell Publishing Company, London, U.K.

Robertson, A. W., Ma, C.-C., Mechoso, C. R., and Ghil, M. (1995a). Simulation of the tropical Pacific climate with a coupled ocean-atmosphere model. I: The seasonal cycle. *J. Climate*, **8**, 1178–1198.

Robertson, A. W., Ma, C.-C., Mechoso, C. R., and Ghil, M. (1995b). Simulation of the tropical Pacific climate with a coupled ocean-atmosphere model. II: Interannual variability. *J. Climate*, **8**, 1199–1216.

Rodbell, D., Seltzer, G., Anderson, D. M., Abbott, M. B., Enfield, D. B., and Newman, J. H. (1999). A 15,000 year record of ENSO-driven alluviation in Southwestern Ecuador. *Science*, **283**, 516–520.

Rooth, C. (1982). Hydrology and ocean circulation. *Progress in Oceanography*, **11**, 131–149.

Ropelewski, C. F. and Halpert, M. S. (1987). Global and regional scale precipitation associated with ENSO. *Monthly Weather Review*, **115**, 1606–1626.

Roulston, M. and Neelin, J. D. (2000). The response of an ENSO model to climate noise, weather noise and intraseasonal forcing. *Geophys. Res. Letters*, **27**, 3723–3726.

Ruddiman, W. F. (2001). *Earth's Climate: Past and Future*. W.H. Freeman and Company, New York, U.S.A.

Ruelle, D. and Takens, F. (1970). On the nature of turbulence. *Comm. Math. Phys.*, **20**, 167–192.

S.-I., A. and Jin, F.-F. J. (2000). An eigenanalysis of the interdecadal changes in the structure and frequency of ENSO mode. *Geophys. Res. Letters*, **27**, 2573–2576.

Saad, Y. (1996). *Iterative Methods for Sparse Matrices*. PWS Publishing Co., Boston.

Sakai, K. and Peltier, W. R. (1995). A simple model of the Atlantic thermohaline circulation: Internal and forced variability with paleoclimatological implications. *J. Geophys. Res.*, **100**, 13,455 – 13,479.

Sakai, K. and Peltier, W. R. (1997). Dansgaard-Oeschger oscillations in a coupled atmosphere-ocean climate model. *J. Climate*, **10**, 949–970.

Sakai, K. and Peltier, W. R. (1999). A dynamical systems model of the Dansgaard-Oeschger oscillation and the origin of the Bond cycle. *J. Climate*, **12**, 2238–2255.

Salmon, R. (1986). A simplified linear ocean circulation theory. *J. Mar. Res.*, **44**, 695–711.

Salmon, R. (1998). *Lectures on Geophysical Fluid Dynamics*. Oxford Univ. Press, 400pp.

Sanders, J. and Verhulst, F. (1985). *Averaging methods in nonlinear dynamical systems*. Springer Verlag.

Sandstrom, J. W. (1908). Dynamische Versuche mit Meerwasser. *Annalen der Hydrographie und der Maritimen Meteorologie*.

Saravanan, R. and McWilliams, J. (1997). Stochasticity and spatial resonance in interdecadal climate fluctuations. *J. Climate*, **10**, 2299–2320.

Saravanan, R. and McWilliams, J. (1998). Advective ocean-atmosphere interaction: an analytical stochastic model with implications for decadal variability. *J. Climate*, **11**, 165–188.

Sato, O. T. and Rossby, T. (2000). Seasonal and low-frequency variability of the meridional heat flux at 36° in the North Atlantic. *J. Phys. Oceanogr.*, **30**, 606–621.

Sausen, R., Barthels, K., and Hasselmann, K. (1988). Coupled ocean-atmosphere models with flux correction. *Clim. Dyn.*, **2**, 154–163.

Schlesinger, M. E. and Ramankutty, N. (1994). An oscillation in the global climate system of period 65-70 years. *Nature*, **367**, 723–726.

Schlösser, P., Bonisch, G., Rhein, M., and Bayer, R. (1991). Reduction of deep water formation in the Greenland Sea during the 1980s: evidence from tracer data. *Science*, **251**, 1054–1056.

Schmeits, M. J. and Dijkstra, H. A. (2000). On the physics of the 9 months variability in the Gulf Stream region: Combining data and dynamical systems analysis. *J. Phys. Oceanogr.*, **30**, 1967–1987.

Schmeits, M. J. and Dijkstra, H. A. (2001). Bimodality of the Kuroshio and the Gulf Stream. *J. Phys. Oceanogr.*, **31**, 2971–2985.

Schmeits, M. J. and Dijkstra, H. A. (2002). Subannual variability of the ocean circulation in the Kuroshio region. *J. Geophys. Res.*, **107**, 3235, doi:10.1029/2001JC001073.

Schmitt, R. W., Bogden, P., and Dorman, C. (1989). Evaporation minus precipitation and density fluxes for the North Atlantic. *J. Phys. Oceanogr.*, **19**, 1208–1221.

Schmitz, W. J. (1995). On the interbasin-scale thermohaline circulation. *Rev. Geophys.*, **33**, 151–173.

Schmitz, W. J. and Holland, W. R. (1982). A preliminary comparison of selected numerical eddy-resolving general circulation experiments with observations. *J. Mar. Res.*, **40**, 75–117.

Schopf, P. and Suarez, M. (1988). Vacillations in a coupled ocean-atmosphere model. *J. Atmos. Sci.*, **45**, 549–566.

Schopf, P. S. and Suarez, M. (1990). Ocean wave dynamics and the time scale of ENSO. *J. Phys. Oceanogr.*, **20**, 629–645.

Semtner, A. J. and Chervin, R. M. (1992). Ocean general circulation from a global eddy-resolving model. *J. Geophys. Res.*, **97**, 5493–5550.

Seydel, R. (1994). *Practical Bifurcation and Stability Analysis: From Equilibrium to Chaos*. Springer-Verlag, New York, U.S.A.

Sheremet, V. A., Ierley, G. R., and Kamenkovich, V. M. (1997). Eigenanalysis of the two-dimensional wind-driven ocean circulation problem. *J. Mar. Res.*, **55**, 57–92.

Shilnikov, L. P. (1965). A case of the existence of a denumerable set of periodic motions. *Sov. Math. Dokl.*, **6**, 163–166.

Siegel, A., Weiss, J. B., Toomre, J., McWilliams, J. C., Berloff, P., and Yavneh, I. (2001). Eddies and coherent vortices in ocean basin dynamics. *Geophys. Res. Letters*, **28**, 3183–3186.

Simonnet, E. and Dijkstra, H. A. (2002). Spontaneous generation of low-frequency modes of variability in the wind-driven ocean circulation. *J. Phys. Oceanogr.*, **32**, 1747–1762.

Simonnet, E., Ghil, M., Ide, K., Temam, R., and Wang, S. (2003a). Low-frequency variability in shallow-water models of the wind-driven ocean circulation. Part I: Steady-state solutions. *J. Phys. Oceanogr.*, **33**, 712–728.

Simonnet, E., Ghil, M., Ide, K., Temam, R., and Wang, S. (2003b). Low-frequency variability in shallow-water models of the wind-driven ocean circulation. Part II: Time dependent solutions. *J. Phys. Oceanogr.*, **33**, 729–752.

Simonnet, E., Ghil, M., and Dijkstra, H. A. (2005). Homoclinic bifurcations of barotropic QG double-gyre flows. *J. Mar. Res.* sub judice.

Sleijpen, G. L. G. and Van der Vorst, H. A. (1996). A Jacobi-Davidson iteration method for linear eigenvalue problems. *SIAM J. Matrix Anal. Appl.*, **17**, 410–425.

Slingo, J., Rowell, D., and Nortley, K. S. F. (1999). On the predictability of the interannual behaviour of the Madden-Julian oscillation and its relationship with El Niño. *Quart. J. Roy. Meteor. Soc.*, **125**, 583–609.

Smith, L., Chassignet, E., and Bleck, R. (2000a). The impact of lateral boundary conditions and horizontal resolution on North Atlantic water transformations and pathways in an isopycnic coordinate ocean model. *J. Phys. Oceanogr.*, **30**, 137–159.

Smith, R. D., Maltrud, M. E., Bryan, F. O., and Hecht, M. W. (2000b). Numerical simulation of the North Atlantic Ocean at $\frac{1}{10}^{\circ}$. *J. Phys. Oceanogr.*, **30**, 1532–1561.

Spall, M. (1996a). Dynamics of the Gulf Stream-Deep Western Boundary Current Crossover. Part I: Entrainment and Recirculation. *J. Phys. Oceanogr.*, **26**, 21522168.

Spall, M. (1996b). Dynamics of the Gulf Stream-Deep Western Boundary Current Crossover. Part II: Low-Frequency Internal Oscillations. *J. Phys. Oceanogr.*, **26**, 21692182.

Spall, M. and Pickart, R. S. (2001). Where does dense water sink? A subpolar gyre example. *J. Phys. Oceanogr.*, **31**, 810–826.

Speich, S., Dijkstra, H. A., and Ghil, M. (1995). Successive bifurcations of a shallow-water model with applications to the wind driven circulation. *Nonlin. Proc. Geophys.*, **2**, 241–268.

Sperber, K. R., Hameed, W. L., Gates, W., and Potter, G. L. (1987). Southern oscillation simulated in a global climate model. *Nature*, **329**, 140–142.

Stammer, D. R., Tokmakian, R., Semtner, A., and Wunsch, C. (1994). How well does a $\frac{1}{4}^{\circ}$ global circulation model simulate large-scale oceanic observations? *J. Geophys. Res.*, **101**, 25,779–25,811.

Stern, M. E. (1975). Minimal properties of planetary eddies. *J. Mar. Res.*, **33**, 1–13.

Steward, W. J. and Jennings, A. (1981). A simultaneous iteration algorithm for real matrices. *ACM Trans. Math. Software*, **7**, 184–198.

Stocker, T. F. and Wright, D. G. (1992). A zonally averaged, ocean model for the thermohaline circulation. II: Interocean circulation in the Pacific - Atlantic Basin System. *J. Phys. Oceanogr.*, **21**, 1725–1739.

Stocker, T. F., Wright, D. G., and Mysak, L. A. (1992). A zonally averaged, coupled ocean-atmosphere model for paleoclimate studies. *J. Climate*, **5**, 773–797.

Stommel, H. (1948). The westward intensification of wind-driven ocean currents. *Trans. Amer. Geophysical Union*, **29**, 202–206.

Stommel, H. (1961). Thermohaline convection with two stable regimes of flow. *Tellus*, **2**, 244–230.

Stommel, H. (1965). *The Gulf Stream: a physical and dynamical description.* University of California Press, Berkeley and Los Angeles, California, USA.

Straughan, B. (2004). *The energy method, stability and nonlinear convection; second edition.* Springer-Verlag, New York.

Stricherz, J., Legler, D., and O'Brien, J. (1997). *TOGA Pseudo-stress Atlas 1985–1994, Volume II: Pacific Ocean.* Florida State University, Tallahassee, FL, USA.

Strogatz, S. H. (1994). *Nonlinear dynamics and chaos: With applications to physics, biology, chemistry, and engineering.* Reading, MA: Perseus Books.

Suarez, M. and Schopf, P. S. (1988). A delayed action oscillator for ENSO. *J. Atmos. Sci.*, **45**, 3283–3287.

Sura, P., Lunkeit, F., and Fraedrich, K. (2000). Decadal variability in a simplified wind-driven ocean model. *J. Phys. Oceanogr.*, **30**, 1917–1930.

Sutton, R. T. and Allen, M. (1997). Decadal predictability in North Atlantic sea-surface temperature and climate. *Nature*, **388**, 563–568.

Sverdrup, H. U. (1947). Wind-driven currents in a baroclinic ocean with application to the equatorial current in the eastern Pacific. *Proc. Natl. Acad. Sci. Wash.*, **33**, 318–326.

Sy, A., Rhein, M., Lazier, J. N. R., Koltermann, K. P., Meincke, J., Putzka, A., and Bersch, M. (1997). Surprisingly rapid spreading of newly formed intermediate waters across the North Atlantic Ocean. *Nature*, **388**, 563–567.

Syu, H.-H. and Neelin, J. D. (1995). ENSO in a hybrid coupled model: Part I: Sensitivity to physical parameterizations. *Clim. Dyn.*, **16**, 19–34.

Syu, H.-H., Neelin, J. D., and Gutzler, D. (1995). Seasonal and interannual variability in a hybrid coupled model. *J. Climate*, **8**, 2121–2143.

Taft, B. A. (1972). Characteristics of the flow of the Kuroshio south of Japan. In H. Stommel and K. Yoshida, editors, *Kuroshio, physical aspects of the Japan current.* Univ. Washington Press, Seattle, USA.

Takayabu, Y., Iguchi, T., Kachi, M., Shibata, A., and Kanzawa, H. (1999). Abrupt termination of the 1997-98 El Nino in response to a Madden-Julian oscillation. *Nature*, **402**, 279–282.

Tang, Y. (2002). Hybrid coupled models of the tropical Pacific. I: Interannual variability. *Clim. Dyn.*, **19**, 331–342.

Tansley, C. E. and Marshall, D. P. (2001). An Implicit Formula for Boundary Current Separation. *J. Phys. Oceanogr.*, **31**, 1633–1638.

Te Raa, L. A. and Dijkstra, H. A. (2002). Instability of the thermohaline ocean circulation on interdecadal time scales. *J. Phys. Oceanogr.*, **32**, 138–160.

Te Raa, L. A. and Dijkstra, H. A. (2003a). Modes of internal thermohaline variability in a single-hemispheric ocean basin. *J. Mar. Res.*, **61**, 491–516.

Te Raa, L. A. and Dijkstra, H. A. (2003b). Sensitivity of North Atlantic multidecadal variability to freshwater flux forcing. *J. Climate*, **32**, 138–160.

Te Raa, L. A., Gerrits, J., and Dijkstra, H. A. (2004). Identification of the mechanism of interdecadal variability in the North Atlantic Ocean. *J. Phys. Oceanogr.*, **34**, 2792–2807.

Terray, L., Thual, O., Belamari, S., Déqué, M., Dandin, P., Delecluse, P., and Levy, C. (1995). Climatology and interannual variability simulated by the ARPEGE-OPA coupled model. *Clim. Dyn.*, **12**, 487–505.

Thompson, C. and Battisti, D. (2000). A Linear Stochastic Dynamical Model of ENSO. Part I: Model Development. *J. Climate*, **13**, 2818–2832.

Thual, O. and McWilliams, J. C. (1992). The catastrophe structure of thermohaline convection in a two-dimensional fluid model and a comparison with low-order box models. *Geophys. Astrophys. Fluid Dyn.*, **64**, 67–95.

Timmermann, A. (2001). Changes of enso stability due to greenhouse warming. *Geophys. Res. Letters*, **28**, 2061–2064.

Timmermann, A., Oberhuber, J., Bacher, A., Esch, M., Latif, M., and Roeckner, E. (1999). Increased El Nino frequency in a climate model forced by future greenhouse warming. *Nature*, **398**, 694–697.

Timmermann, A., Gildor, H., Schulz, M., and Tziperman, E. (2003). Coherent resonant millennial-scale climate oscillations triggered by massive meltwater pulses. *J. Climate*, **16**, 2569–2585.

Toggweiler, J. R. and Samuels, B. (1995). Effect of Drake Passage on the global thermohaline circulation. *Deep-Sea Research*, **42**, 477–500.

Toggweiler, J. R. and Samuels, B. (1998). On the ocean's large scale circulation in the limit of no vertical mixing. *J. Phys. Oceanogr.*, **28**, 1832–1852.

Tomczak, M. and Godfrey, J. S. (1994). *Regional Oceanography: an introduction. Second Edition.* Daya Publishing House, Delhi, India.

Tourre, Y. M., Rajagopalan, B., and Kushnir, Y. (1999). Dominant patterns of climate variability in the Atlantic Ocean during the last 136 years. *J. Climate*, **12**, 22852299.

Trefethen, L. N., Trefethen, A. E., Reddy, S. C., and Driscoll, T. A. (1993). Hydrodynamic stability without eigenvalues. *Science*, **261**, 578–584.

Trenberth, K. E. (1993). *Climate System Modeling.* Cambridge University Press, Cambridge, U.K.

Trenberth, K. E., Olson, J. G., and Large, W. G. (1989). A global ocean wind stress climatology based on ECMWF analyses. Technical report, National Center for Atmospheric Research, Boulder, CO, U.S.A.

Tudhope, A. W., Chilcott, C. P., McCulloch, M. T., Cook, E. R., Chappell, J., Ellam, R. M., Lea, D. W., Lough, J. M., and Shimmield, G. (2001). Variability in ENSO through a glacial-interglacial cycle. *Science*, **291**, 1511–1517.

Tziperman, E. (1997). Inherently unstable climate behavior due to weak thermohaline ocean circulation. *Nature*, **386**, 592–595.

Tziperman, E. (2000). Proximity of the present-day thermohaline circulation to an instability threshold. *J. Phys. Oceanogr.*, **30**, 90–104.

Tziperman, E., Stone, L., Cane, M. A., and Jarosh, H. (1994a). El Nino chaos: overlapping of resonances between the seasonal cycle and the Pacific ocean-atmosphere oscillator. *Science*, **264**, 72–74.

Tziperman, E., Toggweiler, J. R., Feliks, Y., and Bryan, K. (1994b). Instability of the thermohaline circulation with respect to mixed boundary conditions: Is it really a problem for realistic models? *J. Phys. Oceanogr.*, **24**, 217–232.

Tziperman, E., Cane, M. A., and Zebiak, S. E. (1995). Irregularity and locking to the seasonal cycle in an ENSO prediction model as explained by the quasi-periodicity route to chaos. *J. Atmos. Sci.*, **52**, 293–306.

Tziperman, E., Zebiak, S. E., and Cane, M. A. (1997). Mechanisms of seasonal-ENSO interaction. *J. Atmos. Sci.*, **54**, 61–71.

Vallis, G. K. (2000). Large-scale circulation and production of stratification: effects of wind, geometry, and diffusion. *J. Phys. Oceanogr.*, **30**, 933–954.

Van der Ploeg, A. (1992). Preconditioning techniques for non-symmetric matrices with application to temperature calculations of cooled concrete. *Int. J. Num. Methods Eng.*, **35**, 1311–1328.

Van der Vaart, P. C. F. (1998). *Nonlinear Tropical Climate Dynamics*. Ph.D. thesis, Utrecht University, Netherlands.

Van der Vaart, P. C. F. and Dijkstra, H. A. (1997). Sideband instabilities of mixed barotropic/baroclinic waves growing on a midlatitude zonal jet. *Phys. Fluids*, **9**, 615–631.

Van der Vaart, P. C. F., Dijkstra, H. A., and Jin, F.-F. (2000). The Pacific Cold Tongue and the ENSO mode: Unified theory within the Zebiak-Cane model. *J. Atmos. Sci.*, **57**, 967–988.

Van Dorsselaer, J. J. (1997). Computing eigenvalues occurring in continuation methods with the Jacobi-Davidson QZ method. *J. Comp. Physics*, **138**, 714–733.

Van Oldenborgh, G. J. (2000). What caused the onset of the 1997/98 El Nino? *Monthly Weather Review*, **128**, 2601–2607.

Van Oldenborgh, G. J., Burgers, G., Venzke, S., Eckert, C., and Giering, R. (1999). Tracking down the ENSO delayed oscillator with an adjoint OGCM. *Monthly Weather Review*, **127**, 1477–1495.

Vautard, R., Yiou, P., and Ghil, M. (1992). Singular spectrum analysis: A toolkit for short, noisy chaotic signals. *Physica D*, **58**, 95–126.

Vazquez, J., Zlotnicki, V., and Fu, L.-L. (1990). Sea level variabilities in the Gulf Stream between Cape Hatteras and 50°W: A Geosat study. *J. Geophys. Res.*, **95(C10)**, 17957–17964.

Vecchi, G. A. and Harrison, D. E. (2000). Tropical Pacific sea surface temperature anomalies, El Nino, and equatorial westerly wind events. *J. Climate*, **13**, 1814–1830.

Vellinga, M. (1996). Instability of two-dimensional thermohaline circulation. *J. Phys. Oceanogr.*, **26**, 305–319.

Vellinga, M. (1998). Multiple equilibria of the thermohaline circulation as a side effect of convective adjustment. *J. Phys. Oceanogr.*, **28**, 305–319.

Vellinga, M., Wood, R. A., and Gregory, J. M. (2002). Processes governing the recovery of a perturbed thermohaline circulation in HadCM3. *J. Climate*, **15**, 764–780.

Veronis, G. (1966). Wind-driven ocean circulation. II: Numerical solution of the nonlinear problem. *Deep-Sea Research*, **13**, 31–55.

Von Storch, H. (1995). Spatial patterns: EOFs and CCA. In H. Von Storch and A. Navarra, editors, *Analysis of Climate Variability*, pages 227–257. Springer, New York, U.S.A.

Von Storch, H., Buerger, G., Schnur, R., and Von Storch, J.-S. (1995). Principal oscillation patterns: a review. *J. Climate*, **8**, 377–400.

Wakata, Y. and Sarachik, E. (1994). Nonlinear effects in coupled atmosphere-ocean basin modes. *J. Atmos. Sci.*, **51**, 909–920.

Walin, G. (1985). The thermohaline circulation and the control of ice ages. *Paleogeogr. Paleoclim. Paleoecol.*, **50**, 323–332.

Wallace, J., Mitchell, T., Rasmusson, E., Kousky, V., Sarachik, E., and von Storch, H. (1998). On the structure and evolution of ENSO-related climate variability in the Tropical Pacific: lessons from TOGA. *J. Geophys. Res.*, **103**, 14,241–14,260.

Wang, L. and Koblinsky, C. (1995). Low-frequency variability in regions of the Kuroshio Extension and the Gulf Stream. *J. Geophys. Res.*, **100**, 18,313–18,331.

Wang, L. and Koblinsky, C. (1996). Annual variability of the subtropical recirculations in the North Atlantic and North Pacific: a Topex/Poseidon study. *J. Phys. Oceanogr.*, **26**, 2462–2479.

Wang, L., Koblinsky, C. J., and Howden, S. (1998). Annual and intra-annual sea level variability in the region of the Kuroshio Extension from TOPEX/POSEIDON and Geosat altimetry. *J. Phys. Oceanogr.*, **28**, 692–711.

Weaver, A. J. and Hughes, T. M. (1994). Rapid interglacial climate fluctuations driven by North Atlantic ocean circulation. *Nature*, **367**, 447–450.

Weaver, A. J. and Hughes, T. M. (1996). On the incompatibility of ocean and atmosphere and the need for flux adjustments. *Clim. Dyn.*, **12**, 141–170.

Weaver, A. J. and Sarachik, E. S. (1991). The role of mixed boundary conditions in numerical models of the ocean's climate. *J. Phys. Oceanogr.*, **21**, 1470–1493.

Weaver, A. J., Marotzke, J., Cummings, P. F., and Sarachik, E. S. (1993). Stability and variability of the thermohaline circulation. *J. Phys. Oceanogr.*, **23**, 39–60.

Weaver, A. J., Aura, S., and Myers, P. G. (1994). Interdecadal variability in an idealized model of the North Atlantic. *J. Geophys. Res.*, **99**, 12,423 – 12,441.

Webster, P. and coauthors (1996). Monsoons: Processes, predictabilty, and the prospects for predicting. *J. Geophys. Res.*, **103**, 14,451–14,510.

Weijer, W. (2000). *Impact of interocean exchange on the Atlantic overturning circulation.* Ph.D. thesis, I.M.A.U., Utrecht University, Utrecht, The Netherlands.

Weijer, W. and Dijkstra, H. A. (2001). Bifurcations of the three-dimensional thermohaline circulation: The double hemispheric case. *J. Mar. Res.*, **59**, 599–631.

Weijer, W. and Dijkstra, H. A. (2003). Multiple oscillatory modes of the global ocean circulation. *J. Phys. Oceanogr.*, **33**, 2197–2213.

Weijer, W., De Ruijter, W. P. M., Dijkstra, H. A., and Van Leeuwen, P. J. (1999). Impact of interbasin exchange on the Atlantic overturning circulation. *J. Phys. Oceanogr.*, **29**, 2266–2284.

Weijer, W., De Ruijter, W., and Dijkstra, H. A. (2001). Stability of the Atlantic overturning circulation: Competition between Bering Strait freshwater flux and Agulhas heat and salt sources. *J. Phys. Oceanogr.*, **31**, 2385–2402.

Weijer, W., Dijkstra, H. A., Oksuzoglu, H., Wubs, F. W., and De Niet, A. C. (2003). A fully-implicit model of the global ocean circulation. *J. Comp. Phys.*, **192**, 452–470.

Welander, P. (1982). A simple heat-salt oscillator. *Dyn. Atmos. Oceans*, **6**, 233–242.

Welander, P. (1986). Thermohaline effects in the ocean circulation and related simple models. In J. Willebrand and D. L. T. Anderson, editors, *Large-Scale Transport Processes in Oceans and Atmosphere*, pages 163–200. D. Reidel.

Weller, R. and Anderson, S. (1996). Surface meteorology and air-sea fluxes in the western equatorial Pacific warm pool during the TOGA Coupled Ocean-Atmosphere Response Experiment. *J. Climate*, **9**, 1959–1990.

Wiggins, S. (1990). *Introduction to Applied Nonlinear Dynamical Systems and Chaos.* Springer-Verlag, Heidelberg-Berlin, Germany.

Wijffels, S. E., Schmitt, R. W., Bryden, H. L., and Stigebrandt, A. (1992). Transport of fresh water by the ocean. *J. Phys. Oceanogr.*, **22**, 155–163.

Winton, M. (1996). The role of horizontal boundaries in parameter sensitivity and decadal-scale variability of coarse-resolution ocean general circulation models. *J. Phys. Oceanogr.*, **26**, 289–304.

Winton, M. (1997). The damping effect of bottom topography on internal decadal-scale oscillations of the thermohaline circulation. *J. Phys. Oceanogr.*, **27**, 203–208.

Winton, M. and Sarachik, E. S. (1993). Thermohaline oscillations induced by strong steady salinity forcing of ocean general circulation models. *J. Phys. Oceanogr.*, **23**, 1389–1410.

WOCE (2001). *Ocean Circulation and Climate: Observing and Modeling the Global Ocean [Siedler, G. and Church, J. and Gould, J. (eds)].* Academic Press, San Diego, USA.

Woodroffe, C. D., Beech, M., and Gagan, M. K. (2003). Mid-late Holocene El Nino variability in the equatorial Pacific from coral microatolls. *Geophys. Res. Letters*, **30**, doi:10.1029/2002GL015868.

Worthington, L. V. (1976). *On the North Atlantic Circulation.* The Hopkins University Press, Baltimore, U.S.A.

Wright, D. G. and Stocker, T. F. (1991). A zonally averaged model for the thermohaline circulation, Part I: Model development and flow dynamics. *J. Phys. Ocean.*, **21**, 1713–1724.

Wright, D. G. and Stocker, T. F. (1992). Sensitivities of a a zonally averaged global ocean circulation model. *J. Geophys. Res.*, **97**, 12707–12730.

Wright, D. G., Vreugdenhil, C. B., and Hughes, T. M. (1995). Vorticity dynamics and zonally averaged ocean models. *J. Phys. Oceanogr.*, **25**, 2141–2154.

Wright, D. G., Stocker, T., and Mercer, D. (1998). Closures used in zonally averaged ocean models. *J. Phys. Oceanogr.*, **28**, 791–804.

Wu, D.-H., Anderson, D., and Davey, M. (1993). ENSO variability and external impacts. *J. Climate*, **6**, 1703–1717.

Wunsch, C. (1996). *The Ocean Circulation Inverse Problem*. Cambridge University Press, Cambridge, U.K.

Wunsch, C. (1998). The work done by the wind on the oceanic general circulation. *J. Phys. Oceanogr.*, **28**, 2332–2340.

Wunsch, C. (2002). What is the thermohaline circulation? *Science*, **298**, 1179–1180.

Wunsch, C. and Ferrari, R. (2004). Vertical mixing, energy and the general circulation of the oceans. *Annual Review of Fluid Mechanics*, **36**, 281–314.

Wunsch, C., Hu, D., and Grant, G. (1983). Mass, heat and nutrients in the South Pacific Ocean. *J. Phys. Oceanogr.*, **13**, 725–753.

Xie, S.-P. (1994). On the genesis of the equatorial annual cycle . *J. Climate*, **7**, 2008–2013.

Xie, S.-P. and Philander, S. G. H. (1994). A coupled ocean-atmosphere model of relevance to the ITCZ in the eastern Pacific. *Tellus*, **46A**, 340–350.

Yu, X. and McPhaden, M. J. (1999). Seasonal variability in the Equatorial Pacific. *J. Phys. Oceanogr.*, **29**, 925–947.

Zaucker, F., Stocker, T. F., and Broecker, W. S. (1994). Atmospheric freshwater fluxes and their effect on the global thermohaline circulation. *J. Geophys. Res.*, **99**, 12,443–12,457.

Zebiak, S. E. (1982). A simple atmospheric model of relevance to El Niño. *J. Atmos. Sci.*, **39**, 2017–2027.

Zebiak, S. E. and Cane, M. A. (1987). A model El Niño-Southern Oscillation. *Monthly Weather Review*, **115**, 2262–2278.

Zhang, J., Schmitt, R. W., and Huang, R. X. (1999). The relative influence of diapycnal and hydrologic forcing on the stability of the thermohaline circulation. *J. Phys. Oceanogr.*, **29**, 1096–1108.

Zhang, R.-H., Rothstein, L. M., and Busalacchi, A. (1998). Origin of upper-ocean warming and El Niño change on decadal scales in the tropical Pacific Ocean. *Science*, **391**, 879–882.

Zhang, Y., Wallace, J. M., and Battisti, D. S. (1997). ENSO-like interdecadal variability: 1900-1993. *J. Climate*, **10**, 1004–1020.

Copyright Acknowledgements

Albeniz; Title: Asturias; Copyright and Publisher: G. Ricordi and Co., München, 1969; Permission by Albersen Muziek, the Hague, The Netherlands. Fig. 6.1, by Elsevier Science Publishers, (Dickson *et al.*, 1988). ♭ Fig. 6.2 by EGU (Delworth and Mann, 2000). ♯ Fig. 6.8 by Elsevier Science Publishers (Welander, 1982). ♯ Fig. 6.9 by AMS (Tziperman *et al.*, 1994b). ♭ Fig. 6.19 by Gordon and Breach Publishing Group (Thual and McWilliams, 1992). ♯ Fig. 6.20a by Cambridge University Press (Quon and Ghil, 1992). ♭ Fig. 6.29 by AMS (Wright *et al.*, 1998). ♯ Fig. 6.30 by AMS (Vellinga, 1998). ♯ Fig. 6.40-6.44 by AMS (Weaver *et al.*, 1993). Fig. 6.49 by Nature and F. O. Bryan (Bryan, 1986), copyright (1986) Macmillan Magazines Ltd. ♭ Fig. 6.52 by AMS (Marotzke and Willebrand, 1991). ♯ Fig. 6.53 by AMS (Hughes and Weaver, 1994). ♭ Fig. 6.54 by Nature and S. Rahmstorf (Rahmstorf, 1994) copyright (1994) Macmillan Magazines Ltd. ♯ Fig. 6.55 by Nature and S. Rahmstorf (Rahmstorf, 1995b) copyright (1995) Macmillan Magazines Ltd. ♭ Fig. 6.56 - 6.57 by EGU (Mikolajewicz and Maier-Reimer, 1990). ♯ Fig. 6.58 by AMS (Manabe and Stouffer, 1988). ♭ Fig. 6.59 - 6.60 by AMS (Delworth *et al.*, 1993). ■ Chapter 7. Opening; Composer: E. Granados; Title: Danza Española nr. 5; Copyright and Publisher: Union Musical Espagnola, Madrid, 1926; Permission by Albersen Muziek, the Hague, The Netherlands. ♭ Fig. 7.1 - 7.2 by AMS (Ropelewski and Halpert, 1987). ♯ Fig. 7.8, by SCIENCE (Fedorov and Philander, 2000). ♭ Fig. 7.15 by AGU (Neelin *et al.*, 1998). ♯ Fig. 7.19 - 7.20 by AMS (Zebiak and Cane, 1987). ♯ Fig. 7.21 - 7.22 by AMS (Hirst, 1986). ♭ Fig. 7.28 - 7.33 by AMS (Neelin and Jin, 1993; Jin and Neelin, 1993a). ♯ Fig. 7.34 by AMS (Suarez and Schopf, 1988). ♭ Fig. 7.35 by AMS (Jin, 1997a). ♯ Fig. 7.50 by Science and by E. Tziperman (Tziperman *et al.*, 1994a). ♭ Fig. 7.51 - 7.54 by Elsevier Science Publishers and Fig. 7.54 courtesy of David Neelin (Jin *et al.*, 1996). ♯ Fig. 7.57 - 7.61 by EGU (Latif *et al.*, 2001). ■

Index

β-plane approximation, 183

Antarctic Circumpolar Current, 14
Arnold' tongue, 105, 457

baroclinic instability, 198
barotropic instability, 197
barotropic potential vorticity equation, 186
bifurcation
    Bogdanov-Takens, 89
    co-dimension of, 81
    cusp, 89
    cyclic fold, 96
    diagram, 68
    general, 79
    homoclinic, 100
        Shilnikov, 102
    Hopf, 84
        physics of, 113
    imperfection, 86
    limit point, 82
    Naimark-Sacker, 99
    normal form, 82
    period doubling, 97
    pitchfork, 83
        physics of, 112
    reduced equation of, 81
    saddle node, 82
        physics of, 110
    symmetry, 107
    transcritical, 82
        physics of, 111
    turning point, 82
    unfolding, 88
Biot number, 126
boundary layer
    Ekman, 182
    inertial, 203
    Munk, 203
    Stommel, 203
Boussinesq approximation, 39
box model, 64, 65

cellular convection, 122

center manifold, 118
circle map, 103
climate cycles
    Dansgaard-Oeschger, 12
    Milankovitch, 10
climate data
    AVHRR, 171
    ECMWF, 174
    ETOPO5, 176
    GEOSAT, 174
    TOPEX, 174, 256
climate drift
    equatorial, 465
    thermohaline circulation, 351
climate episode
    Eemian, 3
    Holocene, 4
    Last Glacial Maximum, 6
    Little Ice Age, 7
    Younger Dryas, 5
climate index
    AMO-index, 26
    CET, 24
    NINO3, 22, 375
    SOI, 22, 375
    THC, 354
cold wall, 172
complex growth factor, 70
continuation
    bifurcation detection, 134
    branch switching, 136
    method, 130, 158
convective adjustment
    classical, 317
    implicit mixing, 317
convective feedback, 275

Dansgaard-Oeschger cycles, 12
deep decoupling oscillations, 336
delayed oscillator
    BHSS, 434
    coupled wave, 436
    general, 411

recharge, 437
Devil's staircase, 106
Devil's terrace, 460
discretization method, 127
dynamical system, 75

eccentricity, 10
Eemian, 3
eigenvalue solvers
    general, 140
    Jacobi-Davidson QZ-method, 144
    simultaneous iteration method, 141
Ekman number, 179
ENSO-models
    ICM, 370
ENSO-modes
    eastward propagating SST, 424
    mixed SST/ocean dynamics modes, 428
    scatter modes, 427
    stationary SST, 424
    unstable Kelvin wave, 414
    westward propagating SST, 424
equation of state, 39
equatorial heat content, 376
equatorial mixed layer, 399
equatorial ocean 'memory', 399
equatorial radius of deformation, 395
equatorial strip approximation, 415
equatorial thermocline, 372
equatorial thermocline feedback, 403
equatorial upwelling feedback, 404
equatorial warm pool/cold tongue, 371
equatorial wave
    inertia-gravity, 384
    Kelvin, 383
    Rossby, 383
    trapping, 383
    travel time, 387
    Yanai, 384

fast SST limit, 416
fast wave limit, 416, 421
feedback
    ENSO
        thermocline, 403
        upwelling, 404
        zonal advection, 404
    thermohaline circulation
        convective, 275
        salt advection, 274
    wind-driven circulation
        horizontal shear, 197, 201, 205
        vertical shear, 197, 201
fixed point
    general, 75
    hyperbolic, 78
Floquet
    exponents, 96
    multipliers, 95, 99
flushes, 336
flux-correction

equatorial, 439
    procedures, 351
frequency locking, 103, 457
freshwater flux
    surface, 14
Froude number, 179
fundamental solution, 94

geopotential, 38
Great Salinity Anomaly, 269
Gulf Stream, 15, 170
    cold wall, 172, 250
    lateral shifts, 174
    meanders, 171
    recirculation regions, 171
    rings, 173
    separation
        'deflected' pattern, 237
        'separated' pattern, 237
        imperfect pitchfork, 238, 239, 244
    separation point, 171
gyre
    general, 170
    subpolar, 15
    subtropical, 15

heat flux
    surface, 13
Hermite polynomials, 383
high resolution ocean models
    general, 245
    MOM, 246
    POCM, 249
    POP, 246
Holocene, 4
homoclinic orbit, 101
hysteresis, 108

imperfection
    pitchfork, 86
    thermohaline flows
        air-sea interaction, 305
        freshwater flux, 306
        lateral fluxes, 303
    transcritical, 86
incommensurate, 99
instability mechanism
    baroclinic, 198, 201
    barotropic, 197, 201
integral balance
    heat, 46
    mechanical energy, 47, 200
    salt, 46
inverse iteration, 137
isolated branch, 74
isotope ratio
    $\delta^{18}O$, 2

Jacobian matrix, 70, 159

Krylov subspace, 153
Kuroshio, 15

transport, 174
variability, 175

Last Glacial Maximum, 6
linear system solvers
    direct
        frontal methods, 150
        Gaussian elimination, 148
    iterative methods
        BICGSTAB, 155
        Gauss-Seidel, 152
        general, 151
        GMRES, 154
        ILU, 158
        Jacobi, 152
        MRILU, 158
        preconditioning, 157
        projection method, 153
        Successive OverRelaxation, 152
Little Ice Age, 7
loop oscillation, 281, 335
Lorenz equations, 102
low resolution ocean models
    FGM, 327
    LSG, 349
    MOM, 328

matrix pensil, 141
Maunder minimum, 10
Milankovitch cycles, 10
mixed boundary conditions, 274
mixing
    biharmonic, 43
    coefficients, 42
    isopycnal, 43
    Laplacian, 43
monodromy matrix, 95
most dangerous modes, 141
multidecadal mode, 324

neutral curve, 199
Newton-Raphson method, 132
noise, 27
numerical bifurcation software
    ODEsoftware
        AUTO, 120
        CONTENT, 120
        DSTOOL, 120
        MATCONT, 120
    PDEsoftware
        BOOM, 120
        LOCA, 120
        PDECONT, 120
numerical method
    backward Euler, 147
    BICGSTAB, 155
    Crank-Nicholson, 147
    Euler-Newton, 132
    Gaussian elimination, 148
    GMRES, 154
    ILU decomposition, 158

inverse iteration, 137
Jacobi-Davidson QZ, 144
MRILU decomposition, 158
Newton-Raphson, 132
QZ, 141
simultaneous iteration, 142
Nusselt number, 123

obliquity, 9
ocean circulation
    'Ocean Conveyor', 17
    freshwater transport, 18
    heat transport, 18
    thermohaline-driven, 16
    wind-driven, 14
ocean model
    ENSO
        Gill atmosphere model, 393
        reduced gravity ocean model, 380
        two-strip, 431
        Zebiak-Cane, 405
    thermohaline
        Bousinesq model, 282
        zonally averaged model, 316
    wind-driven
        homogeneous, 176
        shallow water model, 183
orbital parameters
    eccentricity, 10
    obliquity, 9
    precession, 9
oscillatory modes
    thermohaline circulation
        deep decoupling, 337
        flushes, 337
        loop, 336
        overturning, 298
    wind-driven ocean circulation
        basin mode, 206
        gyre mode, 207

parabolic cylinder functions, 394
phase portrait, 78
Poincaré map, 92
Poincaré section, 91
polar halocline catastrophe, 336
Prandtl number, 126
precession, 9
proxy
    foraminifera, 2
    ice core, 2
    tree ring, 7

quasi-geostrophic approximation, 189

Rayleigh number, 126
Rayleigh-Bénard convection, 122
regime diagram, 68
Rossby
    basin modes, 196
    equatorial radius of deformation, 395
    external radius of deformation, 186

internal radius of deformation, 204
    number, 179, 185
    waves, 194
Rossby number, 178, 179
rotation number, 104

salt advection feedback, 274
shallow water approximation, 182
spectral radius, 152
stability
    asymptotic, 54
    boundary, 55
    complex growth factor, 70
    linear stability, 70, 76, 158
    neutral curve, 199
state space, 75
Stokes theorem, 47
Stommel-Sverdrup-Munk theory, 190
stress tensor, 43
Sunspots, 10
symmetry breaking
    wind-driven ocean circulation, 203, 204

tensor product
    diadic, 43
    direct, 46
thermal wind balance, 198
thermohaline flow regimes
    convective, 298
    diffusive, 285
thermohaline flow states
    ATH, 308
    NPP, 74, 286
    NPP/TH, 308
    SA, 74, 286
    SPP, 74, 286
    SPP/TH, 308

TH, 74, 286
time integration
    explicit, 147
    implicit, 147
trade winds, 371
trajectory, 66, 76, 333

vector field, 75
vortex
    line, 47
    stretching, 49
    tilting, 50
    tube, 47
vorticity
    equation, 48
    potential, 51, 198
vorticity production
    baroclinic, 50, 188
    friction, 188

watermass
    Antarctic Bottom Water, 17
    North Atlantic Deep Water, 17
weak-coupling limit, 424
western boundary current
    Agulhas, 15
    Brazil, 15
    East Australian, 15
    Falkland, 15
    Gulf Stream, 16
    Kuroshio, 16

Younger Dryas, 5

Zebiak-Cane model, 405
zonal advection feedback, 404

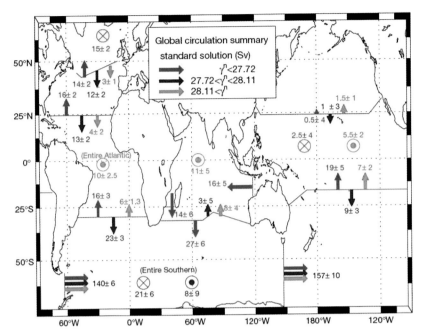

Figure 1.14. Estimated section integrated mass transports as determined in Ganachaud and Wunsch (2000) from the WOCE data. See text for an explanation of the colors and symbols.

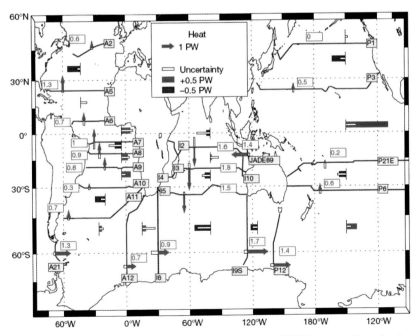

Figure 1.15. Estimated section averaged heat transport over WOCE sections (Ganachaud and Wunsch, 2000); the latter are indicated by their number (such as P12, A8, etc.), and 1 PW = 10^{15} W.

518

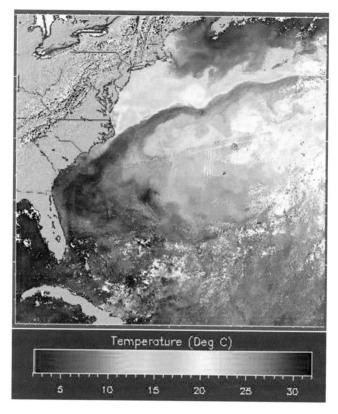

Figure 5.2. Multipass image of the SST field of the Gulf Stream region as determined by the Advanced Very High Resolution Radiometer (AVHRR) in May 1996 (obtained from http://fermi.jhuapl.edu/avhrr/gallery/sst/stream.html).

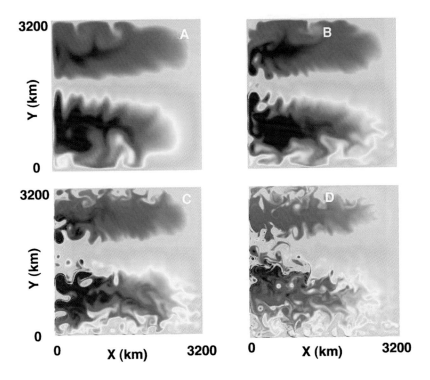

Figure 5.36. Upper layer streamfunction snapshots of the ocean in a 3200 km square basin for varying Reynolds numbers, Re, with $Re = 0.375, 1.5, 6.0, 24$ for the panels (A)-(D), respectively. Here, $Re = UL/A_H$, with $U = 10^{-3}$ ms^{-1} and $L = 3200$ km. The time-mean flow consists of an anticyclonic midlatitude subtropical gyre and a cyclonic subpolar gyre. The resolution in the computations increases from 25 km in (A) to 1.56 km in (D). Note the appearance of coherent vortices throughout the circulation in the highest value of Re results (from Siegel et al. (2001)).

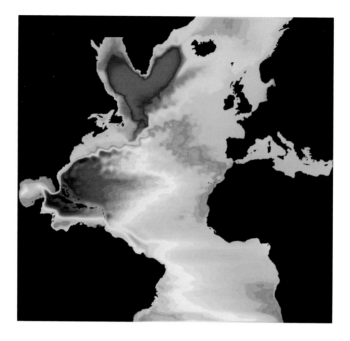

Figure 5.54.   Plot of the two-year mean of the sea-surface height of a high-resolution (1/12°) simulation with MICOM (Chassignet and Garaffo, 2001).

(a)

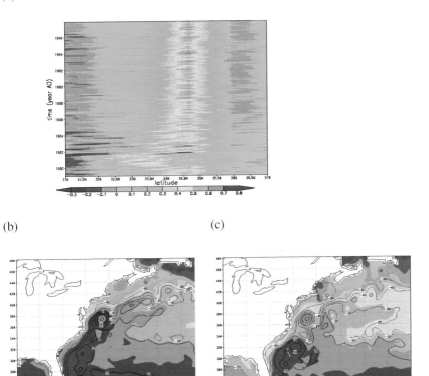

(b)   (c)

Figure 5.55. (a) *Contour plot of zonal geostrophic velocity* (ms^{-1}), *calculated from SSH gradients in POCM, at* $75°$ W *as a function of latitude for the period 1979-1997. Contour plot of monthly mean SSH deviations (cm), superimposed on a shaded contour plot of monthly mean SST for (b) January 1979, representing the 'deflected' Gulf Stream in POCM, characterized by a northerly position of its cold wall at* $75°W$ *and (c) January 1981, representing the 'separated' Gulf Stream in POCM, characterized by a southerly position of its cold wall at* $75°W.$

522

(a)

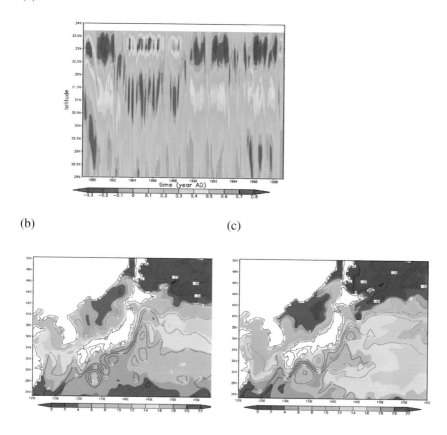

(b)   (c)

Figure 5.56. (a) Contour plot of zonal geostrophic velocity (ms^{-1}), calculated from sea surface height (SSH) gradients in POCM, at 136° E as a function of latitude for the period 1979-1998. (b-c) Contour plot of monthly mean SSH deviations (cm), superimposed on a shaded contour plot of monthly mean SST for (b) January 1988, representing the Kuroshio small meander state in POCM and (c) January 1996, representing the Kuroshio large meander state in POCM.

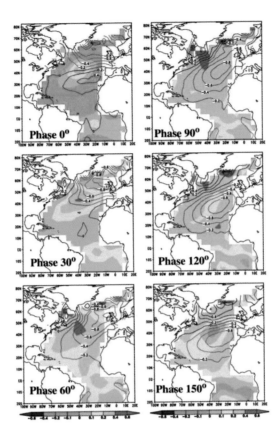

Figure 6.2. Reconstruction of the approximately 52-year signal (panels are about 4.3 years apart) in SST and SLP from Delworth and Greatbatch (2000). Units of SST are in °C (from -0.6°C (blue) to 0.6°C (red)) and that of SLP in hPa.

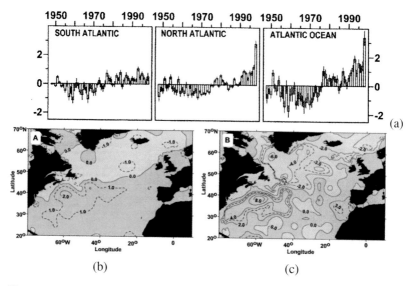

(a)

(b)   (c)

Figure 6.3. (a) *Time series of ocean heat content* $(10^{22} J)$ *in the upper 300 m of the Atlantic for the half-century 1948–1998. For comparison, the climatological range of upper ocean heat content for the North Atlantic is about* 5.6×10^{22} *J. (b-c) Heat storage difference* (Wm^{-2}) *for the North Atlantic between 1988–1992 and 1970–1974 within (b) the upper 300 m and (c) the upper 3000 m; warming is indicated by pink and cooling in light blue (from Levitus* et al. *(2000)).*

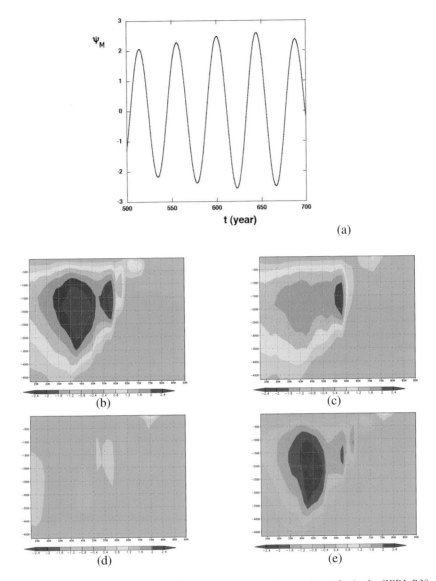

Figure 6.59. Properties of the statistical mode, having a 44-year timescale, in the GFDL-R30 model. (a) Maximum of the meridional overturning streamfunction anomaly (Sv). (b-e) Patterns of meridional overturning streamfunction anomaly (Sv) in the North Atlantic region. The patterns are shown at a 6-yearly interval, starting in model year 600, over about one half-cycle of the oscillation; the other half-cycle is similar but with anomalies of reversed sign.

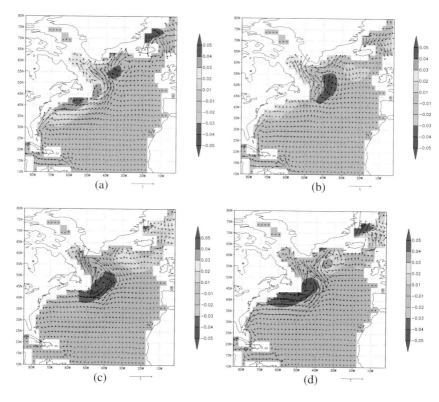

Figure 6.60.   Properties of the statistical mode, having a 44-year timescale, in the GFDL-R30 model. Potential density (kg m^{-3}) and horizontal velocity (cm s^{-1}) anomalies at 680 m depth in the North Atlantic region. The patterns are shown at a 6-yearly interval, starting in model year 600, over about one half-cycle of the oscillation; the other half-cycle is similar but with anomalies of reversed sign.

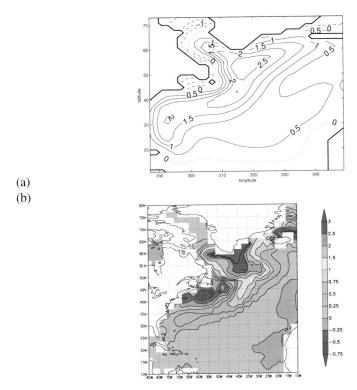

(a)
(b)

Figure 6.62. (a) *Difference in SST fields between maximum and minimum strength of the MOC in the MOM3.1 model. (b) As in (a), but for the statistical 44-year mode in the GFDL-R30 model.*

528

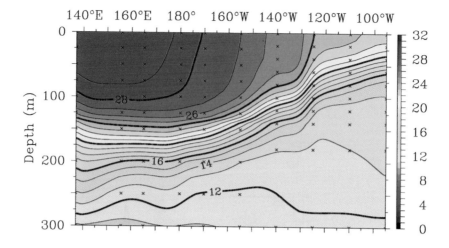

Figure 7.3. Depth-longitude section of the near-equatorial temperature in the Pacific monthly averaged over November 1996. This situation is close to annual-mean conditions for the Tropical Pacific Kessler and McCreary (1995). The crosses in the figure indicate the measurement positions of the TAO-buoys.

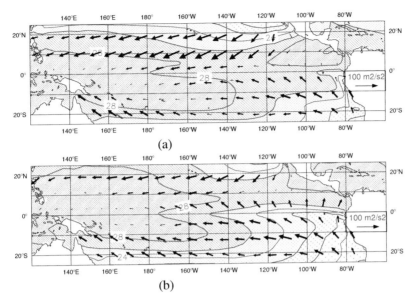

Figure 7.4. SST and wind-stress climatologies for (a) April and (b) October. The contours give the 1961–1990 SST climatology (contour interval 2°C) from NCEP (National Centers for Environmental Prediction), the arrows the 1961-1992 pseudo-wind stress climatology (in $m^2 s^{-2}$) from FSU (Florida State University, (Stricherz et al., 1997). Pseudo-wind stress has the direction of the surface wind and the magnitude of the wind speed squared.

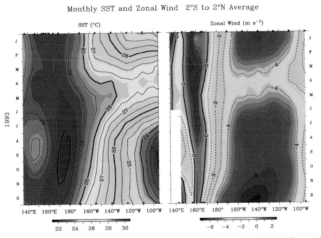

Figure 7.5. Seasonal cycle in 1993 of the monthly averaged equatorial SST (left panel) and zonal wind (right panel). The figure is plotted through the graphics software and data made available through the TAO realtime data-access site at http://www.pmel.noaa.gov/toga-tao/realtime.html. Note that time is downwards.

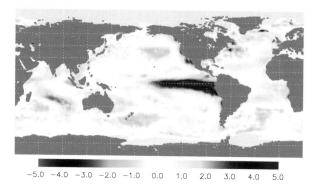

Sea-surface temperature anomaly field of December 1997, at the height of the 1997/1998 El Niño. Data from NCEP.

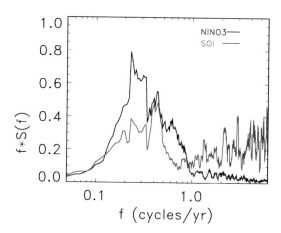

Figure 7.6. Spectra of monthly mean SOI and NINO3 index. Shown are normalized periodograms smoothed over 11 bins, that is over 0.11 cycles per year. Note that x-axis is logarithmic, and $f*S(f)$ rather than $S(f)$ is shown, in order that equal areas make equal contributions to the variance.

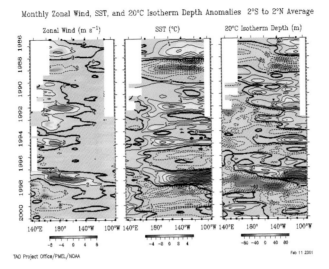

Figure 7.7. *Time-longitude diagrams of equatorial anomalies of zonal wind (left panel), SST (middle panel) and in the right panel the depth of the 20° isotherm (a measure for the thermocline depth). Data are for the period 1986-2000, and measured by the TAO/TRITON array. The plot is made through data and software at http://www.pmel.noaa.gov/tao*

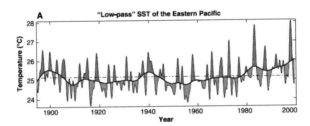

Figure 7.8. *Plot of the SST averaged over the box [120°W, 80°W] × [5°S - 5°N] in the Eastern Pacific on a background of decadal variability, the latter obtained through a low-pass filter (from Fedorov and Philander (2000)).*

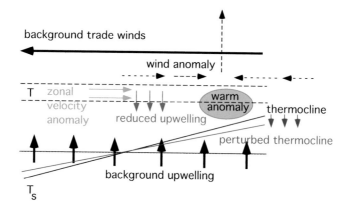

Figure 7.17. Sketch to illustrate the thermocline feedback, the upwelling feedback and the zonal advection feedback as indicated. In each case, a warm SST anomaly induces wind-stress anomalies. (i) Thermocline feedback: the wind anomaly leads to changes in the thermocline slope, which in turn induces — with constant background upwelling — an amplification of the SST anomaly. (ii) Upwelling feedback: the wind anomaly leads to changes in the upwelling which in turn induces — in a background stably stratified temperature field — an amplification of the SST anomaly. (iii) Zonal advection feedback: the wind anomaly induces stronger zonal advection which, if the annual-mean zonal SST gradient is negative, leads to amplification of the SST anomaly.

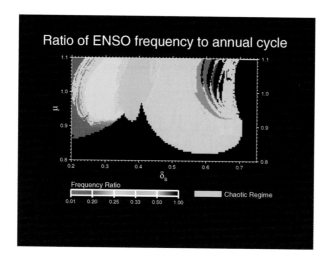

Figure 7.54. The Devil's terrace as computed in Jin et al. *(1996). Colors indicate the frequency locked regimes, with chaotic regimes in between. Along a section for fixed μ, a Devil's staircase, as discussed in section 3.4 appears.*

ATMOSPHERIC AND OCEANOGRAPHIC SCIENCES LIBRARY

1. F.T.M. Nieuwstadt and H. van Dop (eds.): *Atmospheric Turbulence and Air Pollution Modelling.* 1982; rev. ed. 1984
    ISBN 90-277-1365-6; Pb (1984) 90-277-1807-5
2. L.T. Matveev: *Cloud Dynamics.* Translated from Russian. 1984
    ISBN 90-277-1737-0
3. H. Flohn and R. Fantechi (eds.): *The Climate of Europe: Past, Present and Future.* Natural and Man-Induced Climate Changes: A European Perspective. 1984
    ISBN 90-277-1745-1
4. V.E. Zuev, A.A. Zemlyanov, Yu.D. Kopytin, and A.V. Kuzikovskii: *High-Power Laser Radiation in Atmospheric Aerosols.* Nonlinear Optics of Aerodispersed Media. Translated from Russian. 1985
    ISBN 90-277-1736-2
5. G. Brasseur and S. Solomon: *Aeronomy of the Middle Atmosphere.* Chemistry and Physics of the Stratosphere and Mesosphere. 1984; rev. ed. 1986
    ISBN (1986) 90-277-2343-5; Pb 90-277-2344-3
6. E.M. Feigelson (ed.): *Radiation in a Cloudy Atmosphere.* Translated from Russian. 1984
    ISBN 90-277-1803-2
7. A.S. Monin: *An Introduction to the Theory of Climate.* Translated from Russian. 1986
    ISBN 90-277-1935-7
8. S. Hastenrath: *Climate Dynamics of the Tropics*, Updated Edition from *Climate and Circulation of the Tropics.* 1985; rev. ed. 1991
    ISBN 0-7923-1213-9; Pb 0-7923-1346-1
9. M.I. Budyko: *The Evolution of the Biosphere.* Translated from Russian.
    1986
    ISBN 90-277-2140-8
10. R.S. Bortkovskii: *Air-Sea Exchange of Heat and Moisture During Storms.* Translated from Russian, rev. ed. 1987
    ISBN 90-277-2346-X
11. V.E. Zuev and V.S. Komarov: *Statistical Models of the Temperature and Gaseous Components of the Atmosphere.* Translated from Russian. 1987
    ISBN 90-277-2466-0
12. H. Volland: *Atmospheric Tidal and Planetary Waves.* 1988   ISBN 90-277-2630-2
13. R.B. Stull: *An Introduction to Boundary Layer Meteorology.* 1988
    ISBN 90-277-2768-6; Pb 90-277-2769-4
14. M.E. Berlyand: *Prediction and Regulation of Air Pollution.* Translated from Russian, rev. ed. 1991
    ISBN 0-7923-1000-4
15. F. Baer, N.L. Canfield and J.M. Mitchell (eds.): *Climate in Human Perspective.* A tribute to Helmut E. Landsberg (1906-1985). 1991
    ISBN 0-7923-1072-1
16. Ding Yihui: *Monsoons over China.* 1994   ISBN 0-7923-1757-2
17. A. Henderson-Sellers and A.-M. Hansen: *Climate Change Atlas.* Greenhouse Simulations from the Model Evaluation Consortium for Climate Assessment. 1995
    ISBN 0-7923-3465-5
18. H.R. Pruppacher and J.D. Klett: *Microphysics of Clouds and Precipitation*, 2nd rev. ed.
    1997
    ISBN 0-7923-4211-9; Pb 0-7923-4409-X
19. R.L. Kagan: *Averaging of Meteorological Fields.* 1997   ISBN 0-7923-4801-X
20. G.L. Geernaert (ed.): *Air-Sea Exchange: Physics, Chemistry and Dynamics.* 1999
    ISBN 0-7923-5937-2
21. G.L. Hammer, N. Nicholls and C. Mitchell (eds.): *Applications of Seasonal Climate Forecasting in Agricultural and Natural Ecosystems.* 2000   ISBN 0-7923-6270-5

ATMOSPHERIC AND OCEANOGRAPHIC SCIENCES LIBRARY

22. H.A. Dijkstra: *Nonlinear Physical Oceanography.* A Dynamical Systems Approach to the Large Scale Ocean Circulation and El Niño. 2000   ISBN 0-7923-6522-4

23. Y. Shao: *Physics and Modelling of Wind Erosion.* 2000   ISBN 0-7923-6657-3

24. Yu.Z. Miropol'sky: *Dynamics of Internal Gravity Waves in the Ocean.* Edited by O.D. Shishkina. 2001   ISBN 0-7923-6935-1

25. R. Przybylak: *Variability of Air Temperature and Atmospheric Precipitation during a Period of Instrumental Observations in the Arctic.* 2002   ISBN 1-4020-0952-6

26. R. Przybylak: *The Climate of the Arctic.* 2003   ISBN 1-4020-1134-2

27. S. Raghavan: *Radar Meteorology.* 2003   ISBN 1-4020-1604-2

28. H.A. Dijkstra: *Nonlinear Physical Oceanography.* A Dynamical Systems Approach to the Large Scale Ocean Circulation and El Niño. 2nd Revised and Enlarged Edition. 2005   ISBN 1-4020-2272-7 Pb; 1-4020-2262-X

29. X. Lee, W. Massman and B. Law (eds.): *Handbook of Micrometeorology.* A Guide for Surface Flux Measurement and Analysis. 2004   ISBN 1-4020-2264-6

30. A. Gelencsér: *Carbonaceous Aerosol.* 2005   ISBN 1-4020-2886-5